APPLIED CLIMATOLOGY

Climatology has 'come of age' in the last decade or [illegible] ...te consciousness' highlights the vulnerability of society to climatic inconstan[illegible] ...ious that human activities can inadvertently modify the global climatic environment through the combustion of fossil fuels, trace gas pollution and deforestation. At the same time, humans can also deliberately modify the meso-climate through a variety of schemes such as cloud seeding, frost protection, fog dispersal and hurricane moderation.

This book emphasizes the effect of climate on the physical, biological and cultural environments and includes both present-day and future relationships. The book is divided into four main parts. Part 1 examines the current practices/methodological developments which represent the basic 'tools' which underpin applied climatological research (including data collection and statistical representation, modelling procedures and ways of managing atmospheric resources). Part 2 explores the influence of present-day and predicted climates on the functioning of the physical/biological environments including hydrological processes/water resources, glaciers, landforms/geomorphic processes, soils, vegetation and animals. Part 3 concentrates on the relationship between climate and a wide range of human responses/activities, namely comfort/health, building design, industry/commerce, transport systems, agriculture/fisheries, forestry, recreation/tourism, social/legal issues and energy supply/demand. Finally, Part 4 highlights the characteristics and consequences of the changing climatic environment in terms of atmospheric pollution, urbanization, climatic hazards and an historic review of changing human processes in response to (natural) climate change over the past 2000 years.

Applied Climatology is written by 27 authors who are specialists in the topics presented. This up-to-date and authoritative book highlights the importance of applied climatology in understanding the interaction between the physical and human environment within a framework of a changing climate.

APPLIED CLIMATOLOGY

Principles and Practice

Edited by
Russell D. Thompson and Allen Perry

London and New York

First published 1997
by Routledge
11 New Fetter Lane, London EC4P 4EE

Simultaneously published in the USA and Canada
by Routledge
29 West 35th Street, New York, NY 10001

Typeset in Garamond by J&L Composition Ltd, Filey, North Yorkshire

Printed and bound in Great Britain by
Butler and Tanner Ltd, Frome and London

British Library Cataloguing in Publication Data
A catalogue record for this book is available from the British Library

Library of Congress Cataloging in Publication Data
Thompson, Russell D.
Applied climatology: principles and practice/
Russell D. Thompson and Allen Perry
p. cm.
Includes bibliographical references and index.
1. Climatology. 2. Climatic changes. I. Perry, A.H.
(Allen Howard) II. Title.
QC981.T48 1997
551.6–dc20 96–43022
ISBN 0–415–14100–1
0–415–14101–X (pbk)

To Gaynor and Sue for encouraging and enduring our applied climatological practices over the years.

CONTENTS

PART 3: CLIMATE AND THE CULTURAL ENVIRONMENTS

PART 5: OVERVIEW

PLATES

BLACK AND WHITE

COLOUR

FIGURES

TABLES

CONTRIBUTORS

Andris Auliciems	Department of Geography, The University of Queensland, Brisbane, Queensland, Australia
Anthony Brazel	Department of Geography, Arizona State University, PB 870101, Tempe, Arizona 85287–0104, USA
Michael Bridges	International Soil Reference and Information Centre (ISRIC), PO Box 353, 6700 AJ Wageningen, The Netherlands
Howard Bridgman	Department of Geography, University of Newcastle, Newcastle, NSW 2308, Australia
Neville Brown	Mansfield College, Oxford, OX1 3TF, UK
Ian Burton	Environment Adaptation Research Group, Atmospheric Environment Service (AES), 4905 Dufferin Street, Downsview, Ontario, Canada, M3H 5T4
Paul J. Croft	Department of Physics & Atmospheric Sciences, Jackson State University, PO Box 17660, Jackson, Mississippi 39217–0460, USA
Edward Derbyshire	Department of Geography, Royal Holloway University of London, Egham, Surrey, TW20 0EX, UK
Ann Henderson-Sellers	Chancellory, Royal Melbourne Institute of Technology, Melbourne, Victoria, Australia
John E. Hobbs	Department of Geography, University of New England, Armidale, NSW 2350, Australia
David W. Lawlor	Institute of Arable Crops Research (IACR), Rothamsted, Harpendon, Hertfordshire, AL5 2JQ, UK
Sven Lindqvist	Department of Physical Geography, Earth Sciences Centre, Göteborg University, S-41381 Göteborg, Sweden

A.M. Mannion	Department of Geography, The University of Reading, Reading, Berkshire, RG6 6AB, UK
Jonathan Martin	Robert-Jones Associates Incorporated, Phoenix, Arizona 85016, USA
Kendal McGuffie	BMRC, PO Box 1289K, Melbourne, Victoria 3001, Australia
T.R. Oke	Department of Geography, University of British Columbia, Vancouver, British Columbia, Canada, V6T 122
Allen Perry	Department of Geography, University of Wales Swansea, Swansea SA2 8PP, West Glamorgan, UK
Alexander Robertson	Faculty of Engineering and Applied Science, Memorial University of Newfoundland, St John's, Newfoundland, Canada, A1B 3X5
Scott M. Robeson	Department of Geography, Indiana University, Bloomington, Indiana 47405, USA
Marjorie Shepherd	Environment Adaptation Research Group, Atmospheric Environment Service (AES), 4905 Dufferin Street, Downsview, Ontario, Canada, M3H 5T4
Jim Skea	ESRC Global Environmental Change Programme, Mantell Building, University of Sussex, Brighton BN1 9RF, UK
Keith Smith	Department of Environmental Science, University of Stirling, Stirling FK9 4LA, UK
Bernard Stonehouse	Scott Polar Research Institute, University of Cambridge, Lensfield Road, Cambridge CB2 1ER, UK
Ian Strangeways	TerraData, PO Box 48, Wallingford, Oxon, UK
Russell D. Thompson	Department of Geography, The University of Reading, Reading, Berkshire RG6 6AB, UK
John E. Thornes	Department of Geography, The University of Birmingham, PO Box 363, Birmingham B15 2TT, UK
Paul Whitehead	Department of Geography, The University of Reading, Reading, Berkshire RG6 6AB, UK

PREFACE

Applied climatology has indeed 'come of age' in the last decade or so, even though the current 'climate consciousness' emphasizes the vulnerability of society to climatic inconstancy. It is now obvious that human activities can inadvertently modify the global climatic environment through the combustion of fossil fuels, trace gas emissions, aerosol pollution and deforestation. At the same time, humans can also deliberately modify the mesoscale climate through a variety of schemes such as cloud seeding, frost protection, fog dispersal and hail suppression.

It is also apparent that climate clearly controls the development and functioning of the physical, biological and cultural environments, although these relationships have been overshadowed by the global warming euphoria of recent years. However, it must be remembered that these relationships are predicted to change dramatically over the next century or so, if global warming accelerates at the rate suggested by current climate models. This would initiate physical and biological responses which were last evident on Planet Earth many thousands of years ago.

Consequently, it is very timely and appropriate to re-examine these relationships and responses. There is an outstanding case for a modern, applied climatological text which examines current practices/methodological developments and explores these complex climate–environment relationships/responses in such an uncertain and changing situation. There are standard undergraduate texts available in this field but they are all more than 15 years old and are consequently very dated (Griffiths, 1966; Maunder, 1970; Oliver, 1973; Mather, 1974; Smith, 1975; Hobbs, 1981).

There is a clear demand for an up-to-date undergraduate textbook in applied climatology, especially if the material is contributed by authors who are specialized in the wide range of practices and principles to be considered. The standard texts referred to above were written by single authors without overall expertise in the material covered. This textbook incorporates chapters from 27 specialists (including established applied climate names), who are able to present up-to-date, authoritative material.

The 27 contributors represent 21 universities and indeed 12 of the authors are currently, or recently, affiliated to well-known research institutions. These include the UK Institute of Hydrology, the UK IACR, the UK Scott Polar Research Institute, the UK ESRC Global Environmental Change Programme, the US NCAR, the Canadian AES, the Canadian Forest Service and the Netherlands ISRIC. Furthermore, the authors represent a reasonable global coverage, with 12 from the UK, five from Australia, four from Canada, four from the USA and two from Europe.

The text has been written to represent a 'broad-brush' overview of the respective topics and Chapters 6–24 also consider the regional impacts of predicted climate change/global warming. The authors were given complete freedom in the broad way they approached their topics, which is evident from the range of coverage and emphases represented in the 24 individual chapters. The editors were responsible for uniformity of 'house style' and textual consistencies but did not attempt to harmonize the individual contributions in terms of the specific emphasis and the applied climatological content. We hope that we have succeeded in the production of a coherent text with adequate integration through cross-referencing and a substantial overview chapter (25).

Three clear themes have developed from the material presented in this book, which provide challenging discussion topics, as follows:

1 How does climate control the development and functioning of Planet Earth's natural and human-made environments? Are water resources (liquid and frozen), pedo-biogeomorphic processes, animal/human behaviour and human activities/responses controlled unequivocally by climatic elements?
2 How will the natural environments and human responses/activities be modified by proposed climate changes (i.e. global warming) in the next century or so? Will these changes initiate environmental responses which were last evident on Planet Earth many thousands of years ago?
3 What are the evidence and consequences of changing climatic environments due to anthropogenic pollution, urbanization and natural atmospheric disasters? Have human historical processes and (natural) climate change over the past 2000 years suggested that the impact of future global warming will be more minimal than that predicted by current climate models?

We do hope that these three pertinent questions will be answered as you read through the following text.

Russell Thompson and Allen Perry
Reading and Swansea

REFERENCES

Griffiths, J.F. (1966) *Applied Climatology: An Introduction*, London: Oxford University Press.
Hobbs, J.E. (1981) *Applied Climatology*, London: Butterworths.
Mather, J.R. (1974) *Climatology: Fundamentals and Applications*, New York: McGraw-Hill.
Maunder, W.J. (1970) *The Value of the Weather*, London: Methuen.
Oliver, J.E. (1973) *Climate and Man's Environment*, New York: John Wiley.
Smith, K. (1975) *Principles of Applied Climatology*, London: McGraw-Hill.

ACKNOWLEDGEMENTS

The production of this book was greatly facilitated by the assistance and co-operation of the following: Professor Brian Goodall, Head of the Geography Department at Reading University, for the use of facilities and materials; Michelle Norris and, especially, Sonia Luffrum (Department of Geography, The University of Reading), who provided the skills and patience of invaluable typists; and, finally, Guy Lewis (Department of Geography, University of Wales, Swansea) and Heather Browning (Department of Geography, The University of Reading) for all their support and cartographic expertise.

The authors and publishers wish to thank the following, who have supported the submissions and who have kindly given permission for the use of copyright material:

Chapter 4: Funding assistance from the Australian Research Council, the National Greenhouse Advisory Committee and the University of Arizona (NOAA Global Change Program).

Chapter 5: Information and comments from David Etkin, Bob Saunders, Nicola Mayer and Jennifer McKay.

Chapter 7: Figure 7.1(a) with permission of UCL Press, London; Figure 7.2 reprinted from the *Journal of Glaciology* with permission of the International Glaciology Society; Figures 7.3(a) and 7.4(a) and (b) with permission of Butterworth–Heinemann Ltd and W.B. Paterson.

Chapter 9: Facilities provided by ISRIC; assistance from J.H.V. van Baren and Jacqueline Resink.

Chapter 14: Figure 14.1 with permission of the Royal Meteorological Society; Figure 14.2 from Kleinwort Benson Research.

Chapter 16: Figure 16.2 with permission of Martin L. Parry (personal communication).

Chapter 17: Figure 17.1 with permission of *Journal of Tree Physiology*; Figure 17.2 with permission of Roger White and Guy Engelen (personal communication).

Chapter 22: Plates 22.1, 22.2 and 22.3 with permission of Paul Fraser and Clive Elsom, CSIRO Melbourne. Figure 22.1(b) with permission of Carbon Dioxide Information Analysis Center, Tennessee; Figure 22.1(c) with permission of Royal Meteorological Society and *Weather*; Figure 22.4(c) with permission of *Nature* (1990) Macmillan Magazines Ltd, London; Figure 22.4(c) with permission of American Association for the Advancement of Science, Washington; Figure 22.4(c) with permission of the American Geophysical Union, Washington; Figure 22.6(a) with permission of Cambridge University Press.

Chapter 23: Plate 23.1 with permission of the International Federation of Red Cross and Red Crescent Societies.

1

INTRODUCTION: THE EMERGENCE OF APPLIED CLIMATOLOGY AND CLIMATE IMPACT ASSESSMENT

John E. Hobbs

INTRODUCTION: WEATHER, CLIMATE AND EVERYDAY LIFE

We are all affected by the atmosphere and by information about the atmosphere and we react to the atmosphere through our ability to make decisions. Weather and climate are important factors in determining our day-to-day and longer-term activities and lifestyles. Why do we wear clothes if for reasons other than fashion or modesty? Why do we erect buildings and then heat or cool them? These are all adjustments to the atmosphere and the conditions it produces. Such relationships between people and the atmosphere point to the conclusion that in many ways, our lives are in effect governed by applications of our perception and understanding of, and information about, the atmosphere.

The atmosphere provides frequent topics for superficial conversation. How often is a comment about the weather the first thing we say at the start of the day or when we meet someone? A story from a Sunday newspaper magazine about a mother's telephone call from her son in Japan illustrates this well:

> So, what do we talk about? Why, the weather of course. That is what everyone talks about when they call the other side of the world. The topic of climate is such a non-threatening entree and convenient cover-up for the real and big emotions most of us dare not to broach without fervent discussion of low pressure systems.

She suggested that she had friends who get withdrawal symptoms if they have not heard the daily weather report, and that the weather is the reigning monarch of their repertoire of small-talk, representing an opening gambit and a focus of conversation.

The weather is considered a 'safe' topic of conversation and it is seen as neutral ground, where no offence can be given or taken. The weather is something we all have in common since it is in the public domain, is considered by many to be 'safer' than politics and is not at all personal. However, alternatively, others see the weather as intensely personal and as the most mysterious of everyday wonders. There is no question that debate over recent years on matters like the enhanced greenhouse effect and global warming, ozone depletion, acid rain and the nuclear winter has attracted concern about the atmosphere and the weather and climate it produces, at a level involving the whole population and not just the interested scientists.

THE DEVELOPMENT OF APPLIED CLIMATOLOGY

Interest in the atmosphere and an awareness of its impact on human activities is nothing new since the earliest records of weather lore date back at least 6000 years. Humans have always been interested in 'cause and effect' which helps to establish an order to existence. The ancient Egyptians planned their lives on the assumptions embodied in their weather lore and the Greeks, Arabians and Chinese were very much aware of weather and climate. The ancient Romans and Greeks produced descriptions of the climates of Britain, central Europe and Russia which might be considered a form of early applied climatology. The Chinese thought of their environment in terms of the abundance or failure of rains and recorded variations in terms of rainfall that determined harvests and tax revenues (Miller, 1987).

The Greeks gave names and qualities to the winds and their philosophers and writers postulated many ideas concerning the role of climate in the nature of people. Hippocrates, in his *Airs, Waters and Places*, contrasted easy-going Asiatics with penurious Europeans by suggesting that the latter had to be more active to ameliorate their environment. Aristotle wrote of weather lore and his pupil, Theophrastus, devoted a major portion of his life to assembling weather wisdom into orderly presentations. His *Inquiry into Plants and Minor Works on Odours and Weather Signs* contained many weather insights which are as valid today as when they were written in the fourth century BC. Roman writers, like Strabo, suggested that the rise and strength of Rome were due partly to the climate of Italy. With the decline of Rome (see Chapter 24, Into the medieval optimum), explanations of the role of humans in their environment diminished whereas in medieval times, writers were not encouraged to deal with matters that did not conform to biblical teaching (Siddiqi and Oliver, 1987). Outside the Christian world, Arabic writers contributed to a discussion of the climate–human relationship and, for example, Ibn Khaldun divided the hemisphere into seven climatic zones, the extremes of which were uninhabitable. Only the 'middle' zone provided climates in which people were regarded as able to excel in wisdom, being neither too stolid nor too passionate.

A resurrection of classical ideas was apparent in the sixteenth and seventeenth centuries. Montesquieu proposed that people who live in cold climates are stronger, more courageous, less suspicious and less cunning than those in hot regions. Kant, Ritter and Humboldt presented similar views (Siddiqi and Oliver, 1987). Virgil, Chaucer, Shelley and many others wrote of the weather, Shakespeare's plays and sonnets abound with weather insights, and artists such as Constable, Turner and Monet had a clear perception of the atmosphere (Thornes, 1978). As noted by Smith (1987), writers have always commented on the variability of climate and the consequences for human affairs. An understanding of climatic history is central to our understanding of human history and the history of the environment as much as to forward planning and assessment of risks (Lamb, 1969). In a sense, therefore, history, or at least the explanations of history, are part of the realm of applied climatology.

In more recent times, Ellen Churchill Semple and Ellsworth Huntington were closely associated with environmental determinism, the latter being arguably the archetypical climatic determinist. The essence of Huntington's work was that climate change is one of the major factors determining human progress and that there is a climatic optimum for humans regardless of background. His work was not soundly based on scientific evidence and it oversimplified complex variables, but it did represent an early recognition of many aspects of applied climatology. In the light of the above it can be argued that applied climatology has never had to emerge – it has always been with us. This line of thought also supports the proposition that all climatology might be considered as applied or that, at the very least, applied climatology as such should be seen as the pre-eminent branch of the broad field of climatology. Nevertheless it must be recognized that many would disagree with this, instead seeing applied climatology in much narrower terms and of relatively recent emergence.

DEFINITIONS, APPLICATIONS AND DEVELOPMENTS

What is applied climatology? How might it be defined? Do topics presented at conferences on applied climatology define the field? Does the scope of this book define the field? Changnon (1995) has suggested that the field is really much larger than what applied climatologists might define as their domain. Landsberg and Jacobs (1951) defined applied climatology as 'the scientific analysis of climatic data in the light of a useful application for an operational purpose'. Smith (1987) saw applied climatology as the use of archived and real-time climatic information to solve a variety of social, economic and environmental problems for clients and managers, in fields such as agriculture, industry and energy. Marotz (1989) suggested that applied climatology is the scientific use of climatic data and theoretical constructs for the solution of particular problems. It was noted that Oliver (1981) saw applied climatology as one of four major subgroups of climatology together with climatography, physical climatology and dynamic climatology. Smith (1987) used Thomas' (1981) listing of sectors, in which climate has social, economic and environmental significance, as a means of defining the scope of applied climatology. The field of applied climatology can then be considered to encompass four broad groups of problems, as follows:

1 Design and specification of material or equipment.
2 Location and use of equipment or structures.
3 Planning of a particular operation.
4 Climatic influences on biological activities.

Changnon (1995) saw applied climatology as a general term, widely used but often widely misunderstood. He provided a model for applied climatology consisting of a core and two concentric rings (Figure 25.1), all interacting and reflecting an expansion of activity. In his model, the inner core is focused on data (e.g. instruments, collection, transmission, quality assessment, spatial and temporal representativeness, archival and accessing) whereas the inner ring relates to interpretation and generation of climate information from the data core. It includes statistical and physical analyses, performance of specialized studies, generation of publications and a variety of computer-based outputs. A much larger outer circle of activity comprises those who use climate data and information for various purposes, namely scientists, engineers, business people and others in non-atmospheric science disciplines and activities. This user-community includes hydrology, agriculture, ecology, geography, energy supply, medicine, the construction industry, architecture, transportation, leisure and recreation, tourism and government. It thus encompasses most human activity on a day-to-day basis.

According to Changnon, therefore, part of applied climatology is defined by the atmospheric scientists and geographers who practise and identify with it, and part is defined by the many practitioners/users beyond the atmospheric sciences. Applied climatology then begins with the instruments used to sample and measure the atmosphere (Chapter 2) and it includes the acquisition, evaluation and storage of data (Chapter 3). It ends with those who apply such information in their research and decision making as, for example, is represented by Parts 2, 3 and 4 of this book. Such a definition implies that applied climatology did not exist before instrumentation. It also makes an applied climatologist out of, for example, the hydrologist studying the relationship between climatic conditions and surface water or the engineer forecasting public utility loads. Conferences for a wide range of professionals in a variety of fields such as engineering, economics, agriculture, tourism, botany and geography commonly include presentations involving relationships between climate and their specialist areas of study. Changnon pointed to the work of scientists like Horton and Thiessen, who used climatic data for applications in their own fields of interest. Also early efforts to develop climatic classifications, such as those by Koppen and Thornthwaite, were identified as examples of applied climatology, as was the work of Brooks in studies of climatic influences on human behaviour and health.

There is general agreement that the Second World War gave a major impetus to the development of applied climatology (e.g. Changnon, 1995; Haggard, 1982; Marotz, 1989; Miller, 1987; Smith, 1987). Jacobs (1947) noted that applied climatology was born out of the operational requirements of the military during the Second World War. Applications of climatology developed greatly at that time, with more extensive data networks, improved methods for sorting and analysing data and expansion of archives. During the 1940s and early 1950s, the discovery of features like the polar front jet stream and its role in atmospheric circulation led to increased efforts to construct dynamic climatologies which, with the emergence of refined statistical and mathematical approaches, led to new applications and extensions of climatological principles (Marotz, 1989). Hydrologists such as Langbein, Wolman and Ven Te Chow also figured prominently in the post-war development of applied climatology, while researchers like Landsberg, Sewell, Maunder, McQuigg and Kates made significant contributions to the development of climate impact studies.

Smith (1987) suggested that the term 'applied climatology' came into common use in the early 1940s with the realization that meteorological agencies could provide more valuable services to the community than just weather forecasts. He claimed that applied climatology has grown in stature and scope since the 1950s, a period that has been called the 'golden age' of applied climatology (Changnon, 1995). This growth can be attributed partly to the development of specific skills and techniques using modern methods of data handling linked with improved understanding of statistical theory, particularly as related to extreme events and probability analysis (Chapter 3). The problems themselves, while becoming more complex, have also become better defined.

ATMOSPHERIC CONCERN AND AWARENESS

Recent concerns about possible severe climatic impacts on a growing global population have brought the 'discipline' more and more into the public arena. The potential vulnerability of the modern world to climatic variability, growing energy problems and an awareness of environmental consequences of the continued consumption of fossil fuels have led to a 'coming of age' for applied climatology. Recognition over the last 30 years of the need to apply climatic knowledge to societal planning at governmental, industrial and private levels has highlighted the actual and potential roles of the applied climatologist (Haggard, 1982). Hare (1979) even suggested that the 1970s might be remembered as the decade in which climate had a major destabilizing effect on world economies.

Thornes (1978) posed the question 'How does applied climatology fit into the jigsaw of atmospheric studies?' He identified three basic approaches to atmospheric studies: namely, atmospheric science (theoretical), atmospheric perception (experiential) and atmospheric management (behavioural). Atmospheric management, which is equated with applied climatology (Thornes, 1980), is seen as bridging the gap between scientific (atmospheric science) and artistic (atmospheric perception) approaches to the atmosphere. It is concerned with efficient atmospheric resource management (Chapter 5) and atmospheric hazard response (Chapter 23) and requires the merging of physical process–response models with socio-economic, or econoclimatic, models (Perry, 1971). Nearly 30 years ago, Sewell (1968) stressed the need for social science research into atmospheric resource and hazard problems but, despite many similar calls (e.g. Maunder, 1970; Mather, 1974; Smith, 1974; Terjung, 1976), real progress in this regard has been relatively recent. The gap has been generally narrowed between the physical science required to understand the atmosphere on the one hand, and the socio-economic methods needed to understand human interactions with the atmosphere on the other hand.

Physical geographers (as climatologists) have always been prominent in applied climatology but as Thornes (1981) pointed out, there has been little attention to the philosophy of physical geography and atmospheric studies. His view that the history of ideas within atmospheric studies since climatic determinism remains unresearched, is still essentially valid. He suggested that teaching and research in atmospheric studies had been restricted by a positivist paradigm giving little direction on how to apply knowledge produced by an emphasis on measurement and modelling of the atmosphere to facilitate prediction. Again, recent work has made considerable progress with this problem prompted by increased general awareness of a range of global atmospheric issues and stimulated by works such as that by Ausubel and Biswas (1980), which started to bring a level of socio-economic understanding to real atmospheric problems.

Sewell (1968) identified three major problems as having triggered concern about the atmosphere, leading to a growing interest in its management like other resources, which are increasing air pollution, rising losses due to extreme weather events and the increasing ability to modify the weather. Such problems emphasized the fact that knowledge of the human impact of atmospheric variations was limited and that there was a considerable need for research relating to the economic, social, institutional and ecological aspects of atmospheric management (Chapter 5).

The 1972 Stockholm Conference on the Environment stressed the importance of global climate and its variability in determining society's economic and social well-being. The need to understand and apply a knowledge of climatology has been a persistent theme at many conferences since then. In the 1980s, Musk (1983; 1984) identified an international climate of opinion encouraging research in applied and applicable climatology at all levels. He pointed to a growing awareness by politicians, planners and scientists of the role of the atmosphere as a resource needing managing and conserving, and of the importance of climatic variability for weather-sensitive sectors of national economies. Hare (1985) noted an insatiable demand from politicians and the public for predictions of future climate changes and their impacts, a demand which has been sustained to the present and which has been nurtured by the popular media. It has provided a fertile base for much applied climatological work. Musk's reference to 'applied' and 'applicable' climatology raises an interesting distinction, albeit one difficult to define. Chang (1983) exemplified the difference:

> the lack of knowledge of urban planning on the part of urban climatologists is one of the main hindrances to applied research . . . the lack of knowledge of urban climate on the part of urban

> planners is also one of the major reasons why urban climatology has played no important role in urban planning.

A simple view may be to regard 'applicable climatology' as part of the broad field of applied climatology, where the distinction is not worth worrying about.

Commentaries through the 1980s and into the 1990s (e.g. Musk, 1985; 1986; 1987; 1988; Thornes, 1991; 1992; Perry, 1992; 1993; Rogers, 1994; 1995) have discussed and highlighted many areas receiving attention from applied climatologists. These have ranged from, for example, links between variability of weather and climate and demand for power and water, through impacts of weather hazards on transport, problems of acid precipitation, responses of regional food systems to climate changes, to links between many aspects of human health and the atmospheric environment, the biological consequences of stratospheric ozone depletion and to relationships between severe weather and the insurance industry. The full range of topics to which applied climatologists have contributed, and are contributing, is exemplified by this book.

CLIMATE IMPACT ASSESSMENT

Climate impact assessment is an integral part of the broad field of applied climatology. Its special needs have played a major role in the development of applied climatology over the last decade or two, to the extent that it has become arguably the most visible branch of that field. Indeed, some may regard the terms 'climate impact assessment' and 'applied climatology' as synonymous. The ground covered in this book is sufficient to demonstrate that this is in fact a very limited view. Nevertheless, the growing significance of climate impact assessment cannot be denied.

The World Climate Impact Program (WCIP) had distinctive aims to improve our knowledge of the impact of climate variability and change, as follows:

1 To develop knowledge and awareness of interactive relations between these and human socio-economic activities.
2 To improve methodologies employed so as to deepen understanding and improve simulation of such interactions.
3 To determine characteristics of human societies at different levels of development in different natural environments, which might make them resilient to climate variability and change and which also permit them to take advantage of opportunities posed by such changes.
4 To investigate the application of new knowledge of techniques to practical problems of concern to developing countries or which are related to common needs for all humankind. The Scientific Committee on Problems of the Environment (SCOPE) of the International Council of Scientific Unions undertook to prepare an authoritative review of the methodology of climate impact assessment called for in the WCIP. This examined existing methodology, fostered development of new methodological approaches and informed a broad range of disciplines about available concepts, tools and methods beyond SCOPE's own special areas of interest. The initial effort of the WCIP was reported in a study by Kates *et al.* (1985).

Since that report, there have been several important contributions addressing the problems presented by assessment of the impacts of climate change and variability. Parry (1993) noted that it was 50 years since the possibilist reaction to the environmental determinism of the 1930s and considered conceptual models of climate and society. The type of causal explanation, implicit in impact models, was viewed as too simplistic and deterministic, which was characteristic of Huntington's (1907) early studies of the relationship between climate and culture. Parry suggested that 'the premise of a cyclical connection between climate and history still survives, but generally lacks sufficient rigour to be convincing'. Recent work has sought to construct hierarchies of climate–society impact models to trace the effects of a climatic event as it cascades through the physical and social systems. Parry claimed that greater realism can be introduced to climate–society studies by considering adaptation (short term) and adjustment (long term) to climate change. With such interactive models, it is possible to consider factors like the vulnerability or resilience of different systems to an impact, which can in turn affect how societies can and do respond to climate change. This approach also introduces the issue of perception of potential impact, the risks or benefits involved and the opportunities to avoid them or to take advantage of them.

Parry also suggested a need to redefine climate change which can be seen as a change in frequency of extreme events (Chapter 23) and can also be evaluated in human terms as a change in level of risk. Furthermore, if viewing climate as a resource, change can be seen as a shift in resource opportunities or options. He further noted that, as a relatively new field of study, climate impact assessment still has to develop a sophisticated set of analytical tools and more mature concepts, and needs scientists who are able to integrate information from a number of traditionally separate disciplines. Climate impact assessment will require more detailed study of the indirect effects of climate change (e.g. via changes in soil chemistry or in the frequency of outbreaks of pests and diseases). Also, it will need to specify, with greater precision, the interaction between climate and other resources, so that it should then be possible to trace downstream effects of the first-order impacts on an economy. Furthermore, it should focus on direct human concerns such as the role of decision-making, constraints on choice, and the formulation of strategies. Rogers (1995) noted that 'downscaling' is a methodology becoming increasingly popular among applied climatologists, involving translation across spatial scales towards finer resolution. This typically involves the use of large-scale GCM outputs (Chapter 4) to find inter-relationships between large-scale circulation and local and regional climate conditions for present and future scenarios with increased carbon dioxide. Rogers also pointed out that use of 'downscaling' may incorporate neural network approaches to problem-solving in cases where climatic downscaling relationships are non-linear.

The Intergovernmental Panel on Climate Change (IPCC) has played an important role in the development of assessment of climate change and its impacts. Its first assessments were produced in 1990 (Houghton *et al.*, 1990; Tegart *et al.*, 1990) but it was clear that much more work was needed. The reports revealed the difficulties of comparing impacts in different regions and economic sectors assessed using different methods. Subsequent work by an expert group led to two further important reports developing guidelines for the assessment of the impacts of climate change (Carter *et al.*, 1992, 1994). The 1994 report outlines a study framework which will allow comparable assessments to be made of impacts and adaptations in different regions, economic sectors and countries.

CONCLUSIONS

Over 25 years ago, Lamb (1969) suggested that climatology was facing problems not dreamt of a generation before. He saw a demand for forecasts of probable climatic trends over the next decades, and for long-range forward planning in agriculture, industry, trade and government. Changnon (1995) commented on growing concerns, apparent since the 1970s, over severe climatic impacts on a growing global population. Since then we have seen the development of new institutions, supercomputers, better databases, improved statistical techniques, near-real-time products and a much greater public awareness of potential problems associated with changing climates. All of these can be seen to have increased the need for the expertise and vision of the applied climatologist.

On the other hand, Changnon (1991) warned that applied climatology could be in danger of losing its identity, to disappear into allied disciplines like agricultural science, hydrology and engineering, possibly owing to poor cross-disciplinary communications and a relative lack of educational opportunities in applied climatology as such. Applied climatology is a diffuse field that has perhaps not received the attention in universities, or within the funding arena, necessary to ensure its survival as a discrete discipline. Many scientists in other disciplines lack the specific climatic training and understanding to be able to make the most efficient and effective use of climate products in the multitude of weather-sensitive areas of society. Changnon (1995) stressed the needs for education, outreach and delivery to ensure that improvements in the understanding of climate can be used to maximum effect and to enhance climate education in the broad range of disciplines seeking to use and apply climatic principles and information.

There is no question that some key issues face applied climatology including, for example, more effective management of the atmospheric resource, improvement of the global environment and better application of climate data and information to human activities. At the same time, there is commonly a gap between the availability of climatic information and products and their application. Concerns include questions about data quality, about education and training in the application of climate information for decision-making and about the delivery of useful products to those who need them. Despite this, it has been claimed that applied climatology is the oldest and by far the major success story of the atmospheric sciences in service to society. Furthermore, it is the applied climatologist who has gained the greatest credibility within the atmospheric sciences (Changnon, 1995) and evidence to support this claim will be provided by the following 24 chapters in this book.

REFERENCES

Ausubel, J. and Biswas, A.K. (1980) *Climatic constraints and human activities*, Oxford: Pergamon Press.

Carter, T.R., Parry, M.L., Nishioka, S. and Harasawa, H. (1992) *Preliminary Guidelines for Assessing Impacts of Climate Change*, Oxford: Environmental Change Unit, University of Oxford, and Tsukuba: Center for Global Environmental Research, National Institute for Environmental Studies.

Carter, T.R., Parry, M.L., Harasawa, H. and Nishioka, S. (1994) *IPCC Technical Guidelines for Assessing Climate Change Impacts and Adaptations*, London: Department of Geography, University College, and Tsukuba: Center for Global Environmental Research, National Institute for Environmental Studies.

Chang, C.C. (1983) 'Some problems on the study of urban climate', *Acta Geographica Sinica* 38: 73–9.

Changnon, S.A. (1991) 'Applied climatology: atmospheric sciences' biggest success story faces major new challenges', in *Applied Climatological Research: Miscellaneous Papers*, Champaign, IL Illinois State Water Survey.

Changnon, S.A. (1995) 'Applied climatology: a glorious past – an uncertain future', *American Meteorological Society 9th Conference on Applied Meteorology*, Boston: American Meteorological Society.

Haggard, W.A. (1982) 'Applied climatology: some data sources and applications', *Journal of Applied Meteorology* 20: 1412–14.

Hare, F.K. (1979) 'Focus on climate', *Environmental Science and Technology* 13: 156–9.

Hare, F.K. (1985) 'Future environments – can they be predicted?', *Transactions, Institute of British Geographers* 10: 131–7.

Houghton, J.T., Jenkins, G.J. and Ephraums, J.J. (eds) (1990) *Climate Change: The IPCC Scientific Assessment*, Report of Working Group I of the Intergovernmental Panel on Climate Change, Cambridge: Cambridge University Press.

Huntington, E. (1907) *The Pulse of Asia*, New Haven, CT: Yale University Press.

Jacobs, W.C. (1947) 'Wartime development in applied climatology', *Meteorological Monographs* 1: 1–52.

Kates, R.W., Ausubel, J.H. and Berberian, M. (eds) (1985) *Climate Impact Assessment: Studies of the Interaction of Climate and Society*, SCOPE 27, Chichester: John Wiley.

Lamb, H.H. (1969) 'The new look of climatology', *Nature* 223: 1209–15.

Landsberg, H. and Jacobs, W.C. (1951) 'Applied climatology', in T.F. Malone (ed.) *Compendium of Meteorology*, Boston: American Meteorological Society, 976–92.

Marotz, G.A. (1989) 'Current status, trends and problem-solving in applied climatology', in M.S. Kenzer (ed.) *Applied Geography: Issues, Questions and Concerns*, Dordrecht: Kluwer Academic, 99–113.

Mather, J.R. (1974) *Climatology: Fundamentals and Applications*, New York: McGraw-Hill.

Maunder, W.J. (1970) *The Value of the Weather*, London: Methuen.

Miller, D.H. (1987) 'Climatology, History of', in J.E. Oliver (ed.) *The Encyclopedia of Climatology*, New York: Van Nostrand Reinhold, 338–46.

Musk, L.F. (1983) 'Applied climatology', *Progress in Physical Geography* 7: 404–12.

Musk, L.F. (1984) 'Applied climatology', *Progress in Physical Geography* 8: 450–8.

Musk, L.F. (1985) 'Applied climatology', *Progress in Physical Geography* 9: 442–53.

Musk, L.F. (1986) 'Applied climatology', *Progress in Physical Geography* 10: 563–75.

Musk, L.F. (1987) 'Applied climatology', *Progress in Physical Geography* 11: 370–83.

Musk, L.F. (1988) 'Applied climatology', *Progress in Physical Geography* 12: 421–34.

Oliver, J.E. (1981) *Climatology: Selected Applications*, New York: John Wiley.

Parry, M.L. (1983) 'Geographers and the impact of climate change', in R.J. Johnston (ed.) *The Challenge for Geography. A Changing World: A Changing Discipline*, London: Blackwell, 138–47.

Perry, A.H. (1971) 'Econoclimate – a new direction for climatology', *Area* 3: 178–9.

Perry, A.H. (1992) 'The economic impacts, costs and opportunities of global warming', *Progress in Physical Geography* 16: 97–100.

Perry, A.H. (1993) 'Climate, greenhouse warming and the quality of life', *Progress in Physical Geography* 17: 354–58.

Rogers, J.C. (1994) 'Applied climatology', *Progress in Physical Geography* 18: 271–5.

Rogers, J.C. (1995) 'Applied climatology', *Progress in Physical Geography* 19: 555–60.

Sewell, W.R.D. (1968) 'Emerging problems in the management of atmospheric resources: the role of social science research', *Bulletin of the American Meteorological Society* 49: 326–36.

Siddiqi, A.H. and Oliver, J.E. (1987) 'Determinism – climatic', in J.E. Oliver (ed.) *The Encyclopedia of Climatology*, New York: Van Nostrand Reinhold, 382–6.

Smith, K. (1974) *Principles of Applied Climatology*, London: McGraw-Hill.

Smith, K. (1987) 'Applied climatology', in J.E. Oliver (ed.) *The Encyclopedia of Climatology*, New York: Van Nostrand Reinhold, 382–6.

Tegart, W.J.McG., Sheldon, G.W. and Griffiths, D.C. (eds) (1990) *Climate Change: The IPCC Impacts Assessment*, Report of Working Group II of the Intergovernmental Panel on Climate Change, Canberra: Australian Government Publishing Service.

Terjung, W.H. (1976) 'Climatology for geographers', *Annals of the Association of American Geographers* 66: 199–222.

Thomas, M.K. (1981) *The Nature and Scope of Climate Applications*, Canadian Climate Center Report No. 81–5, Downsview: Atmospheric Environment Service.

Thornes, J.E. (1978) 'Applied climatology: atmospheric management in Britain', *Progress in Physical Geography* 2: 481–93.

Thornes, J.E. (1980) 'Applied climatology', *Progress in Physical Geography* 4: 577–87.

Thornes, J.E. (1981) 'A paradigmatic shift in atmospheric studies?', *Progress in Physical Geography* 5: 429–40.

Thornes, J.E. (1991) 'Applied climatology: severe weather and the insurance industry', *Progress in Physical Geography* 15: 173–84.

Thornes, J.E. (1992) 'The impact of weather and climate on transport in the UK', *Progress in Physical Geography* 16: 187–208.

PART 1:

APPLIED CLIMATOLOGY: THE 'TOOLS' OF RESEARCH

The following four chapters (2–5) examine the main methodological practices which underpin applied climatological research. The collection of data through ground and remotely sensed measurements is discussed and statistical analyses are considered, in order to provide a meaningful and useful quantification of the vast quantities of data available. Climate models are examined as essential 'tools' in the practice of applied climatology, particularly in the prediction of climate change. Finally, atmospheric resource management is introduced as a necessary part of applied climatological methodology, required to integrate the assessments of atmospheric impacts on natural and social systems.

2

GROUND AND REMOTELY SENSED MEASUREMENTS

Ian Strangeways

INTRODUCTION: DATA REQUIREMENTS AND SOURCES

Data are required for a variety of applied climatological practices. For example, model predictions of climate change must be checked against reality in the future, if we are to know what climate changes took place in the past or understand the hydrometeorological processes occurring today. They are also necessary in order to manage a country's water resources (Chapter 6), make weather forecasts or study any aspect of the natural environment (Part 2).

Meteorological and hydrological measurements started to be made, in the scientific sense, only during the last few hundred years. For example, Symons did pioneering work on rainfall measurement in the nineteenth century and, throughout the last century and into the twentieth century, instruments were designed for measuring all of the common meteorological and hydrological variables. These developments provide the only direct measurements of past climate events. Before this, there were no instrumental records, although there are examples of rainfall measurement going back millennia in India and the Middle East (Biswas, 1970). Consequently, we are forced to rely instead on techniques such as the interpretation of the contents of ice cores (see Chapter 7, The conversion of snow to glacier ice and the role of climate), to determine what occurred earlier. It might be thought that instruments developed so long ago would now be obsolete. This is not the case since they are still the mainstay of the world's National Weather Service's (NWS's) monitoring networks. They are, therefore, of considerable importance.

In the last 30 years, however, new automatic systems have been developed (Plate 2.1), although they still do not make up the majority of instruments in use today and are only being introduced relatively slowly. However, as they are the key to future improvements in monitoring, they are of great importance. In addition, the new techniques of remote sensing have also become available and provide a further important source of data.

A COMPARISON OF OLD AND NEW INSTRUMENTATION

One of the limitations of the old instruments is their need for operators. Being manually read, or having a clockwork-driven paper chart on which to record a day's data, the instruments are restricted to areas accessible to the observer. In consequence, with but very few exceptions, only the populated parts of the world have been monitored. In contrast, the new instruments can be left unattended for months, even years,

Plate 2.1 An automatic weather station (AWS) operated by the Institute of Hydrology deploying a solarimeter on the top; just beneath there is a solar panel to operate the logger; on the top cross-arm to the left is a wind direction sensor and to the right a cup anemometer; on the lower cross-arm to the left is a net radiometer and to the right a temperature screen. The logger box is just visible to the right on the ground

and so can be installed at the remotest of sites. A further disadvantage of the early designs is that they are less precise than the new instruments. For example, air temperature may only be read once a day (usually at 09.00 hours GMT) with maximum and minimum readings for the previous 24 hours. A modern automatic weather station (AWS), in contrast, will usually take readings every minute or even every second and log hourly means. Ironically, this is often seen as a disadvantage since, by changing to a new instrument (even though it is much better), a discontinuity in the record will occur and this is seen as unacceptable. While it is possible to run the old instrument alongside the new for a time for comparison, a perfect transition is next to impossible to achieve. This applies to instruments for measuring all of the meteorological and hydrological variables. It holds back development quite a lot.

Remotely sensed data from satellites have the important advantage of providing wide areal coverage and of accessing the remotest of locations. However, ground-truth measurements from *in situ* instruments are needed to calibrate the readings and so the two techniques work in partnership not competition.

THE OLD-STYLE INSTRUMENTS

Rainfall

There are two types of conventional raingauges, namely manually read and chart recording. The former are usually read daily at 09.00 hours although, for remote locations, storage gauges are available that collect a week's or a month's rainfall. The collected rain, stored in a bottle within the raingauge, is measured by an observer using a graduated cylinder calibrated in millimetres and tenths. In the case of chart-recording gauges, the water is collected in a container which houses a float that rises as the water accumulates, moving a pen up a chart on a cylinder which is driven by clockwork. When full, the container empties by siphoning with the process starting again.

All rain collectors suffer from errors. There is not the space to go into detail here, but it should be noted that these errors include splashout, insplash, evaporative losses and errors due to the gauge not being levelled accurately. Most serious of all, however, are errors due to the wind, which speeds up as it passes over the gauge, carrying some of the raindrops away that should have fallen into the gauge. The degree of this error is related to the aerodynamics of the gauge and can amount to 30 per cent or more in exposed conditions, especially with drizzle and even more so with snow. In fact, no good way of measuring snowfall yet exists. The best solution to the problem is to expose the gauge in a pit with its orifice at ground level, surrounded by a grating that minimizes eddies and prevents insplash (Rodda, 1967). Such an exposure is not always practical, however, and screens that surround the gauge and deflect the wind downwards have been designed as an alternative. The 'Turf Wall' is another familiar technique used to shelter the gauge, but all such methods are less effective than ground-level exposure.

Temperature and Humidity

Mercury-in-glass thermometers are the 'norm' for temperature measurement and are exposed in wooden, slatted temperature or 'Stevenson' screens which allow the free passage of air while sheltering the sensors from exposure to the sun. The thermometers are normally read daily at 09.00 hours, or, in the case of 'synoptic' stations for weather forecasting by the NWS, every 3 hours during the day and night. They are graduated in 0.1 degrees Celsius.

For relative humidity measurement, a second thermometer, with its bulb covered by a wick dipped into a container of distilled water, gives the 'wet bulb temperature'. Its reading, lower due to evaporation from the wick, is compared with the 'dry bulb', giving the wet bulb depression, from which relative humidity can be calculated. There may also be a thermohygrograph in the screen, an instrument which records a dual trace of temperature and humidity on a paper chart. Temperature is sensed by a bimetallic strip which bends as the temperature changes, while humidity is sensed by a bundle of human hair which changes in length with humidity, both moving pens across the chart. The screen also contains a maximum and a minimum thermometer. Outside of the screen, the night time minimum temperature of short grass and bare soil is also measured, and ground temperatures may also be measured at several depths by thermometers in tubes.

Wind Speed and Direction

The 'cup anemometer' is the most commonly used instrument for measuring wind speed. Developed by Robinson in 1846, it comprises, usually, three 'cups' on a vertical shaft. The pressure on the open side of the cups, being higher than that on the back, causes the shaft to rotate and this is made to turn a mechanical counter which is read at intervals, the change in reading from one observation to the next giving the 'run of wind' or average speed. Wind direction is read visually from a vane which must be exposed, as must the anemometer and raingauge, at a good distance from buildings and trees.

Sunshine Hours

Another long-established instrument is the sunshine-hours recorder, developed by Campbell in 1853 and improved by Stokes in 1880. It comprises a glass sphere mounted so as to burn a trace of the sun on a card as the sun passes across the sky. If the sun is not obscured by cloud, it burns a mark and this can be interpreted to give an estimate of the number of hours the sun was visible. It is still the mainstay of all our solar radiation instrumentation and is widely used throughout the world, even though it gives only a relatively crude indication of solar energy input. Somewhat better is the mechanical chart-recording bimetallic actinograph in which black and white strips of metal are exposed, under a protecting dome, to the sun. Since the black surface absorbs more solar energy than the white, this causes differential heating proportional to the energy received. This moves a pen across a paper chart, the area beneath the curve giving an indication of the total energy received.

Evaporation

Another mainstay of the world's meteorological enclosures is the evaporation pan which is a very imprecise device consisting of a large container filled with water. The water level is measured daily and the change since the last reading, with allowance for rainfall, is an indicator of the amount lost by evaporation. There are different shapes and sizes of pan, which hinders direct intercomparisons.

Barometric Pressure

The traditional mercury barometer has been largely replaced by the aneroid type, which consists of one or more shallow capsules with the air pumped out. As the atmospheric pressure changes, the capsules expand or contract and this movement is magnified mechanically to move either a pen across a clockwork-driven chart or a pointer across a dial.

River Flow

Most of the variables that concern hydrologists are the same as those that interest meteorologists, as described above. However, to these must be added the measurement of river flow. By measuring the depth of water in a river it is possible to estimate its flow, 'rating curves' being obtained to show the relationship between depth and flow. Provided the river bed profile does not change, the relationship is constant and needs only periodic checking. If the river can be contained within a structure, such as a flume or a weir, then the rating curve is predictable and constant. To obtain a rating curve, flow is measured using a meter based on a propeller lowered into the river at different points across its width and at different depths, with pulses from a switch on the propeller shaft being counted on an electromechanical counter. This is done either from a bridge or from a boat pulled across the river on a cableway. Depth is measured manually by observing the height of the water on a graduated-staff gauge fixed into the river bed. Levels can also be recorded automatically with a float that turns a shaft which moves a pen across a clockwork-driven paper chart. The float is housed in a 'stilling well' to provide a surface free from turbulence and the effects of wind. A less common method is the bubble pressure recorder, which is a more complex instrument that operates by bubbling nitrogen from the end of a tube fixed firmly in the river. The pressure, which is proportional to the water depth, is recorded on a chart. It has the advantage of not needing a stilling well but it is more complex and needs a heavy nitrogen cylinder.

THE NEW GENERATION OF INSTRUMENTS

New instruments developed over the past few decades are based on sensors which convert the variable to be measured into an electrical signal and a data logger to record the signal. By replacing chart recorders with a data logger, great benefits result. This section looks first at the more commonly used electrical sensors and then at a typical data logger.

Rainfall

The usual method of measuring rainfall automatically is to collect the water in a funnel and to measure it by a'tipping bucket' – a small, two-sided container which tips from side to side as each side becomes alternately filled and emptied. Other methods, such a weighing the water, are used but they are much less common. The tips of the bucket are detected by a switch, with its brief closures being recorded by a data logger. Such gauges suffer from the same collector errors already described above for the manual gauges together with further errors introduced by the tipping bucket. These include evaporation loss, water sticking to the bucket instead of running off, water continuing to flow into it while it tips and poor adjustment, such that it does not tip for exactly the specified amount of rain (the buckets are generally set to tip for 0.1, 0.2, 0.25, 0.5 or 1 mm of rain). The reading of manual gauges is not of course free from error either. There are other automatic methods that do not collect the rain in a funnel, but instead measure the drops as they fall by detecting the scintillation they produce as they pass through a beam of infrared light. However, this type is not in common use, owing to cost and power requirements.

Temperature and Humidity

The new instruments for measuring temperature differ greatly from the old, the only similarity being that the sensors are still housed in a screen. The screen, however, is generally much smaller than the wooden 'Stevenson' type (typically about 30 cm high by 15 cm in diameter), and so responds more quickly to changes. The most usual sensor for temperature measurement is a platinum resistance thermometer (PRT), which is a small coil of fine platinum wire which varies in resistance with temperature. The logger contains circuits that convert the resistance changes into a voltage for logging. A thermistor may also be used in place of the PRT, which also varies in resistance with temperature but is made from semiconducting material. One disadvantage of measuring humidity by the wet and dry method is that below the freezing point, the method is not practical at automatic stations. Also at automatic stations, the replenishment of water is a problem. Fortunately, over the past few decades an electronic sensor has been developed that replaces the two-thermometer method. It comprises a small thin-film capacitor a few millimetres square which allows water vapour to diffuse into it, changing its capacitance relative to the air's humidity. An electronic circuit converts the small capacitance changes into a voltage for logging.

Wind Speed and Direction

As with the raingauge, a modern wind speed sensor retains the same basic method of detection as the mechanical wind sensor in that it uses the rotating-cup principle. The method of sensing the revolutions of the shaft is similar to the tipping bucket raingauge's magnetic switch, although optical methods are also used, with the logger simply counting and storing the number of pulses. An alternative to cups is a sensor which measures speed using a propeller. Because it must be kept facing into the wind, it is usual to combine the measurements of speed and direction in the one instrument. In the case of cup anemometers, a separate sensor is used to measure wind direction, although the two may often be combined in one housing. There are various methods used to convert the vane position into an electrical signal for logging, ranging from a simple potentiometer to a digitally encoded disc, which is read optically. Other methods have been developed for the measurement of wind, such as the detection of the pressure it exerts on a sensor, but these are rare and are used mostly for special applications.

Solar and Net Radiation

Because of the requirement in some situations to change things as little as possible for the sake of continuity, electronic sensors have been developed to measure sunshine hours, replicating the old Campbell–Stokes recorder as closely as possible. However, it is possible to measure solar energy in a much more precise way by using sensors that detect the actual intensity of solar radiation and not just whether the sun is visible or not. The most commonly used sensor is that based on the exposure of a black disc, protected by a glass dome, to the sun. The energy incident on the disc causes it to warm up and the rise in temperature, above ambient, is measured by a thermopile. The temperature rise is calibrated in terms of incident solar energy, for example in W m^{-2}. Integrated over a period, this gives the total energy received. Recently, light-sensitive diodes have been used to measure solar radiation. They have the advantage of cheapness but the disadvantage of not having such a flat response across the full solar spectrum as does the thermal type.

In some ways more meaningful than solar radiation is net radiation, that is the difference between the total incoming short-wave radiation directly from the sun together with the incoming long-wave infrared radiation from the atmosphere, minus the total outgoing energy both as reflected short-wave radiation and infrared radiation from the ground. This indicates how much energy is available to heat the air, heat the ground and evaporate water. This, of course, is the basis of all physical/biological processes (Part 2) and is indeed the heart of the greenhouse effect. Net radiation is measured by a sensor similar to two (thermal) solar radiation sensors combined in one unit, one facing down, the other up, with the difference in temperature between the downward- and upward-facing black discs giving a measure of net radiation.

Evaporation

By operating temperature, humidity, wind speed and net radiation sensors together as an AWS, evaporation can be estimated using the Penman (1948) equations. These give an infinitely better estimate than that obtained from an evaporation pan. An AWS can also include sensors for wind direction, solar radiation and rainfall, thereby becoming a comprehensive climatological station. Such stations are becoming increasingly common and a typical AWS (Plate 2.1) has many of the sensors described in this section (Strangeways, 1985).

River Flow

As in the case of a mechanical river-level recorder, a float is a very effective way of measuring level electronically, although now the float turns a potentiometer or a digital shaft encoder to produce an electrical signal. Just as in the mechanical era, there is sometimes an advantage in not having to build a stilling well and, as in the mechanical case, the pressure of the water can be measured to give depth. In this case, the pressure is measured by a sensor containing a diaphragm which flexes as the pressure changes, its deflection being measured by strain gauges on its surface. Such sensors have the disadvantage, however, that they are limited to about 0.1 per cent discrimination of the total depth change, and this may not be sufficient if changes are large. For the production of rating curves, flow velocity may still be measured using a propeller flowmeter but the counter is now usually electronic. Flowmeters with no moving parts are also now available, using electromagnetic or ultrasonic techniques to measure flow rates. These are available not only in the form of hand-held instruments, as an alternative to the propeller, but also as large permanent installations set in the river bank or its bed to measure flow continuously across the full width and depth of the river. They are, however, expensive and are not widely used.

Barometric Pressure

A somewhat similar sensor to that for water-level pressure is used to measure barometric pressure, the difference being that in the case of water level, one side of the sensor diaphragm is vented to atmosphere (to compensate for its pressure changes) while the other side is exposed to the water. However, in the case of barometric pressure, one side is open to atmosphere and the other to a vacuum.

Data Loggers

The crucial breakthrough in environmental monitoring came with the development of microelectronics. The first data loggers, capable of unattended, remote operation without the need for mains electricity, were developed in the 1960s, logging their data on quarter-inch magnetic tape. Later, compact cassettes, integrated circuits and printed circuits all improved reliability, reduced size and minimized power consumption. During the 1970s and 1980s, development was rapid and, by 1990, loggers used microprocessors and became programmable by the operator and replaced magnetic tape with solid-state memory. The developments in computing, from the mainframes of the 1960s to the powerful and cheap personal computers (PCs) of today, were also a key factor in the revolution which took place in our ability to measure the hydrometeorological environment with increased precision, by stations which could be left unattended for long periods at remote sites using minimal power. They are programmed by the user via a PC to accept whichever sensors are to be used, along with the ranges of the variables and the frequency of logging. A visit is made to the field site periodically to extract the stored data, the most common way at present being to download the data to a laptop PC. Some loggers also store the data on 'smart card' removable memories, for reading back at base.

Telemetry

The advantages of intelligent data loggers over the old-style mechanical and manual systems are obvious, yet there is an additional technique that enhances the logger yet further. This is telemetry, or the transmission of data from a remote field site to a distant base. Obtaining data in real time is of course essential for various applied climatological practices, weather forecasting and river management, but it also has other advantages. First, it reduces the need for expensive field trips to collect the data and it also gives timely warning of equipment failure, thereby reducing data loss. A telemetering outstation is simply a data logging station with a communications channel added, which can be a telephone line, a ground-based UHF radio link or communication via satellite. Where a telephone line can be installed economically and where the telephone network is reliable, this is generally the best option. Where lines cannot be laid, for whatever reason, but where distances are not too great, UHF radio is usually a good alternative. However, UHF radio links only operate over line-of-sight distances because radio waves in this frequency band behave like light and do not bend far beyond the horizon. In areas lacking a telephone network and where distance or terrain prevents line-of-sight communication, satellite communication now offers an ideal alternative. It also allows the remotest locations to be accessed.

The most commonly used satellites at present for this purpose are the weather satellites (METEOSAT, GOES, GMS and GOMS) operated primarily for weather image collection for remote sensing, but equipped also with the capability of relaying transmissions from 'data collection platforms' (DCPs), which are remote logging stations with a radio transmitter directed at the satellite. Data sent via satellite are most usually received over the communications network that links the world's NWSs, known as the Global Telecommunications System. In some cases, notably with METEOSAT, the user can also receive transmissions directly from the satellite on a small dish antenna and this has many advantages for users not directly involved in weather forecasting. The satellites listed above are in geostationary orbit around the equator and thus appear at a fixed point in the sky, but there is also a network of satellites (ARGOS for example, in low polar orbit), which move across the sky. This orbit produces frequent passes over the polar regions but less frequent coverage the nearer the equator is approached. A detailed discussion of satellite telemetry beyond this brief outline is not possible in the limited space available, but Reynolds (1993) and Strangeways (1990) provide more details.

REMOTE SENSING

All of the instruments so far discussed are ground based, and are in direct contact with the variables they measure. However, from the same weather satellites that are used for telemetry and also from others including Landsat, Insat, Meteor, SPOT and others, images of the earth are being continuously produced. While remote sensing includes observations from aircraft and from the ground, it is those from space platforms that are the most significant. Measurements are made by on-board radiometers, multispectral scanners, thematic mappers and cameras which sense different bands of the electromagnetic spectrum. Observations are either passive, in which natural radiation is sensed, or active, in which radiation (radar for example) is emitted from the platform and the returned signal measured. The sensed spectral bands range from the visible (often subdivided into narrow bands), through the near infrared, far infrared and microwaves.

Depending on orbit and altitude, the resolution (pixel size) obtained varies from 5 km for satellites such as METEOSAT and GOES in geostationary orbits (just beneath the satellite) to the 10, 20 or 30 m of Landsat and SPOT in low polar orbits. Also dependent on orbit is the repeat time of the images, which varies from every 30 minutes for the geostationary satellites to several days for some of the satellites in polar orbits, there being a trade-off between resolution and repeat times. The images can be interpreted to give information on a wide variety of variables, including those sensed by the ground-based instruments, such as temperature, humidity, wind and rainfall as well as cloud, snow and ice cover, the extent of flooding, sea state and currents. However, how this is achieved and the precision attainable are beyond the scope of this chapter and, for detail, readers should refer to, for example, Barrett and Curtis (1993).

Remote sensing has the very considerable advantage of being able to make measurements over the whole of the earth's surface without having to place any instruments on it or to visit any sites. But the precision of the data rarely matches that of *in situ* point measurements. Indeed remotely sensed data need to be calibrated against so-called 'ground-truth' measurements made by instruments on the surface, with the two techniques complementing each other. Remote sensing can also be carried out from the ground, for example radar measurements of rainfall. However, this too requires ground-truth measurements, telemetered in real time from raingauges to act as a check and to calibrate the radar readings. Again, wide areal coverage in real time is the technique's great advantage. 'Sounders' emitting radar and acoustic pulses upwards into the atmosphere can also give information on temperature and winds aloft up to many kilometres, but such instruments are expensive.

CONCLUSIONS

For any understanding of the environment, measurements are needed. Yet today we are still largely dependent on the instruments of a past age. All of our direct knowledge of past climate comes from simple, primitive mechanical and manual instruments, and this continues to this day and will continue for some time yet. Fortunately microelectronics has, in the last 30 years, given us the wherewithal to improve on this and allows us to deploy sophisticated ground-based, precise instruments in the remotest corners of the globe, complemented by remotely sensed measurements from instruments on space platforms. How should we take advantage of these new techniques? For a start, we urgently need a better international consensus on how to deploy the new ground-based instruments in the right places and to ensure, through agreed standards, that the data they produce are directly intercomparable. The WMO is best placed to bring this about, with the co-operation of all countries, and the WMO needs to act urgently over this, perhaps in co-operation with the World Bank. This should include the integration of hydrology with meteorology and oceanography as well as all the other diverse environmental data requirements of today (Strangeways, 1995). Such data are the only hard facts we have about the environment and applied climatological practices and they are absolutely vital.

REFERENCES

Barrett, C.D and Curtis, L.F. (1993) *Introduction to Environmental Remote Sensing*, London: Chapman and Hall.

Biswas, A.K. (1970) *History of Hydrology*, Amsterdam and London: North-Holland.

Penman, H.L. (1948) 'Natural evaporation from open water, bare soil and grass', *Proceedings of the Royal Society*, Series A 193: 120–46.

Reynolds, R. (1993) *MOSAIC Data Collection System*, EUMETSAT Publication EUM UG 02.

Rodda, J.C. (1967) *The Rainfall Measurement Problem. Reports and Discussions*, IAHS General Assembly of Berne, IAHS Publication 78, 113–20.

Strangeways, I.C. (1985) 'Automatic Weather Stations', in J.C. Rodda (ed.) *Facets of Hydrology*, Vol. II, Chichester: John Wiley, 25–68.

Strangeways, I.C. (1990) *The Telemetry of Hydrological Data by Satellite*, Wallingford: Institute of Hydrology Report No. 112.

Strangeways, I.C. (1995) *Merging Meteorological, Hydrological and Environmental Data*, Geneva: *WMO Bulletin* 44(1): 56–60.

3

STATISTICAL CONSIDERATIONS

Scott M. Robeson

INTRODUCTION: THE VALUE OF STATISTICS

Climatologists have vast quantities of data to explore. Daily air temperature and precipitation measurements, for instance, are taken at thousands of locations worldwide, resulting in millions of observations each year. Satellite observations and computer models can produce even larger amounts of information to digest. Even with the vast quantity of data, however, climatologists often cannot produce definitive answers when prompted with seemingly straightforward questions (e.g. 'Have land-use changes made the USA warmer and drier?'). Some of the uncertainty results from a less-than-complete understanding of the climate system and some is associated with data inadequacies, particularly the inability to resolve all of the relevant spatial and temporal scales of climatic variability. While statistical analysis often cannot provide definitive answers, it can be of great help in establishing relationships and quantifying uncertainties within climatic data. Statistical analysis, for instance, provides objective methods (e.g. correlation and regression analysis) for quantifying the relationships between important climatic variables such as solar radiation and cloud cover. Since the climate system is extremely complex and many of its components are not well understood, statistical analysis and models are particularly important in providing meaningful and useful simplifications of system variability.

TIME SERIES ANALYSIS

Nearly all climatological data can be ordered as a time series. This is not the case in some environmental sciences, such as certain areas of geology, where the value of a variable at a given location does not change over human timescales. Since climatological processes operate over a wide range of timescales (e.g. daily, synoptic, annual, etc.), time series analysis provides great opportunities for detecting, describing and modelling climatic variability and impacts. The number of values in a time series (n) and the sampling interval or time step (Δt; e.g. one day) determine the duration of the series ($n\Delta t$) and provide critical limitations on the types of information that can be derived from a time series. In analyses of atmospheric turbulence, for instance, time steps of 0.1 seconds are needed to resolve all of the microscale variability in wind speed and associated variables such as air temperature and humidity. If the appropriate time step is not used, then information that varies at shorter timescales may appear as information at longer timescales – a phenomenon known as *aliasing* (Figure 3.1; Hamming, 1983). Another important property of a time series is whether the series is *stationary* or *non-stationary*. Non-stationary time series

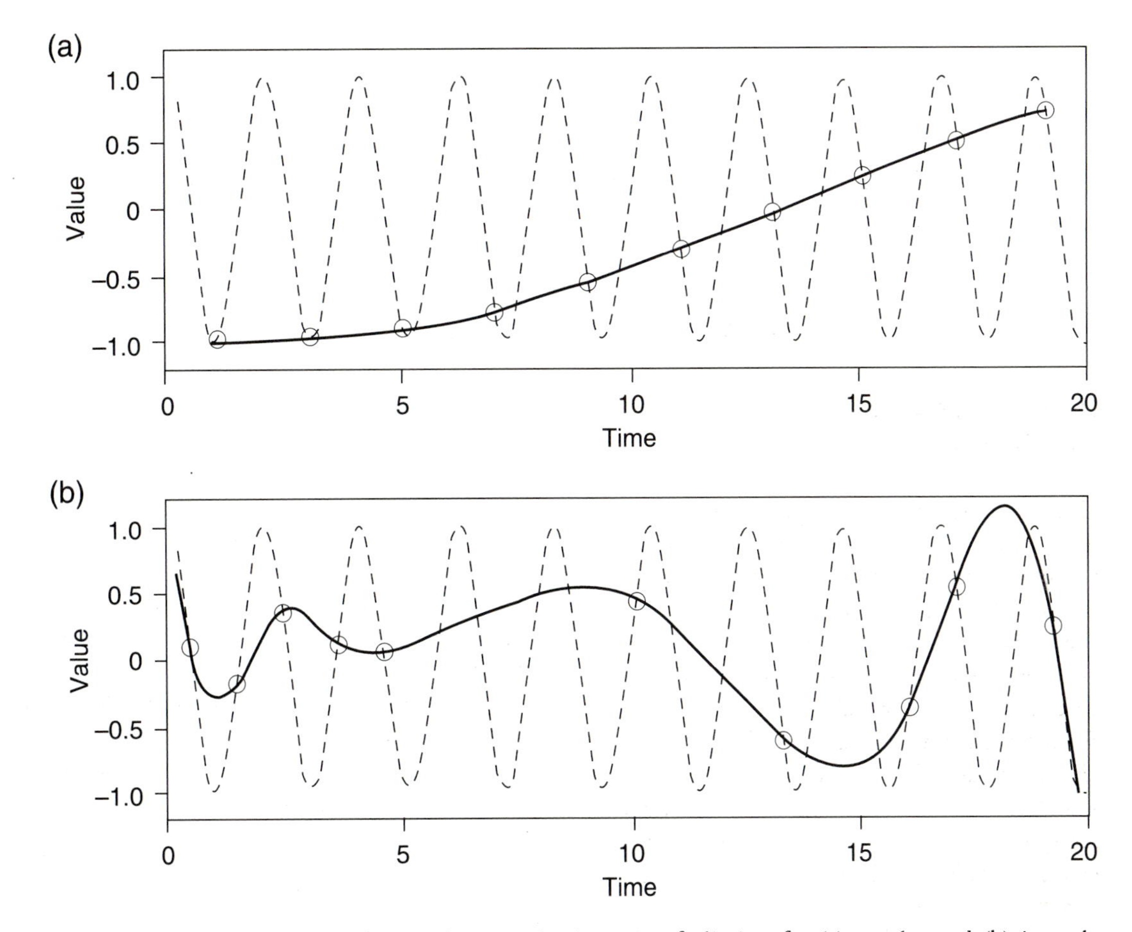

Figure 3.1 A hypothetical time series showing the impacts of aliasing for (a) regular and (b) irregular sampling. Information that is contained at shorter timescales (dashed line) can appear as information at longer timescales (solid line) if the time series is not sampled appropriately (open circles). Although aliasing is usually discussed in the context of time series analysis, it also affects spatial data

have statistical properties (e.g. mean and/or variance) that change through time. Since most climatological time series have diurnal and annual cycles, most contain some amount of non-stationarity.

Graphical Depiction

The most basic, and sometimes most powerful, methods for statistical analysis are graphical ones. A traditional time series plot provides a quick method for visualizing temporal patterns in data, including trends and cycles (Figure 3.2(a)). Traditional time series graphs can also help to identify discontinuities in data that may be the result of non-climatic factors such as station moves (e.g. a change in station elevation often is apparent in both air temperature and precipitation time series) or instrumentation changes. Of course, multiple time series can be plotted on the same graph to examine how variables are related through time (Figure 3.2(b)). Less traditional time series plots, such as a 'state-space' approach (Lorenz, 1993), are

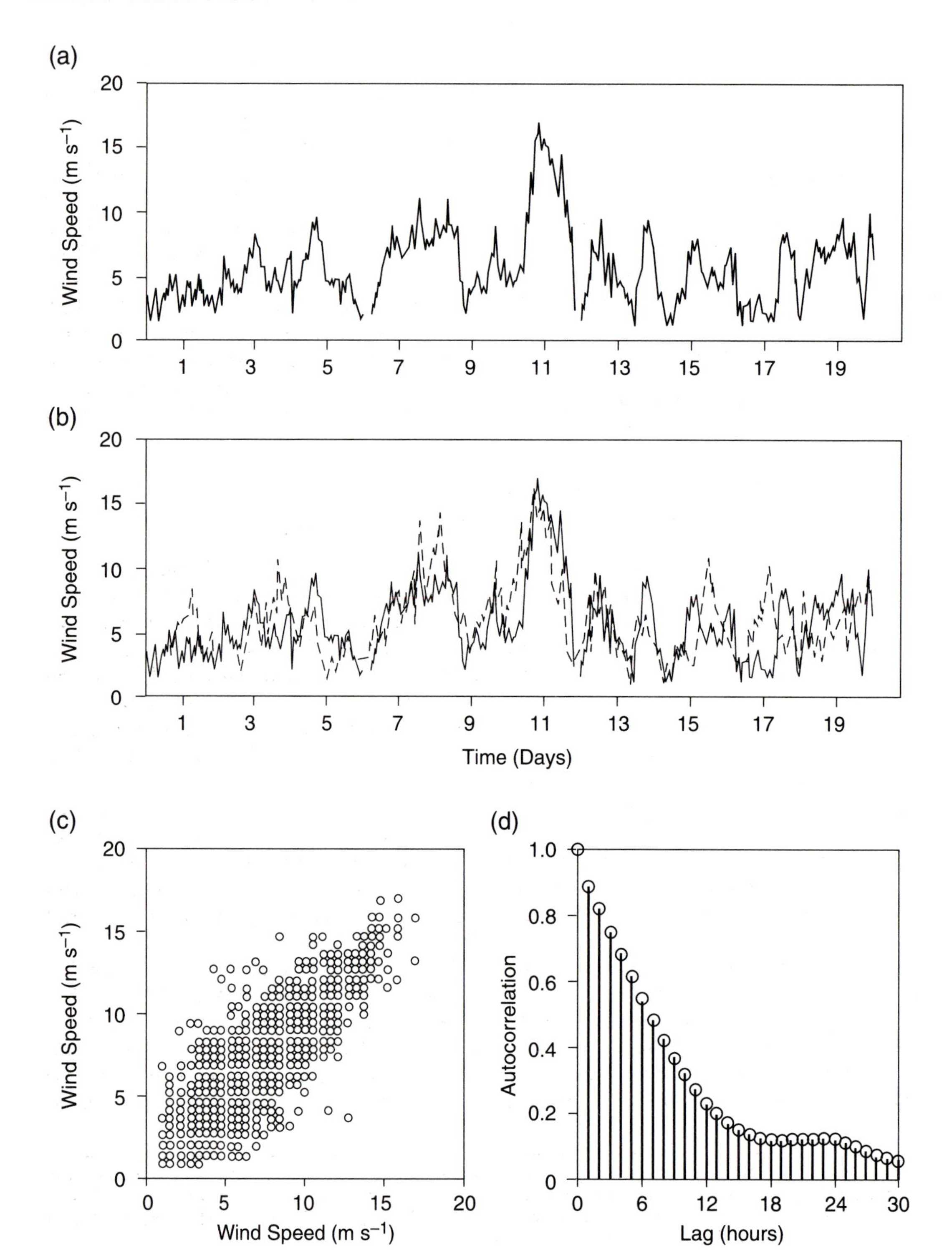
(a)
Wind Speed (m s^{-1})
20
15
10
5
0
1
3
5
7
9
11
13
15
17
19
(b)
Wind Speed (m s^{-1})
Time (Days)
(c)
Wind Speed (m s^{-1})
0
5
10
15
20
Wind Speed (m s^{-1})
(d)
Autocorrelation
1.0
0.8
0.6
0.4
0.2
0
0
6
12
18
24
30
Lag (hours)

helpful in identifying system evolution and hysteresis.

Autocorrelation

Scatterplots of lagged time series components are similar to state-space diagrams and can be used to identify relationships between successive values in a time series (Figure 3.2(c)). Correlation between successive values in a time series is a property known as *temporal autocorrelation* (e.g. today's wind speed is similar to yesterday's wind speed). Most climatic time series exhibit temporal autocorrelation, resulting in a tendency for similar values (above or below the mean) to occur in series. In other words, the climate system exhibits persistence or inertia (e.g. heat waves, rainy periods, etc.). Hourly air temperature exhibits strong autocorrelation; however, the diurnal cycle of air temperature is more appropriately considered a type of non-stationarity since it produces systematic changes in the statistical properties of the time series (the mean value of noon temperature is different from the value for 18.00 hours). In general, non-stationarities such as the diurnal and annual cycle should be removed before estimating autocorrelations. Non-stationarity in the mean can be removed by subtracting the average value for each hour or day from the time series or by fitting and removing an analytical function, such as a series of sinusoids or polynomials.

The most natural way to visualize autocorrelation in a time series is by plotting the autocorrelation as a function of lag time, that is the *autocorrelation function* (e.g. Figure 3.2(d)). An autocorrelation function (acf) clearly shows the amount of persistence in a time series and also reveals non-stationarities (e.g. the rise in the acf near lag 24 in Figure 3.2(d) is caused by the diurnal cycle in wind speed). Nearly all acfs derived from climatological time series exhibit a steady decay with lag time, whereby the amount of time that it takes the autocorrelation function to reach a value near zero indicates how much 'memory' a system exhibits. Wind speeds in many mid-latitude locations often have hourly autocorrelation functions that reach near-zero values after approximately 36 hours (Brett and Tuller, 1991), meaning that (in general) the conditions influencing wind speeds 36 hours ago are nearly unrelated to the current conditions influencing wind speeds.

Time Series Models

Once the variability of a climatological time series has been identified, it is useful to develop an explicit mathematical model for the time series. A time series model provides a concise representation that may contain only a handful of parameters, as opposed to the n original time series values. In addition, the time series model can be thought of as the 'generating process' for the data, allowing unobserved time series values under any conditions to be simulated (Wilks, 1995). The applied climatologist, therefore, can infer statistical aspects of the time series that were not observed (e.g. the probability of having three consecutive days below 0°C in central Florida).

One useful method for time series modelling is *component decomposition*. Most climatological time series can be decomposed into an additive time series that may contain (1) a linear trend, (2) an annual cycle, (3) a diurnal cycle, (4) autocorrelation and (5) a random component.

Figure 3.2 Graphical methods for analysing time series. Hourly wind speeds (in ms^{-1} at 10 m above the surface) (a) at Fargo, North Dakota, during January 1986, (b) at Fargo (solid line) and Minot, ND (dashed line), (c) scatterplot of hourly wind speed at Fargo versus the following hours's wind speed during 1986, and (d) the temporal autocorrelation function of hourly wind speeds at Fargo during 1986

Not all climatic time series contain all of these components, but many do (e.g. hourly air temperature data). Imagine a daily average air temperature time series that contains a long-term linear trend. Most daily air temperature time series also contain a distinct annual cycle associated with the annual cycle of solar radiation. In addition, air temperature time series typically exhibit a large amount of autocorrelation. Below, a daily average air temperature time series from Bloomington, Indiana, is used to illustrate how a time series can be decomposed (over 100 years of daily data are available for the Bloomington climate station; however, only 5 years are used here to simplify the graphs).

The Bloomington air temperature time series (Figure 3.3(a)) exhibits several of the components discussed above: annual cycle, autocorrelation and a random component. The annual cycle is a form of non-stationarity and can be modelled as a series of sine and cosine functions (also known as harmonics in a Fourier series; Figure 3.3(b)):

$$T_t = a_0 + \sum_k \left[a_k \cos \frac{2\pi kt}{n} + b_k sin \frac{2\pi kt}{n}\right] + e_t$$

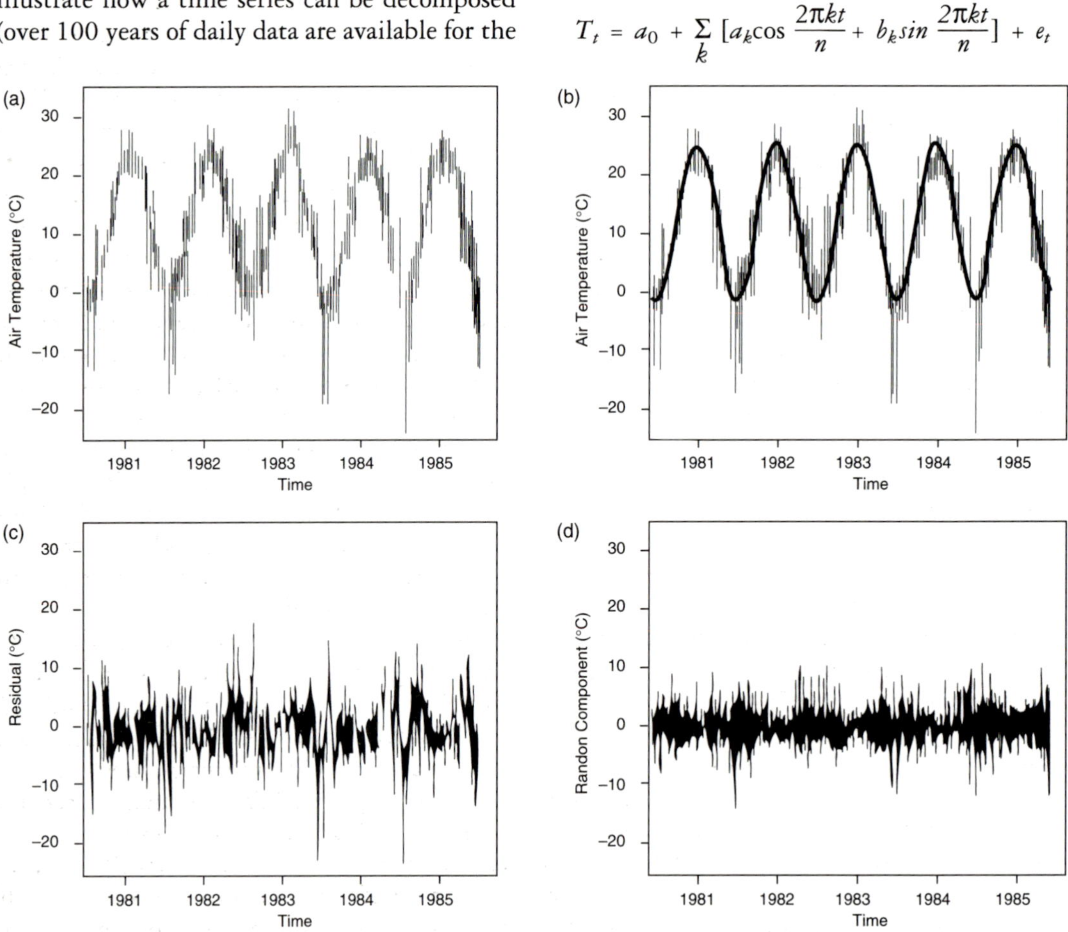

Figure 3.3 Daily mean air temperatures (in °C) in Bloomington, Indiana, during the early 1980s decomposed into a variety of components: (a) original time series, (b) original time series with a series of sinusoids fitted to the annual cycle, (c) residual time series after removing sinusoids, and (d) a random time series resulting from removing an autoregressive component from the residual time series

where T_t is the daily air temperature time series on day t, a_k and b_k are called Fourier coefficients, and e_t is the stationary (or residual) time series. Often, a few harmonics (the index k refers to the harmonic number) are sufficient to model the annual cycle of air temperature (Epstein, 1991). After removing the annual cycle, the e_t series (Figure 3.2(c)) should be stationary; however, more of its temporal variability – specifically, its autocorrelation – can be described and modelled using an autoregressive component, such as

$$e_t = \phi_1 e_{t-1} + r_t$$

where ϕ_1 is an autoregressive coefficient, e_{t-1} is the stationary time series on the previous day, and r_t is a random time series (Figure 3.3(d)). Since only one autoregressive coefficient is used, this is a first-order autoregressive model; however, higher-order models (e.g. incorporating a $\phi_2 e_{t-2}$ term) sometimes are needed (Katz and Skaggs, 1981). If the r_t series is normally distributed (which it often is), then it can be described by two parameters: its mean and standard deviation. The entire time series then can be modelled as follows:

$$T_t = a_0 + \sum_k \left[a_k \cos \frac{2\pi kt}{n} + b_k \sin \frac{2\pi kt}{n}\right] + \phi_1 e_{t-1} + r_t$$

The time series model provides a useful conceptual representation of the variability of air temperature; however, it can also be used to produce probabilistic forecasts and scenarios of air temperature change. Since the r_t time series is random, a random number generator can be used to simulate realistic time series values. By utilizing the other components of the time series along with the randomly generated r_t series, a synthetic time series that has exactly the same statistical properties as the original time series can be produced. The synthetic time series can be of any duration, allowing the applied climatologist to infer statistical properties of the original time series that were not observed. Suppose that a climatologist is interested in studying the probability and impacts of runs of extremely high air temperatures (e.g. air temperatures above 35°C for five consecutive days). A short time series (such as the 5 year one in Figure 3.3) may not contain any events of that magnitude. Using the time series model, however, several thousand years of air temperature variability could be generated and the probability of a variety of extreme events could be easily examined. In addition, the model would provide information on when during the year the events might occur.

Another use of time series models would be to generate scenarios of climate change (Mearns *et al.*, 1984; Chen and Robinson, 1991; Wilks, 1992). Climatic conditions that contain increased variability could be simulated by modifying the standard deviation of the r_t series. Similarly, if the climate of a region is expected to exhibit more persistence under a climatic change scenario, then the autoregressive coefficient could be increased and the statistical properties of the time series could be examined. Historical analogues of future climatic variability (Henderson-Sellers, 1986) also could be incorporated into a variety of elements in the time series model. For example, the statistical properties of years with above-average air temperature could be estimated and time series of these anomalous conditions could be synthesized.

In addition to generating scenarios (or addressing 'what if' questions), time series models also can be used to generate probabilistic forecasts. In studies of air pollution variability, applied climatologists often need to estimate the probability of pollutant concentrations exceeding federal standards. A simple examination of the observed probability density function (pdf) can answer this question, but the pdf groups all data together and does not provide direct information on how likely exceedences are at a given time of year. Stratifying the data by month is one solution, but at some level of stratification, data become limited (e.g.

there are not enough observations to compare conditions on 21 July and 22 July). The time series model can generate a pdf for any day of the year (if hourly data are used in the model, a pdf for any hour of any day of the year could be generated). In addition, the time series model can generate *conditional* pdfs that not only incorporate information on the time of year, but also incorporate information on how past conditions may affect future conditions (Robeson and Steyn, 1989). In studying urban ozone, for instance, a conditional pdf could answer the following question: 'How likely is an ozone concentration of 120 parts per billion on 9 July if the daily maximum ozone concentration on 8 July was 119 parts per billion?' The time series model and conditional pdf, therefore, not only produces a probabilistic estimate of attaining any given ozone concentration, but that estimate is dependent on time of year and on yesterday's ozone concentration. The component decomposition approach is useful for a wide variety of time series variables in applied climatology. One can imagine useful decompositions and simulations that utilize hydrologic variables, solar radiation, wind energy, degree days, and nearly any other climatological variable.

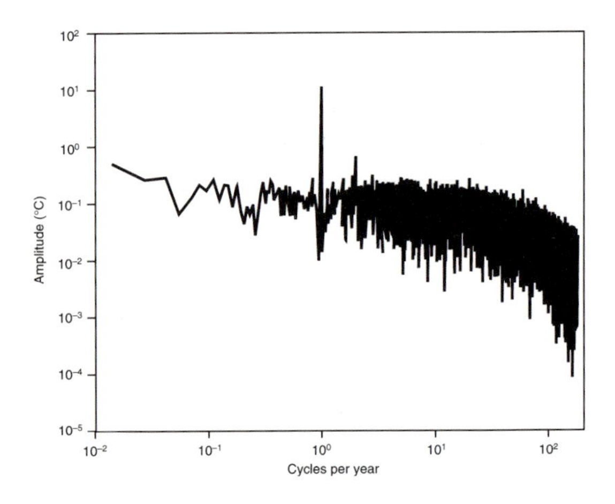

Figure 3.4 Amplitude spectrum of daily air temperatures (°C) in Bloomington, Indiana, for 1901–94. The annual harmonic (at one cycle per year) contains the most information; however, most of the variance in the time series is spread over a wide range of timescales (from one cycle per 94 years to one cycle per 2 days)

Frequency-domain Methods

The other fundamental approach to time series analysis is to analyse data in the frequency domain. The time-domain approaches discussed above retain the data in the form that they were observed (i.e. organized by time). Frequency-domain approaches transform the data into a series of periodic (e.g. sine and cosine) functions, each with a characteristic amplitude and frequency. The frequency of the functions represents the timescale (e.g. one cycle per day) and the amplitude indicates how much information (or variance) is in the time series at that timescale (Figure 3.4). Mathematically, it is straightforward to transform data from the time domain to the frequency domain (and back) using the Fourier transform. Many analyses in applied climatology are naturally done in the frequency domain since many of the dominant forcing processes in climatology are periodic (e.g. the diurnal and annual cycles of solar radiation). In addition, many time-domain calculations (e.g. estimation of the autocorrelation function, smoothing of time series, etc.) are simpler when the time-domain data are transferred to the frequency domain. Bloomfield (1976), Hamming (1983), Holloway (1958) and Rayner (1972) provide further discussion of frequency-domain methods.

SPATIAL ANALYSIS

Although many interesting scientific questions can be addressed via time series analysis, spatial analysis – using data from many different locations – can address an additional range of questions (e.g. 'What locations in the state of

Texas have the greatest potential for solar energy generation?'; 'What locations in the state of Texas have the largest interannual variability in solar energy resources?'; etc.). The spatial organization of data is an important characteristic of most climatological data; however, many statistical methods were not developed with spatial data in mind (Cressie, 1993). As a result, many common statistical techniques (such as significance testing) cannot be applied to spatial data without some modification or qualification. Most climatological data are collected at irregularly spaced locations such as networks of surface climate stations (although gridded data from computer models and satellite-based sensors are also available). Irregularly spaced data are usually interpolated to a regular grid before further spatial analysis. Interpolation is needed both for mapping and to avoid uneven spatial weighting. For instance, if a large number of stations are located in the southern part of a region and few are in the north (e.g. the network of climate stations in Canada), then attaching equal weight to each station would bias any statistical estimates towards the southern part of the region (the statistics probably also would have a bias towards low-elevation locations and urban areas).

Mapping Considerations

Like temporal data, some of the most powerful ways for analysing spatial data are graphical ones. For spatial data, this often translates into creating a map. Climatologists should be familiar with the principles of thematic mapping (Robinson *et al.*, 1978), including map styles, symbolization, and the use of colour. One of the most fundamental cartographic considerations, however, is the choice of map projection. When analysing most maps, climatologists are interested in comparing one region to another; therefore, maps that preserve area relationships typically are most useful. Equal-area map projections include the Albers conical projection for continental-scale maps (e.g. this projection is commonly used for maps of the USA) and the Lambert cylindrical and Molleweide projections for global-scale maps. To illustrate the importance of map projections, consider global climate model simulations of global warming scenarios, which often show the largest changes (from current conditions) in polar regions. A Mercator projection would make the polar regions appear much larger than similarly sized areas elsewhere on the map, exaggerating the impacts of the simulated changes.

Some of the most fundamental spatial calculations that can be performed with climatic data are distance calculations. In many cases, climatological analyses are performed over large areas of the earth's surface. Therefore, the distance (d_{ij}) between two locations i and j (specified by their latitude and longitude: ϕ_i,λ_i and ϕ_j,λ_j) cannot be treated as planar (e.g. as a two-dimensional Euclidean distance) or calculated within a map projection. It must be calculated on the surface of a sphere or as a great-circle distance, as follows:

$$d_{ij} = R \cos^{-1} [\sin\phi_i \sin\phi_j + \cos\phi_i \cos\phi_j \cos (\lambda_i - \lambda_j)]$$

where R is the radius of the earth. Spherical and planar distances diverge with increasing distance between locations. Errors resulting from planar distance calculation have particularly important implications for spatial interpolation (Willmott *et al.*, 1985). Mapping and spatial interpolation, however, are inextricably linked. When mapping continental-, hemispheric- or global-scale climatic data, it is important to perform any spatial operations on the surface of a sphere (or geoid) since operations performed within map projections will be in error. In addition, each map projection will produce a different map (e.g. isolines will be in different locations) since each will have its own characteristic deformation and errors.

Spatial Autocorrelation

One of the fundamental properties of climatological variables is their spatial coherence, that is how rapidly a variable changes with distance from a location. One way to examine spatial coherence is to estimate the correlation of a variable measured at one location with data from a number of locations. Expressing this correlation as a function of distance produces the spatial autocorrelation function. Nearly all climatological variables have a spatial autocorrelation function that decays with distance (Figure 3.5). In other words, nearby locations have more similar climatic variability than distant locations do. How quickly a spatial autocorrelation function decays indicates a good deal about a climatic variable, including (1) the fundamental spatial scales of variability, (2) whether spatial climatic variability is well resolved by the station network, and (3) how reliable spatial interpolation will be. In addition, estimating spatial autocorrelation functions using hourly, daily, monthly and annual data can help to determine the optimal timescale for performing spatial analysis (Figure 3.5).

In climatology, spatial autocorrelation functions are estimated using time series from a number of locations (e.g. correlations might be estimated using monthly air temperature data from a number of climate stations over a 30 year period). If the time series contain non-stationarities (e.g. the annual cycle), then the spatial autocorrelation functions can be biased (Gunst, 1995). For example, two locations might produce high spatial autocorrelations simply because both have high values in July and low values in January. At any given point in time, however, the locations may not be behaving in a similar fashion. Like time series data, spatial data should have all non-stationarities removed before spatial autocorrelation functions are estimated (this has not been done in Figure 3.5 to show the impacts of including non-stationarities). A number of climatological variables also have spatial autocorrelation functions that are dependent on direction (Thiebaux, 1976), a property known as *anisotropy*. Air temperatures, for instance, may be more highly correlated in east–west directions than in north–south directions. Once again, this information provides important insight into how climatologists can best resolve spatial climatic variability (e.g. greater densities of stations may be needed in north–south directions than in east–west directions).

Spatial Interpolation

Spatial interpolation usually is needed to analyse fully spatial variability of climatological observations. Although spatial interpolation frequently is used to generate gridded fields (e.g. for comparison with general circulation model output or for mapping), all isoline maps produced from observational data require interpolation.

In a general sense, all methods of spatial interpolation estimate values of a variable (at unsampled locations) by using a combination of values at sampled points (z_i). As an example, inverse distance weighting explicitly provides the weights (w_{ij}) between sample points i and grid point j as

$$w_{ij} = d_{ij}^{-\gamma}$$

where γ is usually one or two. Again, for most climatological data, the d_{ij} should be estimated spherically. The weights would be used to generated estimated values at any grid point j using.

$$\hat{z}_j = \frac{\Sigma_i w_{ij} z_i}{\Sigma_i w_{ij}}$$

Other methods of spatial interpolation (e.g. thin-plate splines, triangular decomposition, etc.; see Bennett *et al.*, 1984; Lam, 1983) are implemented in different ways, but also are

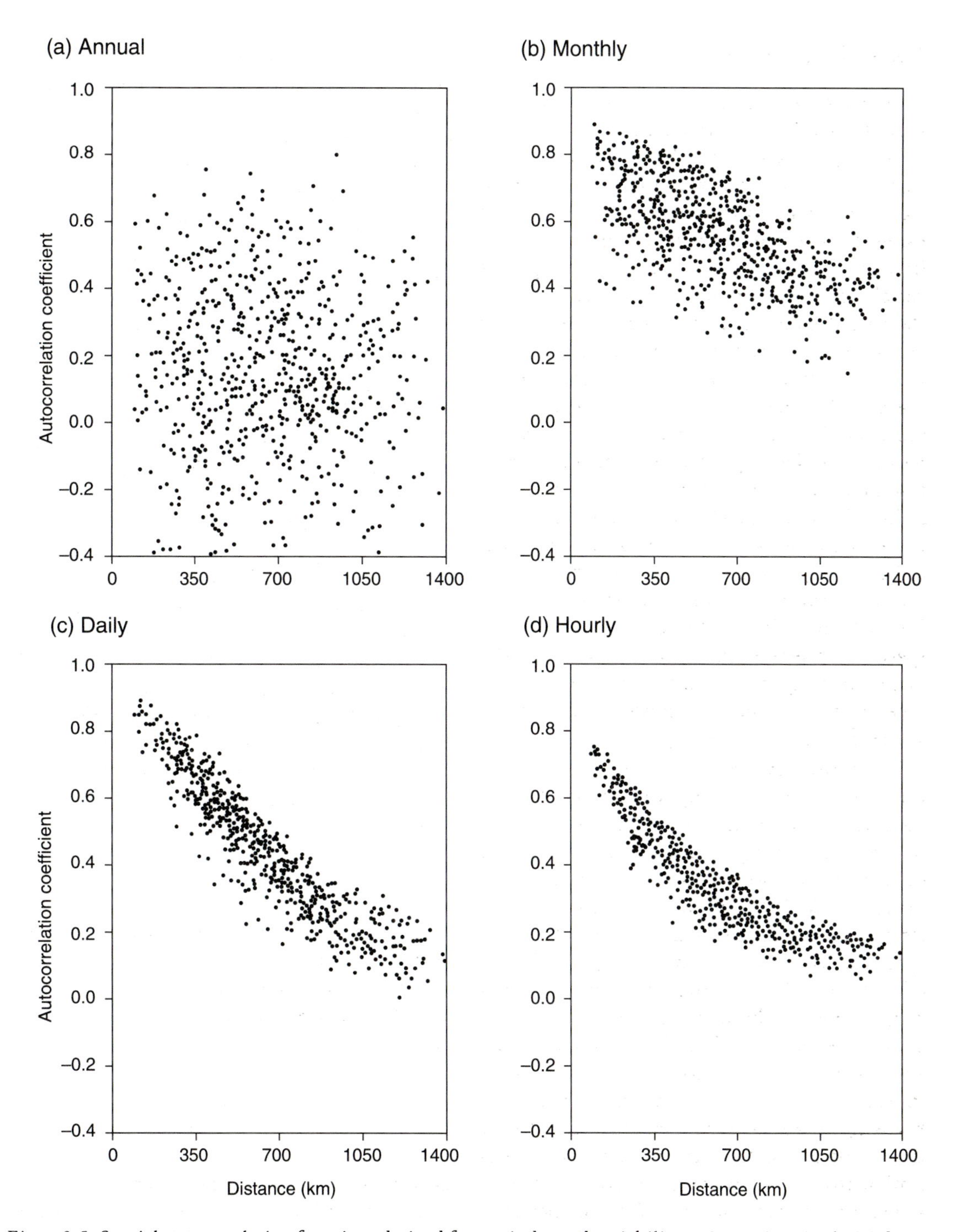

Figure 3.5 Spatial autocorrelation functions derived from wind speed variability at 37 stations in the Midwest during 1961–90, using (a) annual average, (b) monthly average, (c) daily average and (d) hourly data

fundamental functions of distance. Some methods, such as objective analysis and kriging, make explicit use of spatial decay functions (e.g. spatial autocorrelation functions or variograms) to determine optimal spatial weighting (Thiebaux and Pedder, 1987; Cressie, 1993). In data-rich regions, most spatial interpolation methods produce similar results; however, data-poor regions can produce vastly different spatial representations (Robeson, 1994).

As with most methods of data analysis, it is important to express the uncertainty associated with spatial interpolation, although this is rarely done. One method for expressing spatial uncertainty is a resampling method known as *cross-validation*. Cross-validation involves removing a certain portion of the data and trying to estimate their values using the remaining data (Isaaks and Srivastava, 1990; Robeson, 1994). Errors derived from cross-validation, therefore, indicate the uncertainty associated with spatially interpolating data from a given station network. Other resampling methods, such as bootstrap and jack-knife samples (Efron and Gong, 1983), also provide useful measures of uncertainty for spatial climatic data (Robeson, 1995).

Nearly all methods of spatial interpolation used in climatology and elsewhere are univariate, that is they only utilize information regarding the one variable of interest (e.g. sulphur dioxide concentrations). However, many climatological variables are closely associated with other variables that may be observed at higher resolution and, therefore, provide greater spatial detail. One of the most useful ancillary variables for incorporating into spatial interpolation is topography. Air temperature, for instance, is very closely related to elevation; therefore, adding elevation corrections into spatial interpolation can vastly improve spatial interpolation (Hutchinson and Bischof, 1983; Ishida and Kawashima, 1993; Willmott and Matsuura, 1995). Also, if a detailed mean field is available (e.g. long-term averages of air temperature), then anomalies from the mean field for any given month or year can be interpolated reliably (since anomalies are not dependent on elevation) and then added back onto the mean field (Willmott and Robeson, 1995). A similar approach would be to use satellite observations of closely related information to aid spatial interpolation of other surface climate variables.

CONCLUSIONS

The main goals of statistical analysis are to summarize and quantify relationships within data and to specify uncertainty with these relationships. In addition to statistical uncertainty, climatologists also need to consider uncertainties from observational error, such as observer bias, changes in observation practices (e.g. from a midnight-to-midnight to 07.00–19.00 observation schedule for daily maximum and minimum temperature), changes in instrumentation (e.g. switching from mercury-in-glass thermometers to thermistors), and station moves (e.g. from downtown to airport locations). Sometimes these errors can be removed from the data (Mitchell (1953) and Karl *et al.* (1989) provide a discussion of these errors and techniques), but many important changes in observing practices are not well documented.

Other uncertainties with observational data are related to the spatial and temporal representativeness of a particular set of observations. Air temperature observations, for instance, can be described as 'point' data (as opposed to the areal data obtained from satellite observations);

CONCLUSIONS *continued*

however, the air temperature measured at a particular location should represent conditions over some area. Typically, this area increases when air temperatures are averaged through time. Annually averaged air temperatures are representative of larger areas (i.e. they are less spatially variable) than hourly or daily air temperatures. Similarly, the temporal representativeness of data is sometimes difficult to assess. 'Hourly' wind observations in the USA, for instance, are usually taken for 5 minutes at the end of an hour, rather than over the entire hour and averaged. Daily average air temperature often is the simple average of daily maximum and daily minimum air temperature, rather than a true average taken over the course of the entire day. The implications of spatial and temporal representativeness are critical to any statistical analysis in climatology.

Since climatology is a discipline where a large amount of the variability (both spatial and temporal) is governed by well-understood physical laws (e.g. air temperature decreases with elevation owing to well-known thermodynamic principles), scientific considerations should take precedence over statistical considerations when analysing climatological data. The concept of *statistical significance* is less relevant than the issue of *scientific importance*. As an example, if 87,600 (10 years × 365 days × 24 hours) observations of the concentration of a pollutant and wind speed produce a correlation coefficient (r) of 0.1, that correlation would be statistically significant at nearly any level. A correlation of 0.1, however, does not have much scientific importance in climatology since only 1 per cent of the variability in the pollutant can be attributed to variability in wind speed (as measured by r^2). In other words, a statistically significant result is not necessarily a scientifically important result. Conversely, a statistical analysis that does not produce 'significant' results (owing to a very small sample size) may be scientifically important if supported by solid physical reasoning.

Since climatological data do not conform to many standard statistical assumptions, climatologists also need to be careful when applying standard statistical analyses and interpreting the associated results (Bayley and Hammersley, 1946; Wong *et al.*, 1989). Rarely are climatological data either randomly sampled or independent of one another. Since many analyses use historical data or existing data networks, most climatological data are 'samples of convenience' (Freedman *et al.*, 1978) and the sampling process cannot be controlled. Standard statistical methods, therefore, often are not appropriate, making the creative use of graphical data analysis (Tukey, 1977; Tufte, 1983), the use of alternative methods (Efron and Gong, 1983; Mark and Church, 1977; Wong *et al.*, 1989), and the development of new methods (Willmott, 1981; Hanson *et al.*, 1992) critical to the advancement of statistical climatology.

Statistical methods are used to summarize patterns within data; however, the most useful summary is often not a temporal or spatial average. Climatic variability, in the form of temporal fluctuations and spatial variations, is often more important than changes in average values. Much of the global warming debate, for instance, focuses on changes in global average air temperature

CONCLUSIONS *continued*

anomalies; however, there is always important interannual variability, not necessarily systematic change, in air temperature that has important implications for applied climatological research (Katz and Brown, 1992). In addition, global and hemispheric averages disregard the *spatial distribution* of climatic changes and variability. There are often years with very similar global average air temperature or precipitation; however, the spatial distributions of these variables (and their climatic impacts) can be vastly different. When using statistical analysis in applied climatological research, therefore, one must consider not only the 'average' conditions at a given location, but also the variability of important climatological variables over a wide range of temporal and spatial scales.

REFERENCES

Bayley, G.V. and Hammersley, J.M. (1946) 'The "effective number" of independent observations in an autocorrelated time series', *Journal of the Royal Statistical Society* B8: 184–97.

Bennett, R.J., Haining, R.P. and Griffith, D.A. (1984) 'The problem of missing data on spatial surfaces', *Annals of the Association of American Geographers* 74: 138–56.

Bloomfield, P. (1976) *Fourier Analysis of Time Series*, New York: John Wiley.

Brett, A.C. and Tuller, S.E. (1991) 'The autocorrelation of hourly wind speed observations', *Journal of Applied Meteorology* 30: 823–33.

Chen, R.S. and Robinson, P.J. (1991) 'Generating scenarios of local surface temperature using time series methods', *Journal of Climate* 4: 723–32.

Cressie, N. (1993) *Statistics for Spatial Data*, New York: John Wiley.

Efron, B. and Gong, G. (1983) 'A leisurely look at the bootstrap, the jackknife, and cross-validation', *American Statistician* 37: 36–48.

Epstein, E.S. (1991) 'Determining the optimum number of harmonics to represent normals based on multiyear data', *Journal of Climate* 4: 1047–51.

Freedman, D., Pisani, R. and Purves, R. (1978) *Statistics*, New York: Norton.

Gunst, R.F. (1995) 'Estimating spatial correlations from spatial-temporal meteorological data', *Journal of Climate* 8: 2454–70.

Hamming, R. (1983) *Digital Filters*, 2nd edn, New York: Prentice Hall.

Hanson, B., Klink, K., Matsuura, K., Robeson, S.M. and Willmott, C.J. (1992) 'Vector correlation: review, exposition, and geographic application', *Annals of the Association of American Geographers* 82: 103–16.

Henderson-Sellers, A. (1986) 'Cloud changes in a warmer Europe', *Climatic Change* 8: 25–52.

Holloway, J. (1958) 'Smoothing and filtering of time series and space fields', *Advances in Geophysics* 4: 351–89.

Hutchinson, M.F. and Bischof, R.J. (1983) 'A new method for estimating the spatial distribution of mean seasonal and annual rainfall applied to the Hunter Valley, New South Wales', *Australian Meteorological Magazine* 31: 179–84.

Isaaks, E.H. and Srivastava, R.M. (1990) *Applied Geostatistics*, New York: Oxford University Press.

Ishida, T. and Kawashima, S. (1993) 'Use of cokriging to estimate surface air temperature from elevation', *Theoretical and Applied Climatology* 47: 147–57.

Karl, T.R., Tarpley, J.D., Quayle, R.G., Diaz, H.F., Robinson, D.A. and Bradley, R.S. (1989) 'The recent climate record: what it can and cannot tell us', *Reviews of Geophysics* 27: 405–30.

Katz, R.W. and Brown, B.G. (1992) 'Extreme events in a changing climate: variability is more important than averages', *Climatic Change* 21: 289–302.

Katz, R.W. and Skaggs, D. (1981) 'On the use of autoregressive-moving average processes to model meteorological time series', *Monthly Weather Review* 109: 479–84.

Lam, N. (1983) 'Spatial interpolation methods: a review', *American Cartographer* 10: 129–49.

Lorenz, E.N. (1993) *The Essence of Chaos*, Seattle: University of Washington Press.

Mark, D.M. and Church, M. (1977) 'On the misuse of regression in earth science', *Mathematical Geology* 9: 63–75.

Mearns, L.O., Katz, R.W. and Schneider, S.H. (1984) 'Extreme high-temperature events: changes in the probabilities with changes in mean temperature', *Journal of Climate and Applied Meteorology* 23: 1601–13.

Mitchell, J.M., Jr (1953) 'On the causes of instrumentally observed temperature trends', *Journal of Meteorology* 10: 244–61.

Rayner, J. (1972) *An Introduction to Spectral Analysis*, London: Pion.

Robeson, S.M. (1994) 'Influence of spatial sampling and interpolation on estimates of terrestrial air temperature change', *Climate Research* 4: 119–26.

Robeson, S.M. (1995) 'Resampling of network-induced

variability in estimates of terrestrial air temperature change', *Climatic Change* 29: 213–29.

Robeson, S.M. and Steyn, D.G. (1989) 'A conditional probability density function for forecasting daily maximum ozone concentrations', *Atmospheric Environment* 23: 689–92.

Robinson, A.H., Sale, R. and Morrison, J. (1978) *Elements of Cartography*, New York: John Wiley.

Thiebaux, H.J. (1976) 'Anisotropic correlation functions for objective analysis', *Monthly Weather Review* 104: 994–1002.

Thiebaux, H.J. and Pedder, M.A. (1987) *Spatial Objective Analysis: With Applications in Atmospheric Science*, New York: Academic Press.

Tufte, E. (1983) *The Visual Display of Quantitative Information*, Cheshire, CT: Graphics Press.

Tukey, J.W. (1977) *Exploratory Data Analysis*, Reading, MA: Addison-Wesley.

Wilks, D.S. (1992) 'Adapting stochastic weather generation algorithms for climate change studies', *Climatic Change* 22: 67–84.

Wilks, D.S. (1995) *Statistical Methods in the Atmospheric Sciences*, San Diego: Academic Press.

Willmott, C.J. (1981) 'On the validation of models', *Physical Geography* 2: 184–94.

Willmott, C.J. and Matsuura, K. (1995) 'Smart interpolation of annually averaged air temperature in the United States', *Journal of Applied Meteorology* 34: 2577–86.

Willmott, C. J. and Robeson, S.M. (1995) 'Climatologically aided interpolation of terrestrial air temperature', *International Journal of Climatology* 15: 221–9.

Willmott, C.J., Rowe, C.M. and Philpot, W.D. (1985) 'Small-scale climate maps: a sensitivity analysis of some common assumptions associated with grid-point interpolation and contouring', *American Cartographer* 12: 5–16.

Wong, R.K.W., Schneider, C. and Mielke, P.W. (1989) 'Geometric consistency for regression model estimation and testing in climatology and meteorology', *Atmosphere-Ocean* 27: 508–20.

4

CLIMATE MODELS

Ann Henderson-Sellers and Kendal McGuffie

INTRODUCTION: MODELLING THE CLIMATE SYSTEM

Climate Models as 'Tools'

Climate modelling is a young discipline, not yet 30 years old. Despite this youth, it has had great responsibility thrust upon it by the ratification of the United Nations Framework Convention for Climate Change. The ultimate objective of this Convention is to achieve

> stabilization of greenhouse gas concentrations in the atmosphere at a level that would prevent dangerous anthropogenic interference with the climate system. Such a level should be achieved within a time frame sufficient to allow ecosystems to adapt naturally to climate change, to ensure that food production is not threatened and to enable economic development to proceed in a sustainable manner.
>
> (Article 2)

This goal demands the prediction of future climate and of its impacts on natural and human systems. Climate models have become essential tools in the practice of applied climatology.

Forcings and Feedbacks

The climate system is a dynamic system in transient balance. Fluxes (vectors into and out of the system) achieve different budgets over different time and space scales. This concept, which is fundamental to all climate models, can be illustrated in terms of vehicle movement into and out of Manhattan Island, New York. Over time periods greater than a few days, Manhattan has an (approximate) vehicular balance, while over time periods of a few hours there are large negative and positive fluxes of vehicles. If the authorities close all bridges and tunnels or close all the car parks and forbid street parking, the fluxes of vehicles would alter considerably and the net vehicular budget would change in this part of the New York subsystem. However, these drastic changes in Manhattan would have no effect on the vehicular budget of the USA as a whole.

The most important fluxes in the climate system are those of radiant (solar and heat) energy, but the fluxes of water and, to a lesser extent, mass (matter) also affect climate. Net fluxes differ considerably as a function of the time and space period considered and different budgets, the result of the net fluxes, are established when the controlling conditions change. A climate change occurs either because the forcings imposed on the planet change or because the dynamics of the system cause internal variations in fluxes and budgets. Forcings are caused by variations in agents outside the climate system such as solar radiation fluctuations or variations in components of the climate system such as volcanic eruptions, ice-sheet changes, carbon dioxide (CO_2) increases and deforestation. Such alterations in forcings are modified by feedback effects within the climate system.

A feedback occurs when a portion of the output from the action of a system is added to the input and subsequently alters the output. The result can be either an amplification of the process (a positive feedback) or a dampening of the original disturbance (a negative feedback). These feedback effects can be simply illustrated. Someone slightly overweight who eats for consolation can become depressed by their increased food intake and so eat more and rapidly become enmeshed in a detrimental positive feedback effect. On the other hand, perception of a different kind can be used to illustrate negative feedback. As a city grows there is a tendency for immigration but the additional influx of industry, cars and people is often detrimental to the environment so that it may be offset by an outflux of wealthier inhabitants, with a potentially detrimental impact on the economy.

The ice-albedo feedback is often used to illustrate how a perturbation can be modified in the climate system. If something acts to decrease the global surface temperature, then the formation of additional areas of snow and ice is likely. These cryospheric elements are bright and white, reflecting almost all the solar radiation incident upon them. Their albedo (ratio of reflected to incident radiation) is high and, therefore, the surface albedo, and probably the planetary albedo (the reflectivity of the whole atmosphere plus surface system as seen from 'outside' the planet), increases. Thus a greater amount of solar radiation is reflected away from the planet and temperatures decrease further. A further increase in snow and ice results from this decreased temperature and the process continues. This (positive) ice-albedo feedback mechanism is, of course, also positive if the initial perturbation causes an increase in global surface temperatures. With higher temperatures, the areas of snow and ice are likely to be reduced, so reducing the albedo and leading to further enhancement of temperatures. Positive feedbacks always exaggerate the initial change, whatever the direction; negative feedbacks can, at most, return the initial change to zero; they cannot reverse its direction.

Types of Climate Models

The objective of climate modelling is to simulate processes and to predict the effects of imposed changes and internal interactions. It is possible to classify climate models as belonging to one of three types which reflect the Working Groups used by the Intergovernmental Panel on Climate Change (IPCC): global climate models; climate impact models; and integrated assessment models (Houghton *et al.*, 1990; 1992; 1996; Tegart *et al.*, 1990). The latter two model types typically depend upon the results of the global climate models (Figure 4.1). Models describe the climate system in terms of basic physical, chemical, biological and perhaps also social principles. Hence a climate model can be considered as being comprised of a series of equations expressing these laws and, like all models, must be a simplification of the real world. The more complex the representation, the more costly the model is to use, even on the fastest computers, and the results can only ever be approximations.

Assessment of potential effects of climate change can be conducted either using a linear evaluation system comprising an ordered sequence of climate models, impact models and social (e.g. economic) response models or using a single integrated assessment model encompassing all these features and their interactions (Henderson-Sellers and Braaf, 1996). The most critical interface is between the predicted change in the (physical) climate and the societal response. Most assessment processes require that this interface occurs at a spatial

(a)

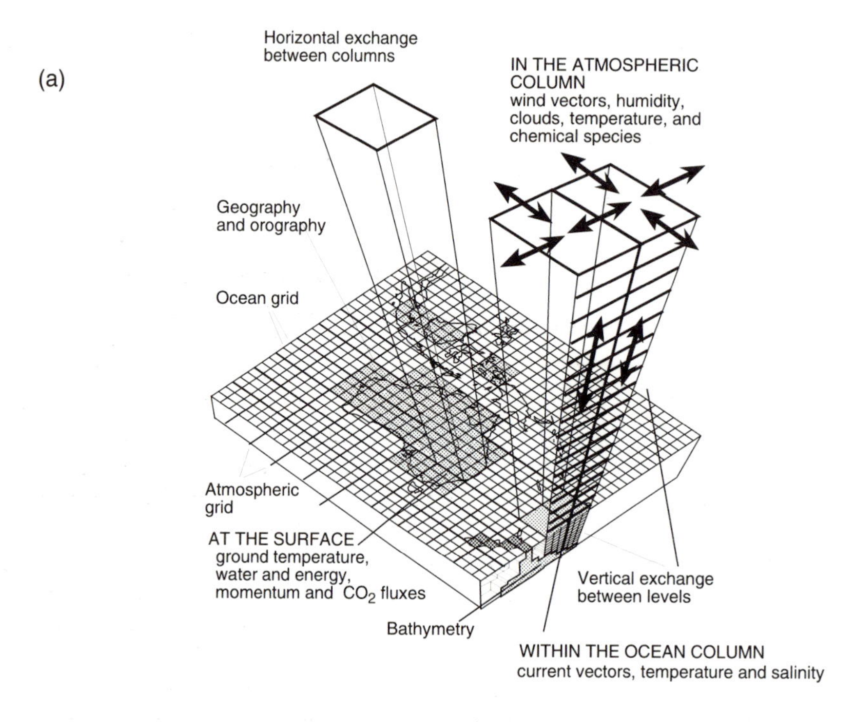
Horizontal exchange between columns
IN THE ATMOSPHERIC COLUMN
wind vectors, humidity, clouds, temperature, and chemical species
Geography and orography
Ocean grid
Atmospheric grid
AT THE SURFACE
ground temperature, water and energy, momentum and CO_2 fluxes
Bathymetry
Vertical exchange between levels
WITHIN THE OCEAN COLUMN
current vectors, temperature and salinity

(b)

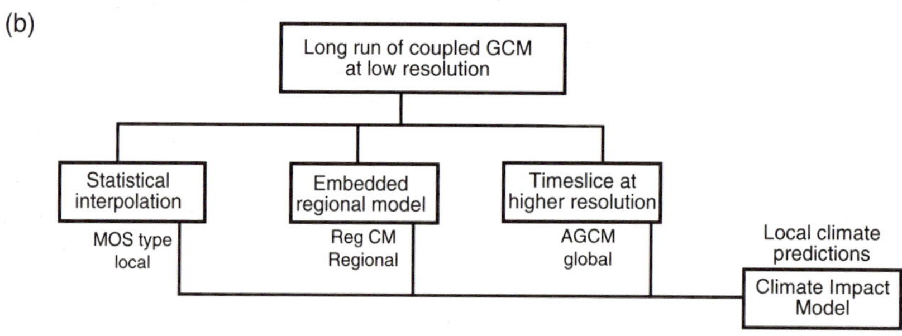
Long run of coupled GCM at low resolution
Statistical interpolation
MOS type local
Embedded regional model
Reg CM Regional
Timeslice at higher resolution
AGCM global
Local climate predictions
Climate Impact Model

(c)

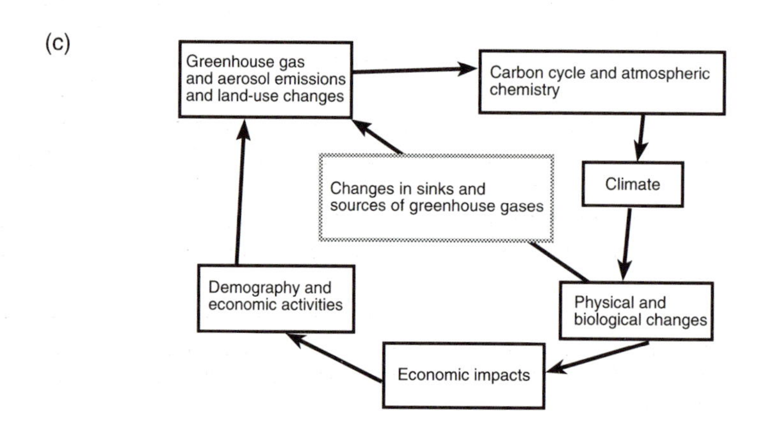
Greenhouse gas and aerosol emissions and land-use changes
Carbon cycle and atmospheric chemistry
Changes in sinks and sources of greenhouse gases
Climate
Demography and economic activities
Physical and biological changes
Economic impacts

scale roughly commensurate with the size of a nation state, a scale often loosely termed 'regional'. Unfortunately, current global climate models (GCMs) have no regional prediction skill (Machenhauer, 1995). The demands made of climate models, for predictions of regional impacts and their consequences, cannot yet be fulfilled because modelling techniques are not yet accurate enough to be used for meaningful impact studies. This chronic lack of skill, reviewed here, is surprisingly overlooked in many climate model applications.

GLOBAL CLIMATE MODELS

Global climate models were derived from weather forecast models, with the modelling communities sharing expertise in fluid dynamics and radiative transfer (Washington and Parkinson, 1986; McGuffie and Henderson-Sellers, 1996). Although numerical climate modelling has, over the last decade, incorporated individuals from other disciplines, it is still true to say that the expertise, reference frame, techniques, enthusiasms and prejudices in this branch of climate modelling derive from mathematics and atmospheric physics, focused particularly on dynamics and radiation (Trenberth, 1992). These climate modellers use high-performance computers as an everyday tool (Semtner, 1995) and they are well aware of the potentially chaotic character of the highly non-linear systems which they model (Lorenz, 1963). They are used to detailed evaluation of their own numerical models, seeing this as a necessary prerequisite to any changes in parameterization. However, the value of intercomparisons amongst numerical model results and group evaluation are only just becoming accepted (Gates, 1992; Henderson-Sellers *et al.*, 1995b).

Fully Coupled Models

There is some ambiguity concerning the meaning of the acronym 'GCM'. Two possible terms are the more recent 'global climate model' and the older 'general or global circulation model'. As the latter also refers to a weather forecast model, in climate studies 'GCM' is understood to mean 'general circulation climate model'. A further distinction is often drawn between oceanic general circulation models and atmospheric general circulation models by terming them OGCMs and AGCMs. The processes that each simulate are coupled to develop a combined ocean–atmosphere global model (OAGCM or CGCM). It has been suggested that as features which are currently fixed come to be incorporated into GCMs, the coupling will be more complete. For example, these include changing biomes (an AOBGCM) or changes in atmospheric, ocean and even soil chemistry (McGuffie and Henderson-Sellers, 1996).

Large-scale global climate models, designed to simulate the climate of the planet, must take into account the whole climate system (Figure 4.1(a)). Simplifications need to be made because interactions operate on different time and space scales (Washington and Parkinson, 1986; Trenberth, 1992). These simplifications can be approached in several ways, resulting in several different types of global-scale climate models. The important components to be considered in constructing or understanding a model of the climate system are: (1) radiation (input and absorption of solar radiation and the emission of infrared radiation); (2) dynamics (movement around the globe by winds and ocean currents

Figure 4.1 Climate model types: (a) schematic of a global climate model (GCM); (b) 'downscaling' from a GCM to a climate impact model; and (c) an integrated assessment model including either a GCM or results from a GCM as one subcomponent

and vertical movements: convection and deep-water formation); (3) surface processes (sea and land ice, snow, vegetation); (4) chemistry (e.g. carbon exchanges between ocean, land and atmosphere); and (5) resolution in both time and space (the time step of the model and the horizontal and vertical scales resolved). Some processes must be treated in an approximate way, either because of our incomplete understanding or because of a lack of computer resources. The relative importance of these processes can be discussed in terms of a 'climate modelling pyramid' (lower part of Figure 4.2).

The aim of GCMs is the calculation of the full three-dimensional character of the climate comprising at least the global atmosphere and the oceans (Figure 4.1(a)). The solution of a series of equations (Table 4.1) that describe the movement of energy, momentum and various tracers (e.g. water vapour in the atmosphere and salt in the oceans), and the conservation of mass, is therefore required. The dynamics are governed by the amount of radiation available at the surface, by the positions and shapes of the continents, and, for the ocean, by the wind stresses imposed by the atmosphere (Figure 4.1(a)). The first step in obtaining a solution is to specify the atmospheric and oceanic conditions at a number of 'points', obtained by dividing the earth's surface into a series of areas, so that a grid results (Figure 4.1(a)). Conditions are specified at each

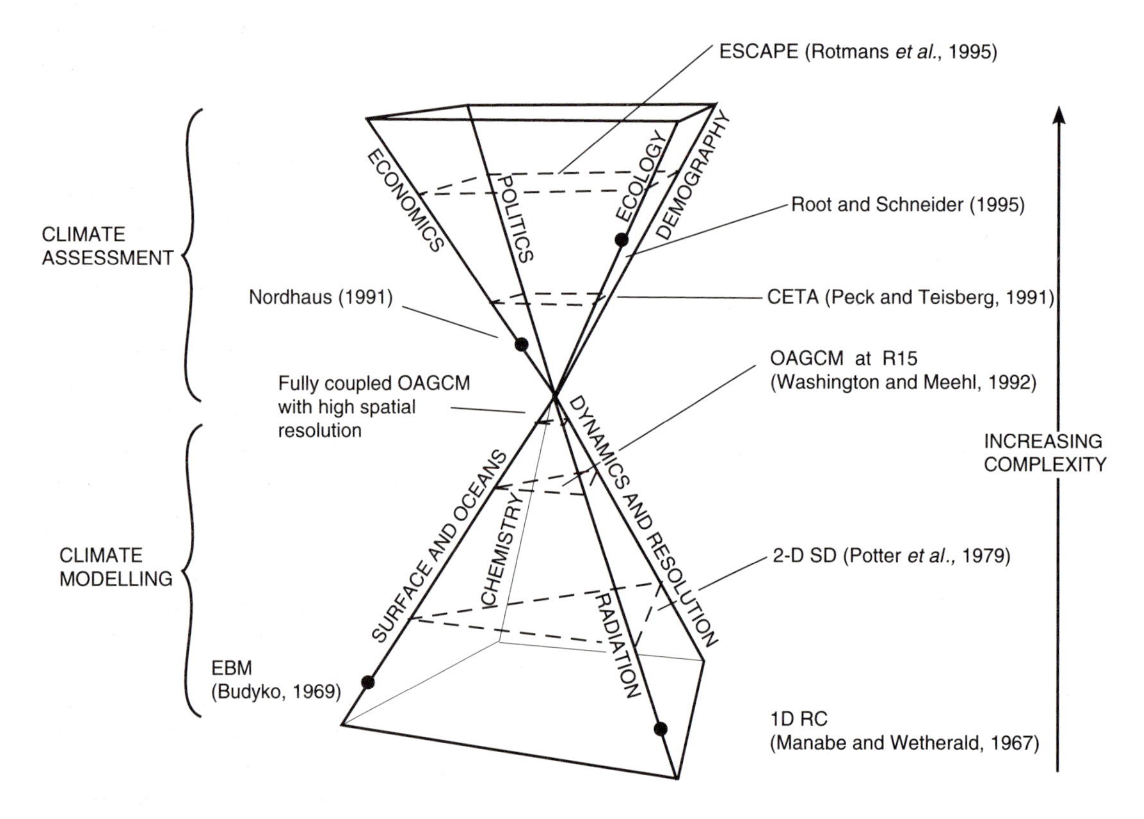

Figure 4.2 The climate modelling pyramid pair: the lower pyramid represents the components of global climate models while the upper pyramid represents aspects of impact and integrated assessment models. The depiction of an apparently unstable balance of much of the latter on an unachieved apex below is intentional

Table 4.1 Fundamental equations solved in GCMs

1 Conservation of energy (the first law of thermodynamics), i.e. input energy = increase in internal energy plus work done
2 Conservation of momentum (Newton's second law of motion), i.e. force = mass × acceleration
3 Conservation of mass (the continuity equation), i.e. the sum of the gradients of the product of density and flow speed in the three orthogonal directions is zero. This must be applied to air and moisture for the atmosphere and to water and salt for the oceans, but can also be applied to other oceanic 'tracers' and to cloud liquid water
4 Ideal gas law (an approximation to the equation of state P atmosphere only), i.e. Pressure × volume = gas constant × absolute temperature

Source: McGuffie and Henderson-Sellers (1996).

grid point for the surface and many layers in the atmosphere and ocean. The resulting set of coupled non-linear equations are then solved at each grid point using numerical techniques. Various techniques are available, but all use a time step approach and an interpolation scheme between grid points (Trenberth, 1992; Semtner, 1995).

An Hierarchy of Global Climate Models

The edges of the lower pyramid in Figure 4.2 represent the basic elements of these GCMs, and complexity is shown increasing upwards. Around the base of the pyramid are the simpler climate models which incorporate only one primary process. For example, the energy balance models (EBMs) are one-dimensional models predicting the variation of the surface temperature with latitude (Sellers, 1969).

Also the one-dimensional radiative–convective (RC) models compute the vertical (usually globally averaged) temperature profile by explicit modelling of radiative processes and a 'convective adjustment', which re-establishes a predetermined lapse rate (Hansen *et al.*, 1981). Two-dimensional (2-D) statistical dynamical (SD) models compute surface processes and dynamics in a zonally averaged framework with a vertically resolved atmosphere (Potter and Gates 1984). More recently, developments have meant that 2-D models can make detailed calculations of chemical reactions in a zonally averaged atmosphere (Garcia and Solomon, 1994). GCMs capture the three-dimensional nature of the atmosphere and ocean and attempt to represent all the climatic processes believed to be important (Washington and Meehl, 1993; Penner *et al.*, 1996).

A similar hierarchy can be seen in ocean climate models. Early ocean models included the 'swamp' model, with no heat storage capacity, and the fixed-depth ocean surface layer (slab) models which have heat capacity but no dynamics. The use of prespecified mixed layer depth (usually between 70 and 100 m), without any allowance for horizontal and vertical motion, is inadequate for the simulation of the annual cycle of zonal heat storage, especially in the tropics. Consequently, most implementations of 'slab oceans' do a very poor job at predicting temperature and sea-ice distributions. An adjustment of surface fluxes at every ocean point can be used as a surrogate for horizontal energy transfer and can greatly improve the seasonality of temperatures and sea ice. The modelling of the oceanic mixed layer has been enhanced by some modellers by the use of more complex treatments of the vertical diffusion of heat away from the surface which have also found application in simple global models such as those used by the IPCC in the Second Scientific Assessment (Wigley and Raper, 1992; Houghton *et al.*, 1996). Full OGCMs calculate the temporal evolution of oceanic variables (velocity, temperature and salinity) on a three-dimensional (3-D) grid of points spanning the global ocean domain

(Bryan, 1969). The formation of oceanic deep water is closely coupled to the formation and growth of sea ice so that full ocean dynamics demand effective inclusion of sea-ice dynamics and thermodynamics.

Limitations of and Constraints on Global Models

Fine-grid models can be used for weather prediction because the integration time is short and simulation is of only the short timescale aspects of the system. GCMs, on the other hand, are designed for long simulations and many GCM integrations must be performed, and their results averaged, to generate an 'ensemble' before a climate 'prediction' can be made. Since the accuracy of the model partly depends on the spatial resolution of the grid points and the length of the time step, a compromise must be made between the resolution desired and the computational facilities available. At present, atmospheric grid points are typically spaced between 2° and 3° of latitude and longitude apart and time steps of approximately 10–30 minutes are used (Gates, 1995). Vertical resolution is between five and 50 levels, with about 20 levels being typical (Figure 4.1(a)). Computational constraints lead to problems for all GCMs. With a coarse grid spacing, small-scale atmospheric and oceanic motions (such as thunder-cloud and eddy formation) cannot be modelled, however important they may be for the real climate dynamics.

Originally, computational constraints dictated that global circulation models could only run for very short periods. For the atmosphere, this meant only simulating a particular month or season, rather than a full seasonal cycle, although now all models include a seasonal cycle and most include a diurnal cycle. For the oceans, restrictions of computer power meant that the models were used before they had fully equilibrated, producing a 'drift' of the ocean climate away from present-day conditions which was often corrected by applying adjusting fluxes at the ocean surface. This is a particular problem for coupled OAGCMs (McGuffie and Henderson-Sellers, 1996; Houghton *et al.*, 1996). The importance of reducing (or preferably removing) such arbitrary adjustments and of including time-dependent features realistically is now well established. Modellers have striven to include increasing numbers of these phenomena, as well as using increasingly available computer power to provide higher-resolution simulations.

CLIMATE IMPACT MODELS

Climate impact modelling is very much more interdisciplinary than the global modelling, encompassing subjects as diverse as ecology, social equity and economics (upper part of Figure 4.2). The climate models used by impact assessors are generally simpler (in mathematical construction) than the GCMs upon whose output they draw. On the other hand, assessment of (un)certainty of impact models, usually based on sensitivity analysis, is probably more fully developed than for GCMs (Shugart *et al.*, 1986; Rind *et al.*, 1992). In contrast to global climate models, impact models are primarily empirically based and hence explicitly (or implicitly) dependent upon long time series information. Application of these data has been carefully evaluated and, hence, is typically regionally or even locally focused (Root and Schneider, 1995). Application to very large areas (say commensurate with the current size of GCM grid elements) raises concerns.

Although the impact modellers initially made rather modest requests of their global modelling colleagues (Robinson and Finkelstein,1991), they are becoming more demanding (EPA, 1989; Henderson-Sellers and Howe, 1996). However, they see themselves, and are seen, as being prepared to use 'old' information and 'making do' with either out-of-date model results or less than complete explanations of output offered (Tegart and Sheldon, 1993; Henderson-Sellers, *et al.*, 1995a). Almost all the impact assessments reported in the First IPCC Scientific Assessment (Tegart *et al.*, 1990; Tegart and Sheldon, 1993) were predominantly single-sector, local to regional and equilibrium assessments only. However, there have also been attempts to assess the possible effects on single sectors at the global scale: for example, forests (Smith and Shugart, 1993); terrestrial ecology (Henderson-Sellers, 1990; Melillo *et al.*, 1993); agriculture (Kane *et al.*, 1992; Rosenzweig and Parry, 1994); and human health (Curson, 1996). There has been relatively little attempt so far to compare the results from different impact models (Rind *et al.*, 1992; Tegart *et al.*, 1990; Tegart and Sheldon, 1993).

One of the major problems faced by impact modellers, wishing to apply GCM projections to regional impact assessments, is the coarse spatial scale of the numerical model results. 'Downscaling', as the production of increased temporal and/or spatial resolution climates from GCM results has come to be termed, has three current forms, namely statistical, regional modelling and time-slice simulations (Figure 4.1(b)). In model-based downscaling, a high-resolution, limited-area model is run off-line using boundary forcing from the GCM. Alternatively, a global atmospheric model is run for limited time 'slices' at high spatial resolution, using sea surface temperatures and sea-ice distributions predicted by a much lower-resolution coupled OAGCM (Giorgi and Mearns, 1991; Mearns *et al.*, 1995; Cubasch *et al.*, 1994). Statistical downscaling can take many forms ranging from a standard interpolation (e.g. Gaussian filtering, kriging) or the application of model output statistics (MOS). Statistical methods include the assignment of the nearest grid box estimate (Croley, 1990); objective interpolation (Parry and Carter, 1988; Cohen, 1991); statistical analysis of local climatic fields (Wilks, 1992; Barros and Lettenmaier, 1993); and the merging of several scenarios based on expert judgement (Pearman, 1988; CIG, 1992). Although GCMs typically have high temporal resolutions, 'downscaling' has also been applied in the time domain because GCM results are found to have rather poor temporal characteristics when examined at timescales less than about a month (Reed, 1986).

All the currently available means of downscaling, and hence of producing climate change scenarios at a 'useful' (regional) spatial resolution, have serious problems. For example, a recent EC-sponsored comparison of regional model simulations from the Hadley Centre (RegCM), the Max Planck Institute (HIRHAM) and Météo-France's variable resolution GCM, with high resolution over Europe, concluded that 'where there were poor regional climates (in the coarse resolution OAGCM) the dynamic embedding made things worse' (Machenhauer, 1995). This evaluation was echoed in 1995 by Leonard Bengtsson (Director of the Max Planck Institute and former Director of ECMWF) who said: 'regional climate prediction requires first and for all good global prediction by coupled climate models – without it regional simulations are useless'.

INTEGRATED ASSESSMENT MODELS

The need for coherence between the social, political and economic and the physical and biochemical aspects of climate change underpins the development of 'integrated assessment models' (IAMs) (Rotmans *et al.*, 1994; Dowlatabadi, forthcoming) (Figure 4.1(c)). IAMs are claimed to have at least two identifying characteristics which distinguish them from other climate model types: they offer greater value than single-disciplinary approaches and they provide information useful to decision-makers. The construction of IAMs ranges from a few equations, representing simple response surfaces, to very large complex sets of equations purporting to capture all the processes (human and physical) involved in climate change. In the former, the climate model input is typically a global mean temperature change following some prescribed forcing, but in the latter, a GCM may become a component of the IAM (Figure 4.1(c)).

It seems likely that integrated assessments have the capability of removing some of the simplistic assumptions underpinning current impact models, for example the neglect of demographics or economics in assessments of human health (Rotmans *et al.*, 1994). In addition, they could be used to try to pinpoint escalation (or de-escalation) in (un)certainty (Dowlatabadi, forthcoming). On the other hand, whilst integrating the effects on natural and human communities, these models have the potential to mask caveats and uncertainties resulting from synergistic interactions. Careless or incomplete identification or balancing of non-linear feedbacks within IAMs may pose a serious problem. This model type also shares with the climate impact models the fundamental difficulty of obtaining local-scale climate predictions from current GCMs.

CLIMATE MODEL EVALUATION

Climate model predictions have been grouped as: (1) unreasonable; (2) reasonable and well known; (3) new, but readily understood once examined; or (4) novel outcomes that challenge current theories (Skiles, 1995). As normal model development would screen out model versions producing unreasonable results and there is little benefit in intercomparison of results which are well known, model intercomparisons and evaluations over the last decade focus on new predictions that are consistent with theory and those which challenge existing ideas. These intercomparisons are contributing to improved assessments of the uncertainties associated with climate model simulation.

International Intercomparisons

Quantitative evaluation of climate model projections is a challenging task. In the past, the most common technique for the evaluation of model simulations was comparison of the results from one model with one observational dataset, often of uncertain validity. Since the development of intercomparison projects (Gates, 1992; Randall *et al.*, 1994; Henderson-Sellers *et al.*, 1995a, b), the practices and techniques have changed. The quality of observational data has increased, as has the quality of correlative information associated with these data. Increasingly, sophisticated techniques and protocols are applied to models and observations so that the understanding of models and the important processes within them continues to develop.

The process of comparison of model predictions and group evaluation is complex, as it has to encompass models and modelling groups from around the world and has to be organized so that similar results are compared. To facilitate the process of group evaluation, the World Climate Research Programme's Working Group on Numerical Experimentation has stratified intercomparisons into three levels (Figure 4.3(a)). Level 1, the simplest, uses only available model results and a common diagnostic set. Until the 1990s, all intercomparisons were conducted at Level 1. The Intergovernmental Panel on Climate Change (IPCC) assessments are Level 1 intercomparisons. Level 2 requires that simulations are made according to prespecified, identical conditions, that common diagnostics are employed and that there is a common dataset against which all the predictions are evaluated. Level 3, the 'best' intercomparison process, requires, in addition to the requirements of the lower two levels, that all the models employ the same resolution and that

(a)

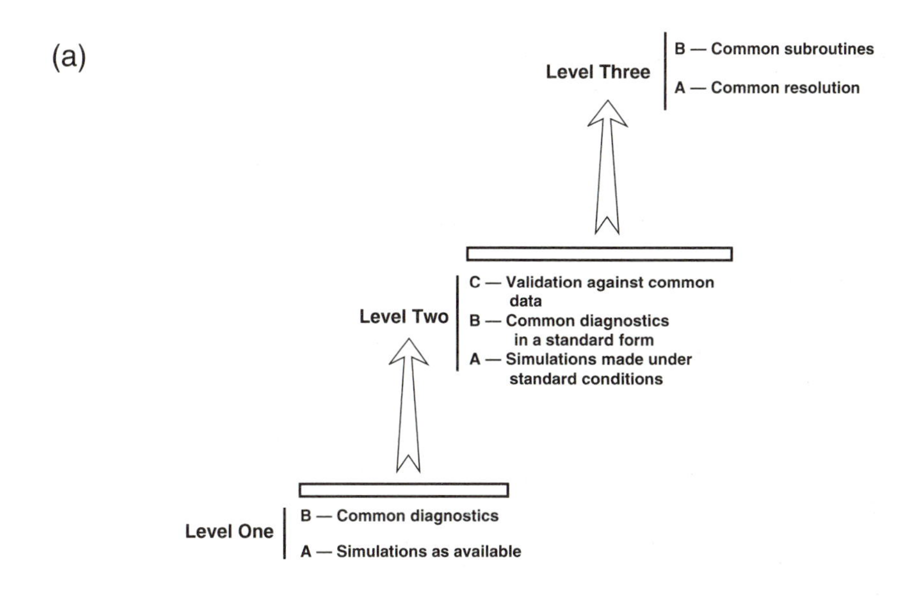

(b)

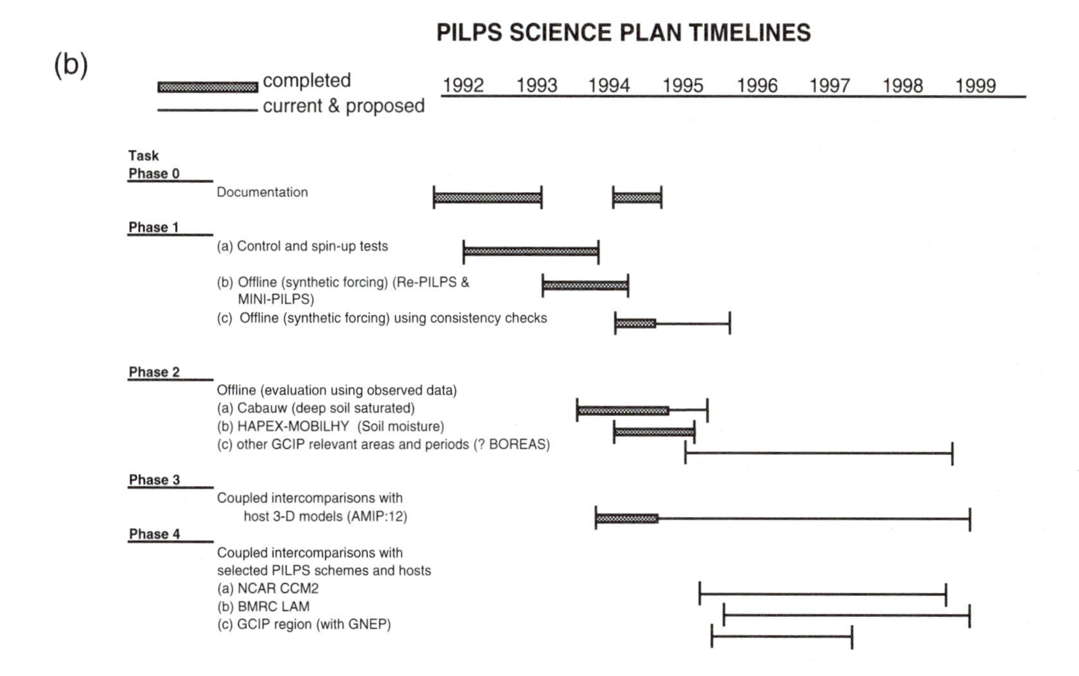

Figure 4.3 (a) Levels of climate model intercomparisons as defined by the Working Group on Numerical Experimentation (WGNE) in support of the World Climate Research Programme (after Gates, 1992). (b) Timelines and lead groups of the Project for Intercomparison of Land-surface Parameterization Schemes (PILPS) (after Henderson-Sellers *et al.*, 1996)

the intercomparison includes the use of common subroutines or code modules.

Among the more important current intercomparisons are AMIP, the Atmospheric Model Intercomparison Project, established in 1989 and now moving into its second stage, the Intercomparison of Radiation Codes for Climate Models (ICRCCM) and the Feedback Analysis of GCMs and Intercomparison with Observations (FANGIO). AMIP focuses on structured (Level 2 in Figure 4.3(a)) intercomparisons of the atmospheric component of GCMs. Participating models use observed ocean surface temperatures and sea-ice extents as well as agreed values of the solar constant (1365 W m^{-2}) and an atmospheric concentration of CO_2 (345 ppmv) as inputs to fixed-length simulations. These operated from 1 January 1979 to 31 December 1988 for AMIP I and 1 December 1978 to 1 December 1995 for AMIP II.

ICRCCM's mandate is to compare results from participating radiation codes in the long- and short-wavelength regions of the spectrum for the cases of clear and cloudy skies. These intercomparisons are then compared with the most detailed radiative transfer calculations available, termed 'line-by-line' calculations, and with observational data from satellite and surface-based field programmes. These include the Earth Radiation Budget Experiment (ERBE), the International Satellite Cloud Climatology Project (ISCCP) and the Surface Radiation Budget Climatology Programme. The FANGIO project seeks to improve understanding of the feedback processes in climate models involving cloud and radiation calculations. FANGIO investigators have calculated the value of a climate model feedback parameter for their models in the cases of clear skies, cloudy skies and for the global response overall (Cess *et al.*, 1993). The range in the clear-sky values is very small, underlining the main conclusion of this intercomparison: that the threefold variation among AGCMs' sensitivity to a prescribed climate change is due almost entirely to cloud feedback processes. The most important outcomes of the international intercomparisons of climate model performance are: (1) the identification of group outliers; (2) the estimation of the range of confidence (or uncertainty) inherent in predictions of any one of the 'reasonable' models; and (3) the development and dissemination of datasets for model evaluation and of tests for climate model results. In the next section, these attributes are illustrated with reference to one intercomparison of particular importance to applied climatology.

The Project for Intercomparison of Land-surface Parameterization Schemes (PILPS)

PILPS is a World Climate Research Programme research activity, jointly sponsored by the Global Energy and Water Cycle Experiment (GEWEX) and the Working Group on Numerical Experimentation (WGNE), which focuses on evaluating and improving land surface schemes for climate prediction. PILPS tests land surface schemes with different datasets representative of different climatic conditions. Documentation (Phase 0) is a continuing activity for PILPS (Pitman *et al.*, 1993; Shao *et al.*, 1994). Phase 1 uses GCM results, Phase 2 employs point-based observed data from a range of sites around the world and Phases 3 and 4 examine the performance of schemes under different types of coupling to other climate model components (Figure 4.3(b)).

Figure 4.4(a) shows the annual mean sensible and latent heat fluxes calculated in PILPS (Phase 2(a), for Cabauw, the Netherlands (51° 58′N, 4° 56′E). Scatter away from the line is associated with differences in the computed value of the surface net radiation, while the scatter along the line reflects differences in the surface energy partitioning between latent heat flux and sensible heat flux. Phase 2(a) shows a range of about 10 W m^{-2} in annual mean net radiation across schemes and a range of about

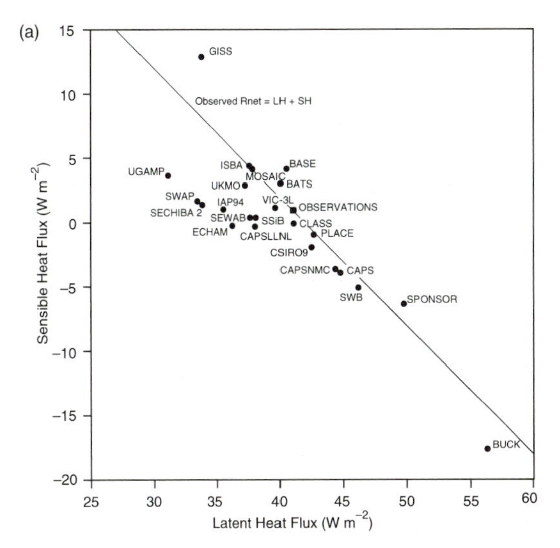

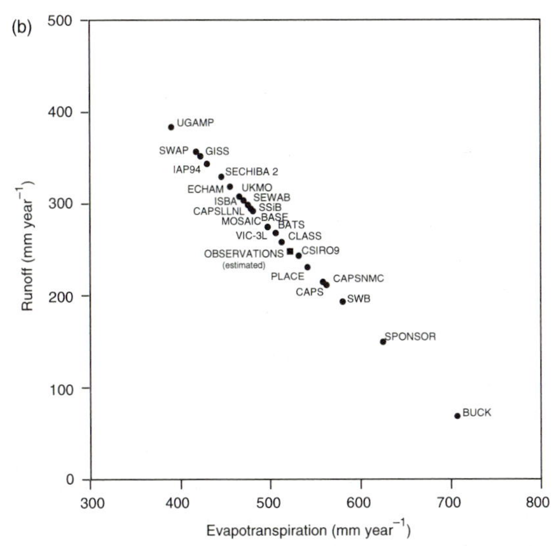

Figure 4.4 (a) Annually averaged sensible heat flux (W m^{-2}) versus latent heat flux (W m^{-2}) for the 23 schemes participating in PILPS Phase 2(a). (b) Annually averaged runoff (mm per year) versus evapotranspiration (mm per year) for PILPS Phase 2(a) (both after Chen *et al.*, forthcoming)

25 W m^{-2} in energy partition. Compared with the observed annual mean it is clear that some schemes (e.g. BUCK) grossly overpredict evaporation, while others (e.g. UGAMP2) allow too little evaporation (Chen *et al.*, forthcoming).

Figure 4.4(b) shows the annual water partitioning of precipitation between runoff and evapotranspiration. All the results lie on the line, an intercept equal to the annual precipitation (776 mm), confirming that the schemes conserve water. The scatter along the line reflects differences in the water partitioning. The range of about 320 mm in both evapotranspiration and runoff (Figure 4.4(b)) is commensurate with the range of 25 W m^{-2} in both annual mean latent heat and sensible heat fluxes (Figure 4.4(a)).

The general conclusions that can be drawn from the results of the PILPS intercomparisons to date (Henderson-Sellers *et al.*, 1996) are as follows:

- Different land surface schemes achieve different annual equilibria when forced with the same atmospheric forcing data and when using the same land surface parameters.
- Different land surface schemes deal with incoming precipitation differently, partitioning it into runoff plus drainage, soil storage and evaporation differently at different times and depending (differently) upon the antecedent conditions.
- Different land surface schemes describe the seasonal cycle of soil moisture and surface hydrology differently with the greatest dispersion occurring when vegetation contributes to the total evaporative flux, there is a great atmospheric demand and the available soil moisture is limited.
- Interactions between land surface and atmospheric models differ as a function of both schemes' characteristics, the location, time of year and antecedent conditions.
- Most land surface schemes can be tuned to observations but no single scheme predicts all the variables describing the land surface climatology and hydrology well. Indeed the consensus (single average) of all the participating schemes generally out-performs all individual schemes.

- Current evaluations of soil moisture are probably unreliable unless care has been taken in specifying land surface parameters and there is accurate atmospheric forcing and, hence, predictions of future soil moisture and drainage are still more uncertain, if for no other reason than because atmospheric forcing will differ.
- The large uncertainty in the simulations indicates that results from 'interactive vegetation' or carbon budgeting models that depend on (uncertain) soil moisture and (dependent) transpiration should be interpreted with great caution.
- There is no single test or evaluation procedure that can demonstrate the 'worthiness' of a land surface scheme's predictions; rather a suite of evaluation procedures ranging from simple conservation requirements through to suite-specific performances is needed.

These conclusions are very similar to the generic summaries being drawn from other international intercomparisons of climate models and their results. They indicate some of the strengths and weaknesses of current climate models.

CONCLUSIONS

This chapter has introduced climate models, the techniques they employ and evaluations of their success. As predictive tools, climate models are remarkably good at some tasks. The global circulation of the atmosphere and ocean is captured as are the seasonal and diurnal signals of temperature. In a known river basin, the change in runoff caused by a large storm can be well simulated and the economic impact of a very cold winter or a poor harvest can be evaluated. These three 'applications' derive from the three types of climate models that are reviewed, namely global climate models, impact models and integrated assessment models. Differences among results arise when these climate models have to be used for tasks for which they were not specifically developed or for which input data are poor (Figure 4.4). A particularly vexing difficulty is the failure of the global models to deliver fine-enough scale predictions upon which impact and assessment evaluations can be based (Figures 4.1 and 4.2). These problems are the focus of current international efforts which will hopefully lead to improvements in all types of climate models and increased confidence in their applications to a wide range of climatological and environmental practices.

REFERENCES

Barros, A. and Lettenmaier, D.P. (1993) 'Dynamic modelling of the spatial distribution of precipitation in remote mountainous areas', *Monthly Weather Review* 121: 1195–213.

Bengtsson, L. (1995) 'On the prediction of regional climate change', *Proceedings of the Third International Conference on Modelling of Climate Change and Variability*, ed. L. Dumenil, Max Planck Institut für Meteorologie.

Bryan, K. (1969) 'A numerical method for the study of the world ocean', *Journal of Computational Physics* 4: 347–76.

Cess, R.D., Zhang, M.H., Potter, G.L., Barker, H.W., Colman, R.A., Dazlich, D.A., DelGenio, A.D., Esch, M., Fraser, J.R., Galin, V., Gates, W.L., Hack, J.J., Ingram,W.J., Kiehl, J.T., Lacis, A.A., Le Treut, H., Li, Z-X., Liang, X.-Z., Mahfouf, J.-F., McAvaney, B.J., Meleshko, V.P., Morcrette, J.-J., Randall, D.A., Roeckner, E., Royer, J.-F., Sokolov, A.P., Sporyshev, P.V., Taylor, K.E., Wang, W.-C. and Wetherald, R.T. (1993) 'Uncertainties in carbon dioxide radiative forcing in atmospheric general circulation models', *Science* 262: 1252–5.

Chen, T.H., Henderson-Sellers, A., Milly, C., Pitman, A.J., Shao, Y., Boone, A., Chen, F., Desborough, C.A., Dickinson, R.E., Ek, M.B., Garratt, J.R., Gedney,

N., Gusev, Y.M., Koster, R., Kowalczyk, E., Laval, K., Lean, J., Lettenmaier, D., Liang, X., Mahfouf, J.-F., Mitchell, K., Nasonova, O.N., Noilhan, J., Polcher, J., Robock, A., Schlosser, A., Schulz, J.P., Shmaking, A.B., Verseghy, D.L.,Wetzel, P., Wood, E.F., Xue, Y. and Yang, Z-L. (forthcoming). 'Cabauw experimental results from the Project for Intercomparison of Land-surface Parameterization Schemes (PILPS)', submitted to *Journal of Climate.*

Climate Impacts Group (CIG) (1992) 'Climate change scenarios for the Australian region', issued November 1992, CSIRO Division of Atmospheric Research, Mordialloc, Victoria, Australia.

Cohen, S.J. (1991) 'Possible impacts of climatic warming scenarios on water resources in the Saskatchewan River sub-basin, Canada', *Climatic Change* 19: 291–317.

Croley, T.E., III (1990) 'Laurentian Great Lakes double-CO_2 climate change hydrological impacts', *Climatic Change* 17: 27–47.

Cubasch, U., Meehl, G. and Zhao, Z.C. (1994) 'IPCC WG 1 Initiative on evaluation of regional climate simulations', Summary report, prepared for IPCC and MECCA, August 1994, 58 pp.

Curson, P. (1996) 'Climate and human health', in T. Giambelluca and A. Henderson-Sellers (eds) *Climate Change: Developing Southern Hemisphere Perspectives*, Chichester: John Wiley.

Dowlatabadi, H. (forthcoming) 'Integrated assessment models of climate change: an incomplete overview', *Energy Policy*.

EPA (1989) 'The potential effects of global climate change on the United States'. Report to Congress (ed. J.B. Smith and D. Tirpak), USEPA, December 1989.

Garcia, R.R. and Solomon, S. (1994) 'A new numerical model of the middle atmosphere, 2. Ozone and related species', *Journal of Geophysical Research* 99: 12937–51.

Gates, W.L. (1992) 'AMIP: The Atmospheric Model Intercomparison Project', *Bulletin of the American Meteorological Society* 73(12): 1962–70.

Gates, W.L. (ed.) (1995) *Proceedings of the First AMIP Scientific Conference, 15–19 May 1995, Monterey, CA*, Geneva: World Meteorological Organization.

Giorgi, F. and Mearns, L.O. (1991) 'Approaches to the simulation of regional climate change: a review', *Review Geophysics* 29(2): 191–216.

Hansen, J.E., Johnson, D., Lacis, A., Lebedeff, S., Lee, P., Rind, D. and Russell, G. (1981) 'Climate impact of increasing carbon dioxide', *Science* 213: 957–66.

Henderson-Sellers, A. (1990) 'Predicting generalized ecotype groups with the NCAR CCM: first steps towards an interactive biosphere', *Journal of Climate* 3: 917–40.

Henderson-Sellers, A. and Braaf, R. (1996) 'Developing new perspectives on climate change, impacts assessment and response', in T. Giambelluca and A. Henderson-Sellers (eds) *Climate Change: Developing Southern Hemisphere Perspectives*, Chichester: John Wiley.

Henderson-Sellers, A. and Howe, W. (1996) 'MECCA achievements and lessons learned', in A. Henderson-Sellers and W. Howe (eds) *Assessing Climate Change – The Story of the Model Evaluation Consortium for Climate Assessment*, Sydney: Gordon and Breach.

Henderson-Sellers, A., Howe, W. and McGuffie, K. (1995a) 'The MECCA Analysis Project', *Global and Planetary Change* 10(14): 3–21.

Henderson-Sellers, A., Pitman, A.J., Love, P.K., Irannejad, P. and Chen, T. (1995b) 'The Project for Intercomparison of Land-surface Parameterization Schemes (PILPS): Phases 2 & 3', *Bulletin of the American Meteorological Society* 76: 489–503.

Henderson-Sellers, A., McGuffie, K. and Pitman, A.J. (1996) 'The Project for Intercomparison of Land-surface Parameterization Schemes (PILPS): 1992–1995', *Proceedings of the Third International Conference on Modelling of Climate Change and Variability*, ed. L. Dumenil, Max Planck Institut für Meteorologie.

Houghton, J.T., Jenkins, G.J. and Ephraums, J.J. (eds) (1990) *Climate Change: The IPCC Scientific Assessment*, Cambridge: Cambridge University Press.

Houghton, J.T., Callander, B.A. and Varney, S.K. (eds) (1992) *Climate Change 1992: The supplementary report to the IPCC Scientific Assessment*, Cambridge: Cambridge University Press.

Houghton, J.T., Mera Filho, L.G., Callander, B.A., Harris, N., Kattenberg, A. and Maskell, K. (eds) (1996) *The IPCC Second Scientific Assessment* Cambridge: Cambridge University Press.

Kane, S., Reilly, J. and Tobey, J. (1992) 'An empirical study of the economic effects of climate', *Climatic Change* 21: 17–35.

Lorenz, E.N. (1963) 'Deterministic non-periodic flow', *Journal of Atmospheric Science* 20: 130–41.

Machenhauer, B. (1995) 'Dynamical down-scaling of climate simulations', *Proceedings of the Third International Conference on Modelling of Climate Change and Variability*, ed. L. Dumenil, Max Planck Institut für Meteorologie.

Manabe, S. and Wetherald, R.T. (1967) 'Thermal equilibrium of the atmosphere with a given distribution of relative humidity', *Journal of Atmospheric Science* 24: 241–59.

McGuffie, K. and Henderson-Sellers, A. (1996) *A Climate Modelling Primer*, Chichester: John Wiley.

Mearns, L.O., Giorgi, F., McDaniel, L. and Brodeur, C. (1995) 'Analysis of the diurnal range and variability of daily temperature in a nested modelling experiment: comparison with observations and $2xCO_2$ results', *Climate Dynamics* 11: 193–209.

Melillo, J.M., McGuire, A.D., Kicklighter, D.W., Moore, B., Vorosmarty, C.J. and Schloss, A.L. (1993) 'Global climate change and terrestrial net primary productivity', *Nature* 363: 234–40.

Nordhaus, W.D. (1991) 'To slow or not to slow: the economics of the greenhouse effect', *The Economic Journal* 101: 920–37.

Parry, M.L. and Carter, T.R. (1988) 'The assessment of effects of climatic variations on agriculture: aims, methods and summary of results', in M.L. Parry, T.R. Carter and N.T. Konijn (eds) *The Impact of Climatic Variations on Agriculture, Volume 1: Assessments in Cool Temperate and Cold Regions*, Dordrecht: Kluwer, pp. 11–95.

Pearman, G. (ed.) (1988) *Greenhouse: Planning for Climate Change*, Melbourne: CSIRO.

Peck, S.C. and Teisberg, T.J. (1991) 'CETA: a model for carbon emissions trajectory assessment', *The Energy Journal* 13: 55–77.

Penner, J.E., Jauman, P., Santer, B., Taylor, K. and Wigley, T.M.L. (1996) 'Anthropogenic sulphate aerosols and climate change: a method for calibrating forcing', in A. Henderson-Sellers and W. Howe (eds) *Assessing Climate Change – The Story of the Model Evaluation Consortium for Climate Assessment*, Sydney: Gordon and Breach.

Pitman, A.J., Henderson-Sellers, A., Abramopoulos, F., Avissar, R., Bonan, G., Boone, A., Dickinson, R.E., Entekhabi, M. Ek. D., Famiglietti, J., Garratt, J. R., Frech, M., Hahmann, A., Koster, R., Kowalczyk, E., Laval, K., Lean, J., Lee, T.J., Lettenmaier, D., Liang, X., Mahfouf, J.-F., Mahrt, L., Milly, P.C.D., Mitchell, K., de Noblet, N., Noilhan, J., Pan, H., Pielke, R., Robock, A., Rosenzweig, C., Schlosser, C.A., Scott, R., Suarez, M., Thompson, S., Verseghy, D., Wetzel, P., Wood, E., Xue, Y., Yang, Z.-L. and Zhang, L. (1993) *Project for Intercomparison of Land-Surface Parameterization Schemes (PILPS). Results from offline control simulations (Phase 1A)*, GEWEX Report, IGPO Publication Series No. 7 (Dec. 1993), 47 pp.

Potter, G.L. and Gates, W.L. (1984) 'A preliminary intercomparison of the performance of two atmospheric climate models', *Monthly Weather Review* 112: 909–17.

Potter, G.L., Ellsaesser, H.W., MacCracken, M.C. and Luther, F.M. (1979) 'Performance of the Lawrence Livermore Laboratory zonal atmospheric model', in *Proceedings of the GARP Joint Study Conference on Climate Models*, GARP Publication Series No. 22 (Vol. 2), Geneva: WMO, pp. 995–1001.

Randall, D.A., Cess, R.D., Blanchet, J.P., Chalita, S., Colman, R., Dazlich, D.A., Delgenio, A.D., Keup, E., Lacis, A., LeTreut, H., Liang, X.Z., McAvaney, B., Mahfouf, J.F., Meleshko, V.P., Morcrette, J.J., Norris, P.M., Potter, G.L., Rikus, L., Roeckner, E., Royer, J.F., Schlese, U., Sheinin, D.A., Sokolov, A.P., Taylor, K.E. and Wetherald, R.T. (1994) 'Analysis of snow feedback in 14 general circulation models', *Journal of Geographical Research* 99: 20757–71.

Reed, D.N. (1986) 'Simulation of time series of temperature and precipitation over eastern England by an atmospheric general circulation model', *Journal of Climatology* 6: 233–53.

Rind, D., Rosenzweig, C. and Goldberg, R. (1992) 'Modelling the hydrological cycle in assessments of climate change', *Nature* 358: 119–22.

Robinson, P.J. and Finkelstein, P.L. (1991) 'The development of impact-oriented climate scenarios', *Bulletin of the American Meteorological Society* 72(4): 481–90.

Root, T. and Schneider, S.H. (1995) 'Ecology and climate: research strategies and implications', *Science* 269: 334–40.

Rosenzweig, C. and Parry, M.L. (1994) 'Potential impact of climate change on world food supply', *Nature* 367: 133–8.

Rotmans, J., Hulme, M. and Downing, T.E. (1994) 'Climate change implications for Europe: an application of the ESCAPE model', *Global Environmental Change* 4: 97–124.

Sellers, W.D. (1969) 'A global climatic model based on the energy balance of the Earth-atmosphere system', *Journal of Applied Meteorology* 12: 241–54.

Semtner, A.J. (1995) 'Modelling ocean circulation', *Science* 269: 1379–85.

Shao, Y., Anne, R.D., Henderson-Sellers, A., Irannejad, P., Thornton, P., Liang, X., Chen, T.H., Ciret, C., Desborough, C., Balachova, O., Haxeltine, A. and Ducharne, A. (1994) *Soil moisture simulation. A report of the RICE and PILPS Workshop*, GEWEX/GAIM Report, IGPO Publication Series No. 14, 179 pp.

Shugart, H.H., Antonovsky, M.Ja., Jarvis, P.G. and Sandford, A.P. (1986) 'CO_2, climatic change and forest ecosystems, in B. Bolin, B.R. Döös, J.Jäger and R.A. Warrick (eds) *The Greenhouse Effect, Climatic Change and Ecosystems*, SCOPE 29, Chichester: John Wiley, pp. 475–521.

Skiles, J.W. (1995) 'Modelling climate change in the absence of climate change data', *Climatic Change* 30: 1–6.

Smith, T.M. and Shugart, H.H. (1993) 'The transient response of terrestrial carbon storage to a perturbed climate', *Nature* 361: 523–6.

Tegart, W.J.McG. and Sheldon, G.W. (eds) (1993) *Climate Change 1992. The Supplementary Report to The IPCC Impacts Assessment*, Canberra: Australian Government Publishing Service, 115pp.

Tegart, W.J.McG., Sheldon, G.W. and Griffiths, D.C. (eds) (1990) *Climate Change. The IPCC Impacts Assessment*, Canberra: Australian Government Printing Service.

Trenberth, K.E. (1992) *Coupled Climate System Modelling*, Cambridge: Cambridge University Press.

Washington, W.M. and Meehl, G.A. (1993) 'Greenhouse sensitivity experiments with penetrative cumulus convection and tropical cirrus albedo feedbacks', *Climate Dynamics* 8: 117–33.

Washington, W.M. and Parkinson, C.L. (1986) *An Introduction to Three-Dimensional Climate Modelling*, Mill Valley, CA: University Science Books, 422 pp.

Wigley, T.M.L. and Raper, S.C.B. (1992) 'Implications for climate and sea level of revised IPCC emissions scenarios', *Nature* 357: 293–300.

Wilks, D.S. (1992) 'Adopting stochastic weather generation algorithms for climate change studies', *Climatic Change* 22: 67–84.

5

ATMOSPHERIC RESOURCE MANAGEMENT

Ian Burton and Marjorie Shepherd

INTRODUCTION: ATMOSPHERIC RESOURCE MANAGEMENT ISSUES

The idea that the atmosphere can be managed as a resource is relatively new. The atmosphere is not a classical resource like minerals, or oil, or forests, or land, or even water. It cannot be appropriated in an exclusive way because it is a fluid medium, constantly in motion and observes no boundaries. The air we breathe circulates freely around the whole globe; hence the atmosphere is classified as a 'common property resource' and is described as part of the 'global commons' – something in which all people now living have a stake and in which future generations also have an interest. For this reason, if atmospheric management is to be at all effective then this can only be achieved through collective action. In the absence of any collective management, the atmosphere is simply a 'free good' that can be used by anyone and in any way they please, with potentially damaging and even catastrophic results. Indeed, the atmosphere has in effect been a 'free good' for most of human history. It is only in recent times with the growth of modern industry and large cities that management has become necessary. An objective of this chapter is to show how the scale of human interference and impact on the atmosphere has expanded over the last half century from the local to the global, and how this has shaped the role and functions of atmospheric management. Some future directions in the evolution of atmosphere management are also discussed.

One legal point of departure for atmospheric management was to limit what could be done on the land in order to protect the rights of others to the use of the atmosphere. In the English Common Law this has traditionally been covered by the law of nuisance. A person is constrained not to use his or her land (e.g. by burning or otherwise emitting noxious smoke or fumes) in such a way that would tend to harm the enjoyment of others of their land, be it on adjacent property or some distance away. With the growth of much greater emissions (pollutants), and the emergence of impacts over larger and larger areas, the law of nuisance has become increasingly inadequate. Although such laws and legal precedents are still used today for some purposes, it has generally become necessary to adopt legislation and a complex set of regulations restricting the use of the atmosphere as a sink for gaseous waste. Regulations within one country are often insufficient and international agreements are necessary for effective atmospheric management.

The atmosphere is a resource in another sense. It is also an asset in terms of the weather or climate which supports a range of human activities and needs, including health, agriculture, forestry and recreation (i.e. Chapters 12, 16, 17 and 18). While it is not possible to deplete the air in a physical sense, it is possible to reduce its quality by affecting its physical and chemical nature from local to global scales (Chapter 22). Poorer air quality reduces health and wealth. Furthermore, because the atmosphere is one continuous medium, all of its properties are interconnected. Air quality, weather/climate and extreme events (Chapter 23) are all attributes of the same atmospheric system. Ideally, management should be recognized and shaped according to this fundamental principle. In practice, however, it has been convenient to segregate the atmosphere into different scientific and policy domains, with each one studied and managed in relative isolation. Such an approach has been justified as a necessary part of the scientific method, and as an effective way of proceeding on the policy agenda one issue at a time.

HISTORY OF ATMOSPHERIC RESOURCE MANAGEMENT

The history of atmospheric resource management reveals two 'polar' perspectives on making use of the atmosphere. One is to try to make the atmosphere behave according to human wishes and the other is to adapt human behaviour to the characteristics of the atmosphere, in order to make use of its benefits and to minimize its adverse effects. These two approaches are the extremes of a spectrum, of a continuously varying pattern of response. The rain dance of the Navajo and other Indian groups of the American southwest is a famous example of an attempt to control the atmosphere. Atmospheric management in modern times has included attempts to disperse fog at airports (London's Heathrow for example) by using heating devices. Cloud seeding enjoyed a period of popularity in the 1950s and 1960s as a more scientific approach to 'rain making'. Many other technologies have been tried, some of which are relatively simple and some 'high tech'. Examples of both are still widely found. Greenhouses and windbreaks are used to modify the microclimate to improve agricultural output. Smudge pots are used to create smoke to protect valuable crops from frost. In these latter examples, the emphasis is shifting from changing the behaviour of the atmosphere *per se* to one of protecting the economic activity. Other protective behaviour includes wearing sun hats, sunscreen lotion, sunglasses or minimizing outdoor activity during episodes of poor air quality. In the recent policy debates over climate change, these two approaches have been enshrined into the United Nations Framework Convention on Climate Change as the 'mitigation' and the 'adaptation' approaches.

In addition to atmospheric control and adaptation to the atmosphere, the third element in atmospheric management is research. Research on the natural science of the atmosphere and the socio-economic causes and consequences of atmospheric change is an essential ingredient of the management process. One important dimension of humanity's relationship to the atmosphere is the changing attitude towards its management. Efforts to control the atmosphere are giving way to the recognition that it is better to adapt to the natural variations in weather and climate, and to attempt to exercise self-control by limiting the anthropogenic sources of air quality problems.

Perhaps the most important dimension of atmospheric resource management in historical terms is the progressive enlargement of spatial scale. The period of early industrialization in Europe and North America was characterized by extremely poor air quality in the inner cities, resulting largely from the burning of coal. Smoke concentrations were high and the fallout of black carbon particulates made the problem quite evident in the densely populated and industrial areas. In the latter half of the twentieth century, coal has been largely replaced in the developed countries by oil and natural gas. However, coal is still used in large coal-fired electrical power generating stations which supply power to industrial, commercial and domestic users. These large power stations typically have tall smokestacks which disperse the emissions higher into the atmosphere, where they are carried long distances (see Acid deposition below and in Chapter 22). A positive result of the 'high stacks' policy, and the more efficient fossil fuel combustion techniques, is that the historical local air quality problems of London and other cities have been greatly reduced. They have, however, been replaced by photochemical air pollution which is aggravated by stable atmospheric conditions and intense sunlight, and for which the major source of pollutants is the very large increase in fossil fuel combustion, particularly transport emissions. It is this photochemical smog which has made cities like Los Angeles, Mexico City and Athens notorious for their air quality problems. Photochemical smog (see Chapter 22, Urban air pollution problems) can have serious health effects especially for those with pre-existing health (especially respiratory) conditions (see Chapter 12, Atmospheric impacts on, performance, and behaviour). The continued expansion of cities and the growth in the number of cars has resulted in poor air quality conditions spreading out from the cities into the surrounding regions where agricultural production can be adversely affected. Major urban areas may also impact regions thousands of kilometres downwind owing to the dispersion of air pollutants during long-range transport of air masses (Figure 22.5).

Regional air quality problems first appeared as photochemical smog spread out from the cities. This was followed by the problem of acidic deposition as a result of the growth in emissions of sulphur dioxide (SO_2) and nitrogen oxides (NO_X) from fossil fuel combustion and smelting activities. The scale of the acidic deposition problem generally exceeds that of photochemical smog. It was shown that SO_2 emissions from the UK and Germany were causing the acidification of lakes in Scandinavia, and Canada protested about damage to lakes and forests in Ontario and Quebec from emissions in the Ohio Valley and the midwest of the United States. The emergence of the acid rain issue (more correctly termed acidic deposition, because snow and fog can be acidic) marks the first recognition of the long-range transport of atmospheric pollutants and the need for an internationally co-ordinated response.

The next atmospheric problem to emerge was the depletion of the stratospheric ozone layer by chlorofluorocarbons (see Chapter 22, Global air pollution problems). These substances (CFCs), which are considered inert gases under tropospheric conditions, have been used in refrigeration and air-conditioning systems since the 1950s. In the late 1980s it was discovered that CFCs had already led to significant depletion of the ozone (O_3) layer over Antarctica (Plate 22.1), and that further depletion could be expected on a wider scale, resulting in increased ultraviolet radiation at the earth's surface and subsequent danger to human health and ecosystems. The problem of climate change resulting from anthropogenic emissions of greenhouse gases (see Chapter 22, Introduction), especially carbon dioxide (CO_2) from the burning of fossil fuels, followed hard upon the heels of ozone-layer depletion. The burning of trees (or wood) also contributes CO_2 to the atmosphere and the indications now are that the level of CO_2 concentrations in the atmosphere will reach double the preindustrial level (1800 AD) during the second half of the twenty-first century, unless greenhouse gas emissions can be stabilized and then substantially reduced.

A new atmospheric resource problem has recently become apparent. This is the problem of global contamination with toxic chemicals transported from their place of use as pesticides, fungicides and other industrial chemicals (including heavy metals), through atmospheric pathways to all regions of the earth. The history of atmospheric resource management has seen the steady expansion in spatial scale of air issues, with the corresponding need to develop systems of management on continental and global scales. An earlier distinction between the management options of mitigation/prevention and adaptation is now breaking down. The reason is that human-induced changes to the atmosphere cannot be studied or managed separately from the natural behaviour of the atmosphere. Since it is now agreed in the scientific community that the earth's climate is being changed by anthropogenic forces, then it follows that any weather-driven event, including hazard events (such as floods, droughts, forest fires, frost, hail, high winds, hurricanes, tornadoes and blizzards, discussed in Chapter 23), cannot be isolated from human activity.

CURRENT MANAGEMENT APPROACHES TO AIR QUALITY PROBLEMS

Urban and Regional Air Quality

Air quality within urban centres and nearby rural areas may be affected by gaseous emissions of nitrogen oxides (NO_X), sulphur dioxide (SO_2), carbon monoxide (CO), volatile organic compounds (VOCs) and particulate matter (see Chapter 22, Urban air pollution problems). The presence of these air contaminants and subsequent reaction products, such as ground-level ozone (O_3), can have serious impacts on human health, vegetation and material surfaces. The major sources of these air pollutants are human activities which require energy, for example electrical power generation, industrial activities and transportation systems that utilize fossil fuels. Approximately 95 per cent of the anthropogenic NO_X in the atmosphere is the result of fossil fuel combustion. Other pollutant sources are industrial process and consumer products such as commercial printing processes, paints and solvents to dry cleaning and personal care products.

Photochemical smog is the most common urban air quality problem. The major component of smog is ground-level ozone (see Chapter 22, Secondary pollutants), which is formed by the reactions of NO_X and VOCs in the presence of sunlight. Ground-level ozone increases respiratory problems in humans. Vegetation effects range from visible leaf damage to reduced yield in agricultural crops and reduced growth in some forest species. Particulate matter, often associated with smog and acidic deposition and the presence of suspended particles in the atmosphere (from 1 μm to 10 μm in diameter), results in health effects which are associated with increased hospital admissions and mortality, and reduced visibility.

The initial management approach to urban air quality was to set ambient air quality guidelines to protect human health and the environment for the individual common air pollutants, namely SO_2, suspended particulate matter, CO, ground-level ozone and nitrogen dioxide. The World Health Organization published Air Quality Guidelines in 1987 for organic and inorganic air pollutants, which were revised in 1995 incorporating ecosystem impacts. These guidelines were then, and are still, used to derive standard emission limits (e.g. Table 22.6) for various pollutant sources on a species-by-species basis, often neglecting the positive or negative consequences of controlling one pollutant on other co-pollutants emitted from the same source.

In the mid-1970s it was recognized that long-range transport of pollutants within Europe and within North America through the atmosphere resulted in damage to ecosystems far from the pollution sources (see Chapter 22, Continental-scale pollution problems). In 1983, the UN ECE Convention on Long Range Transport of Air Pollutants entered into force and became the focal point for determining abatement strategies for reducing the impacts of air pollutants on human health and ecosystems. Under this convention, protocols were signed to reduce SO_2 emissions (1985), NO_X emissions (1988) and VOC emissions (1990), and further SO_2 emission reductions (1994) were made, primarily to reduce ground-level ozone concentrations and acidic deposition. These protocols represent the first co-ordinated international efforts to reduce local and regional air quality problems. Since the development of ambient air quality guidelines, the public (in some countries) has had access to real-time information on local air quality through the daily publication of air quality indices and forecasts. The intent of the advisory air quality information is to encourage behaviour modification from reducing emissions (e.g. encouraging alternative transportation), to taking protective action (e.g. reducing physical activity during bad smog episodes) and staying indoors. Such behavioural responses are particularly important for those at high risk, including young children, older people and those with respiratory illnesses.

Acid Deposition

The acid deposition problem includes wet and dry deposition of acidic compounds, in the form of rain, snow, fog or dry particles (see Chapter 22, Acid deposition). Acid rain or fog was recognized as a problem early on with the switch in principal energy sources from wood to coal. Sulphur dioxide (SO_2) and NO_X produced by metal smelting, thermal electrical utilities and transportation systems account for the majority of anthropogenic emissions causing acidic deposition. Both SO_2 and NO_X are subject to long-range transport through the atmosphere, which means that acid deposition is a continental-scale air quality issue (see Chapter 23, Continental-scale pollution problems). These gases are oxidized in the atmosphere to form primarily sulphuric acid and nitric acid which may be dissolved in rain, snow or fog droplets or form dry particles. When these acidic species are deposited in aquatic and terrestrial ecosystems with a low acid buffering capacity, the habitats are altered enough to affect significantly the fish populations and forest growth. There is also some evidence that acidic particles in the air have an adverse effect on human health. In addition to altering the acidity of the sensitive ecosystems, the higher acidity in lakes often results in the dissolution of toxic metals such as aluminium, cadmium, lead and mercury.

Acid rain was recognized as a significant environmental problem in the 1960s and, within the next ten years, emission guidelines and air quality objectives had been established for SO_2 in North America and Europe. During this period in both Europe and Canada, target loads for sulphate deposition were being defined to protect sensitive ecosystems. This represents the first ecosystem response-based approach to managing an atmospheric issue. Also during this time, under the UN ECE Convention on the Long Range Transport of Air Pollutants, the first SO_2 protocol was signed in 1985 and the first NO_X protocol was signed in 1988. In 1991, Canada and the USA signed the Air Quality Accord Agreement in which the impact of SO_2 emissions on Canada's ecosystems was formally acknowledged, and a framework was established to deal with transboundary air issues more expeditiously. In 1994, the second UN ECE protocol for SO_2 was signed defining further reductions based upon cost-effective approaches.

Ozone-layer Depletion

Life on earth is protected from damaging ultraviolet B (UVB) radiation emitted from the sun, by a thin layer of ozone (O_3) in the stratosphere. Scientists have long understood that any serious threat to the ozone layer could have potentially disastrous consequences for ecosystems and human health (see Chapter 22, Stratospheric ozone depletion). Identified sources of damage to the ozone layer include chlorine from industrial processes, nuclear weapons testing and supersonic jet transport aircraft (SSTs). These risks did not receive much attention until a substantial thinning of the ozone layer over Antarctica was discovered by the British Antarctic Survey in 1985. Subsequent research has identified several CFC compounds that came into widespread industrial use in the 1930s as a substitute for ammonia in refrigeration, as the main source of chlorine that is causing ozone depletion. Because of their distinctive properties (i.e. they are extremely chemically stable compounds), CFCs have found many other industrial applications. It is only when they reach the stratosphere and are exposed to higher radiation levels that they photodissociate and release the reactive chlorine precisely where it can interact with the ozone layer.

Increased exposure to UVB radiation has been linked to an increase in the incidence of skin cancer with the light-skinned populations being most at risk. Other health impacts include damage to the human immune system and the increased risk of eye diseases such as cataracts. Agricultural experiments with crops such as cotton, peas, beans, melons, tomatoes and cabbage show a distinct decrease in yields under higher UVB radiation exposure. Other impacts include the potential loss of micro-organisms in aquatic systems that constitute a threat to the food chain. This suggests that depletion of the ozone layer could have serious consequences for global food production and natural ecosystems as well as human health. Stratospheric ozone depletion is the first atmospheric issue to be recognized as, and addressed as, a truly global concern. Action to reduce the risks of ozone-layer depletion has focused on steps to reduce and eventually eliminate the use of CFCs. Under the leadership of the United Nations Environment Programme (UNEP) negotiations began on the development of an international convention to protect the ozone layer in 1981, and these led to the Vienna Convention of 1985. Specific commitments to reduce emissions of CFCs (and other ozone-depleting substances) came into effect in the Montreal Protocol of 1987 (see Chapter 22, Stratospheric ozone depletion).

The Montreal Protocol was widely greeted as a success story that demonstrated the capacity of the international community and the UN to take action to protect the global (atmospheric) environment. The protocol committed all industrialized nations to reduce their consumption of ozone-depleting substances by 50 per cent of the 1986 level by the year 1999. Developing countries were required to meet similar targets by 2009. In spite of this diplomatic success, it quickly became clear that the measures in the Montreal Protocol were inadequate, based on the rate of change in UVB radiation and stratospheric ozone concentrations, and two subsequent amendments have been agreed in London (1990) and in Copenhagen (1992). Projections of atmospheric chlorine concentrations with and without these actions are shown in Figure 5.1 in relation to a critical level based on reducing health impacts or stabilizing ozone levels. It may be noted that even with this successful action the concentrations of chlorine in the atmosphere will not fall back below the

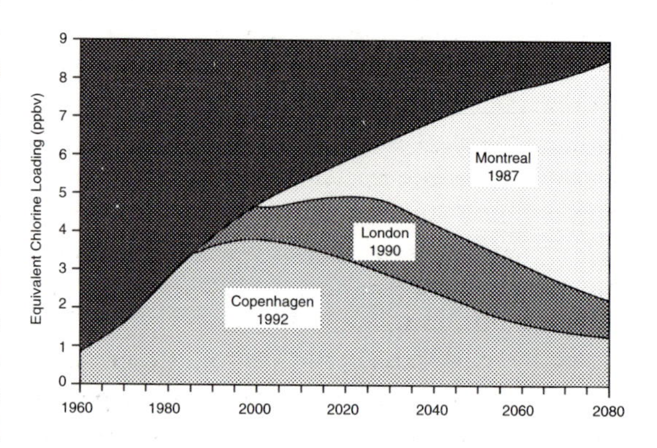

Figure 5.1 Atmospheric chlorine concentrations, 1960–2080

critical level until near 2050 even if the agreements are strictly adhered to. Other, more adaptive responses are available to reduce the impacts of enhanced UVB radiation and these are recommended as temporary expedients pending the successful implementation of the Vienna Convention. Health authorities advise people to avoid exposure to direct sunlight, especially when the sun is high, and to wear protective clothing and a sun-blocking cream.

Climate Change

The temperature of the earth's atmosphere is a result of the amount of incoming radiation, the amount of outgoing (reflected) radiation, and the amount 'trapped' or captured in the atmosphere by water vapour and other heat-trapping gases, primarily CO_2, methane and nitrous oxide. With the exception of water vapour, the amount of gases in the atmosphere is small. These radiatively active trace gases are extremely important, for without them the mean temperature of the atmosphere would be about 15°C cooler than it is today. The most significant trace gas in this regard is CO_2. Preindustrial concentrations of CO_2 were approximately 280 ppm and current levels approximate 355 ppm (Table 22.1).

Atmospheric scientists have suggested for a hundred years that emissions of CO_2 from human activities (see Chapter 21, Impacts of urban climates on GEC) could change the balance of atmospheric radiation. Research in the 1980s with general circulation models of the atmosphere (under assumed conditions of atmospheric forcing by increased CO_2 and other trace gases) suggest that, if current emission rates continue, the mean atmospheric temperature (at ground level) could increase by 2.5°C to 4.5°C by the latter part of the twenty-first century. Carbon dioxide is not solely responsible for radiative forcing. The important role of atmospheric research in atmosphere management is illustrated by Figure 5.2 which shows current estimates of radiative forcing by the family of greenhouse gases (GHGs), tropospheric ozone and solar variation. The height of the bar indicates a mid-range estimate of the forcing whilst the lines show the possible range of values. An indication of relative confidence in the estimates is given below each bar. The contributions of individual greenhouse gases are indicated on the first bar for direct greenhouse gas forcing. The major indirect effects are a depletion of stratospheric ozone (caused by the CFCs and other halocarbons) and an increase in the concentration of tropospheric ozone. The negative values for aerosols should not necessarily be regarded as an offset against the greenhouse gas forcing because of doubts over the applicability of global mean radiative forcing in the case of non-homogeneously distributed species such as aerosols and ozone. There are also 'negative forcings' by stratospheric ozone and tropospheric aerosols. Models of climate impacts (Chapter 4) as a result of global warming have shown the potential for damage and disruption to natural ecosystems, global agricultural production, freshwater distribution and population distributions. A major threat is sea-level rise which could occur in the short to medium term (decades) as a result of the thermal expansion of oceans, and in the longer term (centuries) as a result of melting ice sheets (see Chapter 7, Conclusion). The general circulation models (Chapter 4) project greater warming in the high latitudes and continental interiors, with much less warming over the oceans and in low latitudes (Table 22.3). However, the adverse impacts of climate change are likely to be greatest in those countries where the capacity to adapt to climate change is least – that is, in the low-latitude countries with heavy dependence upon agriculture, forests and water supplies.

The atmospheric management response to the threat of climate change follows three patterns. First, there has been a substantial expansion of atmospheric research and climate

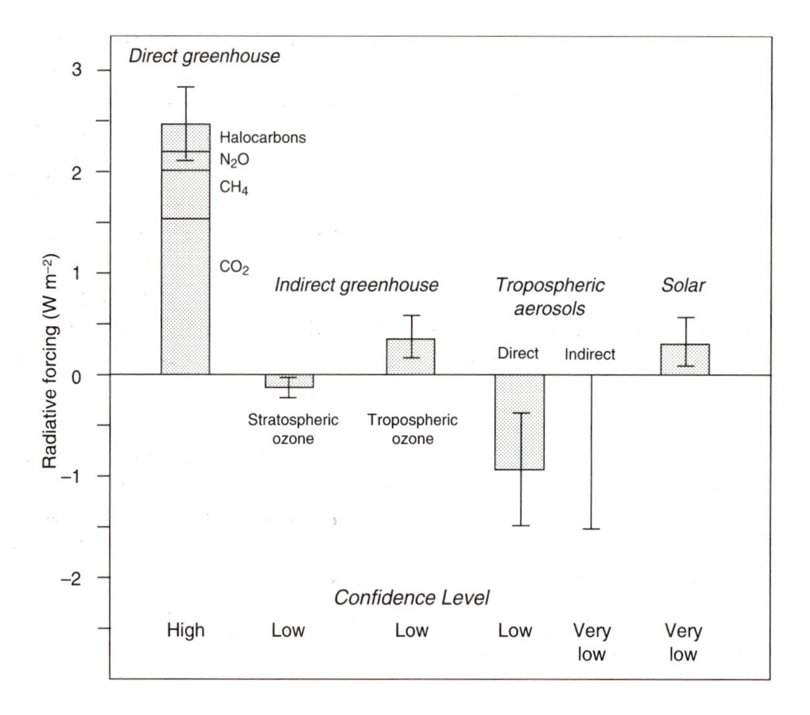

Figure 5.2 Estimates of the globally averaged radiative forcing due to change in greenhouse gases and aerosols from preindustrial times to the present day and change in solar variability from 1850 to the present day

monitoring to strengthen scientific understanding and reduce the uncertainty of current projections. Second, a UN Framework Convention on Climate Change has been adopted (Chapter 22). This convention is analogous to the Vienna Convention on the ozone layer. Negotiations are now underway with a view to the development of specific protocols that will involve international agreements to reduce GHG emissions according to some targets and schedules. In the case of the Montreal Protocol to the Vienna Convention, international agreement was possible in an unusually short time because the adverse effects of ozone-layer depletion could be detected currently along with increased UVB radiation, and because the sources of ozone-layer depletion (CFCs) are limited to a relatively small sector of the economy. In the case of climate change, however, the sources involve the entire energy-intensive industrial economy and domestic lifestyles. Also, curbing emissions of CO_2 and other GHGs will require the participation of more countries. Any climate change mitigative or adaptive strategy must also address the plans of developing countries to expand their use of fossil fuels considerably in the coming decades. Therefore, action to reduce emissions in the industrialized countries will have to be sufficient to accommodate increased emissions from developing countries, if atmospheric concentrations are to be stabilized.

Air Toxics (Persistent Organic Pollutants and Heavy Metals)

Air toxics are also known as hazardous air pollutants. The Organization for Economic Cooperation and Development (OECD) defines them as follows:

> Hazardous air pollutants are gaseous, aerosol or particulate contaminants present in the ambient air in trace amounts with characteristics (toxicity, persistence) so as to be a hazard to human health, or plant, or animal life.

They are of concern because of their persistent and potentially bioaccumulative nature. Air toxics are a global issue, for which the atmosphere is a pathway between the source regions and remote areas, such as the Arctic and Antarctic regions. The means by which toxic species move through the atmosphere to remote regions is called 'global distillation', which describes the selective migration towards colder regions of the globe. This distillation process is the result of revolatilization processes, by which toxic species became vapours under warm conditions and rise up out of the water or soil into the air, until reaching a cold air mass and depositing to the surface. This revolatilization process may happen many times resulting in a global distribution of toxic substances. Air toxics are often defined as persistent organic species including pesticides and heavy metals, such as mercury or lead.

In North America, 362 different contaminants have been detected in the Great Lakes basin, including 32 metals and 68 pesticides. Approximately half of these contaminants are synthetic organic chlorine substances, and one-third of these can have acute or chronic toxic effects on ecosystems, including human health. Many pesticides, which have been banned for nearly 20 years or more, are still observed in the environment, and are abundant in the Arctic food chain. The fact that air toxics do not respect political boundaries has resulted in the establishment of management activities which are international and continental in nature though, to date, many of these agreements and programmes have not been in effect long enough for improvements to be observed in the environment. The circumpolar countries have established an Arctic Environmental Protection Strategy focusing on monitoring and data-gathering programmes to support remediation activities. The UN ECE Convention on the Long Range Transport of Air Pollutants is currently developing protocols for the control and regulation of persistent organic pollutants and heavy metals. UNEP has also initiated a global action plan for hazardous air pollutants. Generally, these international agreements or strategies focus on reducing emissions of the toxic substances at the point of origin into the environment. Air toxics may be managed by two approaches, namely a strategy leading towards virtual elimination from the environment or the management of the toxic substance throughout the entire life cycle to a level of acceptable risk. Substances which are usually targeted for virtual elimination are those substances which are persistent, bioaccumulative and result predominantly from human activity.

Natural Atmospheric Hazards

Long before air quality became an issue, people had to cope with extreme atmospheric variations. Floods, droughts, blizzards, windstorms, hurricanes, fog and frost, for example, have long plagued humanity. The character of atmospheric processes that give rise to these events is such that most of them lie beyond the human capacity to control. There have been numerous efforts to modify these and similar events, however, sometimes with rather perverse results. Efforts to control floods (e.g. by the construction of dams, dikes and other flood control structures) has led to increased use of floodplain lands and has contributed to an increase rather than the expected reduction in flood damage. Previously natural atmospheric hazards have been considered to be 'acts of God' and not therefore the responsibility of any human agent. With the advent of anthropogenic climate change this assumption is no longer so convincing. Climate change will actually be experienced in the changing distribution of weather events, including increased frequency and magnitude of atmospheric extremes (Chapter 23). In a technical sense, therefore, the so-called 'natural' atmospheric hazards can be affected by human activities. In this atmosphere management issue there is a strong case to be made for adaptive response strategies which seek to harmonize human activities with environmental extremes, and to reduce vulnerability.

INTEGRATION OF AIR ISSUES

The previous sections have described the management approaches taken and proposed in relation to each of the separate air issues at local, national, regional and global levels. It is increasingly clear, however, that these issues cannot be effectively managed in isolation. Figure 5.3 shows, in a schematic way, the pattern of interrelationships between some of the air issues. Human activities of many kinds result in emissions into the atmosphere. The level of emissions depends on the scale and the nature of the human activities and the degree of emission content. Many pollutants accumulate in the atmosphere and some interact with each other. Changes in atmospheric physics, density and atmospheric programmes create socially identified problems called 'air issues'. Each of the air issues has an impact on human activities and targets specific vulnerabilities or susceptibilities. In response to the impacts, human activity can adapt in many ways. These include the extent and value of the adaptation of human society which alters the level, and the amount of, emissions produced. For example, sulphate particulates which contribute to acid precipitation also have a local cooling effect on the earth's atmosphere. There is now strong evidence that global warming would have been significantly greater over the past few decades in the absence of sulphur emissions. Policies and programmes introduced to reduce sulphur

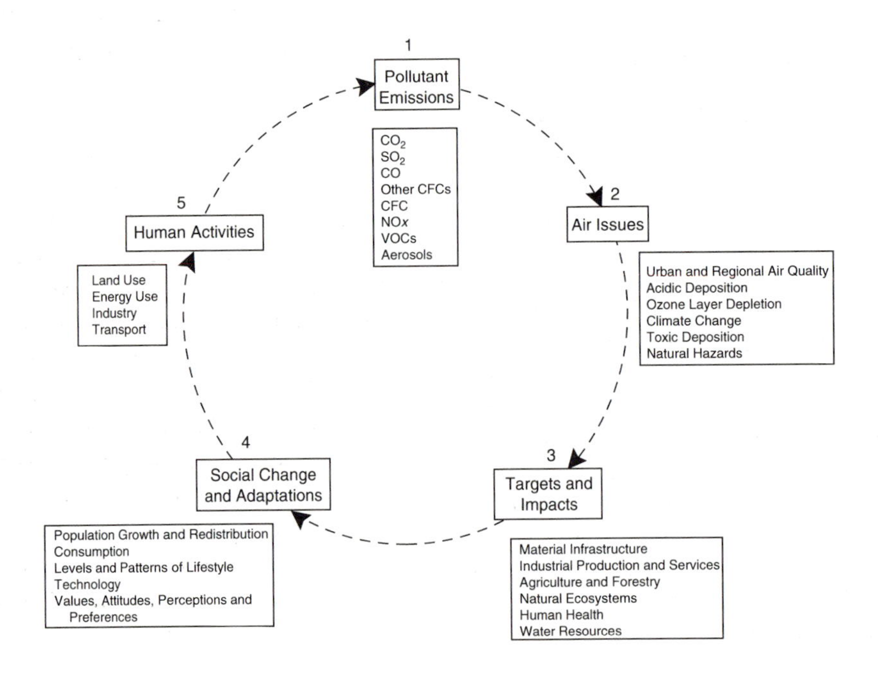

Figure 5.3 Human–atmosphere interactions

emissions therefore, to the extent that they are successful, will tend to exacerbate the global warming problem. On the other hand, CFCs not only deplete the ozone layer, but also serve as greenhouse gases and hence elimination of CFC emissions will contribute to a lessening of the amount of global warming that would otherwise be expected. Economic analysis of the costs and benefits of specific air management proposals should ideally take into account the multiple benefits (and costs) where more than one air issue is affected. For example, reducing NO_X emissions also alleviates acid deposition and smog. Steps to increase energy efficiency can have benefits for urban/regional air quality and acid deposition, as well as climate change.

It is now becoming clear, therefore, that policies directed at one atmospheric management problem cannot be properly assessed or managed in isolation. A policy in one direction may make matters worse in another, and hence the benefits ascribed to that policy may be less than anticipated. On the other hand, a policy that reduces global warming (and at the same time helps to reduce photochemical smog) may have additional benefits which may help to make otherwise less attractive policies seem more economically feasible. A future direction for atmospheric management therefore is to move towards the integration of air issues. This integration, if it is to succeed, should include the integration of the atmospheric research science, the integration of the assessment of impacts of atmospheric changes on natural and social systems and the integration of policies at all levels from local to global. Such an ambition is easily stated, but it is extremely difficult to design and implement. Neither the science, nor the socio-economic studies, nor the policy instruments exist. The most that can be expected is steady progress in the direction of a more integrated approach.

SUSTAINABLE DEVELOPMENT AND THE ATMOSPHERE

Even though the integration of atmospheric science and policy is an important and necessary step that will require a great deal of attention, this will not by itself be enough to resolve the issues of atmospheric resource management. There lies ahead the further requirement to link atmospheric resource management into the agenda for sustainable development agreed by the nations attending the UN Conference on Environment and Development in Rio de Janeiro in 1992. The implications of this agreement are that a new philosophy of management is required and should be worked out and implemented as soon as practicable. The new philosophy leads away from the 'end of pipe' control by technology and regulations (called 'react and cure') to an anticipate and prevent approach. In the case of atmospheric resource management, this means addressing the more fundamental questions of demand for goods and services, full cost accounting and pricing in the marketplace and the development of environmentally friendly technologies. In terms of Figure 5.3, the management interventions should not only include steps to control or reduce emissions, but also address the issues of population growth and redistribution, economic patterns of consumption/lifestyles and environmental technology, along with questions of social values, attitudes, perceptions and preferences. Behind these variables lie questions of ethics pertaining to the equitable distribution of health and wealth among the peoples of the planet, and between generations. Atmospheric resource management has changed substantially from the days when the law of nuisance was sufficient.

CONCLUSIONS

An ultimate objective might be the development of a set of policies at local, national, regional and global levels that work together in a harmonious way to manage the atmosphere for the benefit of all the earth's population, present and future. It is good to have such long-term visions of where management needs to move. For the present it is necessary (if not sufficient) to develop a more integrated understanding of the problems of the atmosphere and to advance its management at the local and national levels. At the same time, it is important to proceed through international negotiations and agreements in order to move the world community towards a common goal, in which atmospheric resource management plays its full part in the move towards sustainable development. While such a vision can be inspiring as a guide to action, it does not command universal acceptance even as a vision. There are limits to what can be achieved by laws, regulations and agreement. Some argue that more market-based economic instruments should play a larger role in atmosphere management. Others propose that neither regulatory nor market-based approaches will suffice, singly or in combination, without a radical change in the willingness of individuals to accept responsibility for the atmosphere.

The atmosphere has traditionally been an unmanaged system. As problems of air quality were identified and grew in scale and complexity, human management systems grew correspondingly in order to cope with the threats to health and economic livelihood. Now the management systems themselves are seen as part of the problem. This has happened in part because more and more divergent interests are at stake and achieving agreement becomes a more and more complicated process. Atmospheric science and the understanding of the economic, social and health significance of the atmosphere are still crucial, but politics and diplomacy are becoming more important. Like the atmosphere itself, 'atmospheric resource management' is in a constant state of flux and both the atmosphere and its management are evolving.

REFERENCES

Auclіems, A. and Burton, I. (1973) 'Trends in smoke concentrations before and after the Clean Air Act of 1956', *Atmospheric Environment* 7: 1063–70.

Bojkov, Rumen D. (1995) *The Changing Ozone Layer*, Geneva: World Meteorological Organization and United Nations Environment Programme.

Bunce, N.J. (1991) *Environmental Chemisty*, Winnipeg: Wuerz.

Burton, I., Kates. R.W. and White, G.F. (1993) *The Environment as Hazard*, New York: Guilford Press.

Elsome, D.M. (1995) 'Atmospheric pollution trends in the United Kingdom', in L. Simon (ed.) *The State of Humanity*, Cambridge, MA: Blackwell, pp. 476–90.

Graedel, T.E. and Crutzen, P.J. (1993) *Atmospheric Change: An Earth System Perspective*, New York: AT&T/W.H. Freeman.

Houghton, J.T., Jenkins, G.J. and Ephraumus, J.J. (eds) (1990) *Climate Change. The IPCC Scientific Assessment*, Cambridge: Cambridge University Press.

National Academy of Sciences (1992) *Rethinking the Ozone Problem in Urban and Regional Air Pollution*, Washington, DC: National Academy Press.

PART 2:

CLIMATE AND THE PHYSICAL/ BIOLOGICAL ENVIRONMENTS

The following six chapters (6–11) examine the influence of climate on the functioning of the physical and biological environments. Hydrological processes/water resources, glaciers, geomorphic processes/landforms, soils, vegetation and animal responses are all considered, with the predicted impact of climate change foremost in the coverage by most of the contributors.

6

HYDROLOGICAL PROCESSES AND WATER RESOURCES

Paul Whitehead

INTRODUCTION: CLIMATE AND THE AQUATIC ENVIRONMENT

In Part 1 of this book the techniques of applied climatology are described particularly within the context of climate modelling and climate impact assessment. In this first chapter of Part 2, the impacts of climate change on hydrological processes and water resources are considered. The hydrological system is particularly complex with many interactions and feedback mechanisms operating. Superimposed on natural processes are human activities which affect the hydrological system (Chapter 21) and enable it to be exploited for potable water supply. Figure 6.1 illustrates the broad interactions between climate and hydrological factors.

Changes in greenhouse gases will increase the counter-radiation flux which will in turn have direct physical effects by changing vegetation response, and hence land use, and by altering temperature and rainfall patterns. These will inevitably lead to changes in evaporation and soil moisture conditions and these will combine to produce changes in run-off, river flows and groundwater levels. Predicting the effects on the hydrological system and water resources is difficult but may be achieved by using a knowledge of the processes involved together with mathematical modelling techniques (Chapter 4).

In the UK there have been several initiatives designed to assess impacts of climate change on hydrology and water resources. The Climate Change Impacts Review Group (CCIRG, 1991) was established by the Department of Environment (DOE) in 1990 to review the impacts of climate change on a wide range of activities. As part of this review the DOE Water Directorate funded a study on the Impacts of Climate Change on Water Resources and the initial results of this programme of research are given by Arnell (1992). Table 6.1 shows the likely effects of climate change on the water environment and, as indicated, a large number of functions of the new Environment Agency will be affected (Arnell and Reynard, 1993). In order to facilitate research on climate change impacts, the DOE has funded the Climate Impact Link Project based at the Climatic Research Unit at the University of East Anglia. The basic aims of this project are to:

1 To liaise with the impacts community to determine the nature of their climate change data requirements
2 To provide information for the impacts community to help them become familiar with the correct interpretation of global circulation model (GCM) data
3 To liaise with the Hadley Centre so that archived GCM data can be tailored to the

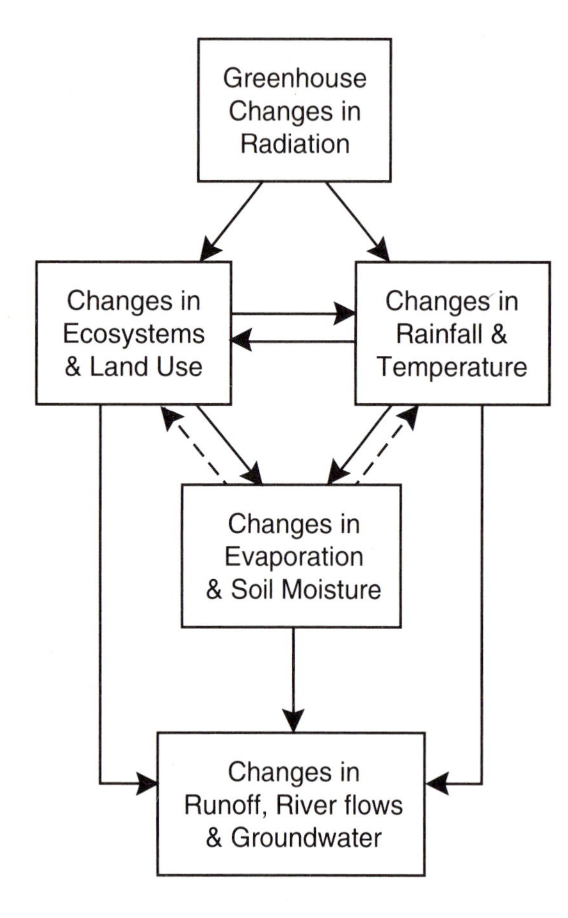

Figure 6.1 Climate change and hydrological impacts

needs of the impacts community (climate model simulations produce vast quantities of output; only a small proportion is stored)

4 To develop and provide a range of climate change scenarios for the UK at time and space scales suitable for use by the impacts community.

This project has been particularly useful in providing scenario data for a wide range of environmental, hydrological and water resource studies and some results are presented in this chapter to indicate likely changes.

HYDROLOGICAL PROCESSES

As indicated in Table 6.1 changes in temperatures and rainfall will have significant impacts on many aspects of the aquatic environment. The degree of effect will vary depending on location, the season, the size of the change and the non-linearity of the processes operating. As indicated in Figure 6.2, the hydrological cycle can be viewed as a series of storages and flows, and changing drying and wetting patterns could alter the system in many ways. It is useful to consider some of the components and processes in turn.

Temperature and Precipitation

Both temperature and precipitation are expected to change in the future. Table 6.2 shows predicted changes adopted in hydrological and water resource studies to date.

The rainfall change scenarios assume increased rainfall in winter and no change in summer rainfall, although as Table 6.1 indicates, there is considerable uncertainty. Spatial patterns of change in rainfall are even more uncertain because of the differences between different climate models, but there are indications that drier summers relative to the present are more likely to occur in the south of the UK than in the north (CCIRG, 1991). The scenarios in Table 6.1 show changes in mean rainfall and temperature. Climate models are not yet sufficiently reliable to estimate with confidence changes in the year-to-year variability in these parameters although the latest runs from GCMs can provide time series of rainfall and temperature data and these are available from the DOE LINK project.

Evapotranspiration

Evaporation is a crucial element in the hydrological cycle. An increase in temperature would be expected to increase the rate of evaporation

Table 6.1 Impacts of climate change on the aquatic environment

	Water resources	*Water quality*	*Flood defence*	*Fisheries*	*Conservation*	*Recreation*	*Navigation*
Change in temperature	×	×		×	×	×	
Change in rainfall	×	×	×				
Change in evaporation					×		
Direct effects of CO_2	×	×			×		
Change in river flows	×	×	×	×	×	×	×
Change in groundwater recharge	×	×					
Change in water chemistry and biology	×	×		×	×	×	
Change in storminess			×		×		
Sea-level rise	×	×	×		×		

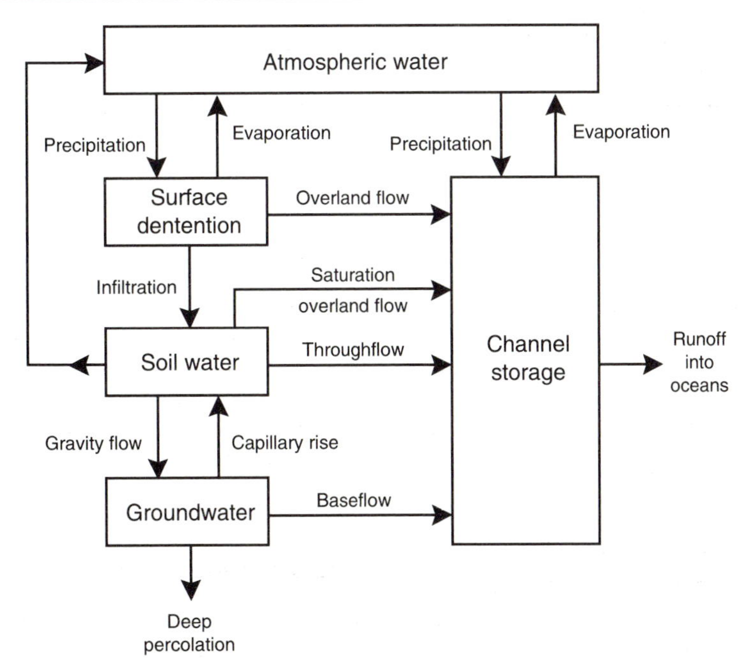

Figure 6.2 The hydrological cycle expressed as a series of water storages and flows

Table 6.2 Climate change scenarios for the UK (CCIRG, 1991), expressed as change from the present climate

	2010		*2030*		*2050*	
	Summer	*Winter*	*Summer*	*Winter*	*Summer*	*Winter*
Temperature (°C)	0.7	0.8	1.4	1.5–2.1	2.1	2.3–3.5
Precipitation % (range)	0 (±5)	3 (±3)	0 (±11)	5 (±)	0 (±16)	8 (±8)

but this change would also be affected by humidity, wind speed, net radiation and water availability. Arnell (1992) and Cole *et al.* (1994) considered the effects of evaporation changes on hydrological response for a range of catchments and found that the hydrology was particularly sensitive to evaporation estimates. Evapotranspiration will also be affected by changes in catchment vegetation resulting from rainfall and temperature variations. A new vegetation mix would alter evaporation totals, and a change in plant growth rates through the year would affect the timing of evaporative demands. Increased plant growth early in spring, for example, would lead to an earlier upswing in the evapotranspiration rate after winter than at present. Another complicating factor is that plant physiology will also change with increased carbon dioxide (CO_2) concentrations perhaps generating increased transpiration. However, this is difficult to predict because different plants will respond in different ways and there will be interactions with temperature, rainfall and nutrient availability which would limit plant growth.

Soil Moisture

As rainfall and evapotranspiration change, so patterns of soil moisture will alter (see Chapter 9, Effect of climate change on soils). Estimates of changes in soil moisture content can be derived directly from climate model simulation output or by applying rainfall and potential evaporation scenarios to a soil moisture accounting model (Whitehead and Calder, 1993). Current climate models already contain a basic soil moisture component as soil moisture content affects land surface energy and water balances. Climate model output can be used to show continental- or global-scale changes in soil moisture, as illustrated in the IPCC report (Houghton *et al.*, 1990), but these outputs are not appropriate for local- or regional-scale assessments because the spatial scale of simulation is very coarse. The Hadley Centre climate model has a grid size of nearly 80,000 km^2, assumes representative values and parameterizes the effect of variability across the cell. Soil moisture will alter under climate change especially if, as expected, evaporation rates increase in the summer. This will lead to drier soils and hence soil cracking which in turn would alter hydrological pathways and, in general, lead to more rapid runoff.

Runoff, River Flows and Groundwater

Runoff and river flow are derived from overland flow, the near-surface saturated areas, flow through the deeper soil horizons or from groundwater inflow. As might be expected, changes in rainfall, evapotranspiration, soil moisture and hence runoff will vary significantly depending on the nature of the catchment, the land use and the particular flow paths operating. Groundwater will also be a significant factor controlling hydrological response in aquifer-fed systems. The recharge of groundwater generally occurs in the winter months once the soil moisture levels have been replenished. Recharge will depend on the soil type and underlying geology and is very dependent on the length of the recharge season. With increased winter rainfall, recharge of groundwater systems will generally improve across aquifer regions despite the increased evaporation during the summer.

IMPACTS OF CLIMATE CHANGE ON HYDROLOGICAL RESPONSE

In order to assess the components of the hydrological cycle shown in Figure 6.2, simulation studies have been undertaken by Arnell (1992) and Whitehead *et al.* (1995). Whitehead *et al.* applied the IHACRES approach (Jakeman *et al.*, 1990) to simulate a range of river basins across the UK and used climate scenarios to assess the changes in hydrological response. Figure 6.3 shows the simulated and observed behaviour respectively for three very different catchments investigated. The highly responsive regime of the River Gretna (Figure 6.3(a)), located in the Pennines and draining Millstone Grit is shown by the peaky nature of the hydrograph. The Fal at Tegony in Cornwall (Figure 6.3(b)) is a moderate to low relief catchment draining Devonian slates, shales and grits and the response indicates a longer recession to the hydrograph. The Coln (Figure 6.3(c)) is a tributary of the Thames and has slowly responding flow dominated by groundwater response from pervious limestone. The simulated and observed behaviour on each graph reflect the accuracy of the IHACRES modelling approach to reproduce the widely differing responses. It is important to realize that the hydrological response is very different in these three catchments and that the prediction of climate change on such catchments is particularly interesting. Figure 6.4 shows the flow duration curves for these three catchments and gives the results of two climate change scenarios on flow duration.

Two high-resolution equilibrium scenarios

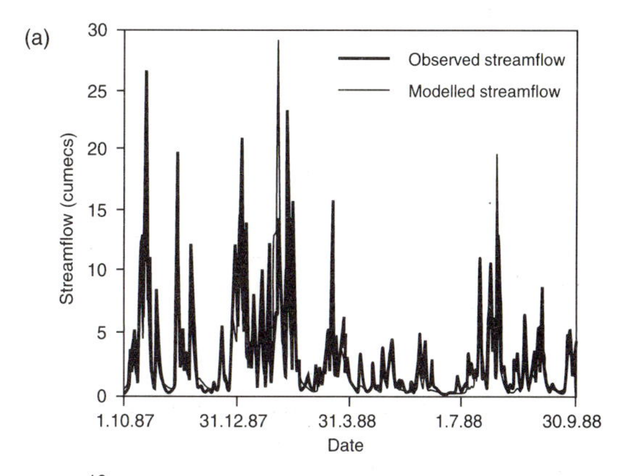

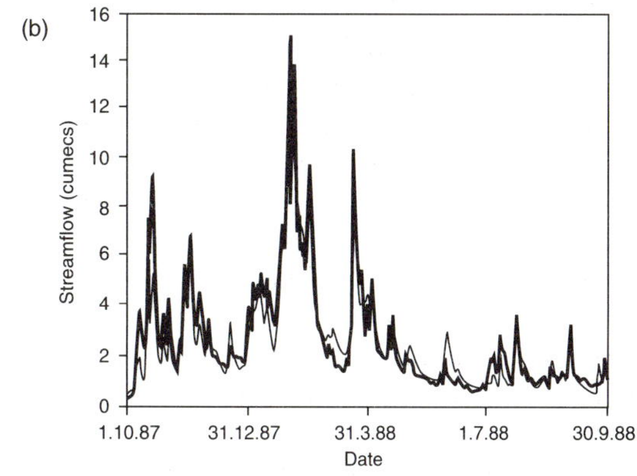

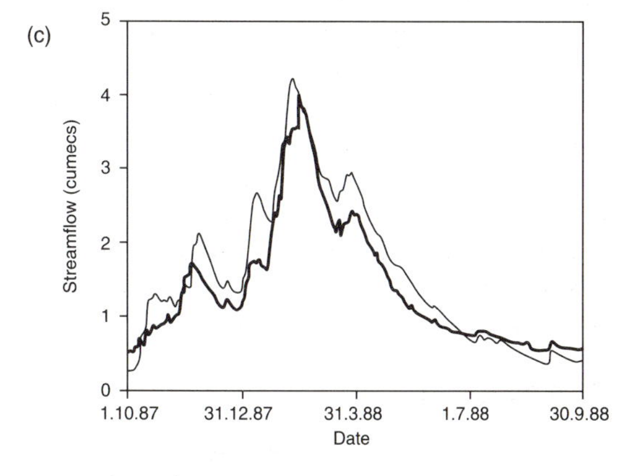

Figure 6.3 Observed and modelled streamflows for calibrations of the (a) River Greta at Rutherford Bridge, (b) River Fal at Tregony and (c) River Coln at Bilbury, in time series form for 1987–8

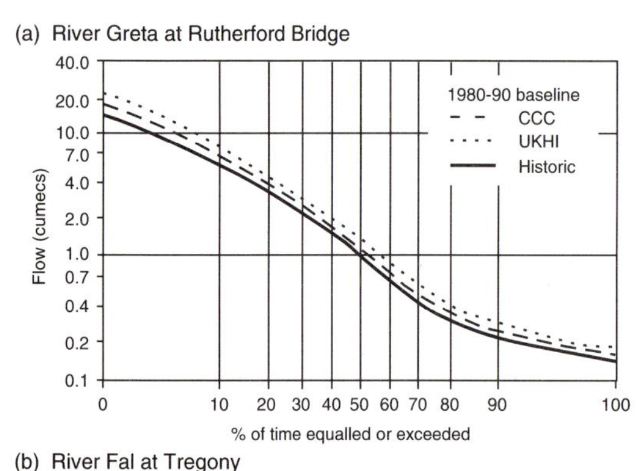

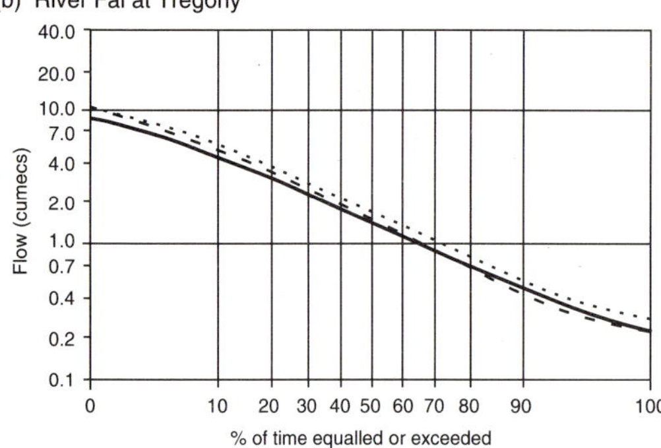

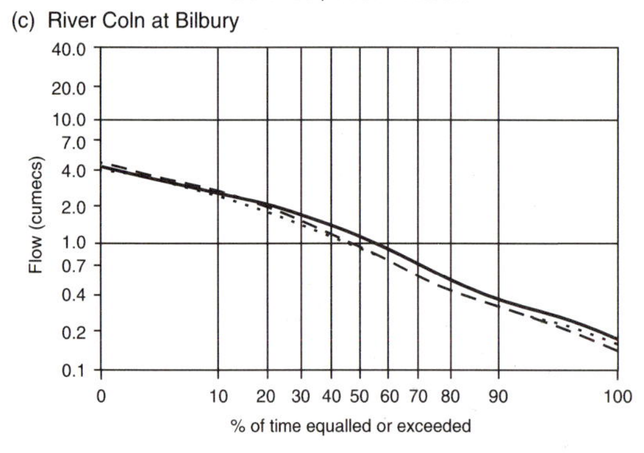

Figure 6.4 Impacts on flow regime of the UKHI and CCC climate change scenarios for a global warming of 2.5°C, due in 100 years' time under the IS92a scenario, without the cooling effects of sulphate aerosols

were applied to the catchments, UKHI and CCC (Hulme *et al.*, 1994). The UKHI scenario was obtained from the Hadley Centre equilibrium experiment in 1989 and the CCC scenario is from the Canadian Climate Center. By equilibrium it is assumed that the atmosphere has shifted to a new CO_2 concentration and the resultant output data from the GCM model is representative of the new climate. As shown in Figure 6.4, flows in the Gretna catchment are generally increased whereas flows in the Coln are decreased. In the River Fal, the flow duration curve is not significantly altered.

A spatial picture of hydrology and hence water resources can be derived from application of similar modelling techniques to a whole range of catchments and Figure 6.5 shows maps of mean flow, 95 percentile flow and mean annual flood for differing climate scenarios in England and Wales. Considering first the impact of CCC on mean flow (Figure 6.5), rivers in the north and west increase in volume, whilst those in central, southern and eastern areas suffer slight decreases. The pattern is repeated under the wetter, warmer UKHI scenario, with greater extremes of wetting and drying seen in the northwest and southeast respectively resulting in closer contours. Clearly the effect of warming dominates over greater rainfall in the southeast, whilst rainfall is the dominant factor in the northwest. In Figure 6.5 this southeast/northwest gradient is significant with increased flow in the north and west and the most severe drying in the south and east. Contours delineating this pattern can be seen to follow the boundary between high- and low-altitude land in England and Wales, separating Cornwall, Wales and the Pennines from the south and east. Southern areas contain the rivers that suffer the greatest reductions in low flows under the scenarios – up to 25 per cent in some catchments. As with the other parameters, the greatest increases in mean annual flood magnitudes are induced by the UKHI scenario in the north and west. Much of the country is predicted as suffering from increased flood levels under a future climate. Further insight into these impacts will be gained when reliable modelling of the temporal redistribution of

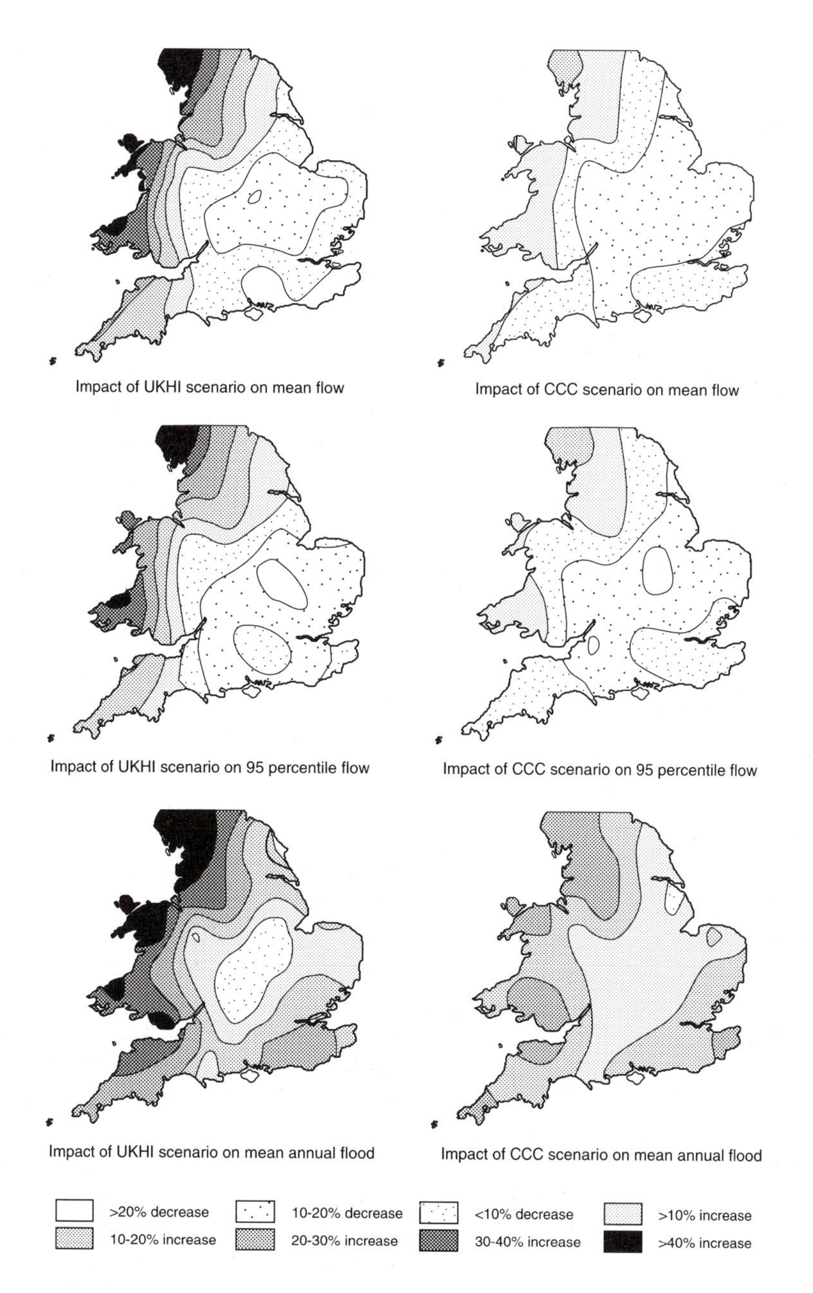

Figure 6.5 Mean flow, 95 percentile flow and mean annual flood for differing climate scenarios in England and Wales

rainfall under climate change scenarios becomes available.

It is clear from the interpolated maps that there is a geographical pattern to climate change response in England and Wales and a key factor in the vulnerability of a river is its location. However, smaller-scale fluctuations in the maps, although they should be interpreted with caution, indicate that location is not the only influencing factor. Such irregularities may indicate that other components of the landscape, such as vegetation, are influencing factors in hydrological response to climate change.

IMPACTS OF CLIMATE CHANGE ON WATER RESOURCES

The climate change scenarios described above indicate that water resources could be affected, particularly in the drier eastern regions of the country. With drier and warmer summers, demand will also increase and public water supplies will have to be increased to meet the demands. The DOE report on climate change impacts (CCIRG, 1991) gives the following summary of likely impacts on water resources:

1 Wetter winters would benefit water resources in general, but warmer summers with longer growing seasons and increased evaporation would lead to greater pressures on water resources, especially in the south and east of the UK. Increased variability in rainfall, even in a slightly wetter climate, could lead to more droughts in any region of the UK.
2 Water demand in warm months may be 25 per cent above average and, locally, a doubling of demand can occur on a hot, dry day. Higher temperatures would lead to increased demand for water and higher peak demands, requiring increased investment in water resources and infrastructure if restrictions were to be avoided.
3 An increase in temperature would increase the demand for irrigation. In times of drought the abstraction for agriculture competes with abstractions for piped water supply by other users.
4 A large range of industries use water in production. Some industrial activity (e.g. chemicals production and food processing) would be severely curtailed if water suppies were interrupted as a result of drought (See Chapter 14, Industry operation).
5 Increases in the frequency of drought could have an impact on public health through interruptions to domestic and other supplies.
6 Climate change may lead to changes in soil structure causing increased leaching, decreased absorption and changes in agricultural applications of fertilizer and pesticides which could affect river, lake and groundwater quality, possibly adversely.
7 Many existing power stations rely on river water for cooling. Climate change may reduce the availability of cooling water, particularly during summer months, with possible implications for the operation of individual power stations.
8 Water quality of rivers and upland catchments could be affected as reduced flows provide less dilution of effluents and also increased temperature is likely to generate mineralized nitrogen from UK soils and affect acidification in the uplands (Ferrier and Whitehead, 1993).
9 Changes in sea water level could also affect aquifer water quality as a rising water level could enhance saline intrusion to coastal aquifers (Cole *et al.*, 1994).

Thus, the effects of climate change on water quantity and quality could be highly significant. The recent situation in Yorkshire of the water authority having to tanker in water from other regions could be a regular sight under a changed climate. It is somewhat surprising therefore that there has been no call for a

National Water Grid (NWG) along the lines of the Electricity Grid to ensure continuity of supply. There are in fact more than adequate resources available across the country as a whole and it is perhaps time that the NWG was put in place to ensure adequate supply.

CONCLUSIONS

There is little doubt that climate change will now occur and the impacts will probably be most significant in the area of aquatic environments. Change in hydrological processes such as evapotranspiration and runoff mechanisms will alter flow duration and the temporal response of catchments. Different parts of the country will be affected in different ways with the dryer southeast being most impacted although increased winter rainfall will probably protect groundwater-fed systems. Water quality will also be affected as lower flows alter pollution–dilution ratios and saline intrusion occurs in areas affected by sea-level rise. Also changed temperature patterns may release nutrients such as nitrogen from land systems exacerbating eutrophication in lowland and upland rivers and lakes. It is particularly important to maintain current monitoring systems in order to have an early warning of hydrological change and long-term experiments such as Plynlimon and Balquhidder should be strongly supported (Whitehead and Calder, 1993).

REFERENCES

Arnell, N. (1992) 'Impacts of climate change on river flow regimes in the UK', *Journal of the Institute of Water Eenvironmental Managers* 6: 433–42.

Arnell, N. and Reynard, N. (1993) 'Impact of climate change on River Flow Regimes in the United Kingdom', *Report to the Department of the Environment*, Water Directorate.

Climate Change Impacts Review Group (1991) *The Potential Effects of Climate Change in the UK*, Department of the Environment, London: HMSO.

Cole, J.A., Oakes, D.B., Slade, S. and Clark, K.J. (1994) 'Potential impacts of climatic change and sea-level rise on the yields of aquifers, river and reservoir source', *Journal of the Institute of Water Environmental Managers* 8: 591–605.

Ferrier, R.C. and Whitehead, P.G. (1993) 'Potential impacts of afforestation and climate change on the stream water chemistry of the Monacyhle catchment', *Journal of Hydrology* V. 145: 453–67.

Houghton, J.T., Jenkins, G.J. and Ephraums, J.J. (eds) (1990) *Climate Change: The IPCC Scientific Assessment*, Cambridge: Cambridge University Press.

Hulme, M., Conway, D., Brown, O. and Barrow, E. (1994) *A 1961–90 Baseline Climatology and Future Climate Change Scenarios for Great Britain and Europe. Part III: Climate Change Scenarios for Great Britain and Europe*, Report accompanying the datasets prepared for the TIGER IV.3a consortium, Climatic Research Unit, University of East Anglia.

Jakeman, A.J., Littlewood, I.G. and Whitehead, P.G. (1990) 'Computation of the instantaneous unit hydrograph and identifiable component flows with application to two small upland catchments', *Journal of Hydrology* 117: 275–300.

Whitehead, P.G. and Calder, I. (1993) 'The Balquhidder catchment and process study'. Special Issue, *Journal of Hydrology* 145: 480 pp.

Whitehead, P.G., Sefton, C. and Jakeman, A.J. (1995) 'Modelling temporal and spatial impacts of climate change on hydrological systems', *Report to TIGER Consortium*, Wallingford: Institute of Hydrology.

7

GLACIERS

Russell D. Thompson

INTRODUCTION: CLIMATE AND GLACIERS

A glacier, ranging in size from the smallest cirque to the largest ice sheet is, in simple terms, an ice mass at the surface of the earth (Souchez and Lorrain, 1991). Flint (1957) was more precise and dynamic when he defined a glacier 'as a body of ice and firn, consisting of recrystallized snow and refrozen meltwater, lying mostly on land and showing evidence of present or former motion'. Earliest studies of glaciers, from the early 1750s to late 1840s (Paterson, 1994), concentrated on glacier movement and the complex climate–glacier relationships were neglected until the pioneer work of Ahlmann (1935) and Sverdrup (1936). Indeed the first textbook dealing entirely with this subject, *Glaciers and Climate*, was published in Stockholm in 1949 (Son Mannerfelt, 1949).

The past 30 years have experienced impressive theoretical developments concerning the relationships between glacier behaviour/flow and climate change.

> It is now possible to predict the future behaviour of a glacier, given sufficient data about its present state and some assumptions about future climate. Such predictions are important when roads, pipelines, mines and hydro-electric plants are being built near glaciers.
>
> (Paterson, 1994)

They are also important when it is realized that glaciers provide enormous reservoirs of pristine water (representing some 75 per cent of available continental water, at present) in an increasingly thirsty world. It is obvious that the usual (liquid!) supplies of fresh, pure water cannot keep pace with the increasing demands and pollution by the current (and forecasted) population explosions. Finally, their importance is paramount to the stability of global sea level since, for example, total ablation of the Antarctic ice sheet could raise this level by some 65 m.

Glaciers are clearly the true 'progeny' of climatic elements and are extremely sensitive to climate changes, with potentially serious consequences for human societies. Furthermore, vast ice sheets and associated cold-air reservoirs influence global climates on a variety of scales. This is especially so in terms of the general circulation of the atmosphere through cyclogenesis at Arctic/Antarctic/polar frontal zones and through the equatorial thrust of polar air aloft, out from the circumpolar vortex in the form of extensive, sluggish, long (Rossby) waves and meridional air flows. This chapter will demonstrate that glaciers interact with climatic elements in a complex cause and effect pattern and will confirm that these relationships are the 'recognized' domain of the applied climatologist.

GLACIER TYPES: PHYSICAL AND THERMAL CHARACTERISTICS

Glaciers are classified according to their physical characteristics (namely, size and surface morphology) and are divided into three main types (Figure 7.1(a)), as follows:

1 Ice sheets are continuous ice masses moving radially in all directions over a vast plateau (East Antarctica, 10,353,800 km^2) or a smaller ice cap occupying the central parts of a mountain range complex (Barnes Ice Cap, NWT, Canada). On coastal margins, ice sheets can extend over the adjacent sea as an ice shelf (Ronne–Filchner, West Antarctica).
2 Mountain glaciers are confined to topographic weaknesses which direct their development and main movement. They are represented as streams of ice flowing down pre-existing river valleys (Tasman glacier, New Zealand), as cirques/corries or pockets of ice on mountainsides (Western Cwm, Mt Everest) and as outlet glaciers, when ice spills over from ice sheets/ice caps through a col or saddle (Meserve glacier, Victoria land, East Antarctica).
3 Piedmont glaciers occupy broad lowlands and coastal plains at the base of mountainous areas where parallel valley glaciers extend beyond the mountains and coalesce into an 'apron' of virtually stagnant ice (western Adelaide Island, Antarctic Peninsula).

The dynamic behaviour of glacial ice is closely related to its thermal characteristics, which are influenced by heat derived from three sources, namely the surface/air, the base and internal friction (Sugden and John, 1976). Surface heating is in the form of conduction from warm air and from the release of latent heat within the glacier, when percolating meltwater refreezes (i.e. 1 g of water freezing warms 160 g of ice by 1°C). Basal heating is due mainly to geothermal heat flows into the base of the ice. This

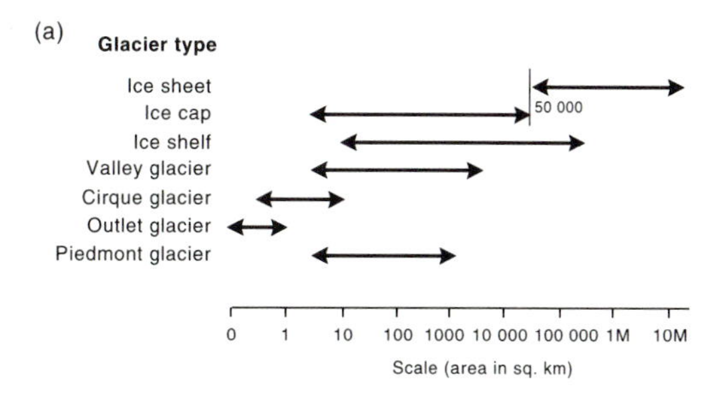

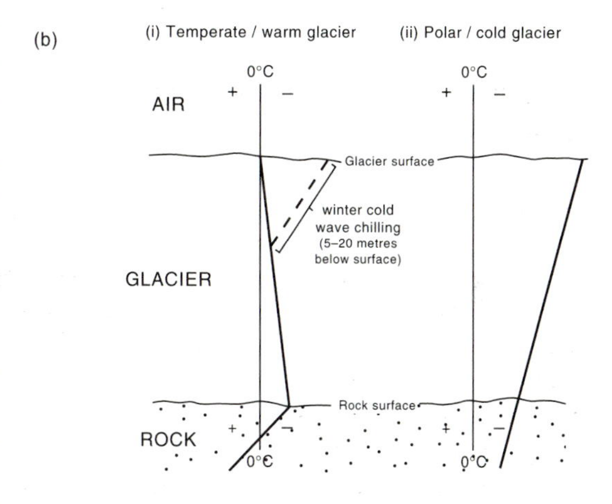

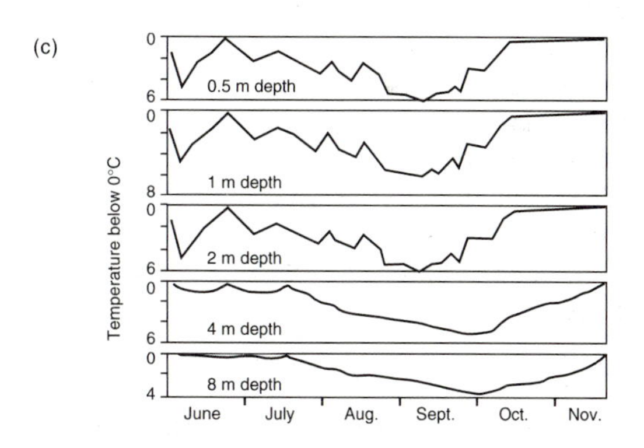

Figure 7.1 (a) Classification of glaciers according to their size and morphology (after Hambrey, 1994); (b) schematic geophysical/glaciothermal classification of glaciers; (c) Orwell glacier: thermal variation with depth

averages 59.9 mW m^{-2} which is enough to melt 6 mm of ice at its pressure melting point each year (Paterson, 1994), although there is considerable spatial variation depending on geological conditions. According to Sugden and John (1976) and Stacey (1969), geothermal heat flow varies from a mean of 90.5 mW m^{-2} in Cenozoic volcanic areas to 38.5 mW m^{-2} in Precambrian shields. Consequently, this heat flux beneath Icelandic glaciers is about twice as high as that beneath the Greenland ice sheet. Internal friction heat is derived from differential movement within and at the base of the glacier and, at a rate of flow of 100 m per annum, it produces five times as much heat as that from the average geothermal heat flow (Sugden and John, 1976).

Glaciers can be classified according to their distinctive thermal characteristics, which develop in response to the above three heat sources. Some 50 years ago, Ahlmann (1948) proposed two main glaciothermal types, namely 'polar' and 'temperate' ice, which have been widely accepted. However, as Sugden and John (1976) emphasized, the terms are confusing since both types of ice occur in polar and temperate areas. Also, entire glaciers have been classified as temperate ('warm') or polar ('cold') when both types of ice can occur in individual glaciers in polar and temperate areas. For example, some subarctic glaciers in Scandinavia and Svalbard are composed of warm ice in their accumulation zones and cold ice in the ablation zones (Paterson, 1972). Temperate glaciers are close to 0°C throughout, owing to the surface heating discussed above, and the actual temperature within the ice equals the pressure melting point corresponding to the depth concerned. This point is actually lowered by 0.0072°C per additional atmosphere pressure although this rate is somewhat misleading because, since ice contains impurities, it does not have a distinct melting point determined soley by pressure (Paterson, 1994).

Even so, it is apparent that the basal layers of a temperate glacier are, on an annual basis, cooler than the more isothermal surface zone (Figure 7.1(b)). For example, at the base of a glacier 1500 m thick, the pressure is great enough to lower the melting point by 1°C (Sharp, 1988). It should be stated that, without an insulating autumnal snow pack in excess of 0.5 m, winter cold-wave penetration can chill the surface/upper layers to a depth of up to 20 m (Figure 7.1(b)). At this time, the surface zone is clearly cold or polar but summer heat sources can soon return conditions to warm with the thaw of the winter snow pack. For example, on the Orwell glacier, Signy Island, West Antarctica, the top 8 m were chilled between −3.7°C and −7.3°C, over a period of 122 days (at 0.5 m) to 172 days (at 8 m) (Figure 7.1(c)). Cold-wave dissipation returned this upper zone to pressure-melting temperatures in only 23 days and 48 days, respectively, at depths of 0.5 m and 8 m (Thompson, 1968; 1977). Alternatively, if a greater thickness of ice is cooled in winter than can be warmed/dissipated in summer, then a cold layer may survive near the surface until the following winter cold-wave penetration. Paterson (1972) and Sugden and John (1976) indicated that this thin layer of cold ice, at a few metres depth below and above warm ice, may be common on many glaciers previously regarded as temperate throughout.

In contrast, polar or cold glaciers develop when surface melting in summer is negligible or absent and the glacier is characterized by temperatures well below the melting point of ice, from surface to basal layers (Sharp, 1988). In these glaciers, the temperature at a depth of 10 m (i.e. below the level of seasonal thermal variations) closely approximates the mean annual air temperature at that site (Souchez and Lorrain, 1991; Sugden and John, 1976). In such a cold ice mass, there is a progressive increase in glacial temperature with depth, towards the source of geothermal heat (Figure 7.1(b)). For example, at Byrd Station, West Antarctica, ice-sheet temperatures ranged from

−28°C near the surface to −1°C at the glacier bed (about 2200 m depth). Similarly, at Camp Century, Greenland, ice-sheet temperatures increased from −25°C at the surface to −11°C at the glacier bed (about 1300 m) (Sugden and John, 1976).

It should be emphasized that the actual rate of temperature increase with depth in cold ice is strongly influenced by the accumulation rate of snow and firn. Since the firn layers are 'cemented' into the ice sheet (see the next section), substantial annual accumulation rates transfer cold conditions into the ice sheet more effectively than small accumulation rates. Consequently, the heavy accumulation can counteract the effect of geothermal heating (quantified above) and reduce the temperature gradient (Sugden and John, 1976). Robin (1955) calculated the relationship between ice temperature and depth for the centre of a hypothetical 3000 m thick ice sheet (assuming negligible/zero flow), in terms of varying values of annual snow accumulation. For example, an accumulation rate of 40 mm (water equivalent) should produce a basal ice temperature some 18°C warmer than one associated with three times that accumulation (Sugden and John, 1976).

Despite the conspicuous warming of cold ice with depth, associated with geothermal heat fluxes, it has been argued that basal melting does not exist since this heat is conducted to the glacier's surface where it is transferred (by conduction/radiation) into the atmosphere. Sharp (1988) unequivocally stated that 'warm glaciers melt at the bottom; cold glaciers do not'. However, this argument ignored evidence in the mid-1970s from beneath Byrd Station in West Antarctica, Vostok Station in East Antarctica and the western parts of the Greenland ice sheet, which confirmed that large basal areas are clearly at pressure melting point. This of course means that water is likely to be present at the ice/rock interface because, once pressure melting point has been reached, only a reduced amount of geothermal heat can be conducted away into the basal ice. The surplus heat is now used to melt a thin layer of basal ice (Sugden and John, 1976). Recently, it has been confirmed that large freshwater lakes are located beneath continental ice sheets. For example, Ellis-Evans and Wynn-Williams (1996) reported the existence of at least 70 subglacial water bodies under the East Antarctic ice sheet. Lake Vostok is the largest of these water bodies, covering some 14,000 km^2 (about the size of Lake Ontario) and up to 500 m deep (similar to Lake Baikal). It exists at a depth of 4 km under the Russian Vostok Station and it is the obvious product of geothermal flux, pressure melting point and the internal frictional heat of sliding.

GLACIERS: CONSTITUENT ZONES AND CLIMATE CONTROLS

A glacier can be considered as an open system with input, storage, transfer and output of mass. It is a system in dynamic equilibrium where the mass balance depends on input (accumulation) versus output (ablation) (see the next section). In very basic terms, the glacier can be divided into two conspicuous parts, namely the area of net accumulation (where there is an excess of input over ablation) and the area of net ablation (where there is an excess of output over accumulation). The boundary between the two zones is termed the equilibrium line (Figure 7.2) where, on an annual basis, ablation equals accumulation (Souchez and Lorrain, 1991). The penultimate section examines in detail the component parts of both accumulation and ablation processes and, in this section, it is useful to discuss the characteristics of these important zones. As Paterson (1994) stated, since conditions vary from one point on a glacier to another and very few glaciers can be fitted into a single category, it is better to

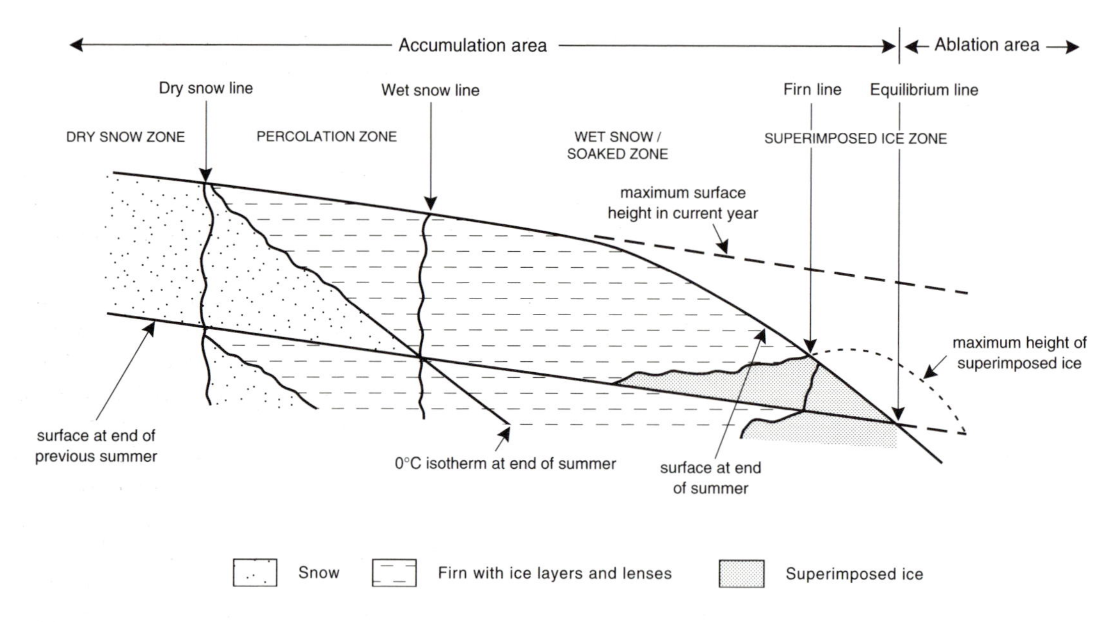

Figure 7.2 Constituent parts of a glacier (after Müller, 1962)

discuss different zones on a glacier than to try and classify entire glaciers.

In the ablation area, output clearly exceeds accumulation and this zone is dominated by progressive glacial thinning of the exposed ice mass. In winter, the entire glacier is covered with the seasonal snow pack, but during summer, the edge of this cover gradually recedes upglacier in the ablation zone, until the highest removal elevation is reached at the firn line (Figure 7.2). Above this line, the previous winter's snow pack remains and is eventually obliterated by the next winter's snow deposition, to become 'old' granular snow called firn. It should be noted that glacier movement is constantly transporting the firn edge downglacier and, despite snow ablation below the actual firn line, on many occasions the firn edge lies at a lower, more advanced position than the snow line on adjacent ground. In contrast, the equilibrium line (Figure 7.2) represents the lower boundary of the 'new' superimposed ice zone, which is part of the net gain of mass over the year (discussed below).

Some glaciers are completely devoid of firn cover and superimposed ice (e.g. the Whakapapanui glacier, Mt Ruapehu, New Zealand), since the entire winter snow pack is normally removed by the summer ablation, together with a substantial amount of 'old' glacier ice. These 'unhealthy' glaciers are thinning and receding at a rapid rate; for example, the Whakapapanui glacier receded 147 m and thinned 5 m in 5 months in early 1970 and had disappeared entirely by the mid-1980s. Conversely, some glaciers are permanently above the firn line (e.g. the Beardmore glacier, flowing into the Ross Ice Shelf, West Antarctica) since the negligible summer ablation fails to remove the winter snow pack deposition (however meagre) from any part of the glacier. These very 'healthy' glaciers are continually gaining firn and glacial ice and are thickening/advancing rapidly over time (e.g. the Beardmore glacier advances at a rate of 1 m per day on average). Despite the simplicity of the basic two-part zonation of a glacier discussed above, the structure of the accumulation zone is much more complicated, and this complexity is illustrated in Figure 7.2. Interestingly, the original Müller zonations of 1962 are still used today (Paterson, 1994) with only very minor changes introduced over the past 34 years. It should be noted though that very few glaciers will show the entire sequence of these distinctive zones and the actual boundaries are very transitional indeed, varying from year to year according to changing weather patterns and climatic elements.

The highest part of the accumulation area is called the *dry snow zone*, where no melting occurs even in summer. In alpine glaciers, this area is likely to be composed of cold ice and it characterizes the middle/upper reaches of ice sheets too. Moving down to the *percolation zone*, this is an area of limited melting which soon refreezes at depth and releases latent heat of fusion, to warm the snow towards 0°C. The lowest division (leading down to the firn line) is the *wet snow or soaked zone*, with excessive surface melting and deep percolation/absorption of meltwater (even into the 'old' firn layers). At the end of the summer, the entire residual snow pack has been warmed to 0°C. In the percolation and soaked zones, the glacial material consists of snow and firn layers with ice bands, deposited by refreezing meltwater or buried rime ice/hoar frost. Around the firn line, percolating meltwater can refreeze in distinctive lenses within the firn itself and along the firn/ice interface. This leads to the deposition of a distinctive *superimposed ice zone*, where the sedimentary ice can survive the summer ablation to be buried by the following winter snow pack. This represents a net gain of glacial mass over the mass balance year so the lower boundary of the zone (where it exists alongside the 'old' glacier ice) is called the equilibrium line. Below this line is the *ablation zone* where all the snow pack, ice bands and superimposed ice are removed over the summer, together with an appreciable amount of 'old' glacier ice (as discussed above).

THE CONVERSION OF SNOW TO GLACIER ICE AND THE ROLE OF CLIMATE

Glaciers are composed of the buried ice bands (discussed above) and mainly recrystallized snow, with each crystal of glacial ice representing thousands of minute snow flakes. These have been welded together into a single, homogeneous crystal structure ranging in size from a sand grain to the human head (Sharp, 1988). Initially, the 'virgin' snow material is a light, loose and fluffy aggregate of new, delicate snow flakes or ice needles which accumulate in distinctive layers, associated with a single storm or different precipitation phases within a storm (i.e. with the passage of warm and cold fronts). Consequently, the accumulated strata display marked differences in density, grain size, porosity, etc., which reflect the varying conditions of precipitation and deposition (e.g. with or without wind packing). The snow deposition also entombs trace gases (especially carbon dioxide, CO_2), pollen, volcanic aerosols and cosmic dust. These deposits provide a record of global climates, tectonic activity and cosmic events, which will become available as evidence of environmental change through the analyses of cores from deep boreholes in the ice mass.

Over a timescale varying from a few years to a few thousands years, the fresh, delicate snow crystals are transformed into granular firn and eventually glacial ice. As Table 7.1 reveals, this transformation is strongly influenced by temperature and proceeds more rapidly in warmer, wet snow environments (e.g. in 3 years and at a depth of 13 m on the Upper Seward glacier, Yukon, and over 3500 years at a depth of 160 m at Plateau Station, East Antarctica). The transformation of glacier material is called 'firnification' and it commences almost as soon as snow is deposited on the glacier surface, in the form of distinctive density increases (Table 7.2). The initial, skeletal snow flakes (density 50–70 kg m^{-3}) are changed into a more dense, settled

Table 7.1 Depth of firn–ice transition (m) and time taken for this transformation

Glacier/station	*Firn temperature (°C)*	*Depth (m)*	*Age (years)*
Upper Seward, Yukon	−1	13	3
Camp Century, Greenland	−24	68	125
S2, Coastal East Antarctica	−19	38	220
Little America, West Antarctica	−24	51	150
Dome C, inland East Antarctica	−54	100	1700
South Pole	−51	115	1020
Vostok, inland East Antarctica	−57	95	2500
Plateau, inland East Antarctica	−57	160	3500

Sources: Gow (1971), Langway (1967), Paterson (1994).

Table 7.2 Typical densities of glacier material (kg m^{-3})

New snow (immediately after deposition in dry, calm conditions)	50–70
Damp, new snow	100–200
Settled, old snow	200–300
Wind-packed snow	350–400
Firn	400–830
Very wet snow/firn	700–800
Glacier ice	830–917

Sources: Seligman (1936), Paterson (1994).

snow pack (density 200–300 kg m^{-3}) and then into small spherical granules of ice called firn (density 400–830 kg m^{-3}) and finally into glacier ice (density 830 kg m^{-3}). This is due to a progressive release of air to the surface from within the glacial material.

For example, 1 kg of firn contains some 63 cm^{-3} of air in the form of interconnecting pores and channels, whereas air bubbles remain in glacier ice (some 10 per cent by volume). Gow and Williamson (1975) observed air bubbles down to 1100 m in the ice at Byrd Station, West Antarctica. They are removed below that depth by compressional creep of the surrounding ice (Paterson, 1994) so that a further increase in density occurs to 917 kg m^{-3} (Table 7.2) with the air now present only in the form of clathrate hydrate (Miller, 1969).

Firnification processes are both variable and complex and since they are the most active in warm, wet snow environments (Table 7.1), it is obvious that refreezing meltwater (see the two preceding sections) has a prominent role to play in removing entrapped air and increasing snow pack density. Conversely, in cold dry snow zones of the highest altitudes and latitudes, snow–firn–ice transformation is due more to pressure sintering/bonding (analagous to 'hot pressing' in ceramics, according to Paterson, 1994). Under pressure, aggregates of particles (including ice crystals) are heated which initiates sintering and creates bonds between them, increasing their size. The space between the particles is reduced (especially by refreezing in the snow pack), air is released to the surface and the density of the aggregate increases. Warm environments are characterized by a high frequency of maritime airstreams and wave depressions, with excessive precipitation, so that snow pack accumulation per annum can be many metres compared with a few millimetres of accumulation on cold, continental interiors. When this accumulation exceeds 10 m, settling and compaction occur with the excessive weight build-up. The air content of the snow pack becomes progressively displaced by slip/rotation of the ice crystals until a high degree of compaction (and density increase) is achieved.

Sublimation occurs in the snow pack because the air content represents an atmosphere of freely mobile supercooled water molecules which surround the ice crystals. The pressure exerted by these water molecules is referred to as the vapour pressure of the ice crystal, expressed in millimetres of mercury. Small crystals in a snow pack have higher vapour pressures than large crystals so vapour tends to flux along the pressure gradient and sublimation (vapour to ice) takes place around the larger crystal, increasing its size. Also, vapour pressure is higher around the sharp points of the skeletal ice crystals than over the smoother parts. Vapour then moves from the sharp projecting points, making them more blunted and filling in intervening hollows by sublimation (Sharp, 1988). Thus, vapour flux/sublimation ultimately produces small, solid grains of recrystallized ice of roughly spherical shape and uniform size, with a marked increase in density (up to 830 kg m^{-3}, Table 7.2). Once again, it must be emphasized that sublimation is accelerated in warm, wet snow zones, since the vapour flux is greatest with higher temperatures. For example, the vapour pressure in ice rises with increasing warmth, equalling 0.1 mm at −40°C and 4.6 mm at 0°C (Thompson, 1977).

Despite the significance of pressure melting/sintering, setting/compaction and sublimation in the firnification process, there is no doubt that rapid transformations from new snow to glacier ice only occur when percolating meltwater is present, which refreezes at depth in warm, wet snow zones. The resultant formation of ice bands/lenses is important enough in densification but the surface tension of the water film also pulls the crystals/grains together, releasing air to the surface. Also, when the meltwater refreezes at depth, it releases sufficient latent heat (see the second section) which extends the warm wave into the glacier (with increased melt/refreezing at colder depths). This thermal amelioration encourages the spherical growth of the firn grains since the crystals melt first at the extremities. In addition, the average grain size increases since the smaller grains melt before the larger ones. The high densities of 760 kg m^{-3} recorded on the Whakapapanui glacier, Mt Ruapehu, New Zealand (Thompson and Kells, 1974), at the end of *one* mass balance year, confirm the importance of meltwater percolation associated with the high frequency of depressions and warm fronts in this temperate location.

GLACIER MASS BALANCE AND CLIMATE CONTROLS

The net mass balance of a glacier (B_n) represents the net loss or gain of snow and ice (in terms of water equivalent), as calculated for one or more balance years (Figure 7.3(a)). 'Such studies form an important link in the chain of events connecting advances and retreats of glaciers with changes in climate' (Paterson, 1994), and have traditionally been the concern of the applied climatologist. Glacial mass balance over the year is determined in a number of ways including the direct measurements of net balance (see the equation below), the photogrammetric method (revealing volume changes over time from photographic evidence) and the hydrologic method (based on measurements of runoff, precipitation and evaporation). The commonly used direct measurement method is based on the following relationship (Paterson, 1994):

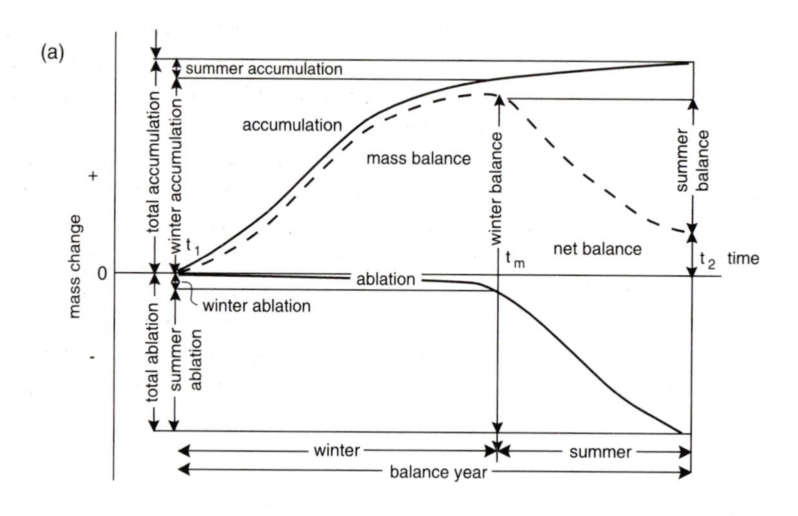

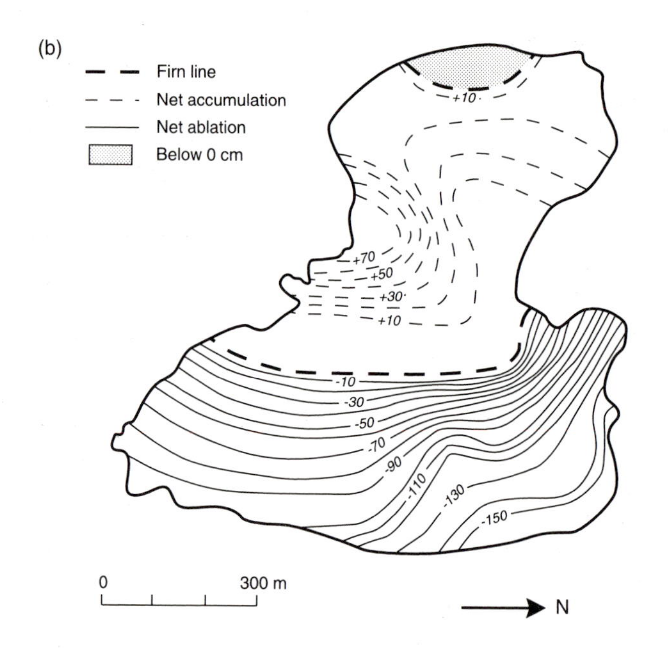

Figure 7.3 (a) Definitions of mass balance terms for the wet zone of a glacier (after Paterson, 1994); (b) Orwell glacier, Signy Island, West Antarctica: mass balance 1961–2

$$B_n = \int_{S_c} b_n \, dS + \int_{S_a} b_n \, dS$$

where: B_n = net balance of a glacier; S_c = accumulation zone; S_a = ablation zone; dS = area of both zones; and b_n = net balance at points in both zones.

This method demands a detailed knowledge

of the water equivalent of net accumulation (above the equilibrium line) and net ablation (below the line). These are acquired from stratigraphic studies of the snow pack (viz. changes in density with depth), from as many points as possible over the glacier, which represent environments of known area related to elevation, aspect and exposure. Pits are dug, or cores taken, at the end of the winter (to determine the gross accumulation from snout to uppermost reaches) and at the end of the summer season above the equilibrium line, to record net accumulation after ablation and firnification changes. Below the equilibrium line, net ablation represents the removal of all the winter gross accumulation together with a quantity of pre-existing glacier ice (measured at stakes drilled into the ice). The total volume gained or lost over the glacier (in cubic metres of water) is obtained when the water equivalent of the net accumulation or net ablation is multiplied by the surface area in each environment, and totalled for the entire glacial system.

Figure 7.3(b) reveals the mass balance of the Orwell glacier, Signy Island, West Antarctica, for the year 1960–1 (Thompson, 1968). It clearly illustrates the ablation zone, equilibrium lines (mid-glacier and ice cap) and accumulation zone for the glacier. The ice-cap equilibrium line is associated with ablation exceeding accumulation due to excessive rates of deflation in this very exposed, windy area. In total net mass balance terms for the Orwell glacier (in water equivalent), net accumulation equalled 182,000 m^{-3}, and net ablation equalled 287,000 m^{-3}, to give a marked deficit of 105,000 m^{-3}. However, single mass balance years do little to indicate the normality of the observed trends and, above the firn line, tunnels excavated into the firn and glacial ice can reveal mass balance trends over past decades. Alternatively, long-term investigations can reveal year-to-year variations in mass balance, which provide far more meaningful climate–glacier relationships.

Figure 7.4 (a) illustrates winter, summer and net balances for Storglaciären, Swedish Lapland, between 1945 and 1975 (Schytt, 1962; Paterson, 1994). The balances confirm the considerable variability during this 30 year period, with fluctuations ranging from −3 to +2 m of water. The net balance appears to be more highly correlated with the summer balance than with the winter balance. This suggests that ablation is more important than accumulation in controlling variations of net balance on the glacier. However, over the investigation period,

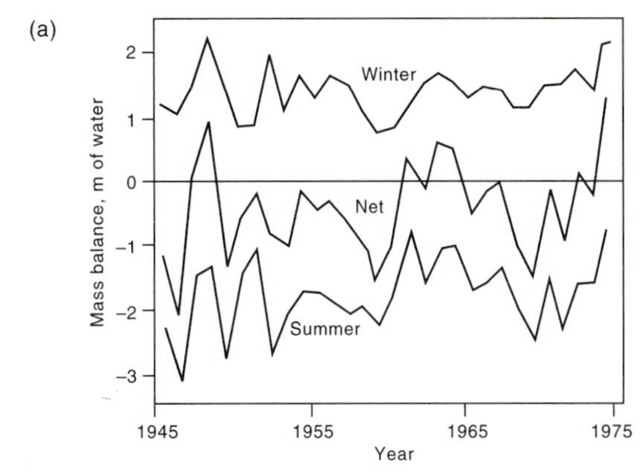

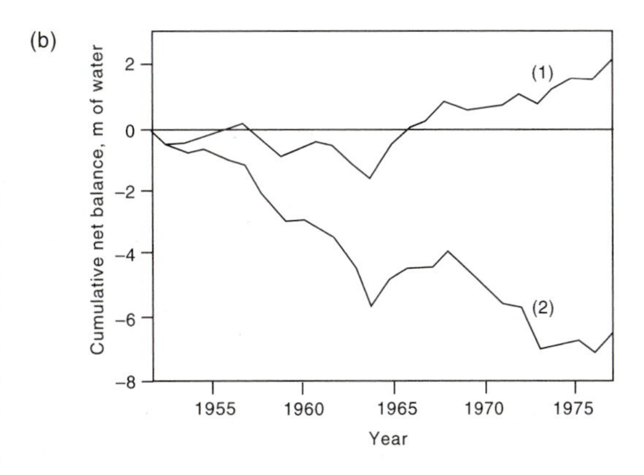

Figure 7.4 (a) Storglaciären, Swedish Lapland: mass balance 1945–75 (after Schytt, 1962; Paterson, 1994); (b) Kesselwandferner (1) and Hintereisferner (2), Austria: mass balance 1952–77 (after Kuhn *et al.*, 1985; Paterson, 1994)

negative balances have been dominant, with a volume decrease of about 12 per cent evident during the 30 year period (Paterson, 1994). The long-term relationship between glacial net balance and climate is also revealed in Figure 7.4(b). Kesselwandferner and Hintereisferner are adjacent glaciers in, presumably, a similar climatic regime. However, between 1952 and 1977, the former glacier gained mass equivalent to 2 m of water, whereas Hintereisferner lost 6.4 m. These significant differences are mainly due to the size and elevation of the glaciers concerned since the mass deficits occurred on the large glacier where 51 per cent of the glacial area is below 3050 m. It appears that the low-lying tongue of Hintereisferner (with high ablation rates) is a relic of the 'Little Ice Age' (1650–1850) and the size of the glacier has not yet adjusted to the current climate (Paterson, 1994).

The net balance of a glacier is controlled by the atmospheric dynamics associated with accumulation and ablation processes. However, it appears that such balances are more sensitive to changes in ablation (associated with melt/runoff and sublimation) than in accumulation. Ablation rates (the mass flux) are influenced by the glacier's energy balance, mainly in terms of solar/net radiation (R_n), sensible heat flux (H) and latent heat flux (LE) expressed as

$$\text{Mass flux} = R_n + H + LE$$

Net radiation effectiveness in the ablation process is minimized by the high albedo values and cloudiness over glaciers, although (according to Paterson, 1994) debris-covered ice albedos (mean 12 per cent) are considerably lower than clean, dry snow (mean 84 per cent). Consequently, ablation rates initiated by net radiation are considerably higher over dirty ice unless, of course, the dirt cover exceed 3 mm and acts as an insulator to minimize heat transfers. Table 7.3 reveals that, in relative terms, net radiation is important in glacial ablation at high elevations, remote from the sea (e.g. Salmon and Barnes glaciers) where sensible/latent heat transfers are negligible.

Sensible heat transfers produce substantial rates of ablation due to the conductive exchange between the glacier surface and an overriding warm airstream, which is accelerated by convection and turbulence. This diffusion removes chilled, stable air from the glacier's surface and prohibits the formation of strong temperature inversions. Ablation from latent heat transfers results when moist, warm air moves across the glacier and is chilled by conduction from the underlying cold material. Cooling air below its dew point will facilitate condensation and the release of latent heat onto the surface. For example, at 0°C, 2.501 mJ kg^{-1} are released per gram of condensing water vapour which is sufficient to melt 7.5 g of glacial material which is close to melting point. This heat transfer is also accelerated by convection and turbulence which continually renew the supply of water vapour to the glacier surface. Table 7.3 confirms the importance of sensible and latent heat fluxes

Table 7.3 Percentage contribution of energy transfers in ablation process of selected glaciers

Name of glacier	*Latitude*	*Longitude*	*Elevation (m)*	*Distance from sea (km)*	R_n	*H*	*LE*
Salmon	56°N	130°W	1700	320	75	15	10
Barnes	70°N	72°W	865	150	68	20	12
Sveanor	79°N	18°E	5	6	24	58	18
Hodges	54°S	37°W	360	1	35	35	30
Franz Josef	43°S	170°W	500	8	10	67	23

Sources: Thompson (1977), Paterson (1969), Ishikawa *et al.* (1992).

in ablation from low-lying glaciers close to the sea (where lapse rates and water vapour content controls are paramount). Consequently, on the Hodges, Sveanor and Franz Josef glaciers, these non-radiative fluxes contribute between 65 and 90 per cent of net ablation. Furthermore, since wastage is considerable in these warm, maritime glaciers (e.g. 12×10^3 m^{-3} of water on the Hodges glacier between 1957 and 1958; Smith, 1960), non-radiative fluxes are also important in absolute terms.

Energy transfer contributions to glacier ablation also vary seasonally with the temperature regime and daily, with changing synoptic conditions. Table 7.4 reveals that R_n dominates ablation early on in the summer season (84 per cent of wastage) whereas non-radiative fluxes (H and LE) are more important in late July–early August (55 per cent of ablation). Table 7.5 illustrates the control of weather types on the energy transfers over the Devon ice cap. Non-radiative fluxes are particularly dominant in warm, turbulent weather (type B) where, collectively, they provide +56 W m^{-2} of energy for substantial melting, compared with a heat loss of −10 W m^{-2} in type A weather, with low temperatures, light winds and little ablation. Hay and Fitzharris (1988) found similar synoptic controls on ablation over the Ivory glacier, New Zealand, when non-radiative fluxes were dominant, causing 60 per cent of the total energy transfer and excessive ablation rates of 62 mm per day, under warm, windy, moist and cloudy weather. This contribution fell to 18 per cent during calm weather and moderate stability, which did not favour convective, non-radiative heat transfers.

Table 7.4 Percentage of seasonal ablation related to energy transfers, Karsa glacier, Swedish Lapland, at 1000 m

Flux component	*Late May–early June*	*Late July–August*	*Late August*
R_n	84	45	58
H	13	40	32
LE	3	15	10

Source: Wallen (1948).

Table 7.5 Surface energy transfers for the Devon Ice Cap Station under different weather types (W m^{-2})

Weather type	*Sky*	*Wind*	*Surface*	R_n	H	LE
(A) Low temperatures	Clear	Light	Frozen	12	13	−5
and little ablation	Overcast	Light	Frozen	23	−4	−6
(B) Warm, vigorous advected air	Overcast	Strong	Melting	45	46	10
(C) Warm	Clear	Light	Melting	44	14	5
(D) Cold	Clear	Light	Frozen	4	16	6

Source: Holmgren (1971).

CONCLUSIONS

The previous sections have clearly emphasized that glaciers are the true 'progeny' of climatic elements and that their thermal characteristics, constituent zones, firnification rates and mass balances will be particularly sensitive to future climate change (as indeed, they have been in the past). Furthermore, it is apparent that glacial responses are controlled more by changes in the rate of ablation (see the previous section). Consequently, it is assumed that the consequences of predicted global warming (due to an enhanced greenhouse effect) will be devastating for the survival of warm glaciers and the stability of global sea levels. However, there is a great deal of controversy associated with the impact of global warming on glacier behaviour and whether ice-sheet growth and decay is driven primarily by temperature fluctuations or precipitation changes. 'A key issue is the question of whether global warming would lead to the decay of ice sheets through increased ablation or their growth through increased accumulation' (Knight, 1992). Furthermore, ice-sheet growth could actually amplify global cooling by removing water vapour from the atmosphere and weakening the greenhouse effect (see Chapter 10, The Pre-Quaternary period).

Greuell and Oerlemans (1989) measured radiation balance, energy transfers and ablation near the 'snout' (Glacstat site) of the Hintereisferner (Austria). They also included a 'best estimate' of the effect on ablation from rising trace gas concentrations, with CO_2 increasing from its 1986 level of 339 ppm to a speculated 449 ppm in 2036. At the end of this 50 year period, ablation would increase by 5.0 per cent due to the CO_2 increases and by 11.3 per cent following increases in *all* trace gases. Braithwaite and Olesen (1990) studied energy transfers for two stations near the western margins of the Greenland ice sheet. They calculated that the ablation rate increased almost linearly with temperature at a rate of about 15 per cent per °C. However, ablation increases would be 'much less' than this at higher elevations above the firn line, because snow melts more slowly than ice at the same temperature. Furthermore, a considerable amount of percolating meltwater in the wet snow zone will refreeze at depth in the snow pack (see the third and fourth sections) as part of the firnification process (Paterson, 1994).

This process seems to be the case at present in the Greenland ice sheet where satellite altimetry indicates current thickening of about 23–28 cm per year (Zwally, 1989; 1990; Zwally *et al.*, 1989). Knight (1992) cites evidence from a variety of unreferenced sources which indicates that, in the past, higher temperatures have correlated with a thicker East Antarctic ice sheet. It is generally accepted that global warming would be associated with an increased moisture capacity of air, greater snowfall and thicker ice in dry, cold zones, but with melting of ice in wetter, warm zones. However, modelling of the Greenland ice sheet suggested that even this melting would be very slow, with a 3°C warming generating only a few millimetres of sea-level rise over hundreds of years (Knight, 1992).

Worldwide flooding does seem unlikely in the twenty-first century even if the predicted global warming is proved to

CONCLUSIONS *continued*

be correct. Meier (1984) concluded that sea-level rises over the past 100 years (about 1 mm per year) were shared equally by the thermal expansion of sea water and the reduction in size of small, warm glaciers. He argued that there was no need to postulate any contribution from the Greenland and Antarctic ice sheets. Media reports of the January 1996 IBG conference (Uhlig, 1996) quoted a paper by David Sugden (University of Edinburgh) which provided evidence that the East Antarctic ice sheet has been stable for at least 16 million years. Consequently, the present cold polar climate persisted throughout the warm Pliocene period, which is assumed to be analogous with the degree of global warming predicted by the year 2050 (i.e. 3°C warmer globally than today). This negates fears of future worldwide flooding since the polar ice sheets seem to be immune from global warming of this magnitude. However, mass balance studies must become more extensive in future decades in order to quantify these assumed generalizations and associations. Climate–glacier studies will become even more important in the next millennium, with applied climatologists leading the way.

REFERENCES

Ahlmann, H.W. (1935) 'The Fourteenth of July Glacier', *Geografiska Annaler* 17: 167–218.

Ahlmann, H.W. (1948) *Glaciological Research on the North Atlantic Coasts*, Royal Geographical Society Research Series: 1.

Braithwaite, R.G. and Olesen, O.B. (1990) 'Increased ablation at the margins of the Greenland ice sheet under a greenhouse effect climate', *Annals of Glaciology* 14: 20–2.

Ellis-Evans, J.C. and Wynn-Williams, D. (1996) 'A great lake under the ice', *Nature* 381: 644–6.

Flint, R.F. (1957) *Glacial and Pleistocene Geology*, Chichester: John Wiley.

Gow, A.J. (1971) 'Depth–time–temperature relationships of ice crystal growth in polar glaciers', *US Army Cold Regions Research Engineering Laboratory*, Research Report 300.

Gow, A.J. and Williamson, T. (1975) 'Gas inclusions in the Antarctic ice sheet and their glaciological significance', *Journal of Geophysical Research* 80: 5101–8.

Greuell, W. and Oerlemans, J. (1989) 'Energy balance calculations on and near Hintereisferner (Austria) and an estimate of the effect of greenhouse warming on ablation', in J. Oerlemans (ed.) *Glacier Fluctuations and Climate Change*, Dordrecht: Kluwer, pp. 305–23.

Hambrey, M.J. (1994) *Glacial Environments*, London: UCL Press, pp. 36–7.

Hay, J.E. and Fitzharris, B.B. (1988) 'The synoptic climatology of ablation on a New Zealand glacier', *Journal of Climatology* 8: 201–15.

Holmgren, B. (1971) *Climate and Energy Exchange on a Subpolar Ice Cap in Summer, Part F*, Uppsala: Meteorologiska Institutionen, Uppsala Universitet, pp. 107–12.

Ishikawa, N., Owens, I.F. and Sturman, P. (1992) 'Heat balance characteristics during fine periods on the lower parts of the Franz Josef glacier, South Westland, New Zealand', *International Journal of Climatology* 12: 397–410.

Knight, P.G. (1992) 'Progress reports: Glaciers', *Progress in Physical Geography* 16(1): 85–9.

Kuhn, M., Markel, G., Kaser, G., Nickus, U., Obleitner, F. and Schneider, H. (1985) 'Fluctuations of climate and mass balance: different responses of two adjacent glaciers', *Zeit Gletscherkd Glazialgeologie* 21: 409–16.

Langway, C.C. (1967) 'Stratigraphic analysis of a deep ice core from Greenland', *US Army Cold Regions Research Engineering Laboratory*, Research Report 77.

Meier, M.F. (1984) 'Contribution of small glaciers to global sea level', *Science* 226: 1418–21.

Miller, S.L. (1969) 'Clathrate hydrates of air in Antarctic ice', *Science* 165: 489–90.

Müller, F. (1962) 'Zonation in the accumulation areas of the glaciers on Axel Heiberg Island, N.W.T., Canada', *Journal of Glaciology* 4: 302–11.

Paterson, W.S.B. (1969) *The Physics of Glaciers*, (1st edn), Oxford: Pergamon.

Paterson, W.S.B. (1972) 'Temperature distribution in the upper layers of the ablation area of the Athabasca glacier, Alberta, Canada', *Journal of Glaciology* 11: 31–41.

Paterson, W.S.B. (1994) *The Physics of Glaciers*, (3rd edn), Oxford: Pergamon/Elsevier.

Robin, G. de Q. (1955) 'Ice movement and temperature distribution in glaciers and ice sheets', *Journal of Glaciology* 2: 523–32.

Schytt, V. (1962) 'Mass balance studies in Kebnekajse', *Journal of Glaciology* 4: 281–8.

Seligman, G. (1936) *Snow Structures and Ski Fields*, London: Macmillan.

Sharp, R.P. (1988) *Living Ice*, Cambridge: Cambridge University Press.

Smith, J. (1960) 'Glacier problems in South Georgia', *Journal of Glaciology* 3: 707–14.

Son Mannerfelt, C.M. (1949) *Glaciers and Climate*, Stockholm: Svenska Sällskapet För Antropologi och Geografi.

Souchez, R.A. and Lorrain, R.D. (1991) *Ice Composition and Glacier Dynamics*, Berlin: Springer-Verlag.

Stacey, F.D. (1969) *Physics of the Earth*, New York: John Wiley.

Sugden, D.E. and John, B.S. (1976) *Glacier and Landscape: A Geomorphological Approach*, London: Edward Arnold.

Sverdrup, H.V. (1936) 'The eddy conductivity of the air over a smooth snow field', *Geofysics Publications* 11: 1–67.

Thompson, R.D. (1968) 'Hydrological budget studies on geophysically temperate glaciers', *Journal of Hydrology* 7(1): 13–23.

Thompson, R.D. (1977) *The Influence of Climate on Glaciers and Permafrost, University of Reading, Geographical Papers No. 57.*

Thompson, R.D. and Kells, B.R. (1974) 'Mass balance studies on the Whakapapanui glacier, New Zealand', in *Role of Snow and Ice in Hydrology*, UNESCO/WMO 1: 383–93.

Uhlig, R. (1996) 'Ice thaw theory gets cold shoulder', *Daily Telegraph* 4 January: 4.

Wallen, C.C. (1948) 'Glacial–meteorological investigations on the Karsa glacier in Swedish Lapland 1942–48', *Geografiska Annaler* 30: 451–672.

Zwally, H.J. (1989) 'Growth of Greenland ice sheet: interpretations', *Science* 246: 1589–91.

Zwally, H.J. (1990) 'Ice-sheet elevation change (Abstract)', *Annals of Glaciology* 14: 336.

Zwally, H.J., Brenner, A.C., Major, J.A., Bindschadler, R.A. and Marsh, J.G. (1989) 'Growth of Greenland ice sheet: Measurement', *Science* 246: 1587–9.

8

GEOMORPHIC PROCESSES AND LANDFORMS

Edward Derbyshire

INTRODUCTION: CLIMATIC GEOMORPHOLOGY

The landforms of the earth are the result of the interplay between internal, or endogenetic processes, and surface, or exogenetic, processes. Endogenetic processes include igneous activity, in which molten rock materials develop and move beneath and upon the earth's surface, mountain building (orogenesis) and the uplift and depression of extensive areas of the surface (epeirogenesis), which occur with varying degrees of tectonic deformation. The exogenetic processes are driven by gravity and by solar radiation as expressed in the earth's bioclimatic diversity.

The idea that assemblages of landforms are the product of particular climatic environments can be traced back to the first formulation of the theory of terrestrial glaciation. Agassiz (1840) clearly recognized that distinctive suites of landforms, including moraines and roches moutonnées, are unique products of modern glaciers and that, where they occur outside presently glaciated terrain, they constitute clear evidence of former, much more extensive glaciers and ice sheets. By the end of the nineteenth century, this relationship had become generalized and a similar case had been made for distinctive landform assemblages in two other climatically defined regions, namely the arid zone and the humid, mid-latitudes (Davis, 1899; 1900; 1905). This approach was further developed by the introduction of the concept of topographical facies arising essentially from differences in surface moisture availability (Martonne, 1913). The view that the major landforms of the earth are the product of endogenetic processes, and that the exogenetic processes (with climate as the foremost influence) are the greatest single influence upon smaller (even individual) landforms, was well established by the first quarter of the twentieth century. It gave rise to the zonal classification of landform associations broadly coinciding with the earth's bioclimatic zonation. This approach, known generically as climatic geomorphology (Büdel, 1948; 1963; 1982; Stoddart, 1969; Derbyshire, 1973; 1976), argues that variations in the climatically driven process complex, involving physical and biochemical reactions at the atmosphere–lithosphere interface, give rise to morphoclimatic regions with distinctive suites of landforms.

The terminology of one of the most influential global morphoclimatic classifications (Tricart and Cailleux, 1972) is pedological and biogeographical rather than strictly climatic or geomorphological, however, so that the landform regions described are based as much upon inferred bioclimatic process systems as upon landform associations. Other zonal classifications of the world's landforms include Büdel's climatogenetic zonation of landform assemblages, based in part on climatic criteria, and Strahkov's (1967) map of landform groupings based on types and rates of geomorphological processes. Thus, all these global zonal classifications of landforms necessarily utilize mixed criteria. Moreover, the application of the zonal factor to the world's landforms is greatly constrained by the scale factor. In common with global- or continental-scale classifications of vegetation associations and soil types, classifications of climatically grouped landforms break down below the regional scale (Eybergen and Imeson, 1989). Finally, the observed fact that landforms inherited from past climatic regimes, and most evidently those of Pleistocene age, are abundant in many parts of the world raises the important question of the temporal framework of much of climatic geomorphology.

The fundamental problems facing climatic geomorphology are, first, to determine the extent to which climate, including its spatial and temporal variations, controls the geomorphic processes and, second, the extent to which these processes create a range of distinctive landforms. These relationships are very difficult to test in practice. Any zonal characterization of landforms on a global scale encounters difficulties arising from the effects of variations in spatial and temporal scale, as well as from uncertainties surrounding the relationship between surface processes and geomorphological form.

CLIMATE AND GEOMORPHOLOGICAL PROCESSES

Various parameters of present-day climates have been used to assess the relative importance and incidence of land surface processes. One of the founding documents of climatic geomorphology uses 'physiographic' (landform) criteria to classify climate. Penck (1910) was concerned with the distinctive hydrological conditions characterizing humid, nival and arid climates. No diagnostic landform types were specified by Penck. Only recently have landforms been grouped together using this rationale (Figure 8.1(a)) and the approach used to construct transects showing morphoclimatic zones in Africa (Figure 8.1(b); Hövermann, 1985; Hövermann *et al.*, 1992). Temperature changes have the most obvious effect around the freezing point of water, with direct responses in the expansion or contraction of the cryosphere (Koster, 1994). However, seasonal (Mediterranean to savanna) landscapes are also vulnerable because of the complex impact upon potential evapotranspiration and vegetation. Changes in precipitation are of major geomorphological importance in most landform realms, and notably in the subhumid to arid range (see below). On a global scale, selected 'threshold' values of temperature (e.g. 0°C and 21°C) and precipitation (50 mm and 127 mm) for annual and seasonal periods have been used to map the global incidence of selected landform-generating processes (Common, 1966). Mean annual temperature and precipitation values were used by Peltier (1950) to graph the differential incidence of physical and chemical weathering of rocks ('weathering regions') and the relative intensity of mass movement, wind action and rainfall erosion as a basis for a simple graph of nine 'morphogenetic' regions. The seasonal effect, in which fluvial erosion and the sediment yield of streams increases with precipitation amount only up to the point where vegetation becomes multilayered (Langbein and Schumm, 1958), has been used to refine this approach. For example, dynamic classifications of climate were used by Wilson (1973) to examine the relationship between climatic regime and the associated climate–process (landforming) systems (Figure 8.2(a)). It was found that fluvial erosion rates are much higher in seasonal, compared with non-seasonal, climates. Peak erosion

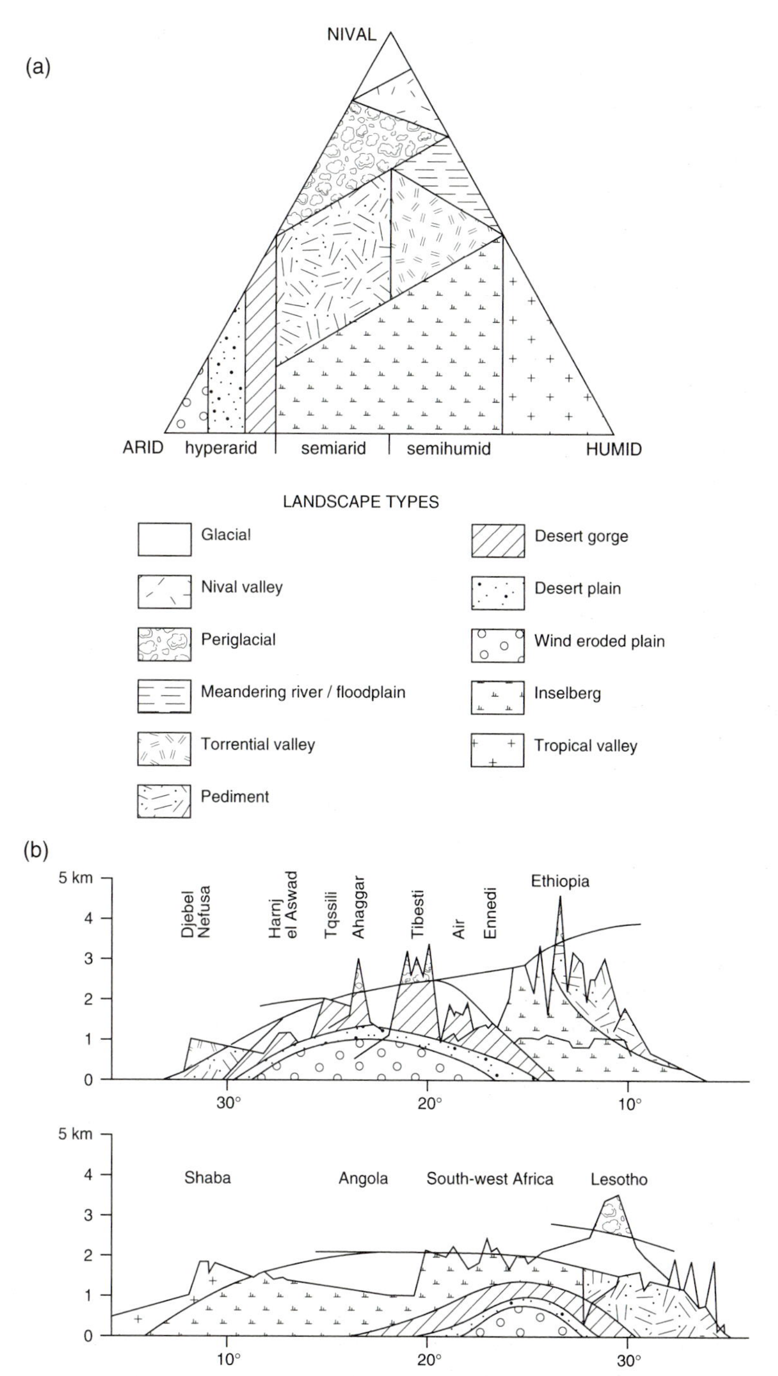

Figure 8.1 A recent approach to the derivation of morphoclimatic regions using hydrological regimes: (a) ternary diagram after Hövermann *et al.* (1992); (b) schematic cross-sections across Africa using the same classification groups (after Hövermann, 1985)

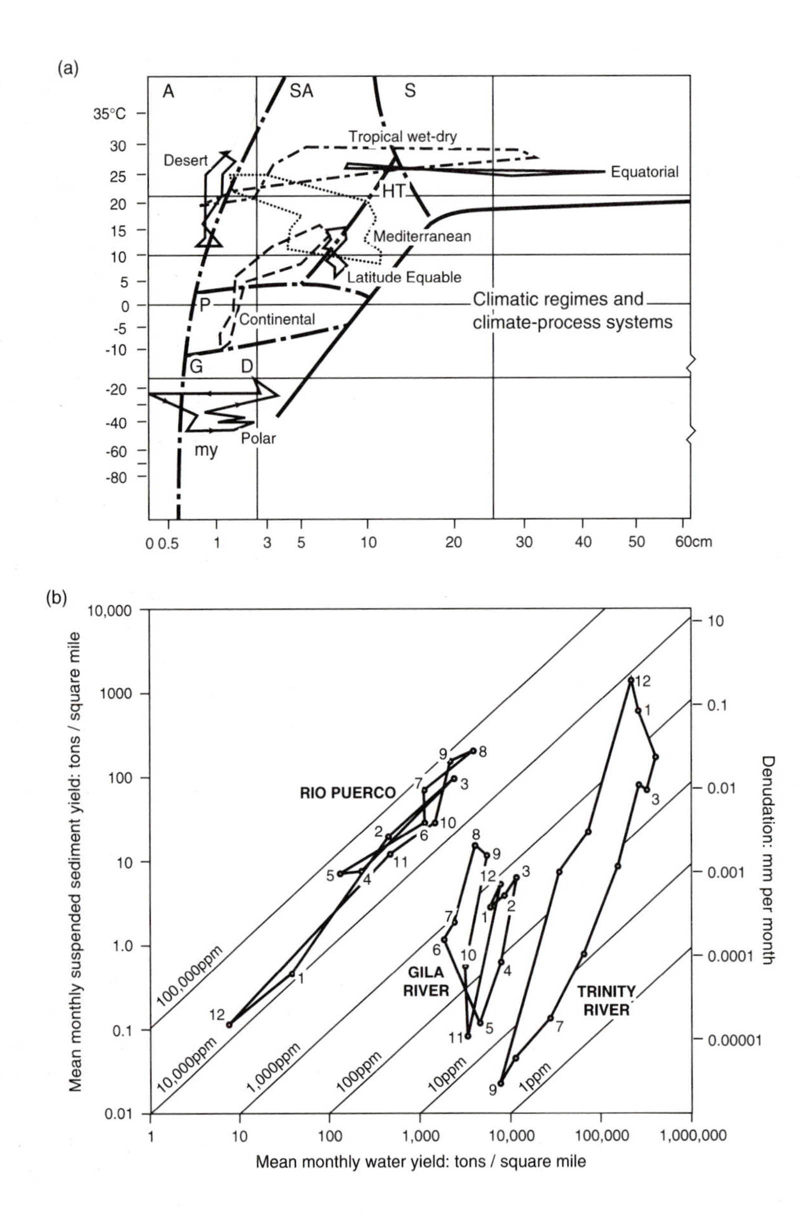

Figure 8.2 Use of mean climatic data, runoff and sediment yield to characterize climate–process systems: (a) climate regimes and climate–process systems, after Wilson (1973): A = arid, SA = semi-arid, S = selva, HT = humid temperate, P = periglacial, G = glacial; (b) sedihydrogram for rivers in three climatic regimes, after Wilson (1972): Trinity River, California (Mediterranean); Gila River, New Mexico (semi-arid continental); Rio Puerco, New Mexico (arid continental)

occurs in dry continental climates at mean annual precipitation values of 250–500 mm, in Mediterranean regimes at 1270–1650 mm and in tropical wet–dry environments at 1780–1905 mm. The seasonal rhythm of fluvial erosion and runoff may be characterized using a co-plot of mean monthly sediment yield and mean monthly water yield (Figure 8.2(b)). On the global scale, Fournier (1960) found a strong correlation between sediment yield patterns and a range of climatic indices, including seasonal and mean precipitation and rainfall frequency. Also on a global scale, Strahkov (1967) claimed that significant shifts in denudation rates occur at certain critical mean values of both temperature and precipitation.

At scales finer than the global or regional, however, generalizations about the role of climatic inputs to the process of landform evolution have proved more difficult to sustain. Assessment of the impact of climate and climate change on landform processes and threshold linkages requires application of climatic data of demonstrable geomorphological relevance based on the concept of magnitude and frequency of events (Ahnert, 1987). There is some evidence to suggest that over shorter timescales (<100 years), in some semi-arid to arid areas, the meteorological regime, vegetation cover, rock type/exposure and sediment yield confuse the generalized relationships characterized by Langbein and Schumm (1958). Indeed, in some cases, they are reversed (Yair and Enzel, 1987). Some specific geomorphological processes, including hillslope change, river channel shift, and soil erosion (Kirkby and Neale, 1987), have been characterized in terms of seasonality and the magnitude and frequency of the occurrence of elements of 'effective' (i.e. process-effective) meteorological parameters.

The relative importance and magnitude of the processes that model hillslopes, for example, are certainly influenced by climate. At a first approximation, and assuming similar conditions of rock type and structure, surfaces in humid climates are weakened by chemical weathering and the products removed by solution, creep and sliding. In dry environments, in contrast, rainwash appears to be the dominant process. However, considerable complexity is introduced by the fact that thresholds for different processes vary with a number of factors including gradient, slope length and vegetation cover. Consequently, in different landform situations, the results of surface processes may differ quite widely within the same climatic regime. Wolman and Gerson (1978) made the general case that clear-cut thresholds for hillslope failure require a continuous mantle of slope debris, steep gradients and intensive rainfall or rapid snowmelt events, and that these are most commonly found in humid mountain environments. Steep slopes and torrential rainfall events occur in arid landscapes, but the process is inhibited by slow rates of rock disintegration except on shale or where debris flows occur on talus slopes.

The effectiveness of major episodic climate-induced events may be gauged by the relative denudation they produce. A graphical plot of data expressed as the ratio between denudation during individual events with different return periods and mean annual denudation (Figure 8.3) suggests that relative denudation in major storms in wet tropical mountains are similar to the mean annual denudation but that such a high-magnitude event may occur almost annually. Also, a rare event in an arid environment may denude many times the mean annual volume of material and the ratio of instantaneous to annual denudation does not vary systematically with climate. Climate–landforming process relationships are subject to considerable and variable modulation by vegetation and soil cover. For example, the role of the soil mantle in influencing the response of slopes to changes in climate has been considered in a recent modelling study. Brooks *et al.* (1995) considered that the enhanced slope activity postulated for

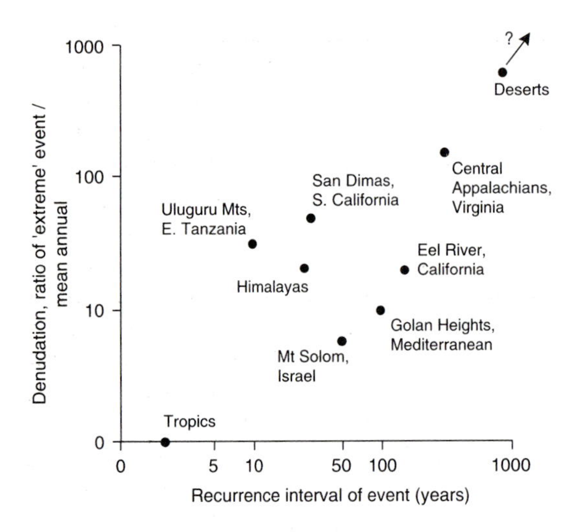

Figure 8.3 Relationship between denudation of hillslopes during large events of different recurrence intervals and mean annual denudation in various parts of the world (after Wolman and Gerson, 1978)

periods of deteriorating climate in the Holocene, such as the Little Ice Age, may have been dependent upon the presence of thicker, more mature and free-draining soils, thinner and less developed profiles, especially under a pine cover, being less vulnerable to higher rainfall totals.

Variations in the magnitude and frequency of climatic or meteorological events may generate a complex of forms making up a spatial and temporal hierarchy (Rapp, 1960). By measuring sediment yields in catchments of different size across a range of climatic types, scale-linked changes in the importance of factors affecting erosion rates may be assessed. Douglas (1976) noted that, although variation at a given scale is often as great as variation between different scale points, at a G-scale (Haggett *et al.*, 1965) of 3–4 the major rivers of western Canada all have similar yields within the range 70–220 m^{-3} km^{-2} per year. However, in Rocky Mountain streams at a G-scale of 4.8–5.6, variability is marked because of differences in lithology, relative relief and the degree of glacier development. With fining of the G-scale down to 7.11–8.7, within-catchment variables such as vegetation and land use become important. At the highest G-scale values, it is the properties of the erodible materials (e.g. particle size) that are most influential. Thus, the status of climatic elements as causal agents of change in geomorphological systems is considered to be heavily dependent on scale, both spatial and temporal.

CLIMATE CHANGE AND LANDFORMS

The presence of landforms in climatic environments in which they could not have originated, as in the classic case of the glacial landforms distributed across extensive areas of lowland Europe, clearly indicates that many landform assemblages retain the imprint of one or more major changes in climate. Whether a particular climate change can be regarded as geomorphologically significant depends not only on the magnitude and duration of the climate change itself, but also on the resistance offered by the landforms. The degree to which a landscape will respond to external stimuli such as climate by changing its forms is known as landscape sensitivity (Brunsden and Thornes, 1979; Thomas and Allison, 1993). The time span between the input of an external stimulus such as a climate change and the beginning of landform response is known as the lag time, and the time required for the input to result in a characteristic landform is the relaxation time. The period over which such a characteristic form persists is known as the characteristic form time (Brunsden, 1980). Highly sensitive landscapes and climate changes of short duration, but high magnitude, may yield a complex of transitional forms but few characteristic landforms. Conversely, landforms of low sensitivity, such as plateaux, will tend to retain their characteristic forms. Some studies have applied this methodology to individual landform types. For example, talus-derived rock glaciers in New Zealand show that, while they all share the same climatic history (having been initiated in the middle Holocene, about 5000 years BP), local differences in morphology and debris supply have controlled the sensitivity of individual features to reactivation during climatic deterioration (Kirkbride and Brazier, 1995).

The complex and subtle relationships between climate change and morphoclimatic landscape elements may be illustrated by studies of Quaternary environmental change (see Chapter 10, The Quaternary period). The alternation of cold (glacial) and warm (interglacial) climates in the recent geological record was first recognized by landform evidence of glaciation and the presence of soils and warm climate fossils between successive (superposed) glacial deposits. This has been confirmed, and an essentially complete record of climatic variation in the Quaternary has been obtained by oxygen-isotope analysis of the calcareous shells of foraminifera in cores from the deep oceans (Shackleton *et al.*, 1988). Oceanic waters become progressively enriched in the heavier isotope of oxygen (^{18}O), but this content declines in warm waters with relative enrichment in the lighter isotope ^{16}O. The oxygen-isotope stages have been dated using the palaeomagnetic timescale and radiometric methods. These stages reflect variations in global ice volumes, variable oceanic temperatures and biological productivity and, by inference, global temperatures. Spectral analysis of data from a number of ocean cores has shown that they contain periodicities similar to the astronomical (Milankovitch, 1930) periodicities (see Chapter 10, The Quaternary period) for solar radiation of about 19–24,000, 41–43,000 and 100,000 years (Figure 8.4) More recently, cores from the great ice sheets (see Chapter 7, The conversion of snow to glacier ice and the role of climate) have contributed data on oxygen-isotope ratios, deuterium–hydrogen-isotope ratios, the carbon dioxide content of air bubbles, and particulate and dissolved matter (Lorius *et al.*, 1985; Jouzel *et al.*, 1989). This has added valuable additional environmental data to the deep-ocean isotope record, including a record of the changing atmospheric content of the greenhouse gases (see Chapter 22, Greenhouse warming), notably carbon dioxide (CO_2) and methane, as an index of the state of the earth's carbon cycle. For example, CO_2 ranged from 190 to 200 ppmv during periglacials to 270–280 ppmv in interglacials (Lorius *et al.*, 1990), enhancing the astronomical forcing of climate (see Chapter 10, The Quaternary period).

Methane shows a similar range of variation, with the evidence from the Vostok ice core suggesting that emissions of methane from low-latitude wetlands were greatly influenced by orbitally controlled changes in monsoonal rainfall (Petit-Maire *et al.*, 1991). Some of the most detailed terrestrial records of palaeoclimatic changes are being provided by analysis of lake levels and basin fills (Street and Grove, 1979; Gasse *et al.*, 1989; Fang, 1993), and the study of the thick accumulations of wind-blown silts (*loess*), particularly those in China and central Asia (Kukla, 1989). Data, such as magnetic susceptibility and the particle size of high-resolution cores and sections, show clear similarities to the oceanic oxygen-isotope records (Figure 8.4). The record of the last 850,000 years shows a dominance of the 100,000 astronomical cycle, the curves suggesting that transitions from glacial to interglacial conditions were relatively rapid events.

As many such records show, the glacial–interglacial cycles (and the many more shorter-term climate changes within these major episodes) had a severe impact upon the landscape, most evidently in changes in plant and animal associations and soil types (Figure 8.5). Micromorphological studies of interglacial and interstadial palaeosols within the Chinese loess–palaeosol sequence, for example, indicate not only variations in pedogenesis under different climatic regimes but different types and

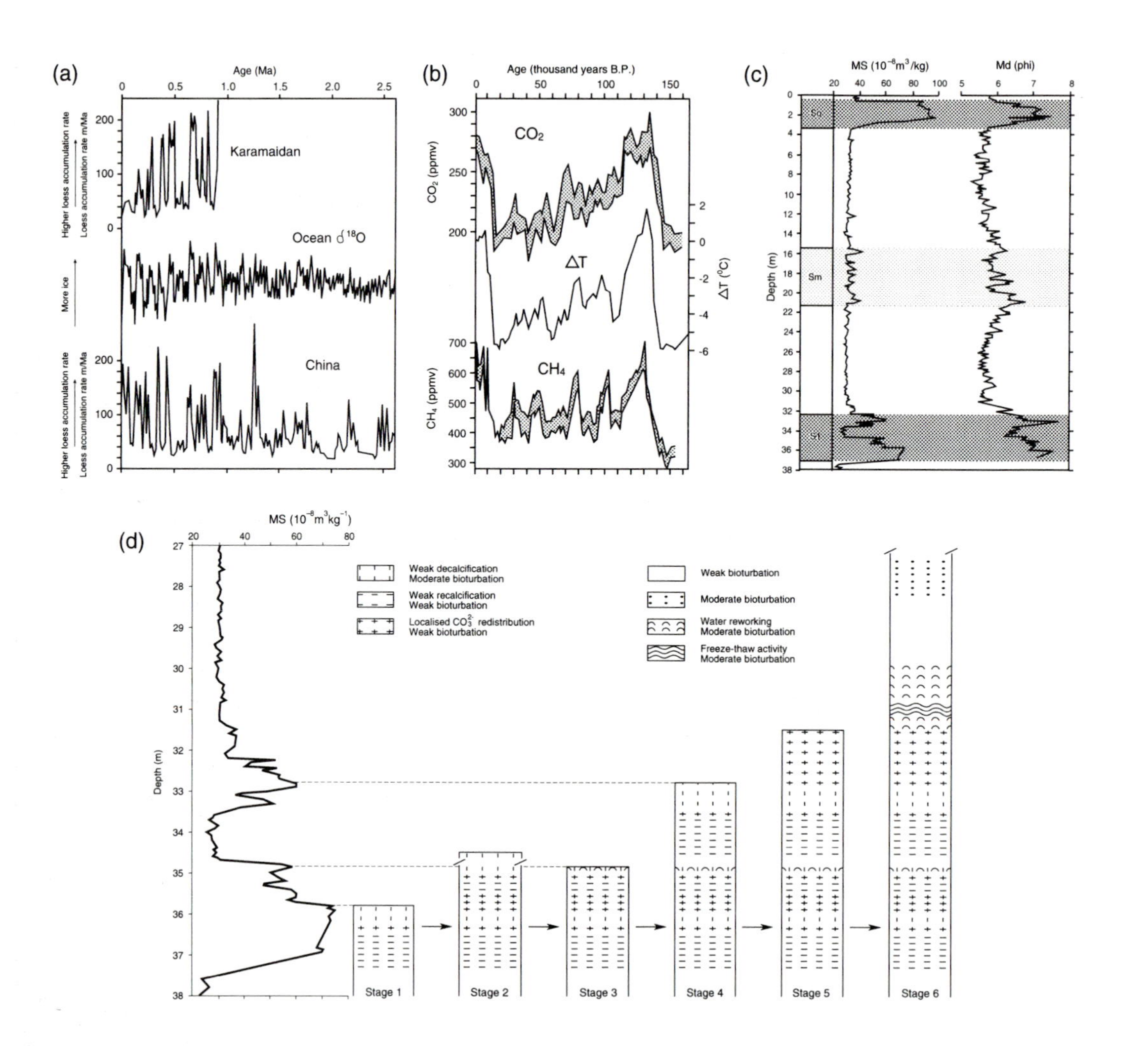

Figure 8.4 Some oceanic and terrestrial proxy indicators of climate change. (a) Loess accumulation rates for Karamaidan (Tadjikistan) and the Loess Plateau of China compared with the $\delta^{18}O$ record (centre) of the SPECMAP stack (to 620 ka) and ODP 677 (older than 620 ka), showing that high loess accumulation rate is associated with increased ice volume (after Shackleton *et al.*, 1995). (b) The Vostok (Antarctica) ice-core record showing variations in carbon dioxide, methane and temperatures during the last glacial–interglacial cycle (after Lorius *et al.*, 1990). (c) Depth functions of magnetic susceptibility (linear SI scale) and median grain size (linear μm scale) showing the mid-Holocene (So), interstadial (Sm) and last interglacial (S1) palaeosols in the 38 m excavation in loess at Gaolanshan, near Lanzhou, western Loess Plateau of China (after Derbyshire *et al.*, 1995). (d) Reconstruction of a sequence of pedosedimentary stages making up the pedocomplex of the last interglacial at Xining, northeastern Tibet (after Kemp *et al.*, 1996)

degrees of soil surface degradation as climatic conditions changed (Figure 8.4: Kemp *et al.*, 1996; Derbyshire *et al.*, 1995). However, changes in landscape *forms* in response to climate changes are more difficult to demonstrate because they are complex systems and the components of which (gradient, rock type and weathering state, soil type, root density, plant

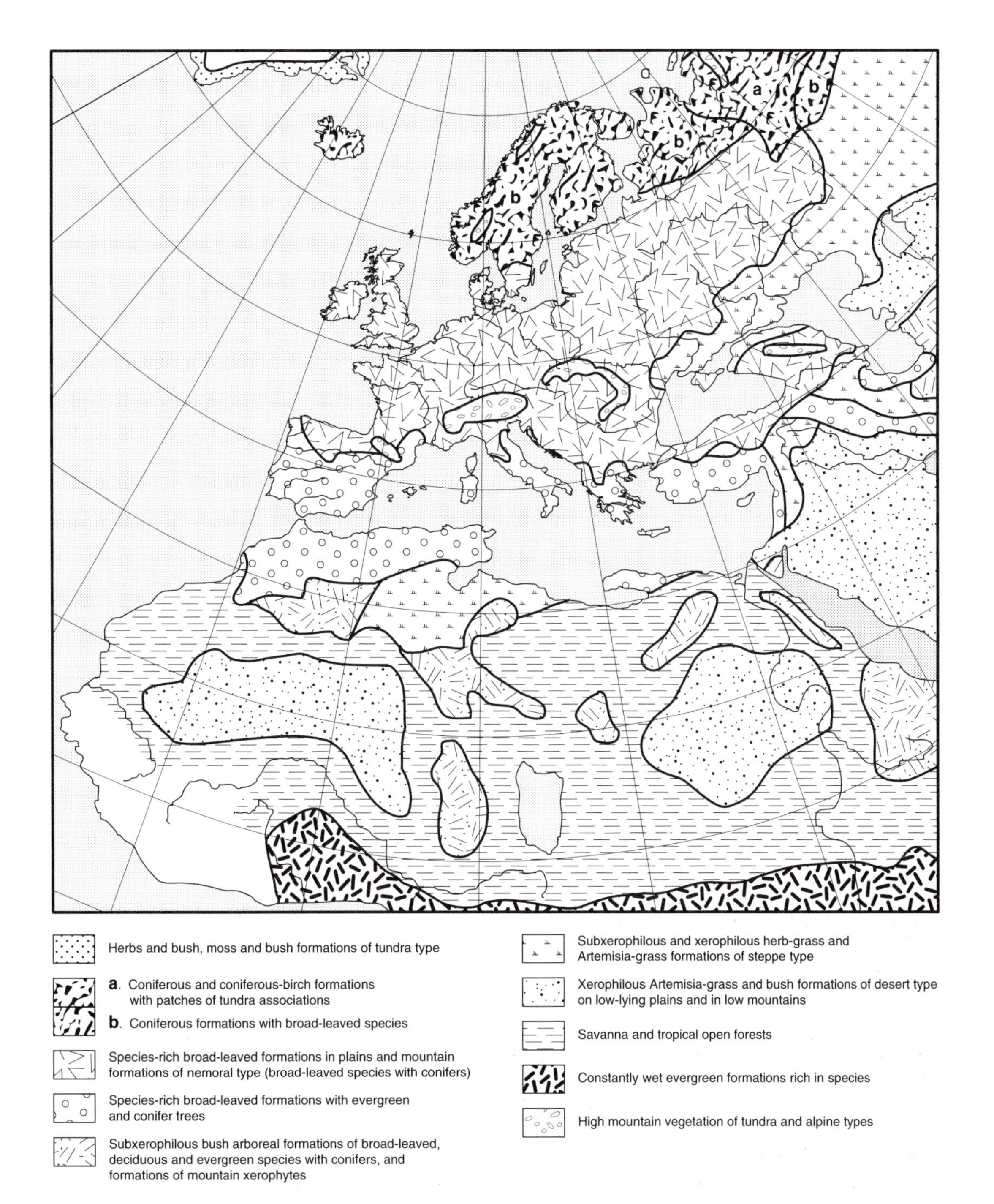

Figure 8.5 Vegetation associations of the last interglacial in Europe and North Africa (after Frenzel *et al.*, 1992)

cover, moisture regime, etc.) have different response thresholds. It is because of their wide range in sensitivity that landforms rarely respond in linear fashion to an external stimulus. This is true in varying degrees at all scales.

Before landforms can be accepted as diagnostic of a former, as against the existing climatic regime in which they are found, it is clearly necessary to establish that they are truly 'relict' by analogy with similar but active modern features. This is not a simple exercise because the specific climatic and process–form characteristics of many landforms have not been demonstrated, a difficulty bound up with the problem of equifinality, that is the derivation of identical landscape forms from quite different initial forms and a number of different processes. Many landform types including, for example, rock stacks (tors), pediments and alluvial fans, occur in a wide range of climatic environments. Nevertheless, there is no doubt that landforms, especially when they are considered together with a correlative sedimentary record (landform–sediment systems), provide valuable evidence of climate change.

Landforms change in response to shifts in the energy balance at the lithosphere–atmosphere interface, with the most influential factors being climate and tectonics. That the landform response to climate may mimic the response to tectonics is a classic geomorphological problem, which is well illustrated by the literature on river terraces, for example. Although river terraces are usually rather simple landscape forms, they frequently contain a rich record of climate change (Figure 8.6). It is important to distinguish between direct and indirect climatic influences on landform change. In the former category is the case of the lowering of world sea levels that broadly coincided with the growth of the Pleistocene continental ice sheets (glacio-eustasy), as well as the loading of the crust by the weight of ice and its subsequent recovery on deglaciation (glacio-isostasy). Relative sea-level rise, whether of eustatic, isostatic or tectonic origin, has a wide-reaching influence upon coastlines, continental shelves, alluvial plains and deltas (Schumm and Brakenridge, 1987). A rising sea level is likely to cause drowning of estuaries and increased sedimentation, at least in the lowermost reaches, so that a suite of distinctive landforms ranging from shoreline bars to extensive deltas may develop. Relict landforms may also be destroyed, such as drumlins and the moraines of glacial cirques at present sea level in, for example, the west of Ireland.

Raised shorelines, often associated with other features including wind-blown sand dunes (Orford *et al.*, 1991), provide the most obvious evidence of relative lowering of sea level, although the effect upon rivers and their valleys by such shifts in base level is more pervasive. On the northwest coast of Alaska, beach ridges provide a 4000 year record of low-frequency storminess (indicated by beach ridge progradation) interrupted by major erosional truncations coinciding with glacier expansion, river terrace aggradation, increased deposition of loess and ice-wedge growth (Mason and Jordan, 1993). Comparable data have been adduced from some large lakes. For example, periods of drought and lowering of the surface of Lake Michigan in the USA have led to the formation of dune-capped beach ridges, while wetter periods with high lake levels have been accompanied by the formation of parabolic dunes (Lichter, 1995). In the simplest case, sea-level lowering increases the relative relief, enhancing the potential energy of the fluvial system and inducing river incision, including incision of the newly exposed continental shelf, the point of incision migrating upstream with the passage of time. Indirect effects of sea-level lowering are numerous, but include changes such as increased continentality inland or increased snowfall on the higher ground following the increase in relative relief.

The difficulty of distinguishing the climatic from the tectonic signal in landforms is well

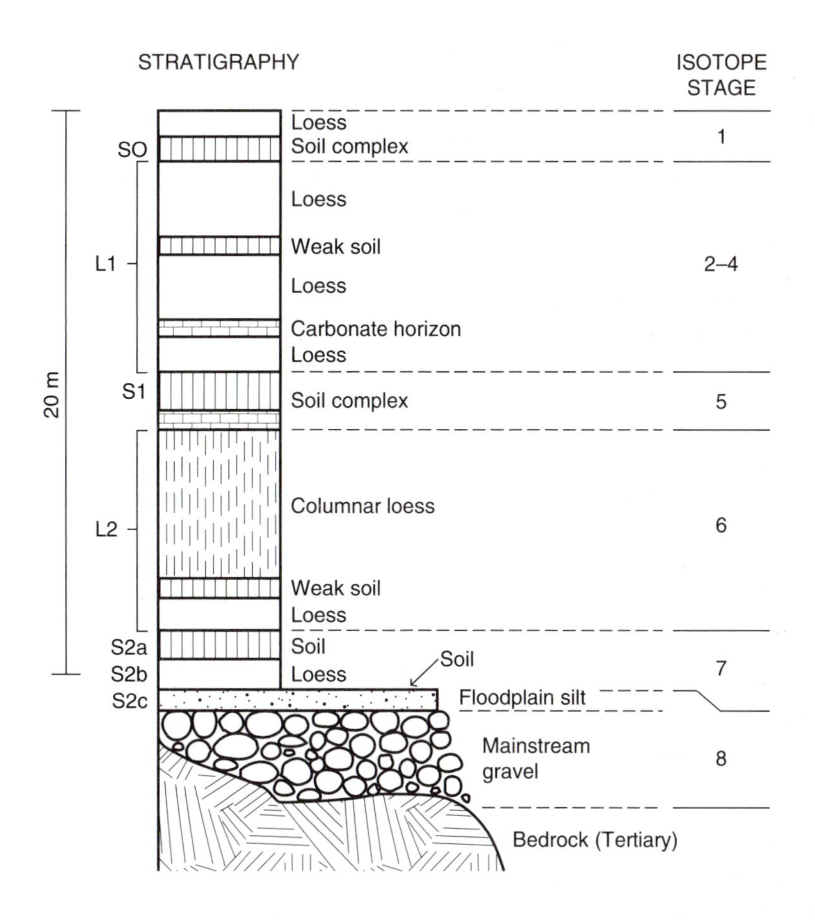

Figure 8.6 Section through a terrace gravel and overlying loess and palaeosol succession along the Ba River near Lantian, Shaanxi Province, China, showing a correlation with the standard marine oxygen-isotope stages (after Porter *et al.*, 1992)

illustrated from the Himalaya (Derbyshire, 1996). These mountains are undergoing rapid uplift, having some of the greatest relative relief on the earth. Also, they underwent multiple glaciation during the Pleistocene, and currently contain the longest glaciers outside the polar regions. It has been suggested that the severe denudation effected by deep incision of initially gently sloping Himalayan valleys by Pleistocene glaciers and their meltwaters would have triggered an isostatic response. Molnar and England (1990) argued that that for every 1 km thickness of crust removed by erosion, reduction in the average elevation would be only about 0.17 km because of the isostatic rebound. Thus, isostatic rebound will produce peaks at elevations higher than those of the preceding landscape, but with a lower mean elevation (i.e. relative relief is greatly increased). Clearly, the relative importance of glaciation *versus* uplift must be determined if the role of climate is to be better understood. This requires evaluation of the role of denudational unloading which, in turn, rests upon other information including estimation of glacial and non-glacial erosion through glacial/

interglacial cycles with appropriate dating control.

In many parts of the Himalayan region, the landform record consists of a complex association of glacial, periglacial, mass movement and fluvial phenomena of different age sensitivity and so at varying stages of modification. For example, widespread erosion and downslope reworking of great volumes of glacial deposits by mass movement has created large sediment storages in the form of huge debris fans on the lower valley side slopes of the Karakoram Range (Figure 8.7; Owen and Derbyshire, 1993). The deposits, ranging in age from Holocene to older than 139,000 years BP (Derbyshire *et al.*, 1984; Shroder *et al.*, 1993), are being eroded by the Hunza and Gilgit rivers (tributaries of the Indus River). Downstream, in contrast, the process combination is rather different. In the middle Indus, the river is incising into bedrock at the very high rates (based on cosmogenic ^{10}Be and ^{26}Al assays of quartz) of between 2 and 12 mm per year. Glacial and related deposits are rare to absent, since landsliding keeps the valley side slopes consistently steep (~32°, approximating the average angle of internal friction of the fractured bedrock debris). It is argued by Burbank *et al.* (1996) that there exists an equilibrium between bedrock uplift and river incision in this part of the Himalaya. It is also believed that the valley sides evolve essentially independently of the erosion rate of the Indus and that a common threshold (angle of internal friction) controls valley side gradients. This is contrary to the widely held view that accelerated denudation results in slope replacement by lengthening of steeper slopes at the expense of gentler surfaces. The result is an extensive geomorphological realm in the middle Indus below the glacial and periglacial realm, and above and essentially independent of the rock-bound fluvial channel (Plate 8.1).

The continued rapid uplift of the Himalaya and Karakoram sustains active (neotectonic) faulting and folding, for example the Karakoram fault and the Rakhiot fault–Liachar thrust below Nanga Parbat. Many of these displacements cut across landforms and deposits of glacial origin, some of which have been loosely correlated with the last glacial maximum (about 20,000 years BP) or the last deglaciation (since 13,000 years BP). Such correlations have assumed synchroneity between these events, derived from the northern hemisphere ice sheets, and glacial events in the Himalaya and central Asia. However, there is some evidence to suggest that at least some of the deposits marking the last glacial maximum in the Himalayas may be significantly older than the maximum of the last glaciation, as indicated by the North American and Fennoscandian ice sheets (Sharma and Owen, 1996). Much-improved glacial chronologies are thus required before the evolution of morphoclimatic terrains in mountain regions can be reliably interpreted in terms of global palaeoclimatic change (Owen *et al.*, 1996).

It is evident from this discussion that the linkages between climate change, valley slopes and river channel morphology are multiple and complex. Despite this complexity and the contentiousness of some of the evidence, the study of fluvial systems has thrown considerable light upon past and present climatic regimes, as illustrated from the arid and semi-arid regions. The work of Schumm (1968) is a classic study, based on contrasts in the sinuosity, channel dimensions, longitudinal gradients, meander wavelengths and sediment bedloads of the present and former channels of the Murrumbidgee River in Australia. This shows that the older palaeochannel was wider, had a much lower sinuosity, a much coarser bedload and a steeper gradient, indicating a much higher peak discharge but an average discharge lower than that of the present, narrower stream. This, in turn, suggests that the older channel was a strongly seasonal river in the drier climate of the Late Pleistocene.

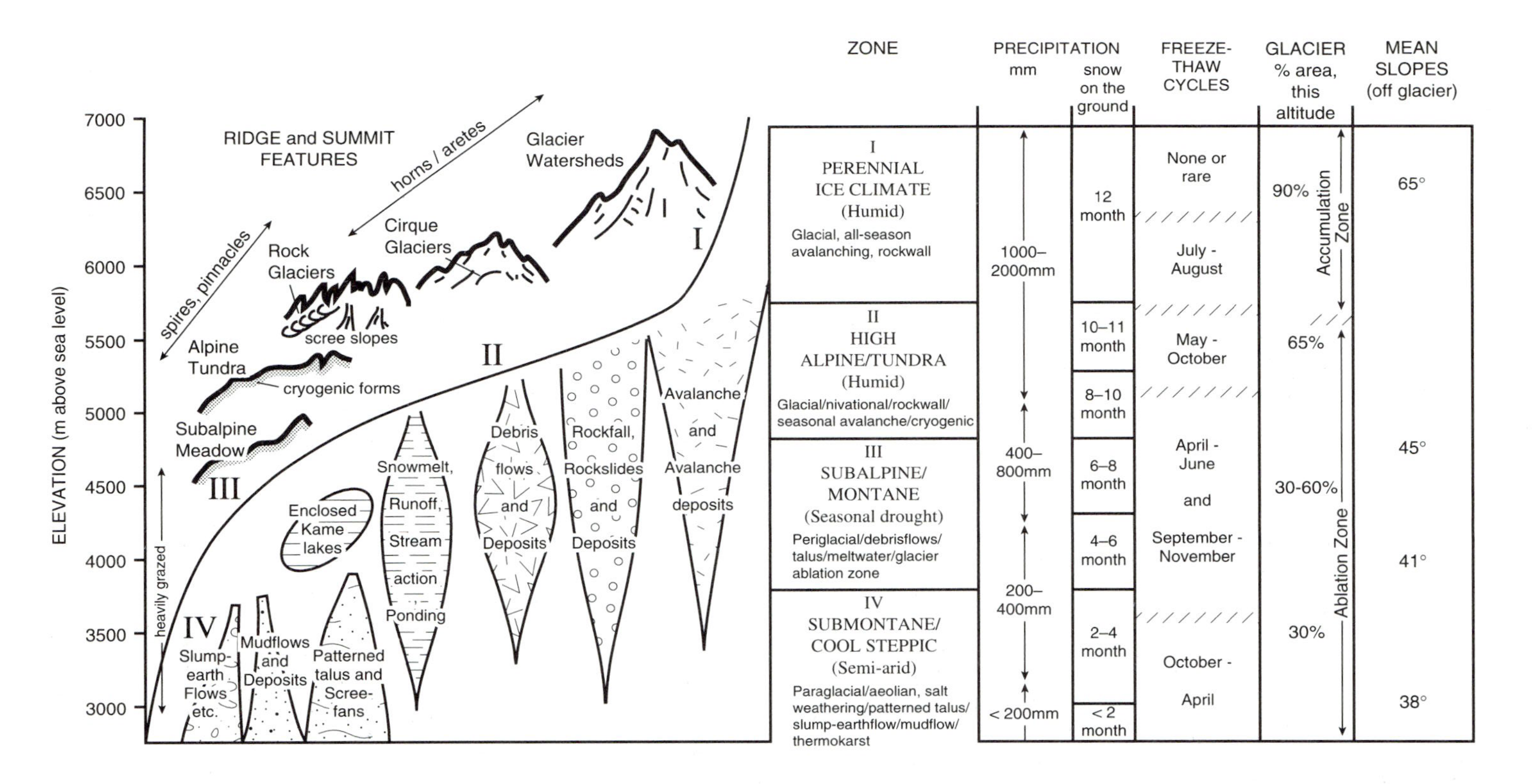

Figure 8.7 Schematic diagram showing altitudinal distribution of environmental variables in relation to currently prevailing geomorphological processes in the central Karakoram (after Hewitt, 1993)

Plate 8.1 Complex debris fan with settlement and glacier meltwater-fed irrigation in the Hunza Karakoram, northern Pakistan. The fan is a composite of glacially-deposited debris (Ghulmet glacier snout upper right), glacial debris reworked by massive debris flows, and some glaciofluvial and fluvial additions. Note that the fan is entrenched both by the Hunza River (foreground) and the meltwater streams

The accumulating evidence that some of the driest regions of the world were once traversed by major rivers includes landform–sediment associations from North Africa and elsewhere. In the eastern Sahara, for example, sinuous, branching gravel ridges up to 10 km in length stand several metres above eroded bedrock surfaces. These are bedload sediments of ancient seasonal rivers (*wadis*) that have become inverted by erosion of the surrounding plain. The raised wadis of Nubia form remnants of an integrated drainage system that was active during the Pleistocene (Butzer and Hansen, 1968; Said, 1975). Further south, in the extremely arid part of northern Sudan, is the Wadi Howar, a 640 km long watercourse that fed into the Nile River between about 9500 and 4500 years BP (Pachur and Kröpelin, 1987). In order to sustain periodic water flow in this very large feature, substantial rainfall would have been needed to recharge the groundwater in a Sahel-type environment that extended at least 500 km further north than it does today. This is broadly consistent with a range of palaeoclimatic evidence derived from studies of former lake basins in the central and western Sahara, implying increased moisture influx from an enhanced African monsoon (Petit-Maire, 1991). Extensive superimposed palaeochannel systems extend for over 200 km across two exhumed fans at the foot of the eastern Mountains of Oman in the Arabian Peninsula. Both inset and raised channels are present, the channel sequence indicating at least 12 successive generations of palaeochannels in at least six major periods of fluvial activity during the Plio–Pleistocene (Maizels, 1990).

The general model linking aridity (general or seasonal) with, on the one hand, reduced vegetation cover, a reduction in stream power and channel aggradation and, on the other, moister climates with channel incision and terracing, has provided a much-used general framework. However, morphoclimatic interpretation of fluvial systems in the drylands is far from straightforward. The highly episodic flow regimes characteristic of many such systems (Baker, 1983) may cause bedload flux to be so irregular that conditions vary from well-graded to blocked channels (Hövermann, 1985). This makes it difficult to determine the form–climate relationship in both modern and ancient arid zone stream systems (Graf, 1983). The channels of wadi systems are frequently poorly integrated, and sometimes partly masked by slope-foot spreads of colluvium. In southern Africa, the dominance of such colluvial deposits has been used to infer semi-arid conditions for the Late Pleistocene (Watson *et al.*, 1985). Consideration of the type (endoreic or exoreic) and hydrological time series of fluvial systems in the Kalahari Desert, led Shaw *et al.*, (1992) to propose that the middle Kalahari valleys were fluvially active between 15,000 and 12,000 years BP (Shaw and Thomas, 1993), and that high-magnitude flood events in the Kuruman River (southern Kalahari) correlate well with southern oscillation high-phase events. Evidence from the lacustrine record over most of Africa's intertropical zone, in contrast, indicates that the last major shift from dry to wet climate occurred at about 12,500 years BP, with rising lake levels and a notable erosional impact (Street-Perrott *et al.*, 1985).

In contrast to the Sahara, where the ergs consist mainly of ancient reworked fluvial sediments, the largest desert in China (the Taklamakan: 337,600 km^2) is being actively replenished with meltwater-borne sediments that provide both valuable oasis water and an abundant sand supply subject to active dune encroachment (Coque *et al.*, 1991). The Taklamakan underwent general flooding up to about 7500 years BP (Jäkel and Zhu, 1991), the Keriya and Tarim rivers having crossed the desert about 2000 years ago, and again during the sixteenth and early nineteenth centuries. Abandoned towns on the lower Keriya River, known from chronicles to have been populous and very prosperous, have been dated by the ^{14}C method at between 2135 and 2684 years BP (the Western Han Dynasty, about 150–250 BC). Maps of Xinjiang dating from the late Qing dynasty (1644–1911 AD) show that, although by then only a seasonal stream, the Keriya still crossed the Taklamakan periodically (Chen, 1991). Overrunning of oases by blown sand is well documented in the Taklamakan. Written records show that some important towns along the Old Silk Road have been lost: for example, Loulan (west of the Lop Nor lake) had over 1600 inhabitants in the Han dynasty (206 BC to 220 AD). However, it was abandoned in 376 AD and is now covered by dunes and yardangs. While there is no doubt that these changes were influenced by warfare and the lowering of meltwater-recharged piedmont water tables by excessive extraction of water by the growing populations (Derbyshire and Goudie, 1996), there is little doubt about the long-term oscillating trend towards arid conditions in the Holocene (Pachur *et al.*, 1995).

In the drylands of central and eastern Asia, sand dunes give way downwind to thick accumulations of wind-blown silts known as loess. Where loess accumulates to considerable thicknesses, a plateau-like landform may result, as in the Loess Plateau of northern China. The alternation of loess and palaeosols in the Chinese Loess Plateau provides a very detailed record of climate change over the past 2.5 million years. Extensive areas of the plateau have been deeply dissected, the resultant valley systems adding a complex but relatively little-studied geomorphological record of environmental change (Derbyshire, 1983). The loess slopes, locally more than 200 m high, are subject to varying degrees of erosion by stream action and mass movement in several climatic regimes from humid to arid. Using an erosional susceptibility coefficient expressive of the kinetic energy input of the runoff water (Ploey, 1990) and a susceptibility coefficient expressing rainwater pressure head (Derbyshire *et al.*, 1995), it has been shown that landslides in the western part of the Loess Plateau indicate some of the highest erosional susceptibilities in the world for non-clay materials, matched only by the tectonically very active Himalaya and New Zealand. The loess fails at thresholds over a range of scales but the failure–threshold relationship is not straightforward. For example, failure at the microscopic scale causes the progressive generation of weak zones that eventually fail in response to trigger mechanisms (climatic, tectonic and artificial), with magnitudes lower than those predicted from loess bulk properties and mean climate data (Derbyshire *et al.*, 1993). Such loess slopes are an important factor in causing the Yellow River to carry one of the highest suspended load concentrations in the world, averaging more than 25 kg m^{-3} throughout its middle and lower reaches (about 2000 km: Derbyshire and Wang, 1993). Thus, the Chinese Loess Plateau, as one of the earth's youngest major landforms, is undergoing very high rates of denudation that reach their peak close to the arid/semi-arid and semi-arid/subhumid climatic transitions.

Climate changes have also been registered using evidence of relict sand dune systems (Grove, 1958). Linear dunes fixed by vegetation, or in various stages of destruction by erosion, have been mapped using remotely sensed imagery in many parts of Africa, Asia, Australia and the Americas (McKee, 1979). Most active dune areas occur in areas with an annual precipitation of less than 250 mm, but relict dunes have been mapped in areas with present-day rainfall totals of more than 1000 mm, implying a shift of dune field margins of up to 1200 km in some regions (Goudie, 1983). The orientation of linear trends within such palaeodune systems has been used to infer past patterns of atmospheric circulation (Madigan, 1936; Mainguet, 1978; Lancaster, 1981; Wells, 1983; Thomas, 1984), although some of the intrinsic assumptions have been questioned (Thomas, 1992; Livingstone and Thomas, 1993). The relict lakes of inland Australia, with their distinctive clay-rich lunette dunes, have provided a detailed record of both landform evolution and human occupation since about 40,000 years BP (Bowler, 1971).

CONCLUSIONS

It is evident from a broad range of phenomena regarded as proxy indicators of climate change that temporal and spatial variation in geomorphological processes affected most climatic zones of the earth throughout the Quaternary. All the existing long-term records indicative of climate change over the past few million years of the earth's history indicate a pattern of global climate change broadly matching the solar orbital and precessional rhythms (see Chapter 10, The Quaternary period). The climatic trend is asymmetrical, changes from warm to cold being gradual with an order of magnitude of 100,000 years, in contrast to which the shifts from cold to warm 'terminations' (Broecker and van Donk, 1970) appear to be much more rapid. Both glacial and interglacial periods included shorter term, including some rapid, climate changes (Street-Perrott and Perrott, 1990) as the curves in Figure 8.4 show. In the tropical zone, including the sensitive subtropical savannas and desert margins, critical changes in erosion rate were controlled by precipitation changes under the influence of variations in the monsoonal circulation (Kutzbach, 1983). The current astronomically forced trend is towards climatic deterioration, with a decline in the summer insolation peak since about 9000 year BP and an expected mimimum in about another 5000 years' time (Berger and Tricot, 1986). Against this trend are the effects of the recent anthropologically induced increase in the greenhouse gases (see Chapter 22, Global air pollution problems), the values for carbon dioxide and methane already exceeding the values for the last interglacial calculated from the Vostok ice core (Chapellaz *et al.*, 1990).

Although all the models (Chapter 4) designed to estimate the amount of global warming likely to eventuate from these opposing trends differ in important particulars (Street-Perrott and Roberts, 1994), there is quite broad agreement that some rise in oceanic temperatures will occur. Rising sea levels (see Chapter 7, Conclusions) and changes in coastal morphology, already documented for parts of the Holocene (Matthews, 1990), are the most likely immediate threat at the regional scale. The estimated model-based global warming of between 2 and 5°C (Table 22.3) is likely to lead to rates of sea-level rise of between four and seven times that of the present (Gornitz, 1995). Neotectonic movements and differences in dynamic effects (wave climates) will minimize the uniformity of sea-level rise (Pirazzoli, 1993), but subsiding coastlines, including the great deltas like the Ganges, will suffer particularly severe changes (Tooley, 1994). Away from the coast, it is known that quite small river systems are highly susceptible to high-frequency climate changes and flood regimes that can be correlated with large-scale upper atmospheric circulation patterns. Rumsby and Macklin (1994) used a study of the last 350 years in the Tyne basin, northern England, as a guide for likely future drainage basin responses to climate change. Since 1700 AD, cool, wet conditions associated with meridional atmospheric circulation patterns have triggered the widespread incision of river channels. Conversely, periods with zonal circulation patterns have been marked by enhanced lateral reworking of floodplains and sediment transfer in the upper reaches of catchments, with channel narrowing and infilling downstream.

CONCLUSIONS *continued*

The effects upon desert-margin landscapes can be sketched out by analogy with changes known to have occurred in the Holocene, although too close an analogy would assume a level of comprehensiveness in palaeohydrological models that has yet to be achieved (Roberts and Barker, 1993). A rise in inter-tropical sea surface temperatures would ameliorate conditions in the semi-arid and arid zones. It has been estimated that parts of the Sahara would see the return of many of the permanent lakes, steppic grasses and animals known to have been present between 9000 and 7000 years ago (Petit-Maire and Riser, 1988). In view of the apparent synchroneity of Holocene monsoon-driven climate changes in dry regions as far apart as tropical North Africa and Tibet (Gasse *et al.*, 1996), any future amelioration would have an impact on large expanses of the world's deserts. Likely hydrological changes would include recharge of currently dry palaeochannels and lake basins. Increased runoff would follow from enhanced glacier melting and replenishment of water tables in the alluvial fan oases in the drylands of central Asia. However, the retreat of glaciers in the more arid mountain systems (such as parts of the Karakoram) might ultimately render many of the directly fed meltwater irrigation networks unworkable, because of a combination of increasing distances from glacier snouts and an enhanced tendency to stream incision.

To summarize, the likely landform response to the proposed climate changes remains difficult to predict because of the sparsity of information on palaeogeomorphological responses to climate change arising, in part, from a terrestrial palaeoclimatic record that is still of rather low resolution. Nevertheless, the present view is that landform response to expected climate changes over the next century or so is likely to be felt first on stable or subsiding low-lying coastlines, some of which have dense population concentrations. Changes in river regimes are also likely to trigger landform changes with potentially severe effects upon well-populated floodplains. Changes of equal severity may occur on the desert margins and in mountainous regions, although the impact may be less evident given the sparser settlement distribution. Taking into account the uncertainties in current models of future global warming (see Chapter 4, Conclusions) and given the complexity of the response rate in geomorphological systems, the rate of change at the landform scale (but excluding, for example, soil erosion), away from the low-lying coastlands, is likely to be measured in centuries rather than decades.

REFERENCES

Agassiz, L. (1840) *Etudes sur les glaciers*, Neuchâtel.

Ahnert, F. (1987) 'An approach to the identification of morphoclimates', in V. Gardiner (ed.) *International Geomorphology*, Chichester and New York: John Wiley, Vol 2, pp. 159–88.

Baker, V.R. (1983) 'Late Pleistocene fluvial systems', in H.E. Wright, Jr (ed.) *Late Quaternary Environments of the United States, Vol. 1 The Late Pleistocene*, Minneapolis: University of Minnesota Press, pp. 115–29.

Berger, A. and Tricot, C. (1986) 'Global climatic changes and astronomical theory of paleoclimates', in A. Cazenave (ed.) *Earth Rotation: Solved and Unsolved Problems*, Dordrecht: Reidel, pp. 111–29.

Bowler, J.M. (1971) 'Pleistocene salinities and climatic change: evidence from lakes and lunettes in south-eastern Australia', in D.J. Mulvaney and J. Golson (eds) *Aboriginal Man and Environment in Australia*, Canberra: Australian National University Press.

Broecker, W.S. and van Donk, J. (1970) 'Insolation changes, ice volumes and the 18_0 record in deep-sea cores', *Reviews of Geophysics and Space Physics* 8: 169–97.

Brooks, S.M., Anderson, M.G. and Collison, A.J.C. (1995) 'Modelling the role of climate, vegetation and pedogenesis in shallow translational hillslope failure', *Earth Surface Processes and Landforms* 20: 231–42.

Brunsden, D. (1980) 'Applicable models of long term landform evolution', *Zeitschrift für Geomorphologie* Supplementband 36: 16–26.

Brunsden, D. and Thornes, J.B. (1979) 'Landscape sensitivity and change', *Transactions of the Institute of British Geographers* 4: 463–84.

Büdel, J. (1948) 'Das System der klimatischen Geomorphologie', *Verhandelingen Deutscher Geographie* 27: 65–100.

Büdel, J. (1963) 'Klima-genetische Geomorphologie', *Geographische Rundschau* 7: 269–85.

Büdel, J. (1982) Climatic Geomorphology, trans. L. Fischer and D. Busche, Princeton, NJ: Princeton University Press.

Burbank, D.W., Leland, J., Fielding, E., Anderson, R.S., Brozovic, N., Reid, M. and Duncan, C. (1996) 'Bedrock incision, rock uplift and threshold hillslopes in the northwestern Himalayas', *Nature* 379: 505–10.

Butzer, K.W. and Hansen, C.L. (1968) *Desert and River in Nubia*, Madison: University of Wisconsin Press.

Chapellaz, J., Barnola, J.M., Raynaud, D., Korotkevich, Y.S. and Lorius, C. (1990) 'Ice record of atmospheric methane over the past 160,000 years', *Nature* 345: 127–31.

Chen, H. 1991 'The change of eco-environment and the rational utilization of water resources in the Keriya River valley', in D. Jäkel and Z. Zhu 'Reports on the 1986 Sino-German Kunlun Shan Taklimakan-Expedition', *Die Erde* Erg.H 6: 133–47.

Common, R. (1966) 'Slope failure and morphogenetic regions', in G.H. Dury (ed.) *Essays in Geomorphology*, London: Heinemann, pp. 53–81.

Coque, R., Gentelle, P. and Coque-Delhuille, B. (1991) 'Desertification along the piedmont of the Kunlun chain (Hetian-Yutian sector) and the southern border of the Taklamakan Desert (China): preliminary geomorphological observations (1)', *Revue de Géomorphologie Dynamique* 15: 1–27.

Davis, W.M. (1899) 'The geographical cycle', *Geographical Journal* 14: 481–504.

Davis, W.M. (1900) 'Glacial erosion in France, Switzerland and Norway', *Proceedings Boston Society of Natural History* 29: 273–322.

Davis, W.M. (1905) 'The geographical cycle in an arid climate', *Journal of Geology* 13: 381–407.

Derbyshire, E. (ed.) (1973) *Climatic Geomorphology*, London: Macmillan.

Derbyshire, E. (ed.) (1976) *Geomorphology and Climate*, Chichester: John Wiley.

Derbyshire, E. (1983) 'On the morphology, sediments and origin of the Loess Plateau of central China', in R. Gardner and H. Scoging (eds) *Mega-geomorphology*, Oxford: Oxford University Press, pp. 172–94.

Derbyshire, E. (1996) 'Quaternary glacial sediments, glaciation style, climate and uplift in the Karakoram and northwest Himalaya: review and speculations', in F. Gasse and E. Derbyshire (eds) 'Environmental Changes in the Tibetan Plateau and Surrounding Areas', *Palaeogeography, Palaeoclimatology, Palaeoecology* 120: 147–57.

Derbyshire, E. and Goudie, A.S. (1996) 'The drylands of Asia', in D.S.G. Thomas (ed.) *Arid Zone Geomorphology: process, form and change in drylands*, 2nd edn, Chichester: John Wiley.

Derbyshire, E. and Wang, J.T. (1993) 'China's Yellow River Basin', in N. Roberts (ed.) *The Changing Global Environment*, Oxford: Blackwell, pp. 417–39.

Derbyshire, E., Li, J.J., Street-Perrott, F.A., Xu, S.Y. and Waters, R. S. (1984) 'Quaternary glacial-history of the Hunza Valley, Karakoram Mountains, Pakistan', in K.J. Miller (ed.) *The International Karakoram Project*, vol. 2, Cambridge: Cambridge University Press, pp. 456–95.

Derbyshire, E., Dijkstra, T.A., Billard, A., Muxart, T., Smalley, I.J. and Li, Y.J. (1993) 'Thresholds in a sensitive landscape: the loess region of central China', in D.S.G. Thomas and R.J. Allison (eds) *Landscape Sensitivity*, Chichester: John Wiley, pp. 97–127.

Derbyshire, E., Kemp, R. and Meng, X. (1995) 'Variations in loess and palaeosol properties as indicators of palaeoclimatic gradients across the Loess Plateau of North China', in E. Derbyshire (ed.) 'Aeolian Sediments in the Quaternary Record', *Quaternary Science Reviews* 14: 691–99.

Douglas, I. (1976) 'Erosion rates and climate: geomorphological implications', in E. Derbyshire (ed.) *Geomorphology and Climate*, London and New York: John Wiley, pp. 269–87.

Eybergen, F.A. and Imeson, A.C. (1989) 'Geomorphological processes and climatic change', *Catena* 16: 307–19.

Fang, J. (1993) 'Lake evolution during the last 3,000 years in China and its implications for environmental change', *Quaternary Research* 39: 175–85.

Fournier, F. (1960) *Climat et Erosion: la relation entre l'érosion du sol par l'eau et les précipitations atmosphériques* Paris: Presses Univ. France.

Frenzel, B., Pecsi, M and Velichko, A.A. (eds) (1992) *Atlas of Palaeoclimates and Palaeoenvironments of the Northern Hemisphere (Late Pleistocene – Holocene)*, Stuttgart, Jena, New York: Gustav Fischer Verlag, and Budapest: Geographical Research Institute, Hungarian Academy of Sciences.

Gasse, F., Ledée, V., Massault, M. and Fontes, J. (1989) 'Water-level fluctuations of Lake Tanganyika in phase with oceanic changes during the last glaciation and deglaciation', *Nature* 342: 57–9.

Gasse, F., Fontes, J.C., Van Campo, E. and Wei, K. (1996) 'Holocene environmental changes in Bangong Co bason (western Tibet). Part 4: discussion and conclusions', in F. Gasse and E. Derbyshire (eds) 'Environmental Changes in the Tibetan Plateau and Surrounding Areas', *Palaeogeography, Palaeoclimatology, Palaeoecology* 120: 79–92.

Gornitz, V. (1995) 'Sea-level rise: a review of recent past and near-future trends', *Earth Surface Processes and Landforms* 20: 7–20.

Goudie, A. (1983) 'The arid Earth', in R. Gardner and H. Scoging (eds) *Mega-geomorphology*, Oxford: Oxford University Press, pp. 152–71.

Graf, W.L. (1983) 'Flood related channel change in an arid region', *Earth Surface Processes and Landforms* 8: 125–40.

Grove, A.T. (1958) 'The ancient ergs of Hausaland, and similar formations on the south side of the Sahara', *Geographical Journal* 124: 528–33.

Haggett, P., Chorley, R.J. and Stoddart, D.R. (1965) 'Scale standards in geographical research', *Nature* 205: 844–7.

Hewitt, K. (1993) 'Altitudinal organization of Karakoram geomorphic processes and depositional environments', in J.F. Shroder (ed.) *Himalaya to the Sea*, London and New York: Routledge, pp. 159–83.

Hövermann, J. (1985) 'Das System der klimatischen Geomorphologie auf landschaftskundlicher Grundlage', *Zeitschrift für Geomorphologie* NF, Supplementband 56: 143–53.

Hövermann, J., Lehmkuhl, F. and Sussenberger, H. (1992) 'Neue Befunde zur Palaoklimatologie Nordafrikas und Zentralasiens', *Abhandlungender Braunschweigischen Wissenschaftlichen Gesellschaft* 43: 127–50.

Jäkel, D. and Zhu, Z. 1991 'Reports on the 1986 Sino-German Kunlun Shan Taklimakan-Expedition', *Die Erde* Erg.H 6, 200 pp and map volume.

Jouzel, J. *et al.* (1989) 'Global change over the last climatic cycle from the Vostok ice core record', *Quaternary International* 2: 15–24.

Kemp, R.A., Derbyshire, E., Meng, X.M., Chen, F.H. and Pan, B.T. (1995) 'Pedosedimentary reconstruction of a thick loess-paleosol sequence near Lanzhou in north-central China', *Quaternary Research* 43: 30–45.

Kemp, R.A., Derbyshire, E., Chen, F. and Ma, H. (1996) 'Pedosedimentary development and palaeoenvironmental significance of the S1 palaeosol on the northeastern margin of the Qinhgai-Xizang (Tibetan) Plateau', *Journal of Quaternary Science* 11: 95–106.

Kirkbride, M. and Brazier, V. (1995) 'On the sensitivity of Holocene talus-derived rock glaciers to climate change in the Ben Ohau range, New Zealand', *Journal of Quaternary Science* 10: 353–65.

Kirkby, M.J. and Neale, R.H. (1987) 'A soil erosion model incorporating seasonal factors', in V. Gardiner (ed.) *International Geomorphology*, Chichester and New York: John Wiley, Vol. 2, pp. 189–210.

Koster, E.A. (1994) 'Global warming and periglacial landscapes', in N. Roberts (ed.) *The Changing Global Environment*, Oxford: Blackwell, pp. 127–49.

Kukla, G. (1989) 'Long continental records of climate – an introduction', *Palaeogeography, Palaeoclimatology, Palaeoecology* 72: 1–9.

Kutzbach, J. (1983) 'Monsoon rains of the late Pleistocene and early Holocene: patterns, intensity and possible causes of changes', in F.A. Street-Perrott, M. Beran and R. Ratcliffe (eds) *Variations in the Global Water Budget*, Dordrecht: Reidel, pp. 371–89.

Lancaster, N. (1981) 'Palaeoenvironmental implications of fixed dune systems in southern Africa', *Palaeogeography, Palaeoclimatology, Palaeoecology* 33: 327–46.

Langbein, W.B. and Schumm, S.A. (1958) 'Yield of sediment in relation to mean annual precipitation', *Transactions of the American Geophysical Union* 39: 1076–84.

Lichter, J. (1995) 'Lake Michigan beach-ridge and dune development, lake level, and variability in regional water balance', *Quaternary Research* 44: 181–9.

Livingstone, I. and Thomas, D.S.G. (1993) 'Modes of linear dune activity and their palaeoenvironmental significance: an evaluation with reference to southern African examples', in K. Pye (ed.) *The Dynamics and Environmental Context of Aeolian Sedimentary Systems*, London: Geological Society Special Publication 72: 91–101.

Lorius, C., Jouzel, J., Ritz, C., Merlivat, L., Barkov, N.I., Korotkevich, Y.S. and Kotlyakov, V.M. (1985) 'A 150,000 year climatic record from Antarctic ice', *Nature* 316: 591–96.

Lorius, C., Jouzel, J., Raynaud, D., Hansen, J. and Le Treut, H. (1990) 'The ice-core record: climate sensitivity and future greenhouse warming', *Nature* 347: 139–45.

Madigan, C.T. (1936) 'The Australian sand ridge deserts', *Geographical Review* 26: 205–27.

Mainguet, M. (1978) 'The influence of Trade Winds, local air-masses and topographic obstacles on the aeolian movement of sand particles and the origin and distribution of dunes and ergs in the Sahara and Australia', *Geoforum* 9: 17–28.

Maizels, J. (1990) 'Raised channel systems as indicators of palaeohydrologic change: a case study from Oman', *Palaeogeography, Palaeoclimatology, Palaeoecology* 76: 241–77.

Martonne, E. de (1913) 'Le Climat – facteur de relief', *Scientia* 1913: 339–55.

Mason, O.K. and Jordan, J.W. (1993) 'Heightened North Pacific storminess during synchronous late Holocene erosion of Northwest Alaska beach ridges', *Quaternary Research* 40: 55–69.

Matthews, R.K. (1990) 'Quaternary sea-level change', in *Sea-Level Change*, Washington, DC: National Research Council, pp. 88–103.

McKee, E.D. (ed.) (1979) 'A study of global sand seas', *United States Geological Survey Professional Paper* 1052.

Milankovitch, M. (1930) 'Mathematische Klimalehre und astronomische Theorie der Klimaschwankungen', in W. Köppen and R.Geiger (eds) *Handbuch der Klimatologie*, 1(A), Berlin.

Molnar, R. and England, P. (1990) 'Late Cenozoic uplift of mountain ranges and global climatic change: chicken or egg?', *Nature* 346: 29–34.

Orford, J.D., Carter, R.W.G. and Jennings, S.C. (1991) 'Coarse clastic barrier environments: evolution and implications for Quaternary sea level interpretation', *Quaternary International* 9: 87–104.

Owen, L.A. and Derbyshire, E. (1993) 'Quaternary and

Holocene intermontane basin sedimentation in the Karakoram Mountains', in J.F. Schroder (ed.) *Himalayas to the Sea: Geology, Geomorphology and the Quaternary*, London: Routledge, pp. 108–31.

Owen, L.A., Benn, D.I., Derbyshire, E., Evans, D.J.A., Mitchell, W.A. and Richardson, S. (1996) 'The Quaternary glacial history of the Lahul Himalaya, northern India', *Journal of Quaternary Science* 11: 25–42.

Pachur, H.J. and Kröpelin, S. (1987) 'Wadi Howar: paleoclimatic evidence from an extinct river system in the southeastern Sahara', *Science* 237: 298–300.

Pachur, H.J., Wünnemann, B. and Zhang, H. (1995) 'Lake evolution in the Tengger Desert, northwestern China, during the last 40,000 years', *Quaternary Research* 44: 171–80.

Peltier, L.C. (1950) 'The geographical cycle in periglacial regions as it is related to climatic geomorphology', *Annals of the Association of American Geographers* 40: 214–36.

Penck, A. (1910) 'Versuch einer Klimaklassification auf physiographischer Grundlage', *Preussen Akademie der Wissenschaft Sitz. Der physikalisch-mathematischen* 12: 236–46.

Petit-Maire, N. (ed.) (1991) *Paléoenvironnements du Sahara: lacs holocènes à Taoudenni (Mali)*, Paris: Editions du Centre National de la Recherche Scientifique.

Petit-Maire, N. and Riser, J. (1988) *Le Sahara à l'Holocène: Mali*, Paris: CCGM, Map at scale 1: 1 million.

Petit-Maire, N., Fontugne, M. and Rouland, C. (1991) 'Atmospheric methane ratio and environmental changes in the Sahara and Sahel during the last 130 Kyrs', *Palaeogeography, Palaeoclimatology, Palaeoecology* 86: 197–204.

Pirazzoli, P.A. (1993) 'Global sea-level changes and their measurement', *Global and Planetary Change* 8: 135–48.

Ploey, J. de (1990) 'Modelling the erosional susceptibility of catchments in terms of energy', *Catena* 17: 175–83.

Porter, S.C., An, Z.S. and Zheng, H.B. (1992) 'Cyclic Quaternary alluviation and terracing in a nonglaciated drainage basin on the north flank of the Qinling Shan, central China', *Quaternary Research* 38: 157–69.

Rapp, A. (1960) 'Recent developments of mountain slope in Karkevagge and surroundings, northern Scandinavia', *Geografiska Annaler* 42: 71–200.

Roberts, N. and Barker, P. (1993) 'Landscape stability and biogeomorphic response to past and future climatic shifts in intertropical Africa', in D.S.G. Thomas and R.J. Allison (eds) *Landscape Sensitivity*, Chichester: John Wiley, pp. 65–82.

Rumsby, B.T. and Macklin, M.G. (1994) 'Channel and floodplain response to recent abrupt climate change: the Tyne basin, northern England', *Earth Surface Processes and Landforms* 19: 499–515.

Said, R. (1975) 'Some observations on the geomorphology of the south Western Desert of Egypt and its relation to the origin of ground water', *Annals of the Geological Survey of Egypt* 5: 61–70.

Schumm, S.A. (1968) 'River adjustment to altered hydrologic regimen – Murrumbidgee River and paleochannels, Australia', *United States Geological Survey Professional Paper* 598.

Schumm, S.A. and Brakenridge, R.A. (1987) 'River responses', in W.F. Ruddiman and H.E. Wright, Jr (eds) *North America and Adjacent Oceans during the last deglaciation*, The Geology of North America K3, The Geological Society of America, pp. 221–40.

Shackleton, N.J., Imbrie, J. and Pisias, N.G. (1988) 'The evolution of oceanic oxygen-isotope variability in the North Atlantic over the past three million years', *Philosophical Transactions of the Royal Society*, London, B 318: 679–88.

Shackleton, N.J., An, Z., Dodonov, A.E., Gavin, J., Kukla, G.J., Ranov, V.A. and Zhou, L.P. (1995) 'Accumulation rate of loess in Tadjikistan and China; relationship with global ice volumes', in E. Derbyshire (ed.) 'Wind Blown Sediments in the Quaternary Record', *Quaternary Proceedings* 4: 1–6.

Sharma, M.C. and Owen, L.A. (1996) 'Quaternary glacial history of NW Garwhal, central Himalayas', *Quaternary Science Reviews* 14: 1–30.

Shaw, P.A. and Thomas, D.S.G. (1993) 'Geomorphological processes, environmental change and landscape sensitivity in the Kalahari region of southern Africa', in D.S.G. Thomas and R.J. Allison (eds.) *Landscape Sensitivity*, Chichester: John Wiley, pp. 83–95.

Shaw, P.A., Thomas, D.S.G. and Nash, D.J. (1992) 'Late Quaternary fluvial activity in the dry valleys (mekgacha) of the Middle and Southern Kalahari, southern Africa', *Journal of Quaternary Science* 7: 273–81.

Shroder, J.F., Owen, L.A. and Derbyshire, E. (1993) 'Quaternary glaciation of the Karakoram and Nanga Parbat Himalayas', in J.F. Schroder (ed.) *Himalayas to the Sea: Geology, Geomorphology and the Quaternary*, London: Routledge, pp. 132–58.

Stoddart, D.R. (1969) 'Climatic geomorphology: review and re-assessment', *Progress in Geography* 1: 160–222.

Strahkov, N.M. (1967) *Principles of Lithogenesis*, ed. S.I. Tomkeieff and J.E. Hemingway, Vol. 1, Edinburgh: Oliver and Boyd.

Street, A.F. and Grove, A.T. (1979) 'Global maps of lake-level fluctuations since 30,000 years BP', *Quaternary Research* 10: 83–118.

Street-Perrott, F.A. and Perrott, R.A. (1990) 'Abrupt climatic fluctuations in the tropics: the influence of Atlantic Ocean circulation', *Nature* 343: 607–12.

Street-Perrott, F.A. and Roberts, N. (1994) 'Past climates and future greenhouse warming', in N. Roberts (ed.) *The Changing Global Environment*, Oxford: Blackwell, pp. 47–68.

Street-Perrott, F.A., Roberts, N. and Metcalfe, S. (1985) 'Geomorphic implications of late Quaternary hydrological and climatic changes in the Northern Hemisphere tropics', in I. Douglas and T. Spencer (eds) *Environmental Change and Tropical Geomorphology*, London: George Allen and Unwin, pp. 165–83.

Thomas, D.S.G. (1984) 'Ancient ergs of the former arid zones of Zimbabwe, Zambia and Angola', *Transactions of the Institute of British Geographers* 9: 75–88.

Thomas, D.S.G. (1992) 'Desert dune activity: concepts and significance', *Journal of Arid Environments* 22: 31–8.
Thomas, D.S.G. and Allison, R.J. (1993) *Landscape Sensitivity*, Chichester and New York: John Wiley.
Thornes, J.B. and Brunsden, D. (1977) *Geomorphology and Time*, London: Methuen.
Tooley, M.J. (1994) 'Sea-level response to climate', in N. Roberts (ed.) *The Changing Global Environment*, Oxford: Blackwell, pp. 172–89.
Tricart, J. and Cailleux, A. (1972) *Introduction to Climatic Geomorphology*, trans. C.J.K. de Jonge, London: Longman.
Watson, A., Price Williams, D. and Goudie, A. (1985) 'The palaeoenvironmental interpretation of colluvial sediments and palaeosols in Southern Africa', *Palaeogeography, Palaeoclimatology, Palaeoecology* 45: 235–49.
Wells, G.L. (1983) 'Late-glacial circulation over central North America revealed by aeolian features', in F.A. Street-Perrott, M. Beran and R. Ratcliffe (eds) *Variations in the Global Water Budget*, Dordrecht: Reidel, pp. 317–30.
Wilson, L. (1972) 'Seasonal sediment yield patterns of US rivers', *Water Resources Research* 8: 1470–9.
Wilson, L. (1973) 'Relationships between geomorphic processes and modern climates as a method in paleoclimatology', in E. Derbyshire (ed.) *Climatic Geomorphology*, London: Macmillan, pp. 269–84.
Wolman, M.G. and Gerson, R. (1978) 'Relative scales of time and effectiveness of climate in watershed geomorphology', *Earth Surface Processes* 3: 189–208.
Yair, A. and Enzel, Y. (1987) 'The relationship between annual rainfall and sediment yield in arid and semi-arid areas. The case of the Northern Negev', in F. Ahnert (ed.) *Geomorphological Models, Theoretical and Empirical Aspects*, Catena Supplement 10: 121–36.

9

SOILS

Michael Bridges

INTRODUCTION: SOIL CLIMATE

In the minds of most people, climate and climatology are associated solely with the atmosphere, its chemical composition, physical structure and the geographical implications of weather systems for any particular place on the earth's surface. However, the atmosphere does not suddenly stop at ground level: it extends into the soil along fissures and pores, well below the surface of the soil, and its effects are felt throughout the body of the soil. This subterranean atmosphere is an important part of the soil environment with its own climate and deserves scientific consideration. Since the beginning of the modern period of scientific soil studies about 100 years ago, the climate above ground has been accepted as one of the main environmental factors of soil formation. Atmospheric temperature determines the speed of chemical reactions, and to a certain extent the amount of biological activity in the soil. The balance of rainfall and evapotranspiration determines how much water percolates through the soil either to remove or redistribute chemical breakdown products. Indirectly, the climate also determines the natural vegetation or crops which may be grown, which in turn influence the nature of the organic component of the soil. During the past decade, it has been realized that there also exists an interaction which works in the opposite direction. That is, soils, or the biochemical reactions which occur in soils, can produce substantial quantities of gases which are significant in the process of global warming. Mindful of this relationship, the position of a chapter on soils is relevant to the subject matter of a book devoted to applied climatology.

THE SOIL

Soils are identified by the appearance of their profile, a section from the soil surface to the underlying parent material. The soil profile is made up of a sequence of horizons, developed in response to the processes of soil formation. Each horizon can be identified by its morphology, physical characteristics, chemical composition and presence of biological activity. Pedologists recognize some half-dozen surface soil horizons (topsoils) which are enriched with organic matter, usually contain the greatest amount of available plant nutrients and have the most biological activity. Below these surface horizons, several distinctive horizons are recognized (subsoils) which plant roots exploit for nutrient and moisture supplies (Bridges, 1990; 1996). These 'diagnostic' horizons are used in the allocation of soils in classification systems in most countries of the world.

Soil is a three-phase system comprising solid, liquid and gaseous components, each of which plays an important part in the soil environment. The solid part of soil is composed of mineral and organic materials. Particles of sand, silt, clay and humus aggregate to form soil structures or peds. In a well-structured soil, these peds are stable units, and between them the voids can amount to about 50 per cent of the volume in a friable topsoil. These voids are occupied either by air or by water. In a saturated soil, water fills the voids and the air is driven out, although bubbles may be trapped in larger sections of convoluted pores. In a freely drained soil, the moisture content between 24 and 48 hours after substantial rainfall (depending upon soil texture) is referred to as field capacity. As a soil dries, it eventually reaches a moisture content which is insufficient to maintain the turgor of plants; this is referred to as the wilting point. Between these two limits lies water available to plants, the available water capacity (AWC), retained in the soil.

Soil moisture was formerly expressed by the tension with which it is held in the soil. Water which was free to move under the influence of gravity was described as gravitational water.. Capillary water is bound to the soil particles and held in the finer pores by capillary forces and is the moisture largely available for plants, but tightly held hygroscopic water is virtually part of the mineral composition and is completely unavailable for plants. Measurement of the tension with which moisture is held in the soil is in terms of a column of water which could be held against gravity. This is effectively zero when the soil is saturated, increasing to 330 cm at field capacity and 15,000 cm at the wilting point. The large range of figures necessitated the use of a logarithmic scale, pF, upon which pF 2.5 equals field capacity and pF 4.2 wilting point. Currently, soil moisture is usually characterized by the matrix potential (psi), described as a measure of the extent to which the free energy of water in the soil (or ability of the water to move) is reduced as a result of gravity, surface attraction and osmotic forces. Plant-available moisture, the water held between field capacity and wilting point, is defined by 33 and 1500 kPa (kilopascals) respectively.

THE SOIL ATMOSPHERE

At the moisture state of field capacity, the volume of the remaining air-filled pores of a soil is referred to as the air capacity. Air within the soil is a natural continuation of the atmosphere above the soil, but although it is similar in some respects, it differs in others. Compared with atmospheric air, soil air is usually saturated with water vapour and it has a greater concentration of carbon dioxide.

The figures given in Table 9.1 indicate that the soil air has slightly less oxygen and more carbon dioxide (eight times) than atmospheric air, but the amounts of both gases vary considerably according to the activity of the micro-organisms living in the soil. Additions of leaf litter or organic manure greatly stimulate the soil fauna, which may result in the depletion of oxygen and an increase in carbon dioxide. The exchange of these gases with the atmosphere above the soil is hindered if the soil pores and fissures are small or limited in number. When the pores are filled with water, fresh oxygen cannot easily diffuse inwards, and such oxygen that may be present in the soil is soon consumed, and anaerobic conditions develop. It is in these conditions that the growth of most plants is inhibited and reducing conditions

Table 9.1 Average composition of soil air (% by volume)

	Oxygen	*Carbon dioxide*	*Nitrogen*
Soil air	20.65	0.25	79.20
Atmospheric air	20.97	0.03	79.00

typical of the process of gleying are brought about.

The movement of gases into and out of the soil takes place by diffusion, the process depending upon the differences in concentration of gases between the soil atmosphere and the free atmosphere above the soil. Even in freely drained soils with wide pores, some water is trapped by constrictions, so the process of diffusion is not straightforward. Changes in atmospheric pressure can cause the soil to 'breathe' and the gustiness of the wind may assist gaseous movement. The passage of water through the soil may also displace air from below the wetting front as it moves downwards. Oxygen diffuses into the soil, where it is taken up by roots and used in microbial activity. During the summer, the demand for oxygen is between 7 and 35 g m^2 d^{-1} and, unless there is a saturated layer, the process of diffusion is usually rapid enough for sufficient oxygen to meet the demand. Conversely, as a result of biological respiration, carbon dioxide diffuses from the soil pores out into the atmosphere, the average rate of diffusion being between 1.5 and 6.7 g m^2 d^{-1}. However, some carbon dioxide also dissolves in water, and it has been predicted that, other than in anaerobic pockets, the concentration of carbon dioxide in soil air would not normally exceed 1 per cent.

SOIL CLIMATE

Although soil scientists remain acutely aware of the importance of climate as an important factor in soil formation, modern soil classification has tended to emphasize the use of 'internal' measurable soil attributes, rather than rely upon the 'external' characteristics of climate. This approach has been followed by FAO (1993) and many national soil survey organizations, avoiding the direct use of non-pedological criteria (i.e. climate) for the classification of soils. However, the concept of soil climate has emerged in which the soil atmosphere is considered to have its own moisture and temperature regimes (Buol *et al.*, 1989). These climatic characters of soils are significant in terms of soil genesis, plant growth and particularly the growth of agricultural crops. This approach is best exemplified in the soil classification system of the USA, *Soil Taxonomy* (Soil Survey Staff, 1975). Abbreviated descriptions of soil moisture regimes are as follows:

Aquic: Saturated, implying anaerobic (reducing) conditions in the soil.

Udic: Moisture usually sufficient for plant growth. Not dry in any part for as long as 90 cumulative days per year in six years out of ten.

Ustic: Moisture is available (but limited) during the warm season. Dry in some or all parts for 90 or more cumulative days per year in six years out of ten.

Xeric: Moisture comes during the cool season. Dry in all parts for more than 45 days in the 4 months following the summer solstice in six years out of ten.

Aridic: Dry in all parts for more than half the cumulative days when the soil temperature at 50 cm depth is above 5°C or is moist for less than 90 consecutive days when the soil temperature at 50 cm depth is above 8°C.

For full details, the reader is referred to *Soil Survey Staff* (1975).

The sun is the source of heat controlling soil temperature, partly influenced by the albedo of the soil. The darker the soil colour, the lower the albedo and the more rapidly heat is absorbed. Soil temperature is critical for seed germination and growth of plants, so it is usually considered in conjunction with soil data when determining the capability of the land to produce crops. In temperate maritime climates, the soil rarely freezes completely, but in interior continental areas, biological activity in the soil during winter is restricted by frost, although the soil may be insulated by a covering of snow. The surface temperature of a bare soil may undergo a wide fluctuation of temperature as it is warmed by the sun or cooled by evaporation, or at night by outward radiation, especially under cloud-free anticyclonic conditions (Figure 9.1). The soil is also an efficient insulating medium, so with depth in the soil, the diurnal temperature wave becomes dampened, until at a metre depth there is little appreciable daily change in soil temperature.

Temperature is used in the definition of lower taxonomic classes in the *Soil Taxonomy*. Abbreviated descriptions of the temperature regimes are as follows:

Pergelic:	Below 0°C mean annual soil temperature.
Cryic:	0 to 8°C mean annual soil temperature.
Frigid:	Below 8°C but difference between mean summer and mean winter soil temperatures is more than 5°C at 50 cm depth.
Mesic:	8 to 15°C but difference between mean summer and mean winter soil temperature is more than 5°C at 50 cm depth.
Thermic:	15 to 22°C but difference between mean summer and mean winter soil temperature is more than 5°C at 50 cm depth.
Hyperthermic:	Above 22°C but difference between mean summer and mean winter soil temperatures is more than 5°C at 50 cm depth.

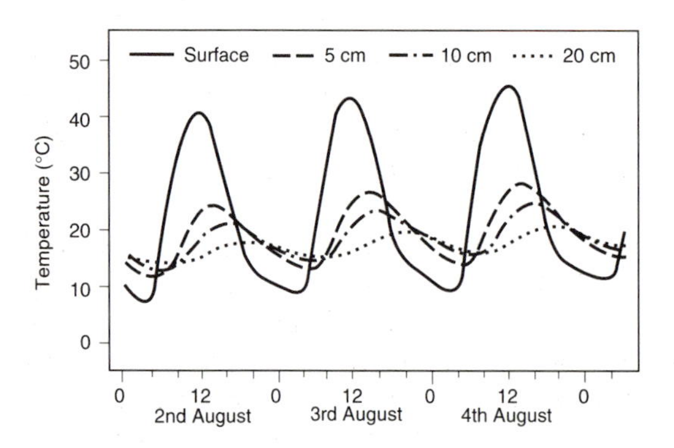

Figure 9.1 The dampening of the daily temperature wave with depth in a bare Rothamsted soil (Russell, 1973)

Where the difference between summer and winter is less than 5°C the prefix iso- is used, for example isothermic, isohyperthermic. For full details, the reader is referred to *Soil Survey Staff* (1975).

Using evapotranspiration and rainfall figures, it is possible to establish an annual budget of soil moisture in a diagrammatic format. The amount of plant-available moisture (AWC) is of critical importance for growing crops and graphical and computer calculations demonstrate the relationships between rainfall and potential evapotranspiration throughout the year (Figure 9.2(a, b)). In the examples shown a surplus (S) of precipitation exists as rainfall exceeds evapotranspiration, but by late spring the position is reversed and utilization (U) of soil moisture reserves takes place leading to a deficit (D) at the end of summer. As potential evapotranspiration falls in autumn, and rainfall becomes more effective, recharge (R) occurs until the annual cycle begins again. Examples of a udic and ustic moisture regime are shown. In the case of the udic regime moisture is used during the summer as demonstrated by the excess of evapotranspiration over rainfall; recharge then occurs in the autumn until the deficit has been removed and a surplus of moisture exists during the spring months. With rainfall and stored water in excess of evapotranspiration, movement of water is downwards through the soil where it effects leaching and redistribution of soil constituents. In a xeric regime, rainfall is lower and evapo-

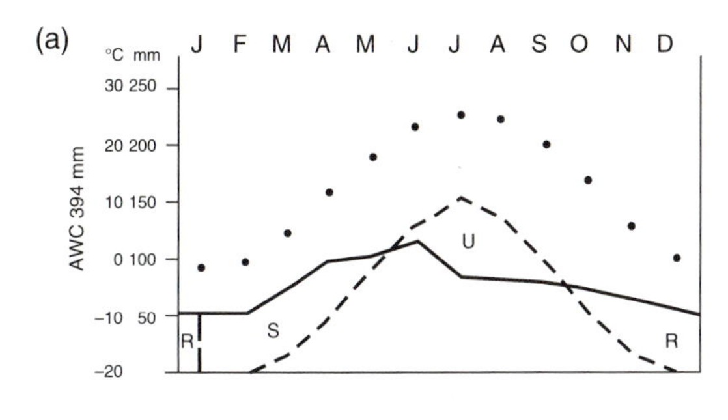

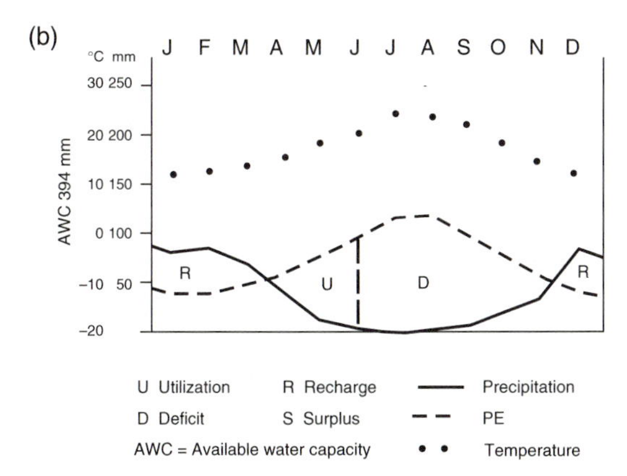

Figure 9.2 Climatic data and soil water balance for a soil (a) with a udic soil moisture regime and (b) with a xeric soil moisture regime (Soil Survey Staff, 1975)

transpiration higher, leading to a larger deficit which is only partly recharged during the winter. Movement of water downwards through the profile may occur during wet periods; some movement may occur upwards by capillary action and, on balance, the soil is not leached.

CLIMATE AND SOIL GEOGRAPHY

Reference to small-scale maps of world soils prepared in the first half of the twentieth century reveals that the compilers drew heavily upon theoretical concepts of what kind of soil should be present, given a particular climate and its associated vegetation. It is little surprising that there was close agreement between maps of climate, vegetation and soils compiled in this way. Subsequently, data gathered in soil surveys between the 1950s and 1980s indicated that, although the earlier assumptions were broadly correct, there were many deviations between the theoretical concepts and the actual soil pattern. This occurs because climate is only one of some five major factors responsible for soil formation; other important influences are the parent material, plant and animal organisms, relief and the time available for soil development (Jenny, 1941). Nevertheless, many correlations have been made between climate and the pattern of soil distribution (Landsberg and Blanc, 1958; Arkley, 1967). The differences are also partly the result of a paradigm shift in soil science which has increasingly come to rely upon soil attributes which can be measured, rather than on unquantifiable concepts of soil genesis for the identification of soil taxonomic units. Not only does the climate of any one place largely determine the natural plant life or crops which can be grown, it also is a dominant factor responsible for the nature of the soil. The two major components of temperature and precipitation, in conjunction with the fertility of the soil, control the productivity of the land. For the pedologist, temperature and moisture have a controlling influence on the chemical processes which are responsible for weathering minerals and the biochemical processes which decompose organic debris. Precipitation also affects losses, redistribution or the accumulation of plant nutrients, soluble salts and other mobile constituents of the soil (FAO, 1993).

Eight climatic zones have been distinguished by FAO on the basis of length of growing period, the incidence of rainfall and the length of frost-free season (Figure 9.3). These 'agro-ecological zones' are now commonly used by scientists involved with soils and agriculture at an international level. However, it must be realized that a zonation based on present-day climate does not necessarily reflect the

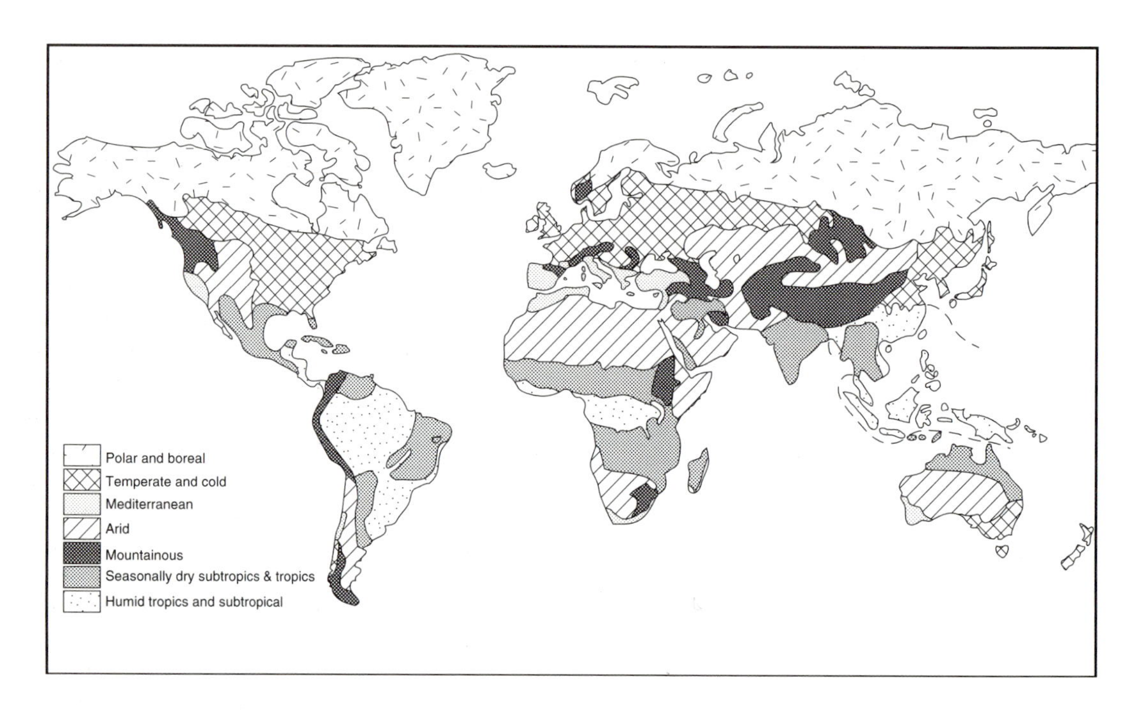

Figure 9.3 Soil ecological zones

conditions under which the soils were originally formed, as these may have been different in the past. Although attributes which are concerned with plant growth are implicit in some soil taxonomic groupings, usually they do not involve purely climatological data. The development of agro-ecological zones has been a result of this gap between the soil and atmospheric sciences. The agro-ecological zones are based on the length of growing period and temperature. The length of growing period (LPG) is defined as the number of days in the year when precipitation exceeds half the evapotranspiration plus a period of time sufficient to remove 100 mm of water from excess water stored in the soil and minus the period of days when the daily mean temperature is less than 6.5°C. The remaining period reflects the time when both water and temperature are available for crop growth. A period of less than 75 days is used to define an arid climate, between 75 and 270 days a seasonally dry climate and more than 270 days a humid climate. The frost-free period is defined as the number of days during the year when the average of the extreme minimum temperatures is above 2°C. Using this definition, the boreal and polar zones have a frost-free period of less than 75 days, the cold zone has a frost-free period of between 75 and 135 days and the temperate climate has more than 135 days with the average of the six warmest months below 25°C. Within the tropical regions, the seasonally dry subtropics are distinguished from the humid tropics by a dry season which extends from 90 and 285 days, whereas the humid tropics have favourable growing conditions all year round. Soil scientists increasingly are using agro-ecological zoning as a practical framework within which traditional soil classification is more meaningful.

SOILS AND THE GREENHOUSE GASES

Considerable energy is currently being exerted by the scientific community in ascertaining the amounts and rates of change of the so-called *greenhouse gases* in the atmosphere (see Chapter 22, Greenhouse warming). As the soil holds large reserves of carbon (globally about twice the amount held in the vegetation) and nitrogen, investigations are taking place to find the potential amounts of carbon dioxide, methane and nitrous oxide which might be released from soils with changing land use and climatic conditions (Bouwman, 1990; Scharpenseel *et al.*, 1990; Rounsevell and Loveland, 1994). From these investigations it has emerged that soils will change under the influence of climate, but because soils are an important factor influencing the amount of trace gases in the atmosphere, they have a reciprocal influence upon the climate (Bridges and Batjes, 1996). When the full impact of soil-emitted greenhouse gases has been evaluated, it may be possible for the global budget of these gases to be balanced.

As it has been realized that soils can affect the composition of the atmosphere, attempts have been made to assemble soil data into a digital form, compatible with modern data-handling equipment. The first attempt was that of Zobler (1986) who took the soil at the centre of each 1° rectangle of the terrestrial area to compile a soil file for global climatology. The International Geosphere–Biosphere Programme (IGBP) has also proposed to collect soil information for a global soils database. Staff of the International Soil Reference and Information Centre have compiled a digital soil database containing details of over 4000 soil profiles on a uniform basis (Batjes and Bridges, 1992; 1994), linked to the digital version of the FAO–UNESCO Soil Map of the World (FAO, 1971–81). Such databases are currently being used to provide a more satisfactory knowledge of the amounts of carbon in soils, distribution of methanogenic soils, and in future will provide a substantial input to global circulation models.

The actions of deforestation, biomass burning and cultivation cause the release of carbon dioxide from soils. Carbon has a natural cycle in the environment with various reservoirs and fluxes between them. Although the oceans and atmosphere are the largest carbon pools, land biota and soils also contain significant quantities: 550 and 1500 Gt of carbon (1Gt = 10^9 metric tons = 10^{12} kg). The content of these terrestrial pools can be changed fairly rapidly by human activities. Until the mid-twentieth century, loss of carbon from deforestation in temperate areas exceeded the amount of carbon released by burning fossil fuels. Since then, the fossil fuel contribution has increased and now stands at 5 Gt of carbon per annum. In 1980, tropical forest clearance resulted in the release of 1.6 Gt of carbon which in the past 10 years has increased to 2 Gt of carbon per annum (Houghton *et al.*, 1990; 1992).

Forest clearance in temperate regions contributed to increasing carbon dioxide (CO_2) concentrations in the atmosphere until the second or third decade of the twentieth century, after which emissions from this source declined. Cultivation of the soil increases the aeration, stimulating the activity of aerobic bacteria and causing a decline in the amount of organic carbon remaining in the soil (Figure 9.4); in some cases this leaves the soils with a weak structure and vulnerable to erosion. Control of groundwater levels in hydromorphic soils is seen as a way in which emissions of carbon dioxide may be minimized and other techniques are used to increase the content of carbon remaining in the soil. Sequestration of carbon by extensive tree planting has been proposed as a partial solution to the increasing atmospheric content of this gas (see Chapters 10 and 17).

Currently, tropical forest clearance and subsequent cropping are both responsible for reductions in soil organic carbon content, most of which is released in gaseous form.

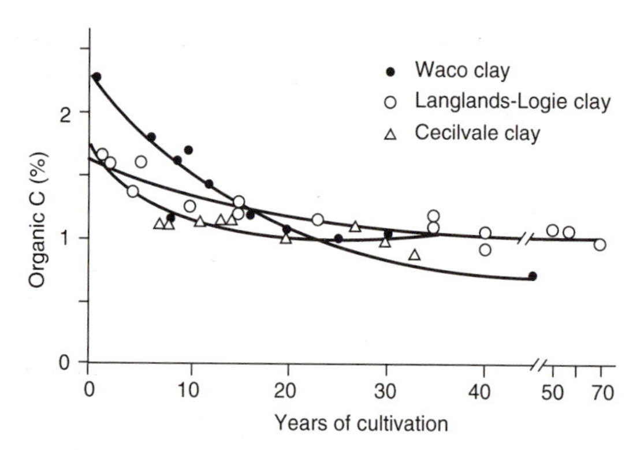

Figure 9.4 Loss of soil organic carbon from selected Australian soils over 70 years (Dalal and Mayer, 1986)

Replacement of natural forest in Kalimantan, Borneo, with cocoa plantations resulted in a fall of 48 per cent in the organic carbon content of the 0 to 10 cm layer of soils, despite a closed canopy and a reasonable litter fall. Aweto (1995), in a study in Nigeria, found similar results. With appropriate management, however, soil organic carbon levels also can be increased as has been the case with some old agricultural soils (Sandor and Easch, 1995). Houghton (1995) gives an indication of the global changes in total carbon content of the soils of major terrestrial ecosystems which suggests an overall reduction of 40×10^{15} g of carbon during the 130 years between 1850 and 1980. The main losses have been from temperate grassland and pastures, tropical seasonal forests and tropical evergreen forests.

Average emissions of methane (CH_4) from wetlands throughout the world amount to between a third and a quarter of the total amount emitted from all sources (560 Tg of CH_4 per year; Tg = teragram; 1 Tg = 10^{12} g). The greater part of this methane comes from emissions by ruminants (80 Tg of CH_4 per year), landfills (30–40 Tg of CH_4 per year), domestic sewage treatment and animal wastes (50 Tg of CH_4 per year), and leakages from natural gas and petroleum industries (100–115 Tg of CH_4 per year) (Houghton *et al.*, 1992; Lelieveld and Crützen, 1993; Adger and Brown, 1993). The residence period of methane in the atmosphere is estimated to be in the region of 10 to 12 years before it is broken down. In the troposphere, methane is oxidized by OH radicals. This results in the destruction of 90 per cent of the methane (455 Tg per year) which is converted to carbon dioxide (enhancing the concentration of carbon dioxide in the atmosphere) and water vapour. A further 10 Tg of CH_4 per year is lost through photochemical reactions in the stratosphere. The flux of methane from rice paddy fields is a significant contributor to atmospheric concentrations and one which is likely to increase as world population and demand for rice grows. Most of the methane escapes into the atmosphere through the stems of the rice plant, rather than directly from the soil by bubbling up through the water, as even a thin aerobic soil layer can cause the oxidation of most of the methane generated. The activity of methanogenic bacteria increases in response to anaerobic conditions when the fields are flooded so methanogenesis can be much reduced by draining the paddy field and then reflooding it during the course of crop growth. Control of emissions can also be achieved by appropriate manurial treatments. The use of different cultivars of rice can also play a part as some varieties are better conductors of oxygen from the atmosphere to the roots where it is released into the soil, counteracting the reducing effect of the flooded soil. In natural poorly drained soils of the subarctic regions, methane production is encouraged by cycles of rising and falling water table, methane being released as the water table falls (Bubier and Moore, 1994).

Nitrous oxide (N_2O) is a greenhouse gas with an atmospheric life of about 130 years. Currently, the mean concentration of N_2O in the atmosphere is about 310 ppbv (parts per billion by volume), which corresponds to a world total amount of 1500 Tg of N, increasing at a rate of

0.2–0.3 per cent per year. Emissions of N_2O are usually given in terms of the amount of nitrogen (N) released. Bouwman *et al.* (1995) estimated the amount of global N_2O emissions from natural terrestrial ecosystems to be 6.7 Tg of N per year; other estimates gave similar values (Khalil and Rasmussen, 1992; Houghton *et al.*, 1992). Emissions which are the direct result of the use of nitrogenous fertilizers in soils amount to between 1 and 2.5 Tg of N per year. Bowden (1986) gave global emission rates of 1.5–15.5 Tg of N per year for temperate hardwoods, 7.4–9.8 Tg of N per year for tropical forests, 4.8–48.0 Tg of N per year for prairie grasslands and 6.0–130 Tg of N per year for wetlands, figures which reflect the variability of measurements recorded. Environmental change, such as reduced rainfall or falling water table levels, could stimulate an increase in emissions by N_2O by 0.03–1.0 Tg of N per year, according to studies of Finnish peatlands (Martikainen *et al.*, 1993). Once in the atmosphere, the major loss of N_2O is by photochemical decomposition in the stratosphere. This loss is estimated to be ± 3.0 Tg of N per year, recently revised downwards to 3.0–4.5 Tg of N per year. Nitrous oxide has an atmospheric lifetime of about 130 years, so unlike methane, once the N_2O is in the atmosphere, it is there for a long period and changing its concentration cannot be effected easily.

Strategies to reduce N_2O emissions are necessary as a growing world population will inevitably lead to an increased demand for food, the production of which will lead to increased emissions of this gas (Janssen *et al.*, 1993; Bouwman, 1994). Applications of fertilizer must be closely matched to the crop requirements and timed to coincide with the needs of the crop. Losses from the soil through leaching must be minimized, particularly in periods when there is no crop on the ground. It is possible to place normal fertilizers in an optimal location for the roots of growing plants, and use of slow-release fertilizers can be encouraged. Animal and crop production systems should be closely integrated so reuse of manure and wastes can be achieved with less loss of nitrogen. Avoidance of anaerobic soil conditions is necessary as this encourages denitrifying bacteria, so control of the soil moisture regime is also critical.

Equally significant is the role of soils in absorption of these trace gases. Under aerobic conditions, some soils may act as a sink of atmospheric methane. However, the role of soils as potential sinks for methane is incompletely known (Mosier *et al.*, 1991). Bacteria capable of oxidizing methane occur in unsaturated soil horizons above the water table in freely drained soils and in the surface layer of rice paddies. Their source of methane is either from organic decomposition deeper in the soil or from low concentrations within the soil pores, developed by decomposition at restricted anaerobic sites. Some carbon is utilized by the bacteria for their growth. Born *et al.* (1990) found that soil absorption of methane is directly related to soil texture through its effect on gas diffusivity. Based on very few data, Schütz *et al.* (1990) estimated that soils absorb 23–56 Tg of CH_4 per year. According to Seiler (1984), global methane consumption in soils is about 20 Tg per year, a figure which was raised by Seiler and Conrad (1987) to 32 ± 16 Tg per year.

Carbon dioxide and some methane are also lost by percolation in water to the subsoil and greater depths; this temporarily reduces the amount which potentially could escape from the soil into the atmosphere, although it may escape later from springs. Although the surface layer of a flooded paddy soil is capable of oxidizing some methane as it is produced, losses directly from the soil are relatively small compared with the amount of methane which passes through the plant stems into the atmosphere. Absorption of N_2O has been observed in soils which are both warm and wet; this could have considerable implications for the global balance, but at the present time the quantities concerned have not been fully ascertained.

Estimates which are available suggest that the global 'sink' of N_2O in soils could be between 1.5 and 3 Tg per year, but further research is required on this topic (Bouwman, 1994).

EFFECT OF CLIMATE CHANGE ON SOILS

Assuming all the global circulation models (GCMs) are correct (see Chapter 4) and climate is changing, what effect will this have on present-day soils? Many soil properties are not in equilibrium with the present environment; some of these are subject to very slow rates of change but others are much more unstable and are liable to change rapidly. With increasing aridity, salt accumulation may occur in many soils currently leached of soluble salts. GCM predictions have suggested that temperature changes will be largest at high latitudes and that loss of organic matter from the organic-rich tundra soils could be rapid. With changing climate, there will be changes in vegetation and this too may change the direction of the soil-forming processes. Considerable areas are likely to be at risk of greater erosion, especially where the vegetation is under heavy grazing pressure (Bouwman, 1990; Rounsevell and Loveland, 1994; Scharpenseel *et al.*, 1990).

Climate is effectively the average of a very variable natural phenomenon and although soil scientists utilize average climatic figures, they are very conscious that occasional extreme events can be more effective than many years of normal climatic activity. One heavy rainstorm in a desert area may leach salts from the soil which remains relatively salt-free afterwards, especially if the water table is well below the soil. Drought conditions in the Sahel of West Africa in recent years have demonstrated how easily overgrazing can occur in fragile ecosystems resulting in desertification and loss of soil productivity as the topsoil is blown away. Even in areas where it was not formerly thought to be a problem, soil erosion is set to increase, such as in the southern UK where drier and warmer summer conditions are predicted (DoE, 1991).

Soil physical conditions also will change with changing climate. Climate changes predicted will affect the soil water balance and soil organic matter contents initially and, through these, the biodiversity of soil-living biota may be affected. Soil stability may well decline as organic matter decreases, leading to capping of the soil in arable lands. This will change the albedo, resulting in slower warming in springtime, and will also increase runoff and erosion. As has been seen in recent dry summers in the UK, clayey soils in some regions have shrunk causing structural damage to property. These deep cracks also may conduct water containing pollutants from the surface rapidly into the aquifers, short-circuiting the normal protective role of soils where contaminated water slowly percolates through and is cleaned in the process (DoE, 1991).

Many examples occur of soils which have developed under climatic conditions which are unlike those of today. Such soils are referred to as palaeosols (Yaalon, 1971). Although the upper horizons of the profile may have been changed in response to current soil-forming conditions, some aspects of their history can be seen in the lower horizons which have not yet been fully modified. In subsoils of many soils in Europe and North America, convoluted materials can be seen which resulted from periglacial activity during the closing phases of the Pleistocene. In other cases, examples of dense soil horizons (fragipans) occur which are brittle when moist, very hard when dry, but slake when placed in water. Formation of such horizons is attributed to repeated freeze–thaw conditions. In desert regions, soils with clay-enriched subsurface horizons occur. Conditions for the development of these soils do not currently exist and therefore these soils relate to an earlier period of a more humid climate. Other examples occur of pockets of soil, with features which indicate deep weathering under warmer conditions during one of the interglacial phases of the Pleistocene. In other cases buried soils indicate changes which have occurred during historical and late prehistorical time. One instance is the onset of podzolization in the soils of many sandy areas in western Europe which has been attributed to human intervention, but which may also be the result of increased rainfall during the Iron Age period of history.

CONCLUSIONS

Soils and climate are intimately interrelated, bringing together two of the most critical environmental parameters of life on earth. This chapter has attempted to show how climate influences soil formation and the effects climate change may have on soils, should the predictions of the GCMs come true. It has also demonstrated how climate can be affected by soils emitting radiatively active gases which are helping to reinforce the greenhouse effect. One significant point which emerges is that soils are dynamic entities which, being the product of multiple factors of soil formation, will themselves adjust to changing climatic conditions. Some of these changes will be rapid, others will take longer. The important message is that human beings must accommodate these changes in their land-use activities and capitalize on the advantages, such as the CO_2 fertilization effect (see Chapter 10, Climate and vegetation in the future) and warmer conditions. However they must realize that it may be necessary to take greater care to prevent soil erosion and to conserve moisture and organic matter, otherwise the current concepts of sustainable agriculture will be jeopardized.

REFERENCES

Adger, W.N. and Brown, K. (1993) 'A UK greenhouse gas inventory: on estimating anthropogenic and natural sources and sinks', *Ambio* 22: 509–17.

Arkley, R.J. (1967) 'Climates of some Great Soil Groups of the western United States', *Soil Science* 103: 389–400.

Aweto, A.O. (1995) 'Organic carbon diminution and estimates of carbon dioxide release from plantation soil', *The Environmentalist* 15: 10–15.

Batjes, N.H. and Bridges, E.M. (1992) *A Review of Soil Factors and Processes that Control Fluxes of Heat, Moisture and Greenhouse Gases*, Technical Paper No. 23, International Soil Reference and Information Centre, Wageningen.

Batjes, N.H. and Bridges, E.M, (1994) 'Potential emissions of radiatively active gases from soil to atmosphere with special reference to methane: development of a global database', *Journal of Geophysical Research* 99 D8: 16479–89.

Born, M., Dörr, H. and Levine, I. (1990) 'Methane consumption in aerated soils of the temperate zone', *Tellus* 42B: 2–8.

Bouwman, A.F. (ed.) (1990) *Soils and the Greenhouse Effect*, Chichester: John Wiley.

Bouwman, A.F. (1994) 'Compilation of a global inventory of emissions of nitrous oxide', unpublished PhD thesis, Agricultural University, Wageningen.

Bouwman, A.F., van der Hoek, K.W. and Olivier, J.G.J. (1995) 'Uncertainties in the global source distribution of nitrous oxide', *Journal of Geophysical Research* 100: 2785–800.

Bowden, W.A. (1986) 'Gaseous nitrogen emissions from undisturbed terrestrial ecosystems: an assessment of their impacts on local and global nitrogen budget', *Biogeochemistry* 2: 249–79.

Bridges, E.M. (1990) *Soil Horizon Designations*, Technical Paper 19, Wageningen: International Soil Reference and Information Centre.

Bridges, E.M. (1996) *World Soils*, 3rd edn, Cambridge: Cambridge University Press.

Bridges, E.M. and Batjes, N.H. (1996) 'Soil gaseous emissions and global climate change', *Geography* 81: 141–56.

Bubier, J.L. and Moore, T.R. (1994) 'An ecological perspective on methane emissions from northern wetlands', *Trends in Ecological Evolution* 9: 460–64.

Buol, S.W., Hole, F.D. and McCracken, R.J. (1989) *Soil Genesis and Classification*, Ames: Iowa State University Press.

Dalal, R.C. and Meyer, R.J. (1986) 'Long-term trends in fertility of soils under continuous cultivation and cereal cropping in southern Queensland; II. Total organic carbon and its rate of loss from the soil profile', *Australian Journal of Soil Research* 45: 281–92.

DoE (1991) *The Potential Effects of Climate Change in the United Kingdom*, London: HMSO.

FAO (1971–81) *Soil Map of the World, 1: 5,000,000*, Vols 1–10, Paris: UNESCO.

FAO (1993) *World Soil Resources: an explanatory note on the FAO World Soil Resources Map at 1: 25,000,000*, Rome: FAO.

Houghton, J.T., Jenkins, G.J. and Ephraums, J.J. (eds) (1990) *Climate Change, The IPCC Scientific Assessment*, Cambridge: Cambridge University Press.

Houghton, J.T., Callander, B.A. and Varney, S.K. (eds) (1992) *Climate Change 1992. The Supplementary Report to the IPCC Scientific Assessment*, Cambridge: Cambridge University Press.

Houghton, R.A. (1995) 'Changes in the storage of terrestrial carbon since 1850', in R. Lal, J. Kimble, E., Levine and Stewart, B.A. (eds) *Soils and Global Change*, Advances in Soil Science, Boca Raton FL: CRC Lewis Publishers, pp. 45–65.

Janssen, L.H.J.M., Gupta, J. and Weenink, B. (1993) 'Policy options for reducing the emissions of methane and nitrous oxides', in A.R. Van Amstel (ed.) *Methane and Nitrous Oxide*, Bilthoven: National Institute of Public Health and Environmental Protection, pp. 365–85.

Jenny, H. (1941) *Factors of Soil Formation*, New York: McGraw-Hill.

Khalil, M.A.K. and Rasmussen, R.A. (1992) 'The global sources of nitrous oxide', *Journal of Geophysical Research* 97: 14651–60.

Landsberg, H.E. and Blanc, M.L. (1958) 'Interaction of soil and weather', *Soil Science Society of America Proceedings* 22: 491–5.

Lelieveld, J. and Crützen, P.J. (1993) 'Methane emissions into the atmosphere, an overview', in A.R. Van Amstel (ed.) *Methane and Nitrous Oxide*, Bilthoven, National Institute of Public Health and Environmental Protection, pp. 17–25.

Martikainen, P., Nykänen, H., Crill, P. and Silvola, J. (1993) 'Effect of lowered water table on nitrous oxide fluxes from northern peatlands', *Nature* 366: 51–3.

Mosier, A.R., Schmiel, D.S., Valentine, D., Bronson, K. and Parton, W. (1991) 'Methane and nitrous oxide fluxes in native and fertilized grasslands', *Nature* 350: 330–2.

Rounsevell, M.D.A. and Lovelend, P.J. (eds) (1994) *Soil Responses to Climate Change*, Berlin: Springer-Verlag.

Russell, E.W. (1973) *Soil Conditions and Plant Growth*, 10th edn, London: Longman.

Sandor, J.A. and Easch, N.S. (1995) 'Ancient agricultural soils in the Andes of Southern Peru', *Soil Society of America Journal* 59: 170–9.

Scharpenseel, H.W., Schomaker, M. and Ayoub, A. (eds) (1990) *Soils on a Warmer Earth*, Developments in Soil Science 20, Amsterdam: Elsevier.

Schütz, H., Seiler, W. and Rennenberg, H. (1990) 'Soil and land use related sources and sinks of methane (CH_4) in the context of the global methane budget', in A.F. Bouwman (ed.) *Soils and the Greenhouse Effect*, Chichester: John Wiley, pp. 269–85.

Seiler, W. (1984) 'Contribution of biological processes to the global budget of CH_4 in the atmosphere', in M.J. Klug and C.A. Reddy (eds) *Current Perspectives in Microbial Ecology*, Washington, DC: American Society for Microbiology, pp. 468–77.

Seiler, W. and Conrad, R. (1987) 'Contribution of tropical ecosystems to the global budgets of trace gases, especially CH_4, CO, H_2 and N_2O', in R.E. Dickinson, (ed.) *Geophysiology of Amazonia, Vegetation and Climate Interactions*, New York: John Wiley, pp. 133–60.

Soil Survey Staff (1975) *Soil Taxonomy*, Agriculture Handbook No. 436, Washington DC: US Department of Agriculture.

Yaalon, D.H. (1971) *Palaeopedology*, Jerusalem: International Soil Science Society and Israel University Press.

Zobler, L. (1986) *A World Soil File for Global Climate*, NASA, Technical Memorandum 87–802.

10

VEGETATION

A.M. Mannion

INTRODUCTION: CLIMATE AND VEGETATION

There has always been a mutually reinforcing relationship between climate and vegetation. As part of the earth's biota, the other components of which are micro-organisms and animals, vegetation has influenced climate via its role in determining atmospheric composition. This it does by its mediation of the global biogeochemical cycle of carbon. Conversely, climate exerts a major control on the distribution of plant species and plant community composition through precipitation and temperature regimes. However, there are other factors which have affected this relationship, especially over the long-term geological past. These factors include plate tectonics and astronomical features which relate to the orbit of the earth around the sun.

Plate tectonics have influenced climate–vegetation relationships in two principal ways. First, the changes in the geography of the continents and oceans have affected the distributions of plants by shifting land masses into new climatic zones and, second, global climatic zones have themselves been affected by mountain building and the changing distributions of the world's oceans. Astronomical features, such as the Milankovitch forcing factors (see The Quaternary period, p. 128), have influenced the amount of solar radiation received at the earth's surface. Plate tectonics have had an impact over hundreds of millions of years and have influenced the climate–vegetation relationship not only directly, but also indirectly. Mountain building and continental subduction, for example, bring about changes in weathering and erosion which in turn influence earth surface processes, including the carbon biogeochemical cycle. Astronomical features, conversely, operate at periodicities of 96,000,000, 42,000, and 21,000 years. Although they have operated throughout earth history, the influence of these factors on global climates is particularly well documented for the last 3 million years when alternating cold and warm periods came to dominate global climates. Thus, in the spirit of the Gaia hypothesis (Lovelock, 1972), the earth and its biota have evolved together in a coupled mechanism. This relationship is likely to continue in the long-term future (Schwartzman *et al.*, 1994). The fact that another agent (i.e. human impact) has come to exert a profound effect on the earth's vegetation will not destroy this relationship but will merely alter its manifestation. The dominance of humankind became apparent about 10,000 years BP as specific plants and animals were domesticated and the first agricultural systems emerged. Ironically, it is possible that this turning point in natural and cultural environmental history was itself caused by the fluctuating climate that characterized the end of the last ice age (Mannion, 1995a; Moore and Hillman, 1992).

The degree of control exerted by humankind on the earth's surface increased gradually over the ensuing millennia. The clearance of natural ecosystems for agriculture, especially the deforestation of Eurasia's temperate zone and the Mediterranean zone, brought about changes in the carbon cycle as carbon dioxide was released from the biomass in soils and vegetation. However, the extent of this control has accelerated since the beginning of fossil fuel use on a large scale some 300 years ago. According to Hannah *et al.*'s (1994) preliminary inventory of human disturbance of world ecosystems, nearly 75 per cent of the earth's habitable land has been disturbed either severely or moderately. Europe is the most severely altered continent and only 15.6 per cent of its area remains undisturbed, 31.8 per cent is partially disturbed and 56.6 per cent is heavily disturbed. The resulting human-dominated environment bears little similarity to the ecosystems it replaced. Moreover, some 60 per cent of the overall decline in global forest and woodland cover has occurred in the last 300 years, much of which is due to the 466 per cent increase in cultivated land between 1700 and 1980, involving an area of about 12 $\times$ 106 km^2 (Meyer and Turner, 1992).

Whilst cultural factors (such as rapidly increasing populations supported by the Industrial Revolution and the migrations of people to the Americas and Australasia) underpinned these land-use changes, their impact has been global. Together with increased fossil fuel use and the expansion of pastureland, this massive change in land use has brought about such substantive changes in atmospheric composition that climate change is likely to ensue (IPCC, 1995). The situation is paradoxical. The reserves of ancient biospheres removed carbon from active circulation (and other substances such as sulphur, the release of which is now causing 'acid rain') which is now being rapidly released back into the atmosphere, with other heat-trapping gases, to alter atmospheric composition. The extent to which this is occurring is beginning to generate positive feedback in the climate system and cause global warming (see Climate and vegetation in the future, p. 135). The resulting patterns of temperature and precipitation regimes will impinge on modern-day ecosystems. Thus, the products of ancient biospheres will significantly affect the modern biosphere, altering both ecosystems (through species extinction and community composition and distribution) and agricultural systems, which are also constrained by climatic characteristics. Humankind will pay a price for its manipulation of the carbon cycle, albeit inadvertent, through agriculture and fossil fuel consumption.

VEGETATION COMMUNITIES IN THE GEOLOGICAL PAST

The relationship between climate and vegetation in the geological past is difficult to examine in detail because it relies on interpreting the fossil and palaeoenvironmental record. The precise nature of the many variables involved, not least of which is the role of life itself, cannot be determined with certainty, nor can their interrelationships be tested. The following exposition is based on an essentially Gaian approach.

The Pre-Quaternary Period

Figure 10.1 illustrates the relationship between atmospheric composition, notably oxygen and carbon dioxide concentrations, major changes in the earth's biota, periods of major ice formation and phases of the supercontinent cycle in relation to the geological timescale. The relationships are highly complex but the overall result has been to maintain temperatures at the earth's surface so that they fluctuate within only a limited range (with an average of 15°C) and one which is conducive to the persistence of life. This maintenance of temperature has been brought about through a number of mechan-

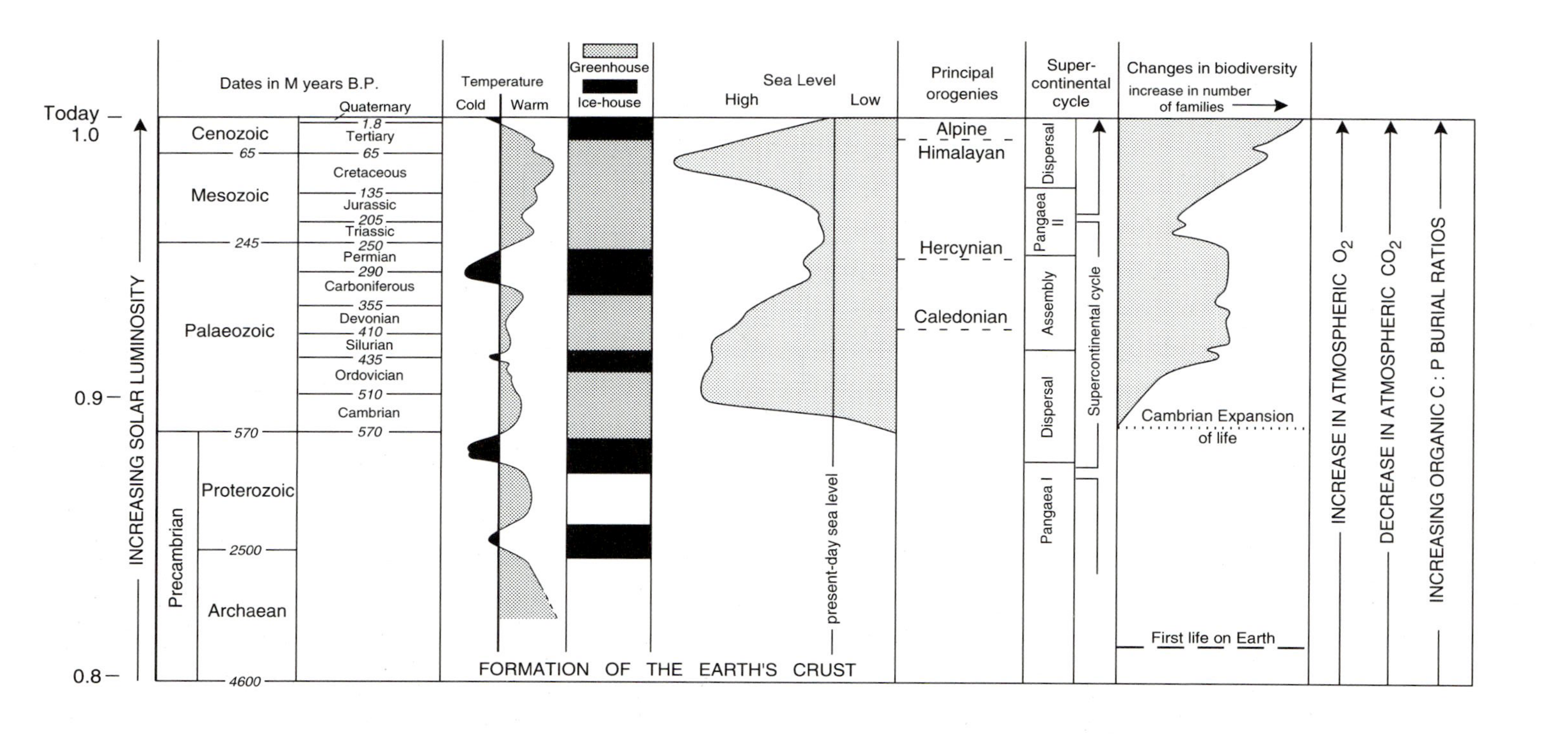

Figure 10.1 The major environmental changes that have occurred throughout earth history (based on sources quoted in the text)

isms involving global biogeochemical cycles which have adjusted to counteract the warming impact of the sun's increasing luminosity. One such mechanism is the silicate–carbonate buffering system. This prevents the build-up of volcanically produced carbon dioxide in the atmosphere. The system operates through the solution of carbon dioxide in water to produce carbonic acid. This in turn reacts with silicate rocks to produce limestone (calcium and magnesium carbonate) and quartz, thus sequestering the carbon and oxygen of carbon dioxide in sedimentary material. The chemistry involved can be summarized by the following equations:

$$CaSiO_3 + CO_2 \underset{}{\overset{\text{chemical weathering}}{\rightleftharpoons}} CaCO_3 + SiO_2$$

(calcium silicate and carbon dioxide) (calcium carbonate and quartz)

$$MgSiO_3 + CO_2 \rightleftharpoons MgCO_3 + SiO_2$$

(magnesium silicate and carbon dioxide) (magnesium carbonate and quartz)

According to Worsley *et al.* (1991), this process causes some 80 per cent of the volcanically produced carbon dioxide to be removed from active circulation. The remaining 20 per cent is removed in a second important buffering mechanism which involves organisms that can photosynthesize (Schwartzman and Volk, 1989). The carbon dioxide is assimilated into organic matter, as represented by the following equation:

$$6CO_2 + 6H_2O \xrightarrow[\text{chlorophyll}]{\text{light energy}} C_6H_{12}O_6 + 6O_2$$

(carbon dioxide and water) chlorophyll (carbohydrate and oxygen)

A proportion of the organic matter so produced will be returned to the atmosphere via respiration but the remainder will be incarcerated in soils, sediments and peats which act as stores of carbon within the lithosphere and biosphere. Moreover, these mechanisms are linked through other biogeochemical components, namely phosphorus and oxygen, and both are involved in the burial of organic matter (Worsley and Nance, 1989). For example, phosphorus is important in the burial of organically fixed carbon (see the equation below) but it can also be a limiting factor to plant growth which 'fixes' the carbon dioxide from the atmosphere and produces oxygen in the process. Adding to the complexity of these relationships is the fact that the availability of phosphorus is itself related to silicate weathering (see the first two equations above):

$$P + {}_n(H_2O + CO_2) \rightarrow P(CH_2O)_n + {}_nO_2$$

(phosphorus, water and carbon dioxide)

where $n = 1 \times 10^2$ to 1×10^3.

Worsley *et al.* (1991) have suggested that the maintenance of a near-constant temperature at the earth's surface is primarily a function of organic carbon removal because both the magnitude of this component of atmospheric carbon dioxide regulation and the mechanisms involved have varied over geological time. Conversely, those involved with the silicate–carbonate system have remained more or less constant. This flexibility is a function of the fact that one phosphorus atom can bury (i.e. remove from circulation) between 100 and 1000 carbon atoms depending on the nature of the burial environment. In addition, the evolution of organisms with higher carbon:phosphorus ratios in their organic matter (e.g. vascular plants), and increased phosphorus-recycling mechanisms in marine algae, probably mediated this increased burial of carbon dioxide. The consequent reduction in carbon dioxide concentrations in the atmosphere would have reduced warming due to the greenhouse effect and counteracted the warming trend caused by the sun's increasing luminosity (Figure 10.1).

The amount of oxygen in the atmosphere is also of prime importance in the relationship between surface temperatures and atmospheric carbon dioxide concentrations. As Figure 10.1 shows, the concentration of oxygen in the atmosphere has increased over geological time. This is also considered to relate directly to the burial of organically fixed carbon dioxide (Worsley *et al.*, 1991). According to Kump (1989), the silicate–carbonate system actually consumes oxygen yet over the past 3000 million years, the amount of oxygen in the atmosphere has increased a thousandfold. Concentrations similar to those of the present day, that is 21 per cent, were attained 250 million years BP (Berner (1987) quoted in Worsley *et al.* (1991)). As the equation above shows, oxygen is produced when organic carbon burial is mediated by phosphorus. Furthermore, the evolution of organisms capable of photosynthesis helped to transform the earth's reducing atmosphere into an oxidizing milieu through the addition of free oxygen produced by photosynthesis (see the third equation above). Such organisms include the Cyanophyta, that is blue–green algae. Subsequently, other forms of life evolved; multicellular organisms evolved and increased in numbers during the Cambrian era (Figure 10.1) for example and, during the Devonian period, vascular plants underwent major expansion about 410 million years BP. The development of an atmosphere containing free oxygen also paved the way for the evolution of the mammals, including humankind. Of particular significance for the global biogeochemical cycle of carbon is the Carboniferous period when vast tracts of forests developed, comprising early vascular plants (e.g. lycopsids, sphenopsids and pteridophytes). These were equatorial zone coal–swamp floras, the remains of which accumulated as land sank creating extensive basins. The deposition of sediments over these organic remains caused them to become compressed into coal, oil and natural gas. As is discussed below, the large-scale use of these fossil fuels is now giving much cause for concern about global climates in the future.

The maintenance of a near-constant surface temperature through the reduction of atmospheric carbon dioxide concentrations has not, however, occurred in a linear fashion through time. As stated in the introduction, there have been other factors at work to cause deviation from the average trend. Figure 10.1 includes reference to tectonic episodes involving the assembly and disintegration of supercontinents. Examples of supercontinents are Gondwanaland, which consisted of modern-day South America, Africa, Antarctica, Australia and India; Euramerica, which comprised modern-day North America and northern Europe, and Siberia (Cox and Moore, 1993). Throughout the last 3000 million years there have been periods when such supercontinents were in existence and other intervals when they have disintegrated. Not only have these tectonic changes involved periods of mountain building but they have influenced sea-level changes. Worsley *et al.* (1991) state that the supercontinent cycle gives rise to various degrees of continental emergence or submergence. This in turn influences the rate of organic carbon burial. Emergent continents, for example, during periods of falling sea level which are characterized by the presence of supercontinents, will enhance the availability of calcium, magnesium and phosphorus. This occurs because of increased chemical weathering and leads to increased carbon burial (see the equations above). Global temperatures will thus decline and concentrations of atmospheric oxygen will increase. Both of these cause what Worsley *et al.* (1991) refer to as 'biotic innovations', which may also be stimulated by new niches created by continental emergence. This enhances the rate of carbon removal through photosynthesis and phosphorus recycling, so reinforcing the existing trend towards lower temperatures. If there are continental land masses in polar positions and if the earth is at its most distant from the sun (Milankovitch's orbital eccentricity cycle; see the next section), then ice-sheet growth may occur and further amplify global cooling by removing water vapour from the atmosphere (see Chapter 7, Conclusions). At various times in earth history, such conditions clearly existed; indeed, this is the present condition of the earth which is in a mild phase typical of interglacial stages of the Quaternary period (see Climate and vegetation today, p. 133).

When the continents became drowned, that is when splitting of the supercontinents occurred, conditions opposite to those described above developed. The earth experienced a warm, or greenhouse, phase. Such conditions occurred during the Cretaceous period which began about 135 million years ago (Figure 10.1). Worsley *et al.* (1991) stated that the average temperature of the earth would be about 10°C higher during a greenhouse phase than during an ice age or 'icehouse' phase. Indeed, there is abundant evidence to show that global temperatures during the Cretaceous period were higher than those of today, for example Spicer *et al.*'s (1993) review of Cretaceous phytogeography. This was a period of angiosperm (flowering plant) diversification (Crane *et al.*, 1995) and increased radiation. As a consequence, the amount of photosynthesis occurring on the earth's surface increased significantly so accelerating the removal of carbon dioxide from the atmosphere. The onset of the Tertiary period was appreciably cooler than the Cretaceous period. Polar ice sheets began to form and seasonality of climate increased considerably by the end of the Oligocene (Figure 10.1).

The Quaternary Period

The Quaternary period began 1.8 million years BP and although it is considered to be synonymous with the ice ages, global cooling and major ice-sheet formation actually began about 3.5 million years BP in the Pliocene epoch. More than half a century ago, Milutin Milankovitch identified various periodicities from his calculations relating to the earth's orbit around the sun (Figure 10.2). In the 1930s, however, there was no possibility of testing Milankovitch's hypothesis through field observations so it remained in obscurity until the 1970s, by which time much research had been undertaken on ocean sediments. Hays *et al.* (1976) recognized the three main cycles of Milankovitch in the oxygen–isotope records of foraminifera in ocean sediments. A synthesis of North Atlantic ocean-core data by Ruddiman and Raymo (1988) has shown that the last 3 to 5

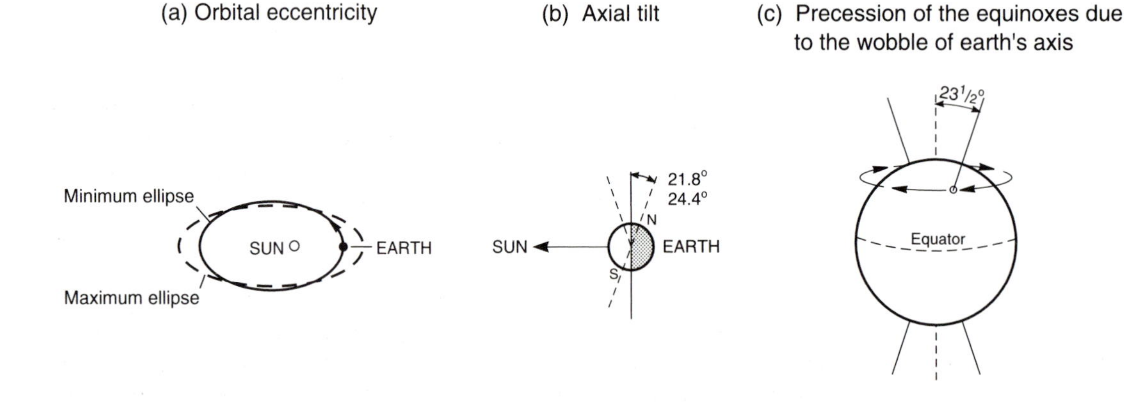

Figure 10.2 The astronomical forcing factors involved in the Milankovitch theory of climate change

million years have been characterized by three distinct climatic regimes. Between 3.5 million years BP and 2.47 million years BP, ice sheets in the northern hemisphere were small, and there were only small-scale climatic oscillations. Between 2.47 million years BP and 0.734 million years BP, rhythmic climatic changes developed and were controlled by the 41,000 year cycle of orbital obliquity (Figure 10.2). However, since 0.734 million years BP, the 100,000 year orbital eccentricity factor has been dominant and caused the onset of intensified glaciation. In line with the views of Worsley *et al.* (1991) discussed above, Raymo and colleagues (reviewed in Raymo, 1994) have suggested that tectonic uplift, notably of the Himalayas, combined with increased weathering could have been responsible for the onset of glaciation by changing atmospheric circulation and removing carbon dioxide from the atmosphere. It is now accepted that during the last 0.7 million years or so there have been numerous phases of ice advance separated by warm intervals known as interglacials. Currently, the earth is in an interglacial phase. Although the change in orbital eccentricity is considered to drive the glacial–interglacial cycle, the reduced amount of solar energy received when the earth is furthest from the sun is unlikely to cause a 10°C drop in global temperatures as the earth enters a glacial phase. Yet this is the difference in temperature which is indicated by fossil faunal assemblages.

Other factors must, therefore, contribute to global cooling and warming either as forcing or reinforcing factors. The heat-trapping gases, notably carbon dioxide and methane, are important in this context. As components of the carbon cycle, changes in the atmospheric concentrations of these gases must be related to carbon storage on land and in the oceans which, inevitably, involves vegetation. Variations in carbon storage during the Quaternary period were first suggested by Shackleton (1977) on the basis of carbon-isotope data from ocean sediments. The trends have since been confirmed by analyses of air bubbles encapsulated in ice cores from the Arctic and Antarctic (see Chapter 7, The conversion of snow to glacier ice and the role of climate). For example, data from the Vostok ice core relating to the transition between the last glacial phase and the current interglacial (the Holocene) show a rise in carbon dioxide from about 190 ppm to 270 ppm (Barnola *et al.*, 1987) and a doubling of methane concentrations (Raynaud *et al.*, 1988); the latter is also registered in the GRIP core from Greenland (Chappellaz *et al.*, 1993). Such changes in the concentrations of heat-trapping gases in the atmosphere would have reinforced global cooling caused by orbital eccentricity. What remains enigmatic, however, is the source and the mechanisms of transfer for this movement of carbon. The removal of carbon (as carbon dioxide and methane) from the atmosphere means that it must be transferred into other storage components of the global carbon cycle. Is it stored in the oceans, on the land, or both? The possibilities have been reviewed in Mannion (1997) and polarize between the oceans, invoking increased phytoplankton (algal) activity, and increased biological productivity in low latitudes on land which was exposed owing to lowered sea levels. Moreover, nutrient availability, oceanic circulation patterns and oceanic heat transfer are likely to be involved in this removal and release of carbon.

There is abundant evidence for environmental change during the Quaternary period. As well as evidence from ocean sediments and ice cores, the deep loess sequences of China (see Chapter 8, Climate change and landforms) and continental palaeobotanical records (notably pollen analytical data) have revealed the interplay between climatic and vegetation changes. The growth and decline of continental ice sheets were paralleled by expanding and contracting vegetation belts, both altitudinally and latitudinally. Altitudinal shifts in vegetation are well exemplified by pollen data from two sedimentary sequences from Funza, near

Bogotá in the Eastern Cordillera (Andes) of Colombia (Hooghiemstra *et al.*, 1993; Hooghiemstra and Ran, 1994). A schematic representation of the vegetation changes that occurred in this region during the Quaternary period is given in Figure 10.3. Some 55 pollen assemblage zones have been identified and related to the well-established oxygen-isotope stratigraphy of ocean sediments. At least 27 climate cycles have been recognized and are reflected in changes in the height of the tree line. During warm (interglacial) phases, forest extended to an altitude of about 3000 m in this Andean region with perennial snow present only above 4500 m. Conversely, during cold (glacial) phases the vegetation belts were depressed; for example, the tree line was some 1000 m lower at 2000 m and perennial snow was present above 3500 m. Between the tree line and snow line, subpáramo, grass páramo and superpáramo (all types of grasslands, with herbs and shrubs) occurred and oscillated altitudinally in tandem with climatic change. The implication of the Funza pollen analytical data is that there was a 9°C temperature difference within a glacial–interglacial cycle at the altitude at which Funza is located, that is 2500 m above sea level. Average temperatures were thus depressed to about 6°C during cold phases and elevated to 15°C during warm phases. Changes of a similar order of magnitude are indicated by data from the Huascarán ice core from Andean Peru (Thompson *et al.*, 1995).

Several other long pollen sequences from middle latitudes reflect corresponding and synchronous changes in vegetation communities, for example Japan (Fuji, 1988), Italy (Follieri *et al.*, 1988) and Israel (Horowitz, 1989). The record from Tenaghi-Philippon in Greece (Mommersteeg *et al.*, 1995), for example, extends back 1 million years whilst that of Ioannina (Tzedakis, 1994), also in Greece, provides a record of vegetation history for the last 430,000 years. This represents stages 1–11 of the marine oxygen-isotope record (Figure 10.3) which includes five interglacials and four cold stages. In terms of vegetation communities, the interglacials were characterized by forests composed of oak, elm, hornbeam, fir, beech and hop hornbeam whilst the cold stages were characterized by forest–steppe and desert–steppe communities. The pollen analytical data from the Jordan–Dead Sea rift valley (Horowitz, 1989) also reflect oscillations of vegetation communities with evergreen oak or deciduous oak forest, sometimes with pine, dominating the glacial phases and desert–steppe communities (i.e. xeric vegetation) dominating the interglacial phases. The relative scarcity of such long sequences, and of continental deposits containing palaeobotanical evidence for Pre-Holocene interglacials, means that it is not yet possible to reconstruct global vegetation communities (i.e. biomes) for these periods.

However, attempts have been made to reconstruct the world's vegetation at the last glacial maximum, about 18,000 years BP. Much of this research has been undertaken in order to understand better the dynamics of the global carbon cycle (see above). This is particularly important because of the reciprocal relationships between the various pools (stores) of carbon in the lithosphere, biosphere and atmosphere (and the interlinking fluxes) and present-day climate. Where the sinks and sources of carbon are located, and which factors influence the flux rates between them, are of major significance for identifying the impact on ecosystems and agricultural systems of likely global warming (see Climate and vegetation in the future). One such attempt to reconstruct vegetation patterns is that of Prentice *et al.* (1993). The results are given in Figure 10.4, which illustrates the projected differences between global vegetation communities 18,000 years ago, at the last glacial maximum (LGM), and those of today. Of particular note are the differences in high and mid to high latitudes where polar desert, polar semi-desert and cool grass–shrublands were much more

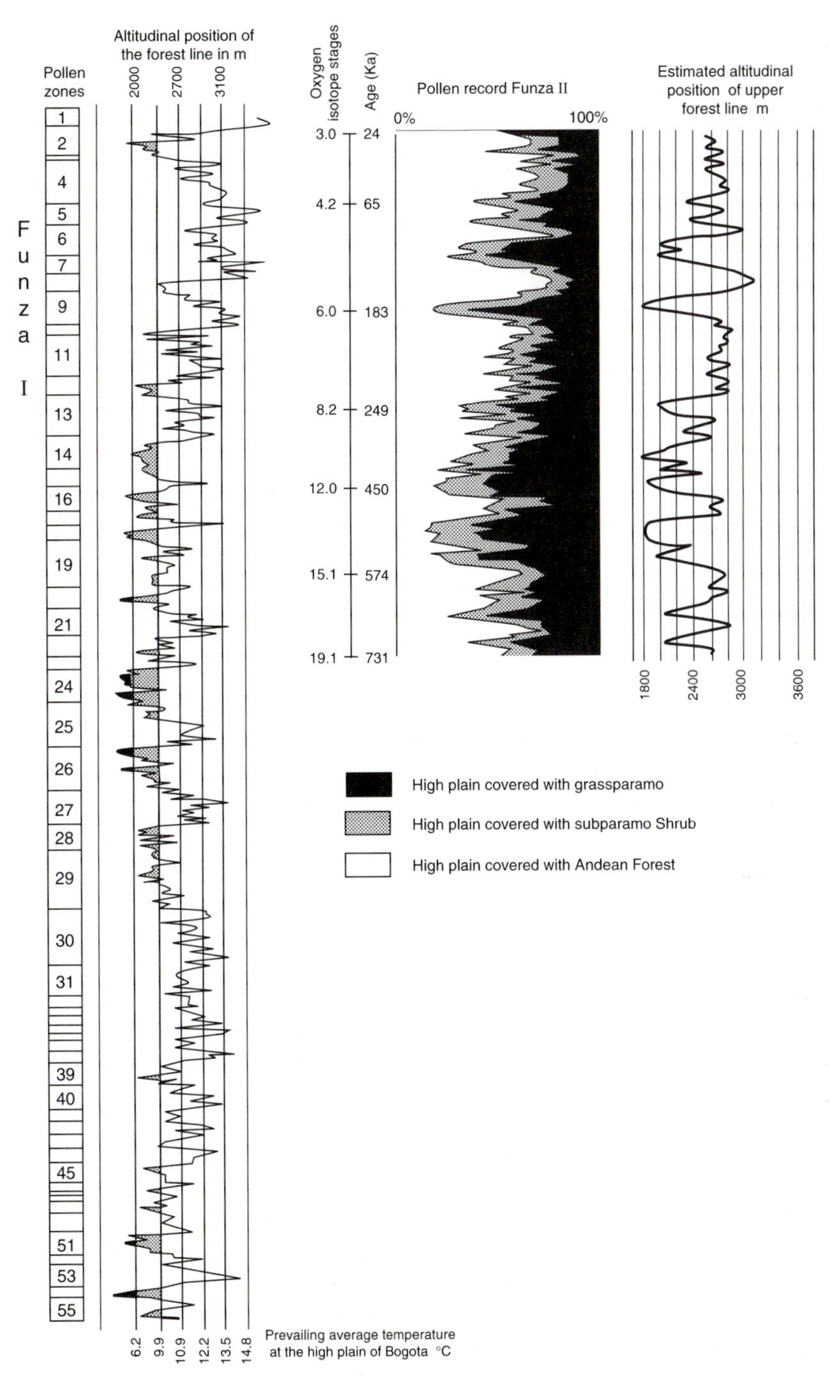

Figure 10.3 The relationship between climate change, as represented by marine oxygen-isotope stages, and changes in the altitudinal position of the tree line in the High Plain of Bogotá in the eastern Cordillera of Colombia, deduced from pollen cores Funza I and II (after Hooghiemstra, 1989; Hooghiemstra and Ran, 1994)

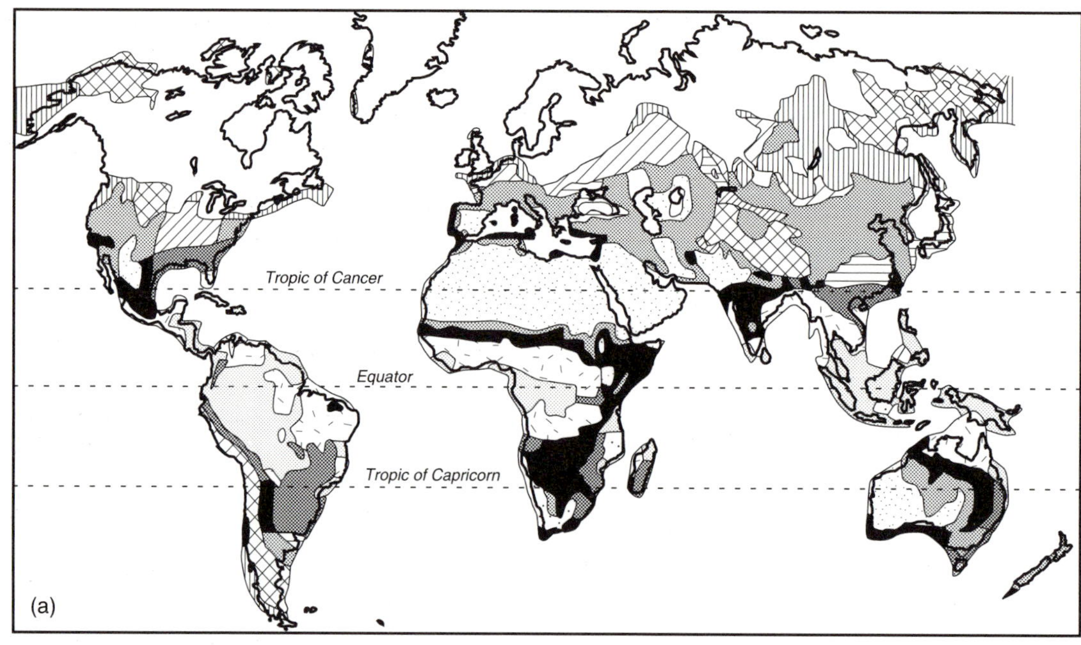
Tropic of Cancer
Equator
Tropic of Capricorn
(a)

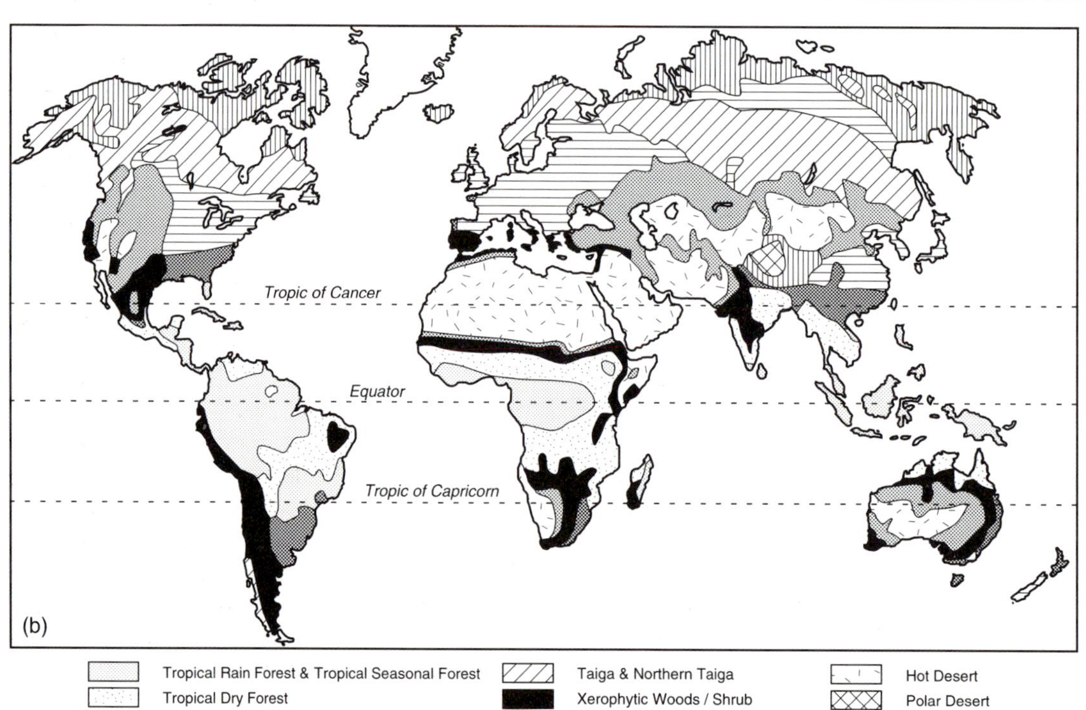
Tropic of Cancer
Equator
Tropic of Capricorn
(b)
Tropical Rain Forest & Tropical Seasonal Forest
Tropical Dry Forest
Evergreen / Warm Mixed Forest
Temperate Deciduous, Cool Mixed, Cool Conifer, Cold Mixed, Cold Deciduous Forest
Taiga & Northern Taiga
Xerophytic Woods / Shrub
Warm & Cool Grass / Shrub
Wooded Tundra & Tundra
Hot Desert
Polar Desert

extensive at the LGM than they are now. Conversely, taiga, temperate and cool/cold mixed forests were less extensive, as was the tundra zone. All of these are important stores of carbon. In addition, tropical rain forest was distributed differently, being reduced in extent and in some areas fragmented, in continental Africa and Amazonia. However, it occupied land in the tropics, especially in southeast Asia, which has since been inundated by rising sea levels consequent on ice-sheet melting that occurred as the last ice age ended and the Holocene began. Prentice *et al.*'s (1993) calculations in relation to changes in carbon storage suggest that after the LGM, terrestrial carbon storage increased by some 300–700 Pg. The implication of this, and the results of another modelling exercise by Adams *et al.* (1990), which gives a value of 1350 Pg, is that the carbon is sequestered in the oceans during ice ages when the amount stored on the continents is considerably reduced.

Two issues arise from this work. First, there are considerable differences in estimates for the amount of increased carbon storage that occurred on the continents as the last ice age ended. The available estimates have been reviewed by Crowley (1995) who points out that the descriptions reflect inadequacies in the models used to generate the values and an imperfect understanding of the fluxes that operate within the global carbon cycle. The latter is of particular concern because it means that prediction of the likely impact of global warming on biological systems is sufficiently inaccurate to be of little assistance for planning purposes. Moreover, there is little evidence to demonstrate that the vast quantity of carbon removed from the atmosphere during glacial phases was in fact incarcerated in ocean waters or sediments as discussed in Mannion (1997). One possibility, however, is that of enhanced marine productivity, stimulated by an increase in windborne nutrients as aeolian erosion removed sediment and soil particles from the continents (Martin, 1990). Given that the global carbon cycle is so important in all aspects of biospheric operation, it is disconcerting that its mechanisms are so poorly understood.

CLIMATE AND VEGETATION TODAY

Figure 10.4(b) illustrates the distribution of the world's major ecosystems today and Figure 10.5 gives the world's major climatic zones. The high degree of correspondence between the two is a prima facie case for a close, causal relationship. In the previous sections, however, the nature of this relationship has been considered only in general terms. The inferences made above about the relationship between vegetation and climate rely on what is known to occur at the present time. Most importantly, annual regimes of temperature and precipitation primarily determine the range of plant species that can grow and persist in any given climatic zone, because of the effect that these factors have on plant physiology. There are, however, other factors that influence the distribution and composition of vegetation communities. Such factors include competition between species, which is not readily quantifiable, and herbivory. A further factor is the presence of biogeographic barriers such as mountain ranges and sea barriers which limit the spread of plant species. Moreover, climate influences vegetation communities through its effects on bedrock, that is rates of weathering, and soils (Chapter 9). In particular, temperature and precipitation regimes influence nutrient availability and biogeochemical or nutrient cycling.

Figure 10.4 (a) Simulated biome distribution for the last glacial maximum (after Pentice *et al.*, 1993); (b) Simulated biome distribution from modern climatology (after Pentice *et al.*, 1993)

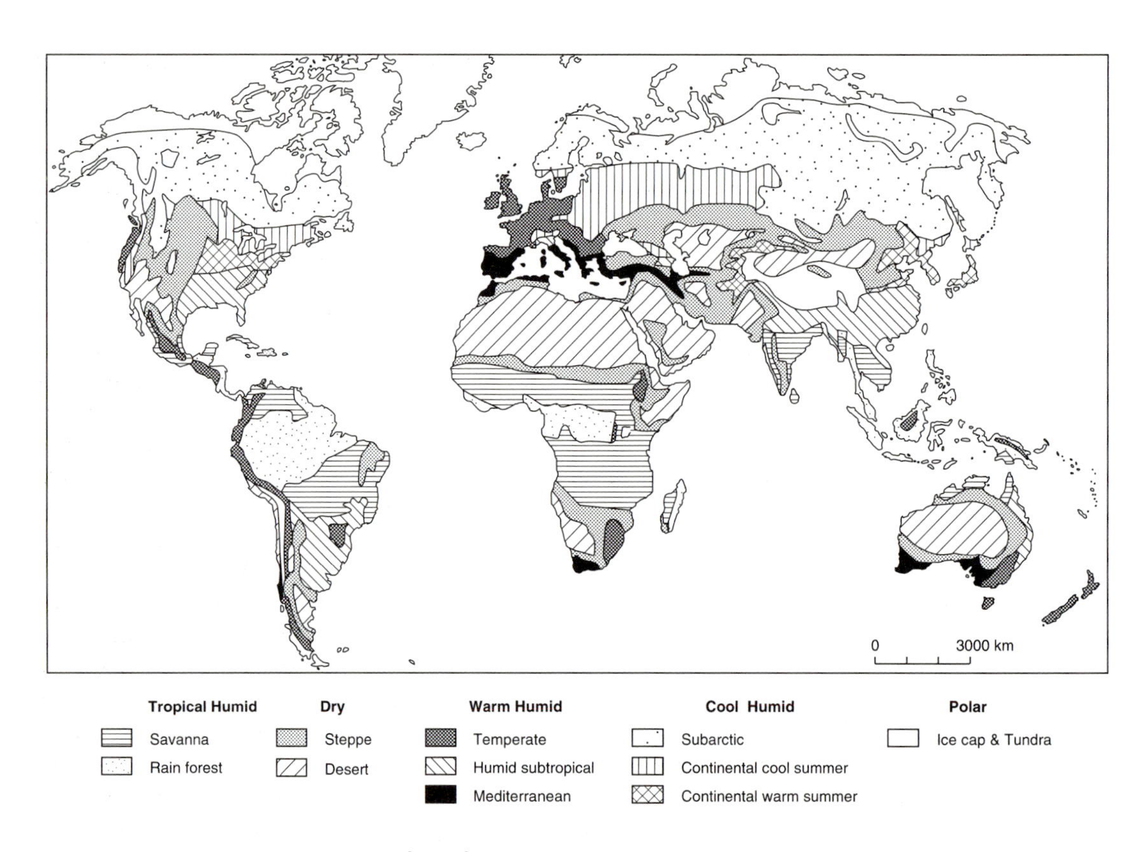

Figure 10.5 World climatic zones (after *The Times*, 1995)

The relationship between plant species and climate is complex and individualistic. It is important to recognize, for example, that when climatic change occurs, individual species respond, rather than entire plant communities. Consequently, plant communities can be considered to be relatively transient or ephemeral. Each species has a particular tolerance to the environmental factors that surround and sustain it and those species with similar tolerances will group together to produce plant formations. These are shown in Figure 10.4 and represent the major biomes of the earth's surface. When climate change occurs, the character of the biomes alters as the species composition alters. The relationship between climate and modern vegetation, which was established a century ago, is reflected in the classic work of Köppen and Holdridge (Archibold, 1995). As well as affecting the actual character of a given biome (in terms of the species it contains) climate also influences the amount of carbon storage. This, in turn, is influenced by management practices and land use which represent ways in which vegetation communities and agricultural systems are manipulated by humankind. For example, the transposition of natural vegetation communities into agricultural systems generally results in diminished carbon storage, especially where woodland or forest was originally present. In addition, climate is the overriding control on plant diversity through its imposition of chilling, drought stress due to frost or aridity and the length of the growing season

(Woodward, 1987). All these factors constrain the amount of carbon that can be 'fixed' via photosynthesis (see the third equation above) through the imposition of limiting factors relating to energy and/or water availability. These factors also influence the distribution and operation of 'plants' in the world's oceans. These are the algae – or phytoplankton – which underpin marine food chains, though nutrient availability is also a major constraint on carbon fixation (the second section above discusses the significance of phytoplankton in the regulation of global climates).

Green plants and algae, because they photosynthesize, are primary producers. This means that, unlike most organisms, they have the ability to trap solar energy through the binding together of carbon dioxide and water (see the third equation). In so doing, they form the basis of food chains and food webs and provide a primary energy source for all those other components of the earth's biota which are unable to produce a source of energy. However, not all of the carbon fixed in photosynthesis enters food chains and webs from whence it is returned eventually to the atmosphere. A proportion is stored in the biomass that characterizes all plant communities. This storage may be in organic matter in the soil or in the living above and below ground components of plants. The largest amount of carbon as primary productivity is fixed and stored in tropical forests (Leith, 1975). Currently, the amount of carbon fixed within the biosphere annually as net primary productivity is in the order of 73.2 Gt (based on Leith, 1975). It is not possible to ascertain what this value was 5000 to 7000 years BP, when Holocene forests were at their maximum extent prior to the widespread initiation of agriculture (Mannion, 1995a). However, it must have been considerably higher in some regions, though possibly not Europe (Peng *et al.*, 1995), if the figures quoted in the introduction for human-induced ecosystem alteration are accurate up to ±20 per cent (Hannah *et al.*, 1994; Houghton, 1994). This means that an important sink for carbon dioxide, that is the living biota, has diminished substantially. Furthermore, in view of the continued increase in world population and the concomitant requirement for increased agricultural productivity and agricultural land, it is likely to decline further. This is despite the fact that there is some evidence for increases in productivity due to the fertilizing effect of the increased carbon dioxide content of the atmosphere (Kohlmaier *et al.*, 1991). In addition, the impoverishment of existing vegetation communities through logging, mineral extraction, overgrazing, etc., may impair carbon fixation and storage by causing a reduction in biodiversity. Naeem *et al.* (1994) have undertaken experiments in an Ecotron, comprising a series of chambers in which growth conditions can be controlled, and have demonstrated that high biodiversity gives rise to enhanced carbon assimilation. These results echo those of Tilman *et al.* (1994) whose work on US prairies indicates that communities with high biodiversity are more resilient to environmental stress, such as drought, than those characterized by low biodiversity. Thus, the maintenance of biodiversity is important. This is not only because it is significant for the regulation of the global carbon cycle, and hence the greenhouse effect, but also because it has the potential to generate other benefits and profits through biotechnology (Mannion, 1995b, c).

CLIMATE AND VEGETATION IN THE FUTURE

The recent report of the Intergovernmental Panel on Climate Change (IPCC, 1995) has acknowledged that anthropogenic carbon dioxide in the atmosphere (see Chapter 22, Greenhouse warming) is contributing to greenhouse gas radiation forcing to the order of 1.56 W m^{-2}. Increases in other heat-trapping gases as a result of human activity are also contributing to radiative forcing but not at the same intensity as carbon dioxide. For example, methane contributes about 0.5 W m^{-2} to radiative forcing, nitrous oxide contributes 0.1 W m^{-2}, halocarbons (e.g. CFCs) contribute 0.3 W m^{-2} and HCFCs and HFCs (alternatives for halocarbons) contribute 0.05 W m^{-2}. In relation to carbon dioxide, this represents an increase in concentration from about 280 ppmv in pre-industrial times (around 1800) to 356 ppmv in 1993. The rate of annual increase is currently 1.5 ppmv (Table 22.1) and is due to three factors: fossil fuel consumption, cement production and land-cover change. This means that 3.2 ± 0.2 Gt of carbon are added to the atmosphere annually in the form of carbon dioxide. Whilst some of this warming is offset by the cooling effect of aerosols (IPCC, 1995), the net effect is positive. Warming trends have also been recognized from other independent sources of evidence, especially in the northern hemisphere. Examples include tree-ring sequences (Briffa *et al.*, 1995), historical data (Pfister, 1992) and sea surface temperatures (Jones, 1995). Overall, the IPCC (1995) reports a global mean warming of 0.45 ± 0.15°C over the past 150 years. Although there are drawbacks to predicting the impact of this and likely future warming because of inadequacies in models, there is no doubt that emissions of heat-trapping gases are continuing to rise. In China alone, coal production has increased from 872×10^6 t in 1985 to 1080×10^6 t in 1990 and whilst the significance of coal as an energy source has actually decreased to about 70 per cent, the deficit is compensated for through increases in oil and natural gas production (Zhao, 1994).

The rate at which heat-trapping gases are emitted as industrialization proceeds, despite reductions by some nations whose energy efficiency is increasing and consumption declining, is only one factor that makes the prediction of warming difficult. Other factors include possible lagged effects within the carbon cycle which may accelerate or mitigate the warming trend. These may involve extant vegetation communities and primary producers in the oceans. Moreover, global warming will affect other biospheric processes which influence the carbon cycle, such as the hydrological cycle. Increased drought stress globally will reduce carbon fixation via photosynthesis, as will the diminished availability of nitrogen. There are also a number of possibilities that may cause the biosphere to act as a reinforcing factor in global warming. In particular, the oxidation of organic matter in soils and/or peats as they dry out will contribute carbon to the atmosphere. To these factors must be added carbon generated from the likely continued diminution of natural vegetation communities, especially tropical forests. These are responsible for 28 per cent of the carbon fixed annually as net productivity in the biosphere, that is 20.4 Gt (Leith, 1975), yet it is not even possible to obtain accurate estimates of rates of deforestation. Grainger (1993), for example, quotes annual rates of between 6 and 20 Mha. Less dramatically, but of no less significance, is the continued extinction of plant species through habitat fragmentation, habitat impairment and inadequate management. As stated in the previous section, this is also likely to reduce the flux of carbon from the atmosphere into the biosphere (Peters and Lovejoy, 1992) whilst rising sea levels consequent on thermal expansion of the oceans and polar ice-cap melting (see Chapter 7, Conclusions) will reduce the extent of coastal wetlands. These are not only important sinks but also economic multipurpose resources.

However, despite these uncertainties, there has been a plethora of papers based on models which have attempted to determine the impact of global warming on the world's ecosystems (Woodward and Lee, 1995; Woodward *et al.*, 1995). Many of these studies have concentrated on North America (Mooney *et al.*, 1993), or on specific ecosystems such as the boreal and tundra zone (Sykes and Prentice, 1995; Kohlmaier *et al.*, 1995). The latter are particularly important biomes because most models predicting global warming, and its spatial variation for the next 50 years or so, indicate that high latitudes will warm relative to low latitudes. The magnitudes of change predicted by GCMs (Chapter 4) vary considerably, ranging from global averages of 2.8°C to 5.2°C for a doubling of pre-industrial carbon dioxide concentrations by the year 2050 (Smith *et al.*, 1992). However, in polar and boreal regions the amount of warming predicted will be above the global average. Increases in temperature alone will affect biome composition and spatial expression but the changes they promote in evaporation rates and soil moisture regimes will also have an impact. Should soils and peatlands dry out, for example, they will become net producers of carbon dioxide rather than sinks. Consequently, global warming will be reinforced. How the boreal and tundra biomes will change will also depend on the rate at which warming occurs. Smith *et al.* (1992) suggest that globally the extent of tundra and desert will decrease while grasslands and forests will expand; tree lines will extend further north and to higher altitudes than at present. There is also the possibility that increased atmospheric carbon dioxide concentrations will give rise to increased fixation of carbon in the biosphere, possibly by as much as 10 per cent (Smith *et al.*, 1992).

However, for any given region, the predicted changes that might occur in vegetation communities and carbon cycle dynamics vary considerably depending on the models being used. This is illustrated by the work of Melillo *et al.* (1995) which compared the results of three models based on biogeographical characteristics with those of three models based on biogeochemical characteristics. The biogeography-based models, all of which proved to be accurate indicators of modern conditions, showed general agreement, although there were considerable variations in the amount of predicted forest cover. The biogeochemically-based models, however, showed wide variation in their prediction of net primary productivity, which increased by between 2 per cent and 35 per cent, and carbon storage, which varied between −33 per cent and +16 per cent. Even larger discrepancies occurred when the biogeochemically-based models were run using the vegetation distributions predicted by the biogeography-based models. Such variations occur because of different sensitivities to the effects of temperature increase on the hydrologic and nitrogen cycles. A further model, the Frankfurt biosphere model (FBM), has been used by Lüdeke *et al.* (1995) to examine the response of northern biomes, in terms of net primary productivity and carbon storage, to enhanced greenhouse warming. Their results show that if warming alone is considered, without the fertilizing effect of increased atmospheric carbon dioxide, there is a 22 per cent decrease in net primary productivity. This is the result of increased respiration by primary producers and water stress brought about by the effect of the temperature increase on water availability. These changes also caused the soils etc. of these regions to become carbon sources, (see Chapter 9, Soils and the greenhouse gases) producing 170 Gt of carbon, rather than carbon sinks. However, when carbon dioxide fertilization is taken into account, there is an increase in net primary productivity of 9 per cent and a total carbon sink of 50 Gt of carbon. Using the FBM in combination with a biogeochemical model, Plöchl and Cramer (1995) have shown that a decrease in net productivity of between 5.4 and 8.8 per cent would occur in high latitudes under CO_2-induced warming. This does not take into account the effect of carbon dioxide fertilization. A comparison of this work with that of Lüdeke *et al.* (1995) indicates that the effect of the latter is a particularly important element in carbon cycle dynamics. These examples indicate not only the difficulties associated with modelling but also the significance of the boreal–tundra zone in global carbon dynamics.

It is equally difficult to determine the effects of global warming on species diversity and extinction. Peters (1992) has pointed out that past climatic changes have brought about changes in biodiversity and prompted extinctions and it is also likely that they have prompted evolution (see the second section). Nevertheless, the rapidity with which global warming is likely to occur means that its impact could be greater than that of past episodes of global warming which occurred more slowly. In addition, the impact is likely to be greater in higher latitudes where boreal species may well expand into tundra areas. Where global warming is predicted to be low (i.e. in the tropics), it may have little impact on biodiversity. Globally, however, the combinations of global warming and habitat loss due to land-cover change will undoubtedly cause biodiversity loss and species extinction. It is also possible that reduced carbon dioxide assimilation, due to loss of biodiversity, will be more than compensated for by the positive effect of carbon dioxide fertilization. Amthor (1995) stated 'On the whole, terrestrial higher-plant responses to increasing atmospheric CO_2 concentration probably act as negative feedbacks on atmospheric CO_2 concentration increases, but they cannot by themselves stop the fossil-fuel-oxidation-driven increase in atmospheric CO_2 concentration'. This does not, however, signify that humankind can afford to lose species through extinction (Mannion, 1995c).

CONCLUSIONS

Climate, vegetation and the global biogeochemical cycle of carbon are intimately and inextricably related, influencing each other through feedback mechanisms. This relationship has existed for most, if not all, of geological time. The gradual diminution of carbon dioxide in the atmosphere has been paralleled by an increase in free oxygen and both have been catalyzed by life, the evolution of which has itself been profoundly influenced by these and other changes in earth surface processes. These three facets of the environment do not operate in isolation; they are affected by the hydrologic and other biogeochemical cycles, incoming solar radiation and astronomical forcing factors. This complexity makes the task of accurate reconstruction of both past and future environmental relationships formidable. At present, there is a rapid release of carbon, so carefully sequestered in sedimentary sinks in the geological past, by humankind's quest for energy. This is beginning to alter the climate–vegetation–carbon relationship in such a way that positive feedback is now thought to be occurring. This in turn, will fundamentally alter the climate–vegetation–carbon system and the people–environment relationships. How this will be manifest is open to conjecture.

REFERENCES

Adams, J.M., Faure, H., Faure-Denard, L., McGlade, J.M. and Woodward, F.I. (1990) 'Increases in terrestrial carbon storage from the last Glacial Maximum to the present', *Nature* 348: 711–14.

Amthor, J.S. (1995) 'Terrestrial higher plant response to increasing atmospheric CO_2 in relation to the global carbon cycle', *Global Change Biology* 1: 243–74.

Archibold, O.W. (1995) *Ecology of World Vegetation*, London: Chapman and Hall.

Barnola, J.M., Raynaud, D., Korotkevich, Y.S. and Lorius, C. (1987) 'Vostok ice core provides 160,000-year record of atmospheric CO_2', *Nature* 329: 408–14.

Briffa, K.R., Jones, P.D., Schweingruber, F.H., Shiyator, S.G. and Cook, E.R. (1995) 'Unusual 20th-century summer warmth in a 1,000 year temperature record from Siberia', *Nature* 376: 156–9.

Chapellaz, J., Blunier, T., Raynaud, D., Barnola, J.M., Schwander, J. and Stauffer, B. (1993) 'Synchronous changes in atmospheric CH_4 and Greenland climate between 40 and 8 K yr BP', *Nature* 366: 443–5.

Cox, C.B. and Moore, P.D. (1993) *Biogeography. An Ecological and Evolutionary Approach*, 5th edn, Oxford: Blackwell.

Crane, P.R., Friis, S.M. and Pederson, K.R. (1995) 'The origin and early diversification of angiosperms', *Nature* 374: 27–33.

Crowley, T.J. (1995) 'Ice-age terrestrial carbon changes revisited', *Biogeochemical Cycles* 9: 377–89.

Foltieri, M., Magri, D. and Sadori, L. (1988) '250,000-year pollen record from Valle di Castighoni (Roma)', *Pollen et Spores* 30: 329–56.

Fuji, N. (1988) 'Palaeovegetation and palaeoclimate changes around Lake Biwa, Japan, during the last ca. 3 million years', *Quaternary Science Reviews* 7: 21–8.

Grainger, A. (1993) *Controlling Tropical Deforestation*, London: Earthscan.

Hannah, L., Lohse, D., Hutchinson, C., Carr, J.L. and Lankerani, A. (1994) 'A preliminary inventory of human disturbance of world ecosystems', *Ambio* 23: 246–50.

Hays, J.D., Imbrie, J. and Shackleton, N.J. (1976) 'Variations in the Earth's orbit: pacemaker of the ice ages', *Science* 194: 1121–32.

Hooghiemstra, H. (1989) 'Quaternary and Upper-Pliocene glaciations and forest development in the tropical Andes: evidence from a long high-resolution pollen record from the sedimentary basin of Bogotá, Colombia', *Palaeogeography, Palaeoclimatology, Palaeoecology* 72: 11–26.

Hooghiemstra, H. and Ran, E.T.H. (1994) 'Late and middle Pleistocene climatic change and forest development in Colombia: pollen record Funza II (2–158 m) core interval', *Palaeogeography, Palaeoclimatology and Palaeoecology* 109: 211–46.

Hooghiemstra, H., Melici, J.L., Berger, A. and Shackleton, N.J. (1993) 'Frequency spectra and palaeoclimatic variability of the high resolution 30–1450 Ka Funza I pollen record (Eastern Cordillera, Colombia)', *Quaternary Science Reviews* 12: 141–61.

Horowitz, A. (1989) 'Continuous pollen diagrams for the last 3.5 M.Y. from Israel: vegetation, climate and correlation with the oxygen isotope record', *Palaeogeography, Palaeoclimatology and Palaeoecology* 72: 63–98.

Houghton, R.A. (1994) 'The worldwide extent of land-use change', *Bioscience* 44: 305–13.

Intergovernmental Panel on Climate Change (1995) *Climate Change 1994. Radiative Forcing of Climate Change and An Evaluation of the IPCC 1S92 Emission Scenarios*, Cambridge: Cambridge University Press.

Jones, P.D. (1995) 'Recent variations in mean temperature and the diurnal temperature range in the Antarctic', *Geophysical Research Letters* 22: 1345–8.

Kohlmaier, G.H., Lüdecke, M., Janecek, A., Benderoth, G., Kindermann, J. and Klaudius, A. (1991) 'Land biota, source or sink of atmospheric carbon dioxide: positive and negative feedbacks within a changing climate and land use development', in S.H. Schneider and P.J. Boston (eds) *Scientists on Gaia*, Cambridge, MA: MIT Press, pp. 223–39.

Kohlmaier, G.H., Hager, G.H., Nadler, A., Wurth, G. and Lüdeke, M.K.B. (1995) 'Global carbon dynamics of higher latitude forests during anticipated climate change – ecophysiological versus biomes migration view', *Water Air and Soil Pollution* 82: 455–64.

Kump, L.R. (1989) 'Chemical stability of the atmosphere and ocean', *Palaeogeography, Palaeoclimatology, Palaeoecology* 75: 123–36.

Leith, H. (1975) 'Primary productivity of the major vegetation units of the world', in H. Leith and R.H. Whitaker (eds) *Primary Productivity of the Biosphere*, New York: Springer-Verlag, pp. 203–16.

Lovelock, J.E. (1972) 'Gaia as seen through the atmosphere', *Atmospheric Environment* 6: 579–80.

Lüdeke, M.K.B., Dönges, S., Otto, R.D., Kindermann, J., Badeck, F.-W., Rarnge, P., Jäkel, U. and Kohlmaier, G.H. (1995) 'Responese in NPP and carbon stores of northern biomes to a CO_2-induced climatic change, as evaluated by the Frankfurt biosphere model (FBM)', *Tellus* 47b: 191–205.

Mannion, A.M. (1995a) *Agriculture and Environmental Change: Temporal and Spatial Dimensions*, Chichester: John Wiley.

Mannion, A.M. (1995b) 'Biodiversity, biotechnology and business', *Environmental Conservation* 22: 201–27.

Mannion, A.M. (1995c) 'Biotechnology and environmental quality', *Progress in Physical Geography* 19: 192–215.

Mannion, A.M. (1997) *Global Environmental Change. A Natural and Cultural Environmental History*, 2nd edn, Harlow: Longman. (In press.)

Martin, J.H. (1990) 'Glacial-interglacial CO_2 change: the iron hypothesis', *Paleoceanography* 5: 1–13.

Melillo, J.M. *et al.* (1995) 'Vegetation ecosystems modeling and analysis project – community biogeography and biogeochemistry models in a continental-scale study of terrestrial ecosystem responses to climate-change CO_2 doubling', *Global Biogeochemical Cycles* 9: 407–37.

Meyer, W.B. and Turner, B.L., II (1992) 'Human

population growth and global land use/cover change', *Annual Review of Ecology and Systematics* 23: 39–61.

Mommersteeg, H.J.P.M., Loutre, M.F., Young, R., Wijmstra, T.A. and Hooghiemstra, H. (1995) 'Orbital forced frequencies in the 975000-year pollen record from Tenagi-Philippon (Greece)', *Climate Dynamics* 11: 4–24.

Mooney, H.A., Fuentes, E.R. and Kronberg, B.I. (eds) (1993) *Earth System Responses to Global Change*, San Diego: Academic Press.

Moore, A.M.T. and Hillman, G.C. (1992) 'The Pleistocene Holocene transition and human economy in southwest Asia: the impact of the Younger Dryas', *American Antiquity* 57: 482–94.

Naeem, S., Thompson, L.J., Lawler, S.P., Lawton, J.H. and Woodfin, R.M. (1994) 'Declining biodiversity can alter the performance of ecosystems', *Nature* 368: 734–6.

Peng, C.H., Guiot, T., Van Campo, E and Cheddadi, R. (1995) 'Temporal and spatial variations of terrestrial biomes and carbon storage since 13,000 yr BP in Europe – reconstructed from pollen data and statistical models', *Water Air and Soil Pollution* 82: 375–90.

Peters, R.L. (1992) 'Conservation of biological diversity in the face of climate change', in R.L. Peters and T.E. Lovejoy (eds) *Global Warming and Biological Diversity*, New Haven, CT, and London: Yale University Press, pp. 15–30.

Peters, R.L. and Lovejoy, T.E. (eds) (1992) *Global Warming and Biological Diversity*, New Haven, CT, and London: Yale University Press.

Pfister, C. (1992) 'Monthly temperature and precipitation in Central Europe 1525–1979: quantifying documentary evidence on weather and its effects', in R.S. Bradley and P.D. Jones (eds) *Climate Since A.D. 1500*, London: Routledge, pp. 118–42.

Plöchl, M. and Cramer, G. (1995) 'Coupling global models of vegetation structure and ecosystem processes: an example from Arctic and boreal ecosystems', *Tellus* 47b: 240–50.

Prentice, I.C., Sykes, M.T., Lautenschlager, M., Harrison, S.P., Denissenko, O. and Bartlein, P.J. (1993) 'Modelling global vegetation patterns and terrestrial carbon storage at the last glacial maximum', *Global Ecology and Biogeography Letters* 3: 67–76.

Raymo, M.E. (1994) 'The initiation of northern hemisphere glaciation', *Annual Review of Earth and Planetary Sciences* 22: 353–83.

Raynaud, D., Chapellaz, J., Barnola, J.M., Korotkevich, Y.S. and Lorius, C. (1988) 'Climatic and CH_4 cycle implications of glacial CH_4 change in the Vostok ice core', *Nature* 333: 655–7.

Ruddiman, W.F. and Raymo, M.E. (1988) 'Northern hemisphere climate regimes during the past 3Ma: possible tectonic connections', *Philosophical Transactions of the Royal Society of London* B318: 411–30.

Schwartzman, D.W., Shore, S.N., Volk, T. and McMenamin, M. (1994) 'Self-organisation of the Earth's biosphere – geochemical or geophysiological?', *Origins of Life and Evolution of the Biosphere* 24: 435–50.

Schwartzman, D. and Volk, T. (1989) 'Biotic enhancement of weathering and the habitability of Earth', *Nature* 340: 457–60.

Shackleton, N.J. (1977) 'Carbon-13 in *Uvigerina*: tropical rain forest history and the equatorial Pacific carbonate dissolution cycles', in N.R. Anderson and A. Malahoff (eds) *The Fate of Fossil Fuels CO_2 in the Oceans*, New York: Plenum, pp. 401–28.

Smith, T.M., Leemans, R. and Shugart, H.H. (1992) 'Sensitivity of terrestrial carbon storage to CO_2-induced climate change: comparison of four scenarios based on general circulation models', *Climatic Change* 21: 367–84.

Spicer, R.A., Rees, P.McA. and Chapman, J.L. (1993) 'Cretaceous phytogeography and climate signals', *Philosophical Transactions of the Royal Society of London* B341: 277–86.

Sykes, M.T. and Prentice, I.C. (1995) 'Boreal forest futures – modeling the controls of tree species range limits and transient responses to climate change', *Water Air and Soil Pollution* 82: 415–28.

The Times (1995) *The Times Atlas of the World*, 7th concise edn, London.

Thompson, L.G., Mosley-Thompson, E.P., Davis, M.E., Lin, P.-N., Henderson, K.A., Cole-Dai, J., Bolzan, J.F. and Liu, K.-B. (1995) 'Late glacial stage and Holocene tropical ice core records from Huascarán, Peru', *Science* 269: 46–50.

Tilman, D., May, R.M., Lehman, C.L. and Nowak, M.A. (1994) 'Habitat destruction and the extinction debt', *Nature* 371: 65–6.

Tzedakis, P.C. (1994) 'Vegetation change through glacial-interglacial cycles: a long pollen sequence perspective', *Philosophical Transactions of the Royal Society of London* B354: 403–32.

Woodward, F.I. (1987) *Climate and Plant Distribution*, Cambridge: Cambridge University Press.

Woodward, F.I. and Lee, S.E. (1995) 'Global-scale forest function and distribution', *Forestry* 68: 317–25.

Woodward, F.I., Smith, T.M. and Emanuel, W.R. (1995) 'Global land primary productivity and phytogeography model', *Global Biogeochemical Cycles* 9: 471–90.

Worsley, T.R. and Nance, R.D. (1989) 'Carbon redox and climate control through Earth history: a speculative reconstruction', *Palaeogeography, Palaeoclimatology, Palaeoecology* 75: 259–82.

Worsley, T.R., Nance, D.R. and Moody, J.B. (1991) 'Tectonics, carbon, life, and climate for the last three billion years: a unified system?' in S.H. Schneider and P.J. Boston (eds) *Scientists on Gaia*, Cambridge, MA: MIT Press, pp. 200–10.

Zhao, S. (1994) *Geography of China*, New York: John Wiley.

11

ANIMAL RESPONSES TO CLIMATE

Bernard Stonehouse

INTRODUCTION: CLIMATE AND LIFE

Life is confined to the thin shell of the earth's surface called the biosphere, extending from 10,000 m down in the ocean to similar heights in the atmosphere. The biosphere captures and distributes solar energy – the radiant energy that flows constantly from the sun, penetrating the atmosphere and impinging on the earth's surface. Plants and animals absorb and hold this energy, exchanging it among themselves, using it for their metabolic processes and creating more of their own kind. The biosphere extends over much of the world with its living components inhabiting a wide range of environments and tolerating climatic conditions that extend from extreme cold to extreme heat.

Life almost certainly began in warm shallow seas, in a watery medium that protected primitive organisms from sudden shifts in temperature and humidity. The chemical processes underlying life developed in stable conditions, subject only to daily and seasonal rhythms of slight warming and cooling. Today, the oceans continue to provide the world's most stable environments. Organisms of shallow water are still well buffered from climatic vagaries; those of deep oceanic waters live virtually free from diurnal or seasonal influences, in conditions that vary little over tens, hundreds or thousands of years.

Emergence from water to land provided daunting physical and biochemical challenges. Atmospheric climates become significant at the intertidal zone, and dominant in terrestrial environments. Over the billions of years since life emerged onto land, terrestrial organisms have adapted to climates ranging from equatorial to polar, often with strong diurnal and seasonal variations. Continental heartlands experience hot summers and bitingly cold winters; tropical mountain tops suffer extreme heat each day and extreme cold each night. Animals have adapted both to cope with gentle, stable climates, and also, more remarkably, with almost every extreme of climate that the world provides.

ELEMENTS OF CLIMATE

The standard climatic elements that most strongly affect animals are ground and air temperatures, humidity, precipitation and wind. All depend directly or indirectly on solar radiation, the mainspring that drives both climatic systems and life itself. Temperature is a key element in life because body temperature controls both the possibility of life and the pace at which it runs. A mean value of atmospheric temperature immediately above the earth's surface, averaged over all latitudes, altitudes and seasons, is about 14–15°C. Extreme temperatures measured under standard conditions range from below −80°C on the high plateau of Antarctica to +50°C and more in the arid tropics. However, most animals live within an overall ambient temperature range of about 50°C, approximately from −5 to 45°C. This appears to be the range within which life processes work most effectively. Organisms seek constantly to maintain their bodies at temperatures that allow their metabolic processes to work efficiently. Every animal needs to take into account heat from two sources, namely the environment and their own body chemistry, and to maintain a balance between them.

Atmospheric humidity affects water relationships throughout the animal body. While aquatic animals usually need to protect themselves from absorbing too much of their environment, subaerial creatures need defences against excessive loss of water to an often thirsty atmosphere. Conversely, the latent heat of vaporization is high, and controlled evaporation from both lungs and skin is for some animals an important means of maintaining heat balance.

Winds are atmosphere in motion, driven by geographical inequalities of solar heating and modified in direction by the earth's rotation and other influences. By disturbing the boundary layer immediately adjacent to the body, winds prevent organisms from cocooning themselves in a stable atmosphere, and so intensify the effects of temperature and humidity. Warm winds increase evaporation, cooling animal bodies but also drying them out, whereas cold winds accelerate heat losses from warm bodies. Hence they feel colder than still air at the same temperature, giving rise to the 'wind-chill factor' (Figure11.1). Winds have important secondary effects, carrying sound and chemical stimuli, distributing plant propagules and small animals, and helping or hindering long-distance migrants.

Climatic data collected under standard conditions at meteorological stations (Chapter 2) have the purpose of recording weather and long term climatic changes. The parameters measured, called macroclimate, are of little use in studying animal responses to climate. In a

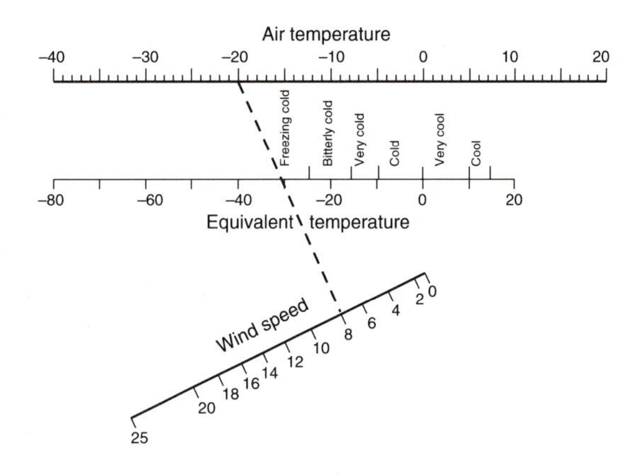

Figure 11.1 Nomogram for determining wind-chill. The dashed line shows an example of its use. A wind speed of 8 m s^{-1}, coupled with an air temperature of −20°C, yields a wind-chill temperature of −32°C. This would be judged a condition of 'freezing cold', in which there would be danger of frostbite to exposed skin (after Rees, 1993)

world in which insects and nematode worms are said to outnumber all other forms of life, most animals are small (i.e. less than 2 cm long) and live close to the ground. Temperatures, humidities and winds impinging directly upon them are different from those recorded in ventilated screens and on masts high above the ground. More immediately relevant is the microclimate at the air–ground interface, close to where the animals are living.

Microclimate is important alike to the small animals and to the biologists who study them. Diurnal extremes of heat and cold measured in deserts and on tropical mountain tops, for example, may be avoided altogether by animals that live in burrows half a metre underground. Lemmings and other small mammals of the tundra avoid extreme winter cold by living at the snow–ground interface. These local conditions cannot readily be deduced from standard meteorological data, and are not easy to measure. Macroclimatic records are to some degree relevant for the larger animals that live on or above the ground. However, they usually lack such important parameters as radiation data, and provide only a rough guide to true environmental conditions.

BIOCLIMATIC ZONES

The earth's surface is divided into latitudinal climatic zones, in a pattern that depends primarily on amounts of solar radiation received. Ecologists always, and climatologists often, tend to characterize the zones by the vegetation growing within them. This is not surprising since temperature, precipitation and evaporation (the key factors in climatology) are equally key factors in the growth of the vegetation that dominates animal habitats.

Köppen's system of classification (Köppen and Geiger, 1936), as elaborated by Thornthwaite (1948), is the system most generally favoured by ecologists. It distinguishes five major types of world climate, each defined primarily by temperature and rainfall, but strongly reflecting a predominant vegetation, as follows:

A: equatorial and tropical rain climates
B: dry climates of the arid zone
C: temperate climates of the broad-leaved forest zone
D: cold temperate forest climates
E: treeless polar climates.

Each of the warmer climates is divided into subtypes according to the amount and seasonality of rainfall. Polar climates tend to be subdivided into zones according to temperature, defined by isotherms for particular months (Stonehouse, 1989). However, either way, what is usually being defined is the length of plant-growing season. The consequences of seasonal plant growth, more than any other factor, determine the distribution of animals within the biosphere.

While both plants and animals show preferences for particular climatic zones, plants on the whole are static and have only passive means of coping with fluctuating environments. Most animals have the option of moving about and have developed patterns of behaviour that optimize their ability to search both for food and for the environmental conditions that best suit them. What they usually seek is stability; whether hot or cold, wet or dry, steady environmental conditions are easier to manage than those that vary unpredictably.

Very few animals live the whole of their lives in the adult environment. Parental doorsteps quickly become overcrowded and small animals usually require different food from adults. Creatures that are sedentary as adults tend to travel during early stages of their lives, wandering through several different climatic phases or regimes before settling. Fish and bottom-living invertebrates, such as clams and polychaete worms that live as adults in the darkness of the the ocean depths, cast their eggs into sunlit surface waters, where food for larvae is more plentiful. The offspring travel and adventure widely before descending to the placid adult environment. Insects that mate in aerial swarms high above ground may spend juvenile months in the very different environment of ponds or damp soils.

Similarly many larger animals, birds and mammals especially, use their powers of locomotion to migrate, that is spend successive seasons in different bioclimatic zones (Cloudesley-Thompson, 1978). Geese hatched and reared on the Arctic tundra fly south to winter on temperate shores, returning annually to the Arctic each spring for breeding. Arctic reindeer migrate from boreal forest wintering grounds to summer grazing on the taiga or open tundra. Insectivorous swifts, swallows and martins commute annually between temperate and tropical regions. Herds of herbivorous mammals in East Africa follow traditional migration routes that take them from areas of wet-season to dry-season grazing. Alpine birds breed in the mountains in summer, and winter over the warmer plains.

Though numbers of species fall towards either end of the range, both polar and tropical regions are surprisingly well stocked with animal life. Polar seas, especially, are rich in species across a wide zoological spectrum, ranging from protozoa to whales, all tolerating temperatures close to the freezing point of sea water (−1.8°C). Polar lands in comparison are relatively poor, though summer visitors to the Arctic are often impressed by the ecological diversity and wealth of insects, birds and mammals on tundra and taiga (Chernov, 1985). The warmest tropical seas, though often poor in biomass (total mass of living material), are usually well endowed with species. Equatorial rain forests are exemplars of diversity of species and complexity of ecosystems. Hot deserts, by contrast, have relatively little animal life and the limiting factor is likely to be want of water rather than excess of heat.

CLIMATIC INSTABILITY

Individual animals adapt to diurnal and seasonal changes and, in the longer term, animal species are subject to secular changes in climate that demand a further order of adaptability. The first two centuries of climatic studies led to the assumption that there was an inherent stability in climatic systems. Now we think more in terms of widely fluctuating climates, at least in middle to high latitudes and at least over the most recent 30 to 40 million years from the Mid-Cenozoic era onward. This period, in which climate change rather than constancy has been the rule, followed a much longer spell of climate stability, which may well be more representative of world climates in general.

The causes of long-term climate change are dealt with more fully in Chapters 8, 10, 21 and 22 and include, as important factors, changes in the intensity and distribution of solar radiation due to shifts in the earth's orbital parameters, movements of continents and consequent changes in ocean circulation, and clouding of the atmosphere due to volcanic action and other major disturbances (Lockwood, 1979; World Meteorological Organization, 1975). Animals appear to have been able to cope with these changes and, almost three decades ago, a leading taxonomist described as 'staggering' the amount of current diversity in the living world (Mayr, 1969). It was estimated that about 1 million species of animals and half a million species of plants had already been described, and between 3 and 10 million living species still awaited description. Other authorities estimate far more diversity and the 15th edition of *Encyclopaedia Britannica*, for example, estimates between 2 and 4.5 million plant and animal species, and others again write of 1 million species of arthropods alone. Whichever figure is chosen indicates that many species survived a long spell of climate instability.

The fossil record is too patchy and inaccurate for precise comparisons to be made with more stable eras of the past. However, Mayr (1969) cited an estimate of half a billion extinct species, while *Encyclopaedia Britannica* quoted 'authoritative estimates' of 15 million to 16 billion extinct species. Again whichever figure is chosen, many of these would have disappeared owing to environmental changes, notably shifts of climate. The demise of the large Mesozoic reptiles, typified by the dinosaurs, is often quoted as a consequence of climatic upheaval. The warm, placid climates of the Late Mesozoic encouraged their spread and diversification and the cooling and instability that accompanied the start of the Cenozoic witnessed their dramatic departure. That two alternative major groups, the birds and the mammals, were able to replace them may well be a testimony to the effectiveness of endothermy (see below) in a changeable environment.

Human beings are the products of an ice age, which was a period of major instability which started some 20–30 million years ago (Andrews, 1978) and was characterized by geological upheaval, mountain building and the development of polar ice caps. Climate change, and more specifically 'global warming', that have recently caught the attention of scientists and the media, are symptoms of this same instability, possibly (though by no means certainly) accentuated by human interference.

SOURCES OF ENERGY

Practically all energy available at the earth's surface comes from the sun. The solar spectrum includes a wide range of frequencies of radiant energy, of wavelengths 0.2–5 μm (Lockwood, 1979). Of these, varying amounts are absorbed selectively in their passage through the atmosphere, the longer ones mainly by carbon dioxide and water, the shorter ones by oxygen and ozone. Of the radiation that reaches the earth's surface, a proportion is reflected back into space, and the balance is absorbed by rocks, plants and animals. The longer waves (of frequencies 0.4 μm and above) provide light, warmth and energy for the chemical bonding of photosynthesis, on which ultimately all life depends. Shorter waves, in the ultraviolet band of 0.2–0.4 μm, tend to be harmful to life, penetrating and disrupting living tissues. Without the atmospheric filter, the terrestrial life that we know today might never have come into existence. The atmosphere and earth's surface are both warmed by solar radiation and, in turn, radiate energy at longer wavelengths within the 5–50 μm or infrared band. Thus an animal standing in the sun (Figure 11.2) is subject to three sources of short-wave radiation (direct, which is received directly from the sun; diffuse, which is received from the atmosphere; and that reflected from the ground and other surfaces), and several sources of long-wave radiation from the atmosphere, ground and nearby objects. The sum of energy that the animal absorbs at its surface is transformed to heat and circulated within the body.

Animals maintain life by feeding since food is both their prime source of raw materials and an important alternative source of energy. Food energy is of course solar energy in another form, which is essentially the energy that plants or other animals have stored in their own bodies for their own purposes, and lost to grazing or predation. Digestion breaks down the food into its chemical components and further metabolic processes (in muscles, liver and other organs) release the chemical energy stored within them. In most of these reactions, heat is released as a by-product. Herbivores and carnivores alike rely on this 'stolen' energy, either for immediate use in their metabolic processes, or to enhance their own energy stores. Animals constantly exchange energy through their surface, gaining or losing heat by conduction, convection, radiation and evaporation of water from internal or external surfaces. Gains and losses of heat within an animal body, like those of an inanimate object, may be expressed simply as follows:

$$M = C + Cv + R - E$$

where: M is the rate of metabolic heat production; C, Cv and R are gains or losses due respectively to conduction, convection and radiation; and E is the loss due to evaporation. M is always positive and E is always negative, but C, Cv and R may have positive or negative values according to the direction of energy flow. All values are expressed as rates of energy exchange per unit of surface, for example as watts per square metre (W m^{-2}).

When the two sides of this energy equation are out of balance, energy is either stored in the body or lost from it. Energy stored in the short term is manifest as heat and results in a rise of body temperature, whereas energy lost results in body cooling. How much energy an animal

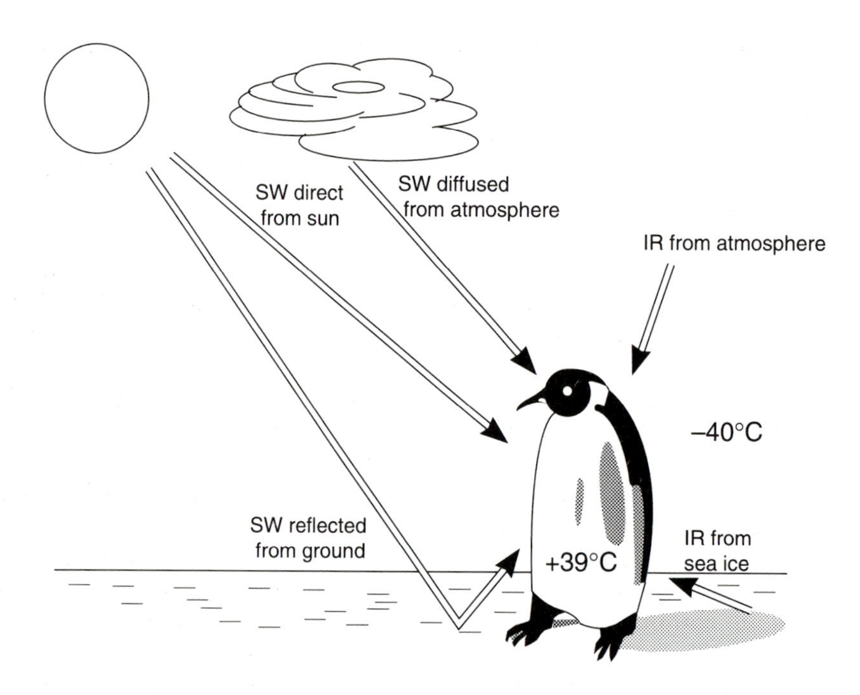

Figure 11.2 Radiation environment of an emperor penguin standing on sea ice. The penguin receives direct, diffuse and reflected short-wave radiation and long-wave radiation from the sea ice and atmosphere. It will itself be radiating to the atmosphere in the long-wave band, and losing heat by conduction through its feet, convection to the atmosphere (possibly enhanced by wind) and evaporation from its lungs. Despite the steep gradient between body and atmospheric temperatures, insulation ensures that it loses little energy (after Stonehouse, 1989)

derives from each of these sources depends on the immediate circumstances and varies from minute to minute. Thus, an animal that increases *M* by exerting muscular activity, or *R* by sitting in the sun, may exceed its own capacity for losing heat, and so raise its body temperature. An animal that falls into cold water will rapidly increase conductive losses whereas one that sits in a draught may increase *E*. However, if other variables remain constant, both will tend in consequence to cool. In general terms, all animals generate some of their own heat, and use behaviour to seek the immediate climatic surroundings that will maintain them in a suitable state of thermal balance. To achieve this important balance, animals have adopted two distinct strategies, the second quite clearly an evolutionary development of the first, with corresponding advantages and penalties. These strategies, called *ectothermy* and *endothermy*, are discussed below.

BIOCHEMICAL ACCLIMATIZATION

Few animals survive long at temperatures above 45°C, probably because particular organic substances vital to life (e.g. enzymes and lipids) are inactivated by high temperatures. Few live actively at temperatures below −5°C, mainly because they start to freeze, and the formation of ice within cells damages delicate membranes and upsets critical concentrations of cell fluids. Within this overall range, life processes are restricted to much narrower spans of temperature, to which the cell chemistry of the species or individual is adapted. Some species have evolved a chemistry that is efficient only within very narrow thermal ranges – some polar fish, for example, within the 1–2°C to which they are normally subject in life. Others work within wider spans of 5–10°C, consistent with diurnal ranges, and given time, they can adjust their spans to accommodate even greater seasonal ranges.

Changes in body temperature have two major effects on cell chemistry (Hochachka and Somero, 1984). The first is the rate at which chemical reactions take place since practically all reactions work faster at high temperatures than low, and the chemistry of metabolism that underlies living processes is no exception. The increase in velocity of a reaction over a span of 10°C, expressed as the ratio of rate at $t + 10^{\circ}$ divided by the rate at t°, is called the Q_{10} value. For most biochemical reactions, Q_{10} values lie between 2 and 3 indicating that the reaction rate doubles or trebles for each 10°C rise. A mean value of 2.5 indicates an increase in reaction rate of 9.6 per cent for each 1°C rise in temperature (Morris, 1974). Values of Q_{10} remain constant or increase slightly within rising bands of temperature over the critical 50°C temperature range for living organisms.

The second important effect of temperature is to affect the points of delicate chemical equilibria that stabilize membranes and other structures within cells. This is the most likely reason why cells work effectively within particular narrow ranges of temperature, and lose efficiency outside that range. It may explain why finding an optimal thermal environment, and remaining in it, is so important to so many organisms, and why endothermy (see below) confers singular advantages on those that possess it. Temperature acclimatization represents physiological adaptation to working at higher or lower temperatures (whether seasonal, diurnal or on a much longer evolutionary timescale) and is thus a matter of biochemical adjustments within the cells.

ECTOTHERMY AND ENDOTHERMY

For a long time, a distinction has been made between animals that feel warm to human touch (called *warm blooded*), and those that feel cold (*cold blooded*). When cold-blooded animals were perceived not to be perpetually cold, but to warm up and cool down with their environment, they became known as *poikilotherms*, a name derived from a Greek root expressing changeability. Warm-blooded animals, known to maintain a constant and relatively high body temperature, became *homeotherms* (alternatively *homoiotherms*), indicating sameness or stability. This is still a useful distinction, although modern terminology reflects a more fundamental truth. Cold-blooded animals are as warm or as cold as their environment because they have little capacity for generating heat internally or retaining it in their bodies. Warmed mainly from outside, they are called *ectotherms*. Warm-blooded animals are warm because they possess much higher metabolic rates and better insulation. Generating much of their heat internally, they are called *endotherms* (Hardy, 1979).

Figure 11.3 models the essential differences between the two strategies where every animal finds a range of environmental temperatures tolerable. Outside that range, conditions are either too cold or too hot, incommoding the animal and putting it at risk. Within the zone of tolerance, ectotherms conform to changing conditions and their body temperatures rise or fall in close approximation to environmental temperature. Endotherms regulate their internal temperatures metabolically, to a high and constant level throughout the tolerance zone. Both kinds of animals have limited capacity to survive outside the zone, and both use behaviour, simple or complex, to keep themselves as much as possible within it. Should this fail,

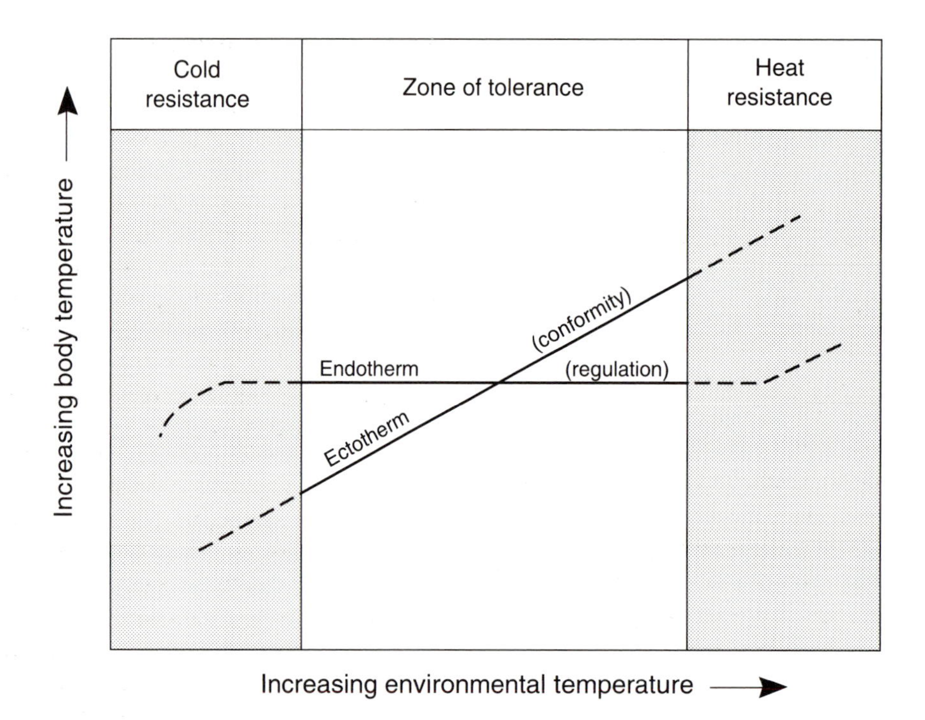

Figure 11.3 Physiological compensation as seen in ectotherms and endotherms. In response to increasing environmental temperature, the body temperatures of ectotherms rise in conformity; those of endotherms remain constant by regulation. Both are capable of extending their activities marginally into zones of intolerance on either side (after Richards, 1973)

they shift ground and, for example, animals that are prominent in summer disappear in winter, by migrating, hibernating or simply 'going to ground' to avoid extreme cold.

THE ECTOTHERM WAY

Ectotherms were for a long time the only kind of animal in the world and are currently by far the majority. With relatively low rates of metabolic heat production, they rely heavily on external energy to keep them within the zone of tolerance. In terms of the energy equation above, M is insignificant in comparison with the environmental variables, and body temperature is stabilized only by maintaining balance between the terms C, Cv, R and E. The balance is maintained largely by relatively simple behavioural responses to environmental stimuli. Ectotherms are in no sense second-rate animals: theirs is a hugely successful stratagem which allows a host of animal species to work economically within the wide range of environmental temperatures that the world provides (Schmidt-Nielsen, 1972). They penetrate the coldest and the warmest seas and they are found in the coldest of polar deserts, on the most frigid mountain tops and flourish abundantly in tropical rain forests. They use resources (especially food and water) with commendable economy and when environmental conditions become intolerable, they retire to states of dormancy and wait for recovery.

Ectotherms' body chemistry works sluggishly at the lower end of their zone of temperature tolerance. Slight warming, perhaps from basking in the sun, raises both body temperature and metabolic rate. This is often manifest in increased activity since ants and beetles run faster at mid-day than in the evening or early morning and lizards and snakes, that are torpid when cold, liven up after a short spell of basking. Activity produces further heat and a further rise in body temperature, bringing the animal to efficient running condition. Most ectotherms have poor insulation and cannot retain heat for long. Furthermore, an effective insulation would not have been advantageous owing to their low metabolic rates and reliance on external warming. Better insulation (which would necessarily work both ways) would inhibit the inward flow of heat on which their strategy depends. However, they could have evolved a more effective metabolism but only at the expense of burning more fuel, compromising the economy that is a key element in their strategy.

The most advanced ectotherms, in fact, raise their body temperature by metabolic action and maintain it several degrees above ambient for as long as their activity continues. Large flying insects, lizards and large, fast-moving fish (e.g. members of the tuna family) are among those that achieve high and relatively stable body temperatures. Most live in temperate or warm environments that minimize heat losses, and show patterns of behaviour that take full advantage of environmental heating and cooling. They tend to be active and their size is significant, since the larger the animal, the greater the body mass (from which internal heat is generated) in proportion to surface area (through which heat is lost). Some of the largest Mesozoic reptiles, living in warm, stable environments with plenty of food, may well have maintained high and relatively constant body temperatures in a physiological condition not far removed from endothermy.

THE ENDOTHERM WAY

Almost exclusively birds and mammals, endotherms have developed relatively high metabolic rates, and rely on food and metabolic heat production to maintain body temperature. Though the environmental terms C, Cv, and R still represent powerful external moderators, M has now become the key component in maintaining internal stability, with E an important adjunct for shedding excess heat.

Among prototherian or primitive mammals, monotremes maintain a core temperature (i.e. that of the main body mass) of about 30°C, and marsupials average about 35°C, although the temperatures of both tend to oscillate. Metatherian (placental) mammals run at 37–38°C and birds 1–2°C higher, both maintaining temperatures that remain constant within 1 to 2°C. This surprising degree of uniformity in temperature suggests that endotherms adopt the highest temperatures at which most body systems can be made to work. Indeed, 39°C is slightly higher than optimal for some body processes, for example sperm production. Hence, there is a need for the testes of most mammals to be excluded from the body core, and maintained at a slightly lower temperature in a scrotal sac, while those of most birds are cooled by direct contact with air sacs within the respiratory system. There is a further need in all endotherms for heat-shedding devices, to ensure that core temperatures do not rise above critical limits. For reviews of the many anatomical, physiological and behavioural adaptations and strategems adopted by endotherms in extreme heat and cold, reference should be made to (for example) Musacchia and Jansky (1981), Lyman *et al.* (1982) and Folkow (1992). Responses of human beings to extreme heat and cold are well reviewed in Edholm (1978) and are covered in Chapter 12.

Creating within themselves an optimal, stable temperature environment demands a particular combination of qualities, of which the first and most obvious is efficient insulation. This is achieved externally by fur or feathers, and internally by well-vascularized subdermal tissues which radiate heat when blood floods through them, but insulate when the flow is restricted. Fur and feathers are more than mere blanketing. Many mammals in seasonal climates moult from thin summer to thicker winter fur, which may also involve protective coloration. The polar bear's white fur proves to be a mobile greenhouse, trapping the energy of weak polar sunshine to warm the body within.

Insulation that is efficient when the animal is at rest may be too efficient while it is active and generating internal heat. This demands rapid cooling mechanisms, especially in warm climates, for endotherms run at body temperatures that are perilously close to their lethal temperatures. Most mammals sweat and all mammals and most birds have patches of bare skin that can be exposed to cooling air. Birds have an internal system of sacs that allows cooling air to permeate every part of the body. Both birds and mammals pant, increasing heat losses through the lungs, and birds flutter their throat pouches for the same effect. Most well-covered endotherms show areas of bare skin with a network of near-surface blood vessels that can be turned on like radiators to shed heat in an emergency.

Enhanced metabolism depends on highly efficient lungs, which provide the flow of oxygen required for heat-releasing oxidative metabolism. Blood flow is increased by a four-chambered heart, providing a high-pressure double circulation which ensures supplies of oxygen and metabolites to the muscles, liver and other metabolically active tissues. Muscular tremors (shivering) and brown fat, a special tissue rich in mitochondria and oxidative enzymes (Smith and Horwitz, 1969; Horwitz, 1989), generate internal heat rapidly when required. Small endotherms, whose heat losses are great, require much higher metabolic rates than larger ones. Many endotherms have blood systems with local carotid heat exchanges that monitor and moderate the temperature of the blood supply to the brain, protecting the hypothalamus which plays a key role in temperature control. From this centre are mediated the many delicate feedback mechanisms, involving advance endocrine and autonomic nervous systems, that ensure the endotherm's remarkable level of constancy of internal temperature.

The environmental advantage of endothermy is a degree of independence from external temperatures. The penalty is a high energy intake since a mammal living in a temperate climate requires three to seven times as much food as a similar-sized reptile. A very few mammals, and even fewer birds, avoid the constant expense of endothermy by allowing their body temperature to fall to environmental levels, for example during winter when food is scarce. While many ectotherms (including both invertebrates and fish, amphibians and reptiles) become inactive in these circumstances, torpor and hibernation in endotherms is more dramatic, involving temporary loss of thermoregulation (Lyman *et al.*, 1982). True hibernation is restricted to small mammals, for example dormice and ground squirrels, with already high metabolic rates, which would find it impossible to maintain themselves in cold winters. They seek a safe place, their temperature drops and they become lethargic and eventually moribund, with radically reduced heartbeat, respiration and metabolism. From this condition, small mammals can quickly recover, though the underlying mechanisms are by no means clear. Larger animals, for example bears, undergo torpor rather than true hibernation, letting their core temperatures fall to intermediate levels from which they too can make a quick recovery. Despite its cost, endothermy seems to confer many benefits beyond a wider choice of environments. It was almost certainly a prerequisite for the elaboration of the brain and central nervous system, and the development of complex behaviour, that contribute so strongly to the evolutionary success of birds and mammals.

CONCLUSIONS

This book considers possible consequences of climate change on world systems. If climate is in fact changing – if for example global warming or ozone depletion are actually occurring – what are the likely consequences for animals? The first consideration must be what magnitude of change is postulated. For global warming, recent atmospheric circulation models (Stauffer *et al.*, 1989, reviewed by Cattle, 1991) suggest that in the Arctic, where changes may be greatest, we should anticipate possible increases in mean annual temperature of up to 3°C within the next 50 years, and up to 7°C within a century. Corresponding values for the equator and southern hemisphere are unlikely to exceed 2°C and 3°C respectively.

As has been noted, today's animals represent stocks that have evolved during a period of chronic environmental instability. Their fairly recent ancestors are likely to have coped with temperature changes that far exceeded those predicted in both rate and magnitude. Only a few animals (ectotherms that live in the most constant marine environments) show little individual ability to survive small shifts in temperature. Most animals, as individuals and species, would find the postulated increases – even if doubled – well within their zones of tolerance. They would be unlikely even to detect, and far less respond to, mean annual changes occurring at these rates.

Effects of increased ultraviolet radiation at the earth's surface, owing to depletion of stratospheric ozone (Chapter 22), are more difficult to quantify and assess. The dramatic rise in atmospheric transparency that occurs each spring over high latitudes of both hemispheres may let through enough additional UVB radiation to damage both terrestrial and aquatic surface organisms, though this has yet to be demonstrated. Levels of ultraviolet radiation are chronically above the mean in high plateau and mountain areas. The overlying atmosphere is thin, the radiation shield weak, but all kinds of animals go about their business undismayed.

Should any of these postulated changes result in shifts of bioclimatic zones, then we might expect marginal changes in the distributions of some animals, and possible radical changes in distributions of others that, for whatever reason, are currently circumscribed by temperature-related constraints. However, most ecologists recognize that their discipline seldom models so precisely (Stonehouse, 1991) and only rarely can a single effect be ascribed to a single cause. It is known that environments change constantly, and that animal life is well adapted to change. The changing scene is watched with continuing interest, tinged only faintly with alarm.

REFERENCES

Andrews, J. (1978) 'The present ice age: Cenozoic', in B.S. John (ed.) *The Winters of the World*, Newton Abbot: David and Charles, pp. 173–218.

Cattle, H. (1991) 'Global climate models and Antarctic climate change', in C. Harris and B. Stonehouse (eds) *Antarctica and Global Climate Change*, London: Belhaven Press, pp. 21–34.

Chernov, Y.I. (1985) *The Living Tundra*, Cambridge: Cambridge University Press.

Cloudesley-Thompson, J. (1978) *Animal Migration*, London: Orbis.

Edholm, O.G. (1978) *Man-hot and cold*, London: Edward Arnold.

Folkow, L. (1992) *Aspects of Thermoregulation in Terrestrial and Aquatic Arctic Mammals*, Tromsø: University of Tromsø.

Hardy, R. N. (1979) *Temperature and Animal Life*, London: Edward Arnold.

Hochachka, P. W. and and Somero, G. N. (1984) *Biochemical Adaptation*, Princeton, NJ: Princeton University Press.

Horwitz, R.A. (1989) 'Biochemical mechanisms and control of cold-induced cellular thermogenesis in placental mammals', in L.C.H. Wang (ed.) *Advances in Comparative and Environmental Physiology. 4. Animal Adaptations to Cold,* Berlin: Springer-Verlag, pp. 83–116.

Köppen, W. and Geiger, R. (eds) (1936) *Handbuch der Klimatologie*, Berlin: Verlagsbuchhandlung Gebrüder Borntraeger.

Lockwood, J.G. (1979) *Causes of Climate*, London: Edward Arnold.

Lyman, C.P., Willis, J.S., Malan, A. and Wang, L.C.H. (1982) *Hibernation and Torpor in Mammals and Birds*, New York: Academic Press.

Mayr, E. (1969) *Principles of Systematic Zoology*, New York: McGraw-Hill.

Morris, J.G. (1974) *A Biologist's Physical Chemistry*, 2nd edn, London: Edward Arnold.

Musacchia, X.J. and Jansky, L. (eds) (1981) *Survival in the Cold: Hibernation and Other Adaptations*, New York: Elsevier.

Rees, W.G. (1993) 'A new wind chill nomogram', *Polar Record* 29(170): 229–34.

Richards, S.A. (1973) *Temperature Regulation*, London: Wykeham.

Schmidt-Nielsen, K. (1972) *How Animals Work*, Cambridge: Cambridge University Press.

Smith, R.E. and Horwitz, B.A. (1969) 'Brown fat and thermogenesis', *Physiological Review* 49: 330–425.

Stonehouse, B. (1989) *Polar Ecology*, Glasgow: Blackie; New York: Chapman and Hall.

Stonehouse, B. (1991) 'Polar ecosystems, management and climate modelling', in C. Harris and B. Stonehouse (eds) *Antarctica and Global Climate Change*, London: Belhaven Press, pp. 147–54.

Stauffer, R.J., Manabe, S. and Bryan, K. (1989) 'Interhemispheric asymmetry in climate response to a gradual increase of atmospheric CO_2', *Nature* 342: 660–2.

Thornthwaite, C.W. (1948) 'An approach toward a rational classification of climate', *Geographical Review* 38: 55–80.

World Meteorological Organization (1975) 'The physical basis of climate and climatic modelling', *GARP Publication Series* 16, Geneva: World Meteorological Organization.

PART 3:

CLIMATE AND THE CULTURAL ENVIRONMENTS

The following nine chapters examine the impact of climate on the development of the cultural environment, through its influence on human behavioural patterns (namely, comfort/health and social responses) and a wide range of human activities, including the built environment and a host of commercial–economic considerations. These include industrial operations, transport systems, agriculture, fisheries and forestry, recreation/tourism and fuel/power supplies. Once again, the issue of climate change is discussed by most authors (to varying degrees).

12

COMFORT, CLOTHING AND HEALTH

Andris Auliciems

INTRODUCTION: HUMAN ADAPTABILITY AND MICROCLIMATE MANAGEMENT

In Part 2 of this book, the impacts of the past, current and predicted climates were discussed within the context of the functioning of the physical and biological environments. In this first chapter of Part 3, the impacts of climate on human comfort, clothing and health are considered. The proliferation of western life-styles, clothing, technology in building construction and microclimate control have tended towards homogenizing atmospheric environments to which humans are exposed. Irrespective of market forces that drive these developments, in a global ecosystem increasingly threatened by environmental degradation and anthropogenic climate change, such specialization in adaptation needs to be examined in terms of (1) sustainability over the longer term, and (2) the human species overall 'biological fitness', or adaptability (Medwar, 1957). Here, we need to be mindful of the general principle that, within a changing environment, survivability is greater among the adaptable than the adapted (Sargent and Tromp, 1964), and ask which trend is technological development favouring?

This chapter attempts to examine human homeothermic adaptation strategies and their quantification, and to synthesize management issues relating the achievement of comfort, health and the microclimate envelopes in which we reside. Inevitably such integration is bound to be biased towards the particular disposition of the writer, in this case being that of a biometeorologist–environmental geographer. Not surprisingly, the thrust is towards an argument that favours physiological and psychological adaptability over technology that utilizes energy-wasteful methods. Inevitably also, the materials reviewed are highly selective and obviously fail to do justice to the vast amount of research in the fields covered.

HOMEOTHERMY AND ADAPTATION

A precondition to the well-being of all homeotherms, including humans, is a successful maintenance of a constant core temperature near 37°C. This requires continuous physiological and behavioural adjustments to balance energy exchanges between the body and the environment. Indeed, the course of human evolution and distribution on the earth's surface may be charted by the organism's responses to prevailing thermal environments. Physiological adaptation and technological development have gone hand in hand to permit human settlement in climate zones well beyond the likely original warm human birthplace (Burton and Edholm, 1955; Sargent, 1963; Scholander, 1955).

In the first half of the twentieth century, many authors believed that heat constituted a major constraint to human well-being in the low latitudes (for general arguments see Mills, 1946; Taylor, 1959; Huntington, 1926; Missenard, 1957; and a review of specific psychological and physiological concerns in Sargent, 1963). Markham (1947), and manifestly the various heating, ventilating and air-conditioning (HVAC) engineering fraternities, believed that the technology of microclimatic control, in particular, was the key to the success of human beings in these environments. Despite the seeming triumph of this latter philosophy in modern cities throughout the world, nowadays overly deterministic conclusions are tempered by a recognition that some of the behavioural patterns, at times mistakenly viewed as symptoms of cultural debilitation, actually represent patterns and processes of specialized adaptations (Wulsin, 1949; Sargent, 1963).

Diversity in adaptations may be attributed to particular combinations and weightings within seven interlaced groups of adaptation strategies to atmospheric, especially thermal, stimuli, as follows:

1 Physiological adjustments, ranging from minor vasomotor to major sweating and metabolic responses.
2 Acclimatization (including habituation) of both physiological and psychological mechanisms by periodic exposure to thermal stimulus.
3 Food energy intake and dietary alterations.
4 Metabolic alterations in scheduling of activities, selection and curtailment of particular tasks or their sequencing.
5 Migration, either temporary or permanent avoidance of particular stress conditions.
6 Clothing and building fabric interposition between the source of stress and the organism.
7 External energy generation for space heating and cooling.

Depending upon atmospheric signal strength and particular circumstances of the adapter, these processes may be largely involuntary (especially in the above categories 1, 2 and probably 3) or voluntary (strategies 4 to 7). In the behavioural adaptation categories, however, there may be no deliberate decision-making involved in the response (e.g. in weather-related recreation, or largely unconscious adjustment of clothing). Overall, a precondition to human well-being in terms of both productivity and health appears to be the achievement of a harmonious balance between minimization of physiological responses in category 1 above, and maximization of acclimatization in strategy 2. Environmentally the least demanding categories are 1 and 2, and the most destructive 7. The overreliance on the latter strategy, especially air cooling, in conjunction with ill-designed forms in strategy 6, is the concern of this chapter.

THE HUMAN ENERGY BALANCE

Modern studies of thermoregulatory responses are based on Gagge's (1936) energy-budget model. Over time, it is assumed that the net heat generated within the body will be balanced by the several fluxes as shown in the following equation:

$$(M + Q) \pm R \pm C - E = \pm S \quad (\mathrm{W\ m^{-2}})$$

where M is metabolic heat minus energies used in work; R, C and E are long-wave radiation, convection and evaporation, respectively; S is storage within body tissues, which in the maintenance of homeothermy must equal zero over time. To cater for dealing with outdoor conditions, Q represents short-wave solar radiation income, conveniently expressed as part of the total energy input that needs to be eliminated. Total metabolic rates generated during particular work categories are shown in Table 12.1.

Given that the thermal condition is also a function of the clothing insulation variable, the equation of homeostasis needs to be quan-

Table 12.1 Metabolic rates for selected activities in W m^{-2}

Category	*Range*	*Examples*
Minimal activity	(<100)	Sleeping 40, sitting 60, standing relaxed 90
Light activity	(100–200)	Standing at attention 100, strolling (3 km h^{-1}) on level 120, light assembly work 130, walking (5 km h^{-1}) on level 160, bicycling 6 km h^{-1}, bricklaying 160, walking gradient 5 % 170, digging soft earth 180, mixing cement 190, volley ball 200
Moderate activity	(210–200)	Walking (5 km h^{-1}) on 5% gradient 240, carrying 20 kg load 250, sawing softwood 260, ploughing with tractor 270, gardening 280
Heavy activity	(>300)	Pushing wheelbarrow 340, jogging 10 km h^{-1} on level 350, walking 3 km h^{-1} on 25 % gradient 380, carrying 20 kg load 400, pick and shovel work 500, running 14 km h^{-1} 560, running 16 km h^{-1} 680, sawing hardwood by hand 750, running 19 km h^{-1} 900

Sources: Categories based on Christensen (1953) 'work' categories; metabolic rates approximated and extracted from Christensen (1953), Fanger (1970), Fourt and Hollies (1969) and Passmore and Dumin (1955).

tified by two personal and four atmospheric parameters; as follows:

- Metabolic rate
- Clothing insulation
- Air temperature
- Radiant temperature of surroundings
- Rate of air movement
- Atmospheric humidity.

Quantification has variously involved monitoring the atmospheric parameters and simultaneously symptoms of stress, through either physiological or verbalized responses. Alternatively, simple instrument readings of dry bulb and globe thermometers, or the rates of heat loss by kata thermometers or from physical manikin models, have been used as simulations of the human body. All of these physical assessments have needed to translate the readings into human responses and make assumptions about body geometry, its thermoregulatory state, metabolic heat production and what represents a realistic environment. Many empirical measurements of heat exchange algorithms are available and a selection is shown in Table 12.2.

THERMAL STRESS ESTIMATES

Depending upon application and, particularly, the environmental conditions encountered, the expressions within the thermal balance have been detailed or simplified, and the basic equation has assumed a variety of forms. Most often, the value of the $\pm S$ term in the equation above itself has been assumed to be directly proportional to the degree of bodily strain likely to be experienced in an environment, and particular assemblages of several of the algorithms, predicting such strain, have been used as unitary indices of the thermal stress properties of environments.

Indices need to assume different mathematical functions in cold and warm conditions. In cold environments, evaporative–expired air energy losses can be treated as nearly constant, at about a quarter of the metabolic rate, and therefore relatively simple assumptions can be made about increased elevations in metabolic rates necessary to make up any shortfall in the thermal balance. The fluxes are assumed to be linear to the thermal gradient between the body and the environment. In warm environments,

Table 12.2 Summary of heat-loss equations

Radiation	R	$= e f_{eff} f_{cl} h_r (T_{cl} - T_r)$	$W m^{-2}$
	h_r	$= 4.6(1+0.01 T_r)$	$W m^{-2} K^{-1}$
	f_{eff}	$= 0.72$	
	f_{cl}	$= 1 + 0.15 I_{clo}$	
	e	$= 0.95$	
Convection	C	$= h_c (T_{cl} - T_a)$	$W m^{-2}$
	h_c	$= 8.3 \sqrt{v}$ $v > 0.2 ms^{-1}$	$W m^{-2} K^{-1}$
	h_c	$= 4.0$ $v < 0.2 ms^{-1}$	$W m^{-2} K^{-1}$
Evaporation			
Regulatory	E_{max}	$= h_c (p_{ssk} - p_a)$	$W m^{-2}$
	h_e	$= 1.65 h_c$	$W m^{-2} mb^{-1}$
	E	$= w E_{max}$	$W m^{-2}$
Insensible			
Skin diffusion	E_{is}	$= 4 + 0.12(p_{ssk} - p_a)$	$W m^{-2}$
Respiration (latent)	E_{res}	$= 0.0017M(59 - p_a)$	$W m^{-2}$
Respiration (dry)	C_{res}	$= 0.0014M(34 - T_a)$	$W m^{-2}$
Conduction through clothing	$R + C$	$= K = (T_{sk} - T_{cl})/(0.155 I_{clo})$	$W m^{-2}$
Evaporation through clothing	E_{max}	$= f_{pcl} h_c (p_{ssk} - p_a)$	$W m^{-2}$

Where e = emissivity, f_{eff} = effective radiation area, f_{cl} = clothing area factor, h_r = radiative transfer coefficient, T_{cl} = surface temperature of clothed body (in °C), T_r = radiant temperature (in °C), I_{clo} = clothing insulation in clo units, h_c = convective transfer coefficient, T_a = air temperature (in °C), v = air speed (in m s^{-1}), E_{max} = maximum evaporative loss from wet skin per unit area, p_{ssk} = saturated vapour pressure (mb), p_a = partial vapour pressure (mb), h_e = evaporative transfer coefficient, w = skin wettedness, E_{is} = evaporative loss by diffusion through skin, E_{res} = evaporative loss through respiration, C_{res} = dry respiration loss, T_{sk} = skin temperature (in °C), and f_{pcl} = permeation efficiency factor.

Source: McIntyre (1980).

however, variable sweating occurs which progressively leads to both increased metabolic rates as well as decreasing efficiency of cooling by sweat covering the evaporative surfaces, which makes calculations exceedingly complicated. The inefficiencies incurred in the sweating response have been estimated in laboratory experiments in which direct measurements of physiological responses (e.g. heart rates, metabolism, sweat secretion) have been translated into unitary indices such as the predicted 4 hour sweat rate (McArdle *et al.*, 1947), the heat strain index (Belding and Hatch, 1955), the relative strain index (Lee and Henschel, 1963) and the index of thermal stress (Givoni, 1963).

Szokolay (1985) has differentiated between these physiologically determined indices and those which are empirically established using social survey methods, such as the comfort–vote technique using the Bedford scale (Table 12.3), including effective temperature (Houghton and Yaglou, 1923), equivalent warmth (Bedford, 1936), operative temperature (Winslow *et al.*, 1937), and others. Effective temperature (ET) and its derivative corrected effective temperature (CET) are probably the best known and the latest variant, the new effective temperature ET* (Gagge *et al.*, 1971) is used in the ASHRAE Comfort Standard 55–74 (1981). Simply described, ET* is the temperature of an isothermal environment at moderate

Table 12.3 Verbal scales used in thermal comfort research and likely physiological and satisfaction responses

Numerical code	*ASHRAE scale*	*Bedford scale*	*Physiological responses*	*Comfort level*
+3	Hot	Much too warm	Profuse sweating	Unacceptable
+2	Warm	Too warm	Onset of sweating	Uncomfortable
+1	Slightly warm	Comfortably warm	Vasodilation	Acceptable
0	Neutral	Comfortable	Minimal	Maximum
−1	Slightly cool	Comfortably cool	Vasoconstriction	Acceptable
−2	Cool	Too cool	Onset of body cooling	Uncomfortable
−3	Cold	Much too cold	Shivering	Unacceptable

humidity, in which a lightly clothed individual would exchange the same total heat as in the atmospheres in question.

The much publicized PMV or predicted mean vote, however, comes from measurement of both the fluxes and subjective responses to stress questionnaires within the controlled laboratory studies of Fanger (1970). The PMV equations essentially are empirical translations of the heat exchanges described in the equation above but, in addition to the achievement of homeostasis, conditions of comfort need to satisfy the criteria of comfort in skin temperature and an appropriate rate of sweating according to given metabolic rates. The PMV equation indicates comfort for the sedentary samples of young Americans wearing light clothing at $-0.5 < \text{PMV} < +0.5$. Together with the standard effective temperature (SET), which allows for further variations in ET* as well as in metabolic rate, clothing amounts, and air movement (Nishi and Gagge, 1977), PMV and SET indices represent the most comprehensive mathematical models of human physiological responses to particular environments. Table 12.4 shows McIntyre's (1980) comparison of the main characteristics of some of the indices mentioned.

Outdoor calculations present particular sets of problems, and a large variety of models have been developed, including those that attempt to incorporate the complicated urban fluxes (Jendritzky and Nuebler, 1981; Hoeppe, 1993). Similarly complicated, but using aggregated data, have been the classification schemes of Terjung (1966) and Auliciems and Kalma (1979). For most everyday applications, using commonly forecasted synoptic parameters, the traditional indices suffice. These include the wind-chill index for cold K_0 (Siple and Passel, 1945) and the temperature humidity index (THI, or the US Weather Service's named 'discomfort index' or DI) for hot conditions (Thom, 1959), expressed as follows:

$$K_0 = 11.62\,(v^{0.5} + 1.45 - 0.1v)\,(33 - T_a)$$
$$\text{THI} = 0.72\,(T_a + T_d) + 40.6\ (^\circ\text{C})$$
$$\text{THI} = 0.4\,(T_a + T_d) + 15\ (^\circ\text{F})$$

where v is wind velocity in m s^{-1}, and T_a and T_d are air temperature and dew point, respectively. K_0 actually is an assessment of heat losses in W m^{-2}, and thus can be directly taken as an objective measure of stress, while THI (being dimensionless) has been interpreted by the US Weather Service. A value of THI = 75 is thought to cause discomfort in 50 per cent of people, while values in excess of 85 is cause enough to dismiss workers (Mather, 1974). Both indices have seen considerable developments for particular applications, with particular attention paid to time resolutions (for K_0 see Steadman, 1971; Coronato, 1995; and for THI see Steadman, 1984; Kalkstein and Davis, 1989).

Table 12.4 Comparison of comfort indices

Index	*Variables**	*Range*	*Comments*
Effective temperature (ET)	T_a, RH, v Two levels clo	0 < ET < 45°C $v < 2.5$ m s^{-1}	Major index, now superseded. Overestimated effect of humidity at low temperatures
Corrected effective temperature (CET)	T_a, T_r, RH, v		As above. Used in British armed forces
Resultant temperature (T_{res})	T_a,T_{wb}, v	$20 < T_{res} < 40$°C $v < 3$ m s^{-1}	Similar to ET, but more accurate. Separate charts for nude–sedentary and clothed–active
Equivalent temperature (T_{eq})	T_a,T_r, v	$8 < T_{eq} < 24$ $v < 0.5$ m s^{-1}	Originally defined as eupatheoscope reading. Later defined by Bedford's regression equation
Fanger's comfort equation,	All	Comfort condition	
Predicted mean vote (PMV)	All	2.5 < PMV < 5.5	No range stated, but PMV becomes less certain away from thermal comfort
Standard effective temperature (SET)	All	Shivering to upper limit of prescription zone	Based on model of physiological response. Most general index
New effective temperature (ET*)	T_a, RH	Sedentary, light clothing only	Used by ASHRAE for indoor comfort zone
Subjective temperature (T_{sub})	T_a,T_r, v, H,I_{clo}	Near comfort	Simple index used to predict comfort conditions, based on Fanger's equation

*T_{wb} = wet bulb temperature; other terms are defined in Table 12.2.
Source: McIntyre (1980).

THE CLOTHING ENVELOPE

The energy flow estimates in the preceding sections are based either on controlled laboratory studies, or some assumed average human state. However, humans mostly do not reside in a free atmosphere, but rather within a microclimate of their own creation. This atmospheric environment can be conceptualized as a series of insulating envelopes, where each consecutive envelope is enclosed by physiological or technologically created physical boundaries. The well-being of the organism is determined by the successful maintenance of the whole envelope series and neither clothing, nor the surrounding building fabric envelopes, can be entirely thought of as independent artefacts, or even in isolation from the prevailing climatic or cultural determinants. As in the case of transplanted technocultural patterns of migrant groups moved to climatic zones which are considerably different compared with those in their original homelands (as heatedly debated in places such as Australia, for example Griffith Taylor (1959) and Cilento (1925) and reconciliation by Sargent (1963)), failure to adapt has led to considerable thermal stress, ill health and social inequalities (Auliciems and Deaves, 1988).

Estimates of the insulating properties of clothing, very aptly described by Burton and Edholm (1955) as 'the private climate', require the calculation of the resistance to thermal transfer (I_{cl}, in °C W m^{-2}) from the following equation:

$$I_{cl} = \frac{T_s - T_a}{M_T} - \frac{[M_T + (Q + q)/A_D]}{hM_T}$$

where T_s is the temperature of the body surface, T_a is the dry bulb temperature, M_T is the net metabolic rate, $(Q + q)$ is incoming direct and diffuse short-wave radiation, A_D is the Dubois and Dubois (1916) surface area of the body as calculated from body mass and height. The term h is dry heat flux, or the inverse of Burton and Edholm's (1955) 'insulation of air' as determined by convection and radiation: $h = 6.6 + 8.7\ v^{0.5}$ (Monteith and Mount, 1974), where v is wind velocity in m s^{-1}.

The insulation properties of clothing have received much detailed attention and are discussed in Burton and Edholm (1955) and Fourt and Hollies (1969) and have been used to describe a variety of climates (Anstey, 1966; Auliciems and Kalma, 1981; de Freitas, 1979; Lee and Lemons, 1949; Siple, 1949). Particularly elegant has been the translation of I_{cl} into clothing or clo units (Gagge *et al.*, 1941), as graphically portrayed in Figure 12.1. Typical insulation values of I_{cl} appear in Table 12.5. Owing to the availability of new thermally efficient fabrics, and changing customs in clothing tradition, the facility for alterations of garments could be the simplest and technologically the 'most appropriate' response to climate variability. Maximization of choice in clothing fabric, fibre and insulation as ever represents a major survival strategy.

THERMAL COMFORT

Thermal comfort is the mental state achieved when (1) physiologically the thermoregulatory mechanisms are minimally activated, and (2) psychologically the perceiver is satisfied with

Figure 12.1 Clo unit representation. The numbers show the value of clo units corresponding to the ensembles depicted (Auliciems and Hare, 1973)

Table 12.5 Typical insulation values in clo units (1 clo unit is equivalent to 0.155°C per W m^{-2} or approximately 60 mm thickness of clothing including air spaces)

Clothing ensemble	*clo units*
Nude	0
Shorts	0.1
Light dress, pantihose, bra and panties	0.3
Shorts, short-sleeved shirt, sandals	0.4
Blouse, slacksdress, pantihose, bra and panties, shoes	0.5
Long lightweight trousers, short-sleeved shirt, socks, shoes	0.5
Sweater, skirt, blouse, slip, pantihose, bra and panties, shoes	1.0
Business suit with usual undergarments and footwear	1.0
1 clo plus light overcoat	1.5
1 clo plus substantial overcoat	2.0
Long underwear, flannel shirt, sweater, heavy trousers, heavy coat, woollen socks, boots	3.0–3.5
Specialized polar or alpine ensembles	4.0–4.5

Sources: ASHRAE (1981), Fanger (1970), Fourt and Hollies (1996).

the thermal environment. Roughly speaking these conditions are likely to be achieved when $S = 0$ in the first equation. However, it cannot be assumed that the achievement of either $S = 0$, or any other particular values of thermal stress indices, can be anything more than gross predictors of human comfort and well-being. $S = 0$ 'over time' does not adequately reflect the dynamic nature of thermoregulation, and temporal deviations from $S = 0$, or even the threat of such deviations, become stimuli for the organism to fine-tune its own responses and to ameliorate its environment.

There has been considerable debate about the semantics involved in the assessment of comfort (Auliciems, 1981; Braeger *et al.*, 1993; de Dear, 1985; 1993; Rohles and Laviana, 1985; Oseland and Humphreys, 1993), However, subjective responses to either the Bedford (1936) or ASHRAE scales, as shown in Table 12.3, are regarded as fair indicators of the degree of comfort experienced. Each of the central response categories is about 2°C wide, and the three central category responses are assumed to constitute the comfort zone for an individual. For public spaces, in practice, no particular thermal level is likely to satisfy more than 80 per cent of the population and the limits of the comfort zone need to be set within ±1.5 votes of the central category. The comfort zone is that within which physiological activity is restricted to vasomotor responses (dilatation and constriction of surface blood vessels in warm and cool environments respectively) and postural adjustments. The main mechanisms of physiological responses beyond this zone are those of sweating in heat and thermogenesis (increased metabolism by muscular tension and shivering) in cold.

Within the range of moderate environments, as usually employed in laboratory studies, very few comparisons are available to test for differences in physiological responses and preferences in the various peoples of the earth. This lack of information has been at the basis of an oversimplified but convenient assumption of a causal chain of events that determines response to thermal stimulus: thermal stimulus → increased thermoregulatory activity → reduced thermal comfort → increased behavioural response → increased modification of ambient warmth → reduced thermal stress. These convenient assumptions of simplicity and universality of response have been endorsed by the dominant building codes of the last decades of the twentieth century, and have created a demand for standard products and services within the HVAC marketplace and especially amongst the engineering proponents of air conditioning.

Integrations of field studies of comfort surveys (Humphreys, 1975; 1976; Auliciems, 1981; 1983) have increasingly dispelled this oversimplified constancy or static hypothesis. Reviews of field comfort surveys around the world, using naturally acclimatized subjects and mainly the Bedford (1936) scale, have produced distributions as shown in Figure 12.2. Several regressions of group comfort values (T_φ) on outdoor mean monthly temperatures (T_m) and indoor mean monthly temperatures (T_i) have been established, including the equations below (Auliciems, 1983). The possible range of

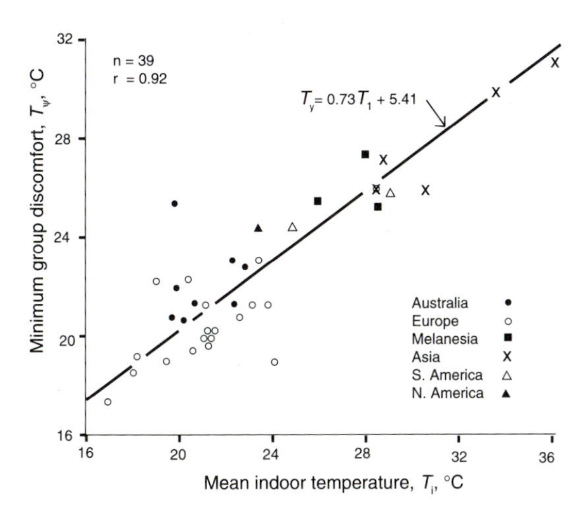

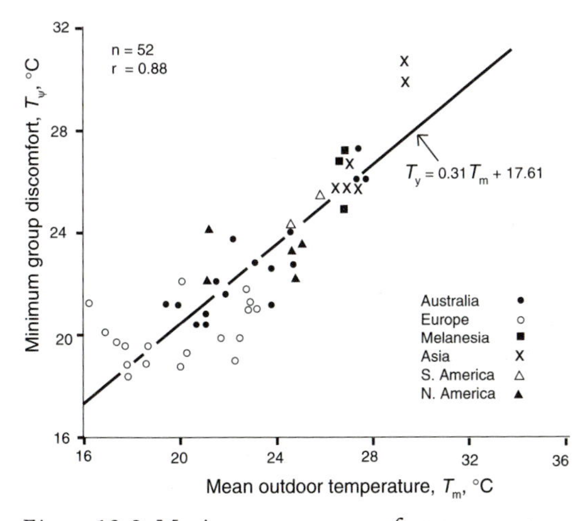

Figure 12.2 Maximum group comfort temperatures against mean monthly indoor and outdoor temperatures. Observations are based on comfort voting for a quarter of a million individuals using the Bedford scale. Minimum discomfort can be observed to range between about 16 and 32°C. The results show acclimatization to prevailing levels of warmth (Auliciems, 1983)

T_φ appears to be some 10°C in width, ranging upwards from about 17°C.

$$T_\varphi = 0.31T_m + 17.6 \qquad (r = 0.88)$$
$$T_\varphi = 0.73T_i + 5.41 \qquad (r = 0.92)$$
$$T_\varphi = 0.48T_i + 0.14T_m + 9.22 \qquad (r = 0.95)$$

For closely climate-controlled buildings (those having centralized heating and ventilating systems) on the other hand, a curve rather than a straight line produces the best fit (Humphreys, 1976). The explanation for such differences in response is given by de Dear (1985), de Dear, *et al.* (1991) and de Dear (1993): people within closely controlled microclimates, especially within air-conditioned ones, lose their adaptation capacities to thermal variability. These real-life observations have necessitated the development of a more complex adaptive model of thermoregulation (Figure 12.3) that recognizes that both group differences in thermal thresholds at which physiological processes are activated, and the thermopreferendum, are related to personal and cultural experiences (Auliciems, 1981; 1983; de Dear, 1985; 1993; Braeger *et al.*, 1993; Heijs, 1993; Cena, 1993; Schiller *et al.*, 1988). These thresholds and preferences are enhanced by maximized acclimatization (strategy 2), which, together with clothing (strategy 4), provide the most effective strategies for coping with atmospheric variability, with minimum cost to the environment. Not surprisingly, within some vested interest groups such findings are anathema.

ACCLIMATIZATION

Acclimatization (including habituation) is discussed in Adolph (1947), Glaser (1966), Nadel *et al.* (1974), Hellon *et al.* (1956), Scholander (1955), Sargent (1963), Fox (1974), Gonzalez (1979), Weihe (1992) and Wyndham (1970). It is a complex set of physiological and psychological readjustments that take place when the

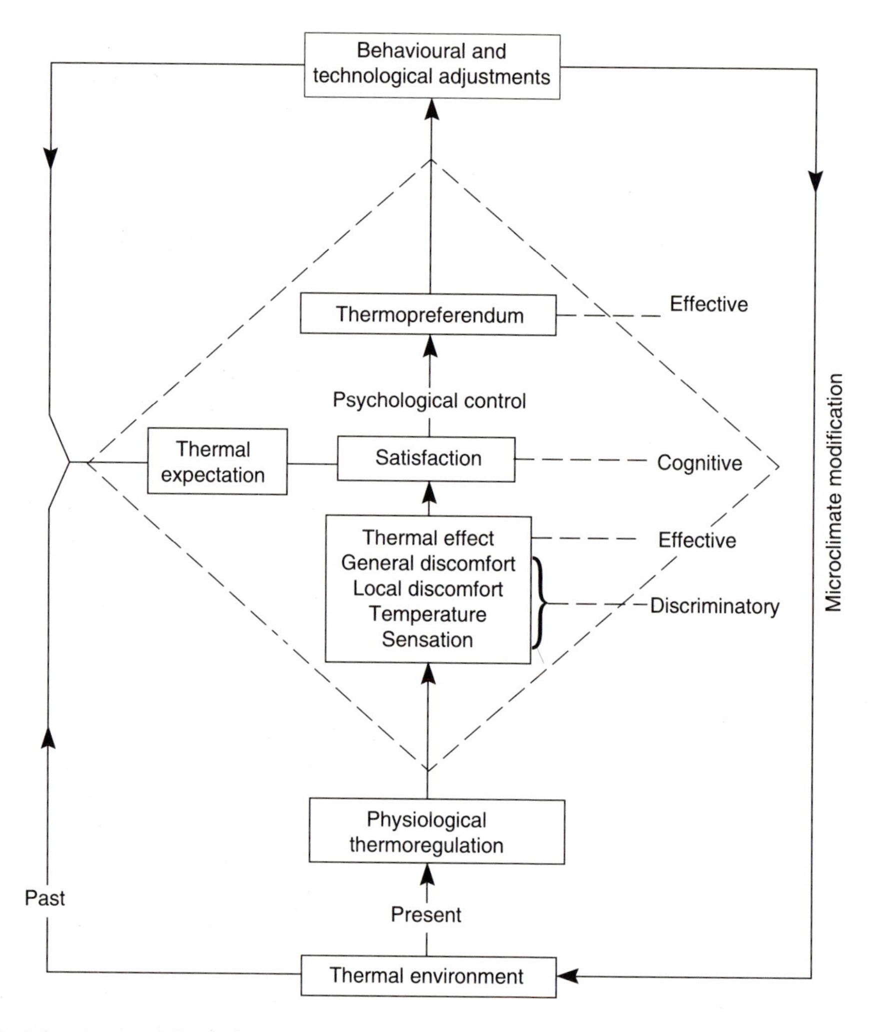

Figure 12.3 Adaptive model of thermoregulation. Behavioural and technocultural responses to thermal discomfort are modified by acclimatization processes that determine thermal expectation. The model places emphasis on psychological control that is absent from more traditional thermophysiological models (Auliciems, 1983)

organism is exposed to stress. This exposure acts to reset organic and cellular thresholds and rates of function by cardiovascular, pulmonary, endocrinal, digestive and nervous systems. The manifestations of heat acclimatization include an enhanced efficiency of heat loss by sweat, both in terms of volumes and composition and, in accord with adaptation theory (Helson, 1964; Wohwill, 1974), a readjustment of temperature preference towards the stress stimulus, as discussed in an earlier section. The tangible results include decreased sensations of discomfort, improved work performance and in general an increased well-being. Acclimatization probably begins to occur within days of exposure to the stimulus but, in general, it is a prolonged seasonal process where its full attainment results from everyday thermal experiences. Acclimatization is speeded up in people whose work is sufficiently vigorous to elevate metabolic heat production.

Although considerable field research is still necessary it would seem obvious that, in everyday life, acclimatization frequently takes place through people's recreational activities in outdoor environments. While escape from air conditioning may increase such outdoor exposure (Hirokawa and Horie, 1985), such avoidance may become a stress response in itself, and one that is not likely to lead to vigorous activities. For voluntary exposure to outdoor variability, indoor conditions need to be harmonious with those outdoors (Hirokawa and Horie, 1985; de Dear, 1985; 1993; Bromberek, 1996). That is, both satisfaction and opportunity for acquiring and maintaining acclimatization are likely to be enhanced when there is inducement for exposure to natural variability in atmospheric stimuli, and when people are free to choose between the higher strategies (3 to 7).

ATMOSPHERIC IMPACTS ON PERFORMANCE AND BEHAVIOUR

Acclimatization to severe stimuli, such as extreme temperatures or synoptic-scale weather changes, is not always possible, resulting in discomfort and dysfunction. In the case of extreme geophysical phenomena, or hazard events (Chapter 23), the impact also very much depends upon societal adjustment processes, including their perceptions and attitudes towards the event and their own responses. Large numbers of studies have been conducted within this field (as for example the series of papers in the Natural Hazards Research series at the universities of Toronto and Chicago) but, for a synopsis, reference should be made to Burton, *et al.* (1985).

Within the area of human performance, there is some evidence to suggest that moderate thermal stress may actually lead to improved performances in schools (Pepler and Warner, 1968) and this applies within factories, with heat-acclimatized workers (Wyon *et al.*, 1982). However, in general, exposure to discomfort leads to loss of capacities for physical and mental work. This has been observed in tasks of vigilance (Mackworth, 1950), motor co-ordination and dexterity (Teichner and Wehrkamp, 1954) and ability or perhaps willingness to concentrate (Auliciems, 1972). Significantly, in recent years, attention has turned towards productivity as related to the total work environment 'sick building syndrome' (Table 12.6). In general, and not surprisingly, it may be concluded that comfortable environments and a 'fit' workforce are likely to be the most productive.

Changes in more complex human behavioural patterns, in association with short-term temperature and weather variability, also have become documented, including moods (Auliciems, 1981), traffic accidents (Jendritzky, *et al.*, 1979), prison order (Haertzen *et al.*, 1993), street riots (Maunder, 1970), sexual aggression (Cohn, 1993) and domestic violence (Rotton and Frey, 1985; Auliciems and DiBartolo, 1995). The direct causes for such changes in behavioural patterns are not always obvious and law enforcement officials readily point out the likely connection of antisocial behaviour to drug (including alcohol) abuses, which in themselves may be equally influenced by atmospheric parameters (see Chapter 19, Social aspects). While the results of the behavioural studies represent statistical associations only, the common trend indicates that antisocial behaviour is a function of environmental warmth, especially elevations of temperature above likely comfort ranges.

Table 12.6 Causes, symptoms and costs of the sick building syndrome (SBS)

SBS causes	*SBS symptoms*	*SBS costs*
Environmental control	**Mucous membrane irritation**	**Job satisfaction**
HVAC or naturally ventilated	Irritation of eyes, nasal	
Lighting control	congestion,	**Self-esteem**
Cellular or open-plan office	dry throat and cough	
Type of organization		**Productivity**
	Neurotic disorders	
Human parameters	Headaches,	**Absenteeism**
Duration of occupancy	dizziness,	
Density of occupancy	fatigue,	**Personal turnover**
Computer usage	nausea,	
Age, sex, status	irritability,	**Medical**
Job satisfaction	lack of concentration,	
Environmental satisfaction	malaise	**Building management**
Ambient conditions	**Skin irritation**	**Perceived value of**
Inadequate fresh air	Skin rash,	**property**
Chemicals pollutants,	itchiness	
bioaerosols,		
inorganic dust,	**Odour**	
airborne pollutants	Odour and taste complaint	
Lack of thermal stimulus		

Sources: Hedge *et al.* (1989), Hodgson (1992), Jaakola (1993), Rowe (1994).

ATMOSPHERIC IMPACTS ON MORBIDITY AND MORTALITY

Beyond behavioural dysfunction, exposure to environments that repeatedly demand the more extreme responses may prove to be damaging to human health, either directly or indirectly, or by debilitation of capabilities in the resistance to micro-organism infection. In general, seasonal increases in morbidity and mortality may be expected in human populations, especially in the elderly (Hsia and Lu, 1988; Fox *et al.*, 1973; Collins, 1979), the poor and very young (Kalkstein, forthcoming), and when the stress risk is unexpected and sudden (Weihe and Mertens, 1991). Amongst the most notable direct effects are the impacts of the passage of mid-latitude low- and high-pressure 'weather phases' (Ungeheuer, 1955; Bucher and Haase, 1993) and the consequences of heat and cold upon the impaired or diseased thermoregulatory systems. These may result in cardiovascular, cerebrovascular, respiratory, endocrinal, renal, rheumatic and consumptive diseases (Maarouf, 1992).

Whatever the method of describing the atmospheric condition, not surprisingly, the relationship between atmospheric factors and mortality has received even more attention. Over the past two decades, analysis has been carried out in many locations both for (1) synoptic time series (Driscoll, 1971; Ellis and Nelson, 1978; Kalkstein and Valimont, 1986; Lye and Kamal, 1977) and (2) aggregated meteorological data (Hsia and Lu, 1988; Kalkstein and Davis, 1989; Langford and Bentham, 1995; States, 1977). Because of its dominance in the categories of causes of death, and its thermoregulatory connotations, particular attention has been paid to death resulting from heart diseases (Al-Yusuf *et al.*, 1986; Bull, 1973; Frost and Auliciems, 1993; Keatinge *et al.*, 1986; Mukammal, *et al.*, 1984; and Sarna, *et al.*, 1977).

The common observation in these studies is that death rates are elevated by both thermal extremes, the response being in part relative to the nature of the certified cause of death (Kalkstein and Davis, 1989; Ellis and Nelson, 1978; Hsia and Lu, 1988; Langford and Bentham, 1995). Most affected appear to be those related to the cardiovascular system and particularly notable are the differences in temperatures thresholds at which death rates accelerate. These vary according to prevailing warmth of locations, and may be related to particular thermoregulatory thresholds (Figure 12.4). In general, the percentage variability in death rates explicable by atmospheric factors increases with the aggregating period employed, but in general the percentages are approximately 10 per cent for daily periods, 50 per cent for weekly periods, 70 per cent for fortnightly and monthly time resolutions and, depending upon the statistical approaches used, over 90 per cent for annual periods (Auliciems *et al.*, forthcoming). Irrespective of such statistics, however, the direct causation of ill health, as in the case of morbidity and behavioural effects, is far from self-evident (Driscoll, 1992). Impacts may result from complex interactions at several levels with complicated feedbacks and controls, any of which may be affected by atmospheric factors with different weightings within individual parameters. To attribute morbidity and mortality to a specific parameter would be erroneous, and the phenomena need to be treated as parts of complex biological–environmental interactions.

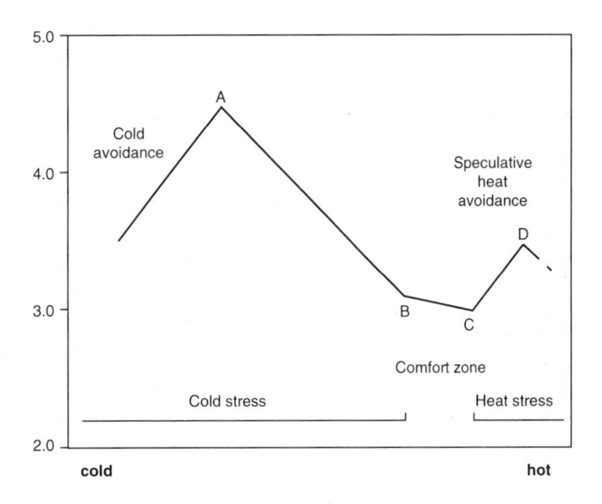

Figure 12.4 A postulated general relationship between mortality and temperature. The thermal range is divided into a narrow 'comfort zone', bounded on either side by stress zones. At both extremities, amongst the more prosperous societies, susceptible individuals are likely to avoid thermal stress and death rates may fall. Point A, cold maximum death temperature; point B, acceleration temperature for cold-related mortality; point C, minimum death temperature probably coinciding with maximum comfort; point D, hot maximum death temperature (speculative). Since the positions of thermal responses and comfort sensations are relative to exposure, the thermal abscissa is shown as a cold → hot continuum. The death rates on the ordinate are approximations in rates per million of all ages of populations (based on Auliciems, 1972; Frost and Auliciems, 1993)

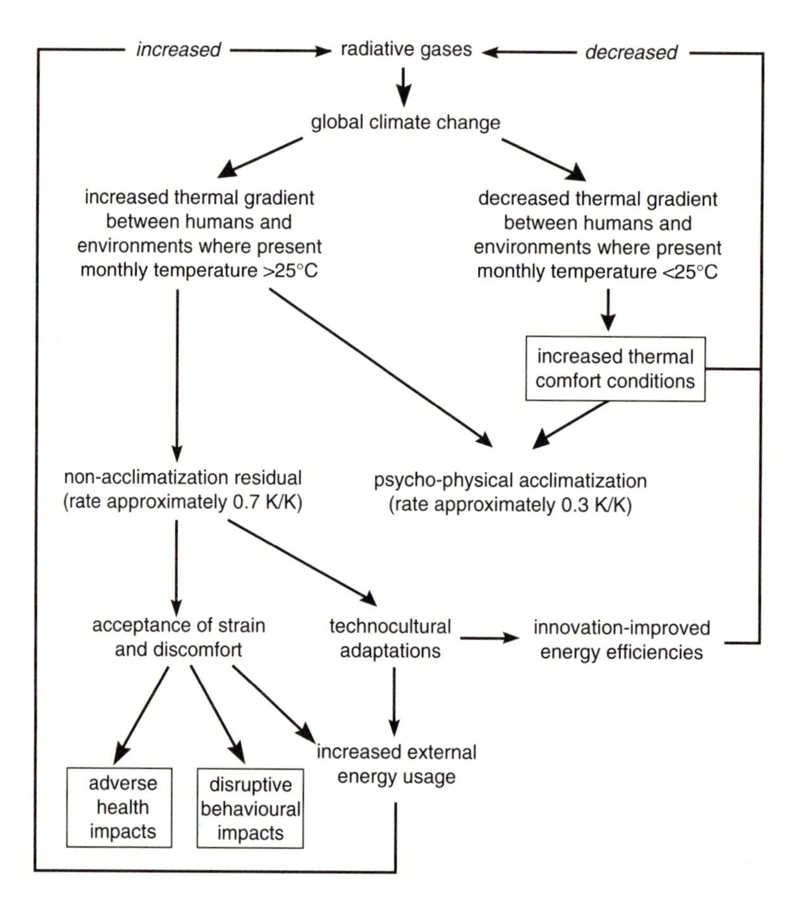

Figure 12.5 Some theoretical relationships between acclimatization, thermal comfort, health and behaviour impacts, and the human generation of greenhouse gases. Critical to the feedback loop is the capacity for acclimatization and the nature of technocultural innovation

CONCLUSIONS

While thermally comfortable indoor environments are taken for granted, especially by urban dwellers, the increasing dependence on equable indoor warmth must also be examined in the light of the global need to conserve energy resources and reduce the emissions of radiative gases. It should be remembered that thermal comfort is remarkably expensive: in the latter part of the twentieth century, globally it probably consumes about a quarter of all energy conversions. Looking towards the future, with typical doubled CO_2-equivalent scenarios as generated by computerized GCMs (Chapter 4), even with maximum acclimatization and energy-efficient designs, achievement of homeostasis would be more difficult for some 20 per cent of the world's population the whole year round, and heat stress would be an increased seasonal problem for another 65 per cent (Auliciems, 1994).

CONCLUSIONS *continued*

Increasing educational efforts designed to wean people away from unnecessary space heating, and especially air cooling, seems to be an essential policy. The role of homeothermy in potential global warming is illustrated in Figure 12.5. Clearly, desirable processes are the functions of acclimatization and adaptation on the right hand side of the diagram that tend to reduce radiative gas emissions.

Physiologically and psychologically well-adapted people are likely to be those achieving an effective balance between the gaining of maximum acclimatization and normally residing within comfortable environments of their own choosing. That is, sensible management strategies would provide amenities and environments to encourage physical exercise with adequate exposure to thermal stimulus, but otherwise provide freedom from thermal stress during everyday work and rest periods. Adopting approaches that focus on adaptability, especially acclimatization and increased reliance of clothing variation, and less reliance on active energy utilization, would promote a more harmonious synchronization between the psycho-physiological functions of people and the natural environment. The use of variable indoor comfort concepts provides increased scope for natural alterations of clothing, and would reduce the average temperature gradient between indoors and outdoors by about a third of a degree per degree of gradient (see the equations above). Such a change would imply a directly proportional 30 per cent reduction in the need for active energy utilization for space heating and cooling, without reduction in comfort, but probably with considerable improvement in ambient air quality and human health.

Neglecting at this time the educational problem of how such change can be instituted within free societies, it seems desirable that for populations already residing in buildings with close climate controls, a programme of deregulation could be instituted by allowing a gradual but progressively increasing indoor temperature drift towards that of outdoors. Depending upon available data and the type of building, the general target indoor level of warmth could be determined by empirical relationships such as those in the equations above. For example, in old buildings, providing that adequate temperature information exists, the last two equations would be preferable; in new buildings the third last equation would be appropriate. With further verification of short-term weather effects on indoor comfort, other equations could be established and utilized depending upon the facility of building technology.

During the past decades, there have been considerable advances in architectural science principles based on human requirements (Ruck, 1989) for 'passive' energy utilization, 'smart building' designs and microprocessor developments. Innovative cybernetic approaches have been developed for the provision of environments for individual needs and specialist tasks (Duffy, 1993), and indoor environments can be made to respond interactively with dynamic diurnal and seasonal changes within the atmosphere. Perhaps the simplest approach is to readjust again the already versatile thermostat, already variously modified to anticipate changes in heat flows to and from the outdoors, to set back to reduce levels of

CONCLUSIONS *continued*

heating and cooling at specified times, and in general to achieve an increased precision in operations. What needs to be further altered, either by designated chip or by suitable interactive programme, is its 'stat' functions to dynamic ones (for discussion of this thermomobile, or more simply 'thermobile', concept and design see Auliciems, 1984).

Apart from utilizing smart building designs and individual control capacities, there is reason to reappraise the passive energy technologies of traditional designs that, in conjunction with lifestyles to suit particular climatic rhythms, have coped well with thermal inclemency. Here, in the main, indoor thermal regulation has depended on natural ventilation, shading, sizes and positions of openings, orientation materials, and thermal mass – the very elements essential for the enhancement of outdoor exposure and acclimatization. There exists more than just prima facie evidence to suggest that, at least in some instances, a return to such microclimate control methods may be environmentally and socially the most appropriate approaches in many locations. The alternative of a smart building–cybernetic approach, based on adaptation rather than some current mechanistic close control 'zookeeper' (Cooper, 1982) methods of microclimate management, may produce five major groups of benefits, as follows:

1 Enhancement of overall comfort levels, and increase of individual capacity for microclimate adjustment.
2 Reduction in the incidence of the 'sick building syndrome'.
3 Encouragement of adjustment by individual clothing changes and the wearing of ensembles that better reflect weather conditions outdoors.
4 Reduction of outdoor–indoor temperature gradients and thus the thermal 'shock' of moving from one environment to another, promotion of outdoor activities and improvement in 'physical fitness' and the facility for seasonal acclimatization.
5 Reduction in the need for HVAC technologies and fossil energy requirements.

Irrespective of the strategies adopted, the rationale and economy of air cooling in latitudes north and south of 35° needs to be put under close scrutiny. What is questioned here is not innovation in HVAC technologies, nor air cooling when thermal stress is otherwise unavoidable in the workplace, but air cooling's uncritical and wasteful usage in large volumes and at times when other technologies or behavioural patterns are available. In the larger sense, whatever may be the economic arguments in favour of air-conditioning developments, the proliferation of this technology is likely to increase the production of greenhouse gases, and lead to a progressive degeneration of thermoregulatory adaptability in people living in closely controlled environments. Should disruptions occur in their maintenance, as could take place with global warming, populations are likely to become increasingly susceptible to impairment in thermoregulatory mechanisms, social behaviour, productivity and health.

REFERENCES

Adolph, E.F. (1947) *Physiology of Man in the Desert*, New York: John Wiley.

Al-Yusuf, A.R., Kolar, J., Bhatnaga, S.K., Hudak, A. and Smid, J. (1986) 'Seasonal variation in unstable angina and acute myocardial infarction: effect of hot dry climate on the occurrence of complications following acute myocardial infaction', *Journal of Tropical Medicine and Hygiene* 89: 157–61.

Anstey, R.L. (1966) 'Clothing almanac for Southeast Asia', *Technical Report 66–20 ES*, US Army Nattick Laboratories, MA.

ASHRAE (1981) *Standard 55–81: Thermal Environmental Conditions for Human Occupancy*, ASHRAE, Atlanta.

Auliciems, A. (1972) *The Atmospheric Environment. A Study of Comfort and Performance*, Toronto: University of Toronto Press.

Auliciems, A. (1981) 'Towards a psycho-physiological model of thermal perception', *International Journal of Biometeorology* 25: 109–22.

Auliciems, A. (1983) 'Psycho-physiological criteria for global zones of building design', in O. Orerdieck, J. Muller, and H. Lieth (eds) *Proceedings of the 9th International Society of Biometeorology Conference, Part 2, Biometeorology 8*, Lisse: Swets and Zetlinger, pp. 69–86.

Auliciems, A. (1984) 'Thermobile controls for human comfort', *Heating and Ventilating Engineer* 58: 31–4.

Auliciems, A. (1994) 'Thermoregulatory adaptation to global warming – winners and losers', in A.R. Maarouf, N.N. Barthakur, and W.O. Haufe (eds) *Biometeorology*, Part 2, Vol. 1, *Proceedings of the 13th International Congress of Biometeorology, 12–18 September 1993, Calgary*, Downsview: Environment Canada.

Auliciems, A. and Deaves, S. (1988) 'Clothing, heat and settlement', in R.L. Heathcote (ed.) *The Australian Experience. Essays in Australian Land Settlement and Resource Management*, Melbourne: Longman Cheshire, Ch. 8.

Auliciems, A. and DiBartolo, L. (1995) 'Domestic violence in a subtropical environment: police calls and weather in Brisbane', *International Journal of Biometeorology* 39: 44–9.

Auliciems, A. and Hare, F.K. (1973) 'Visual presentation of weather forecasting for personal comfort', *Weather* 28: 478–80.

Auliciems, A. and Kalma, J.D. (1979) 'A climatic classification of human thermal stress in Australia', *Journal of Applied Meteorology* 18: 616–26.

Auliciems, A. and Kalma, J.D. (1981) 'Human thermal climates of Australia', *Australian Geographical Studies* 19: 3–24.

Auliciems, A., Frost, D. and Siskind, V. (forthcoming) *The Time Factor in Mortality: Weather Associations in a Subtropical Environment*.

Bedford, T. (1936) 'The warmth factor in comfort at work', *Report of the Industrial Health Research Board* 76, London: HMSO.

Belding, H.S. and Hatch, T.F. (1955) 'Index for evaluating heat stress in terms of resulting physiological strain', *Heating Piping and Air Conditioning* 27: 129–36.

Braeger, G.S., Fountain, M., Benton, C.C., Arens, E.A. and Baumen, F.S. (1993) 'A comparison of methods for assessing thermal sensation and acceptability in the field', in N.A. Oseland, and M.A. Humphreys (eds) *Thermal Comfort: Past, Present and Future, Proceedings of a conference held at the Building Research Establishment, Garston, 9–10 June 1993*, Watford: Building Research Establishment.

Bromberek, Z. (1996) 'Passive climate control for tourist facilities in the coastal tropics (Far North Queensland)', unpublished PhD thesis, University of Queensland.

Bucher, K. and Haase, C. (1993) 'Meteorotropy and medical-meteorological forecasts', *Experientia* 49: 759–68.

Bull, G.M. (1973) 'Meteorological correlates with myocardial and cerebral infarction and respiratory disease'. *British Journal of Preventative and Social Medicine* 27: 108–13.

Burton, A.C. and Edholm, O.G. (1955) *Man in a Cold Environment*, London: Edward Arnold.

Burton, I., Kates, R.W. and White, G.F. (1985) *The Environment as Hazard*, New York: Oxford University Press.

Cena, K. (1993) 'Thermal and non-thermal aspects of comfort surveys in homes and offices', in N.A. Oseland and M.A. Humphreys (eds) *Thermal Comfort: Past, Present and Future, Proceedings of a conference held at the Building Research Establishment, Garston, 9–10 June 1993*, Watford: Building Research Establishment.

Cilento, R.W. (1925) 'The white man in the tropics', *Department of Health Service Publication (Tropical Division), No. 7*, Melbourne: Commonwealth of Australia, Govt. Printer.

Cohn, E.G. (1993) 'The prediction of police calls for service: the influence of weather and temporal variables on rape and domestic violence', *Journal of Environmental Psychology* 13: 71– 83.

Christensen, E.H. (1953) 'Physiological evaluation of work in Nykroppa Iron works', in W.F. Floyd, and A.T. Welford, (eds) *Symposium on Fatigue*, London: Lewis, pp. 95–107.

Collins, K.J. (1979) 'Hypothermia and thermal responsiveness in the elderly', in P.O. Fanger, and O. Valbjorn (eds) *Indoor Climate: Proceedings of the First International Indoor Climate Symposium*, Copenhagen: Danish Building Research Establishment.

Cooper, I. (1982) 'Comfort theory and practice: barriers to the conservation of energy by building occupants', *Applied Energy* 11: 243–88.

Coronato, F.R. (1995) 'Windchill influence on thermal conditions in North Patagonia', *International Journal of Biometeorology* 39: 87–93.

de Dear, R.J. (1985) 'Perceptual and adaptational basis for the management of indoor climate', unpublished PhD thesis, University of Queensland.

de Dear, R.J. (1993) 'Outdoor climate influences on indoor comfort requirements', in N.A. Oseland and M.A. Humphreys (eds) *Thermal Comfort: Past, Present and Future, Proceedings of a conference held at the Building*

Research Establishment, Garston, 9–10 June 1993, Watford: Building Research Establishment.

de Dear, R.J., Leow, K.G. and Foo, S.C. (1991) 'Thermal comfort in humid tropics: field experiments in air conditioned and naturally ventilated buildings in Singapore', *International Journal of Biometeorology* 34: 259–65.

de Freitas, C.R. (1979) 'Human climates of northern China', *Atmospheric Environment* 13: 71–7.

Driscoll, D.M. (1971) 'The relationship between weather and mortality in ten major metropolitan areas in the United States, 1962–1965', *International Journal of Biometeorology* 15, 23–40.

Driscoll, D.M. (1992) 'Everyday weather and human response: should we be concerned?', in A.R. Maarouf (ed.) *Proceedings of the Weather and Health Workshop, 19–20 November, Ottawa, Canada*, Downsview: AES, pp. 30–9.

Dubois, D. and Dubois, E.F. (1916) 'A formula to estimate approximate surface area if weight and height are known', *Archives of Internal Medicine* 17: 863–71.

Duffy, F. (1993) 'Designing comfortable working environments based on user and client priorities', in N.A. Oseland, and M.A. Humphreys (eds) *Thermal Comfort: Past, Present and Future, Proceedings of a conference held at the Building Research Establishment, Garston, 9–10 June 1993*, Watford: Building Research Establishment.

Ellis, F.P. and Nelson, F. (1978) 'Mortality in the elderly in a heat wave in New York City, August 1975', *Environmental Research* 15: 504–12.

Fanger, P.O. (1970) *Thermal Comfort*, Copenhagen: Danish Technical Press.

Fourt, L. and Hollies, N. (1969). 'The comfort and function of clothing', *Technical Report 69–74 CE*, US Army Natick Laboratories, MA.

Fox, R.H. (1974) 'Heat acclimatization and sweating response', in J.L. Monteith, and L.E. Mount (eds) *Heat Loss from Animals and Man: Assessment and Control*, London: Butterworths.

Fox, R.H., Woodward, P.M., Exon-Smith, A.N., Green, M.F., Donjnison, D.V., and Wicks, M.H. (1973). 'Body temperature in the elderly: a national study of physiological, social and environmental conditions', *British Medical Journal* 1: 200–6.

Frost, D. and Auliciems, A. (1993) 'Myocardial infarct data, the population at risk and temperature regulation', *International Journal of Biometeorology* 37: 46–51.

Gagge, A.P. (1936) 'The linearity criterion as applied to partitional calorimetry', *American Journal of Physiology* 116: 656–68.

Gagge, A.P., Burton, A.C. and Bazett, H.C. (1941) 'A practical system of units for the description of the heat exchange of man with his thermal environment', *Science* 94: 428–30.

Gagge, A.P., Stolwijk, J.A.J. and Nishi, Y. (1971) 'An effective temperature scale based on a simple model of human physiological regulatory response', *ASHRAE Transactions* 77: 247–62.

Givoni, B. (1963) *Man, Climate and Architecture*, Amsterdam: Elsevier.

Glaser, E.M. (1966) *The Physiological Basis of Habituation*, London: Oxford University Press.

Gonzalez, R.R. (1979) 'Role of natural climatization (cold and heat) and temperature: effect of health and acceptability in a built environment', in P.O Fanger, and O. Valbjorn (eds) *Indoor Climate*, Copenhagen: Danish Building Research Institute, pp. 737–52.

Haertzen, C., Buxton, K., Covi, L. and Richards, H. (1993) 'Seasonal changes in rule infractions among prisoners: a preliminary test of the temperature-aggression hypothesis', *Psychological Reports* 72: 195–200.

Hedge, A., Burge, P.S., Robertson, A.S., Wilson, S. and Harris-Bass, J. (1989) 'Work-related illness in offices: a proposed model of the sick building syndrome', *Environment International* 15: 143–58.

Heijs, W.I.M. (1993) 'The dependent variable in thermal comfort research: some psychological considerations', in N.A. Oseland and M.A. Humphreys (eds) *Thermal Comfort: Past, Present and Future, Proceedings of a conference held at the Building Research Establishment, Garston, 9–10 June 1993*, Watford: Building Research Establishment.

Hellon, R.F., Jones, R.M., MacPherson, R.K. and Weiner, J.S. (1956) 'Natural and artificial acclimatization to hot environments', *Journal of Physiology* 132: 559–76.

Helson, H. (1964) *Adaptation-Level Theory*, New York: Harper and Row.

Hirokawa, Y. and Horie, G. (1985) 'Effect of air conditioned indoor environment in summer on the living behaviors in the immediate outdoor surroundings', in P.O. Fanger (ed.) *Clima 2000, Indoor Climate Vol. 4*, Copenhagen: VVS Kongres-VVS Messe.

Hodgson, M. (1992) 'Field studies on the sick building syndrome', *Annals New York Academy of Science* 641: 21–35.

Hoeppe, P.R. (1993) 'Heat balance modelling', *Experientia* 49: 741–6.

Houghton, F.C. and Yaglou, C.P. (1923) 'Determining lines of equal comfort', *ASHVE Transactions* 29: 163–75.

Hsia, L.B, and Lu, J.K. (1988) 'Association between temperature and death in residential populations in Shanghai', *International Journal of Biometeorology* 32: 47–51.

Humphreys, M.A. (1975) 'Field studies of thermal comfort compared and applied', *UK Dept of Environment, Building Research Establishment current paper 76/75*, Watford: BRE.

Humphreys, M.A. (1976) 'Comfortable indoor temperatures related to the outdoor air temperature', *UK Dept of Environment, Building Research Establishment note PD117/76*, Watford: BRE.

Huntington, E. (1926) *The Pulse of Progress*, New York: Simpson.

Jaakola, J.J.K. (1993) 'Sick building syndrome in relation to room temperature, relative humidity and air exchange', in N.A. Oseland, and M.A. Humphreys (eds) *Thermal Comfort: Past, Present and Future, Proceedings of a conference held at the Building Research Establishment, Garston, 9–10 June 1993*, Watford: Building Research Establishment.

Jendritzky, G. and Nuebler, W. (1981) 'A model analysing the urban environmental and physiologically significant terms', *Archives for Meteorology, Geophysics, and Bioclimatology* 29: 313–26.

Jendritzky, G., Stahl, T. and Cordes, H. (1979) 'The influence of weather on traffic accidents', in S.W. Tromp, and J.J. Bouma (eds) *Biometeorologcal Survey, Vol. 1 1973–1978*, London: Heyden, pp. 198–202.

Kalkstein, L.S. (forthcoming) *Climate and human mortality: relationships and mitigating measures*.

Kalkstein, L.S. and Davis, R.E. (1989) 'Weather and human mortality: an evaluation of demographic and interregional responses in the United States', *Annals of the Association of American Geographers* 79: 44–64.

Kalkstein, L.S. and Valimont, K.M. (1986) 'An evaluation of summer discomfort in the United States using a relative climatological index', *Bulletin of the American Meteorological Society* 67: 842–8.

Keatinge, W.R., Coleshaw, S.R., Easton, J.C., Cotter, F., Mattock, M.B. and Chelliah, R. (1986) 'Increased platelet red cell counts, blood viscosity, and blood cholesterol levels during heat stress, and mortality from coronary and cerebral thrombosis', *American Journal of Medicine* 81: 795–800.

Langford, I.H. and Bentham, G. (1995) 'The potential effects of climate change on winter mortality in England and Wales', *International Journal of Biometeorology* 38: 141–7.

Lee, D.H.K., and Henschel, A. (1963) 'Evaluation of thermal environment in shelters', *Division of Occupational Health Technical Report 8*, Cincinnati: US Dept. of Health Education, and Welfare.

Lee, D.H.K. and Lemons, H. (1949) 'Clothing for global man', *Geographical Review* 34: 181–213.

Lye, M. and Kamal, A. (1977) 'The effect of a heat wave on mortality rates in elderly impatients', *Lancet* 1: 529–31.

Maarouf, A.R. (ed.) (1992) *Proceedings of the Weather and Health Workshop, 19–20 November, Ottawa, Canada*, Downsview: AES.

Mackworth, N.H. (1950) 'Researches on the measurement of human performance', *Medical Research Council Special Report No. 268*, London: HMSO.

Markham, S.F. (1947) *Climate and the Energy of Nations*, London: Oxford University Press, 105–8.

Mather, J.R. (1974) *Climatology: Fundamentals and Applications*, New York: McGraw-Hill.

Maunder, J. (1970) *The Value of Weather*, London: Methuen.

McArdle, B., Dunham, W., Holling, H.E., Ladel, W.S., Scott, J.W., Thompson, M.L. and Weiner, J.S. (1947) 'The predication of physiological effect of warm and hot environments, *Medical Research Council Report 47–391*, London: HMSO.

McIntyre, D.A. (1980) *Indoor Climate*, London: Applied Science.

Medwar, P.B. (1957) *The Uniqueness of the Individual*, London: Methuen.

Mills, C.A. (1946) *Climate Makes the Man*, London: Gollan.

Missenard, A. (1957) *In Search of Man*, New York: Hawthorn Books.

Monteith, J.L. (1973) *Principles of Environmental Physics*, London: Edward Arnold.

Monteith, J.L. and Mount, L.E. (1974) *Heat Losses from Animals and Man: Assessment and Control*, London: Butterworth.

Mukammal, E.I., McKay, G.A. and Neumann, H.H. (1984) 'A note on cardiovascular diseases and physical aspects of the environment', *International Journal of Biometeorology* 28: 17–28.

Nadel, E.R., Pandolf, K.B., Roberts, M.F. and Stolwijk, J.A.A. (1974) 'Mechanisms of thermal acclimatization to exercise in heat', *Journal of Applied Physiology* 37: 515–20.

Nishi, Y. and Gagge, A.P. (1977) 'Effective temperature scale useful for hypo- and hyperbaric environments', *Aviation Space Environmental Medicine* 48: 97–107.

Oseland, N.A. and Humphreys, M.A. (eds) (1993) *Thermal Comfort: Past, Present and Future, Proceedings of a conference held at the Building Research Establishment, Garston, 9–10 June 1993*, Watford: Building Research Establishment.

Passmore, R. and Durnin, J.V.G.A. (1955) 'Human energy expenditure', *Physiological Reviews* 35: 801–40.

Pepler, R.D. and Warner, R.E. (1968) 'Temperature and learning: an experimental study', *ASHRAE Transactions*, 74: 211–19.

Rohles, F.H. and Laviana, J.E. (1985) 'Indoor climate: new approaches to measuring how you feel', in P.O. Fanger (ed.) *Clima 2000*, Vol. 4, Copenhagen: VVS Kongres-VVS Messe, pp. 1–6.

Rotton, J. and Frey, J. (1985) 'Air pollution, weather and violent crimes: concomitant time series analysis of archival data', *Journal of Personality and Social Psychology* 49: 1207–20.

Rowe, D. (1994) 'Sick building syndrome: the mystery and reality', *Architectural Science Review* 37: 137–48.

Ruck, N.C. (ed.) (1989) *Building Design and Human Performance*, New York: Van Nostrand Reinhold.

Sargent, F. (1963) 'Tropical neurasthenia: giant or windmill?', in *Arid Zone Research. Environmental Psychology and Physiology, Proceedings of the Lucknow Symposium*, Paris: UNESCO.

Sargent, F. and Tromp, S.W. (eds) (1964) 'A survey of human biometeorology', *WMO Technical Note No. 65*, Geneva.

Sarna, S., Romo, M. and Siltanen, P. (1977) 'Myocardial infarction and weather', *Annals of Clinical Research* 9: 222–32.

Schiller, G.E., Arens, E.A., Benton, C.C., Bauman, F.S., Fountain, M.E. and Doherty, T.J. (1988) 'A field study of thermal environments and comfort in office buildings', *Final Report 41.62*, Atlanta: ASHRAE.

Scholander, P.R. (1955) 'Evolution of climatic adaptation in homeotherms', *Evolution* 9: 15–26.

Siple, P.A. (1949) 'Clothing and climate', in L.H. Newburgh (ed.) *Physiology of Heat Regulation and the Science of Clothing*, New York: Hafner, Ch. 12.

Siple, P.A. and Passel, C. (1945) 'Measurements of dry

atmospheric cooling in sub-freezing temperatures', *Proceedings of the American Philosophical Society* 89: 177–99.

States S.J. (1977) 'Weather and deaths in Pittsburgh, Pennsylvania: a comparison with Birmingham, Alabama', *International Journal of Biometeorology* 21: 7–15.

Steadman, R.G. (1971) 'Indices of windchill of clothed persons', *Journal of Applied Meteorology* 10: 674–83.

Steadman, R.G. (1984) 'A universal scale of apparent temperature', *Journal of Climate and Applied Meteorology* 23: 1674–87.

Szokolay, S.V. (1985) 'Thermal comfort and passive design', in K.W. Boer (ed.) *Advances in Solar Energy 2*, New York: American Solar Energy Society.

Taylor, G.T. (1959) *Australia: A Study of Warm Environments and Their Effect on British Settlement*, 7th edn, London: Methuen.

Teichner, W.H. and Wehrkamp, P.F. (1954) 'Visual motor performance as a function of a short duration ambient temperature', *Journal of Experimental Psychology* 47: 447–50.

Terjung, W.H. (1966) 'Physiologic climates of the conterminous United States: a bioclimatic classification based on man', *Annals of the Association of American Geographers* 56: 141–79.

Thom, E.C. (1959) 'The discomfort index', *Weatherwise* 12: 57–60.

Ungeheuer, H (1955) *Ein meteorolgischer Beitrag zu Grundproblemen der Medezin-Meteorologie*, Bad Kissingen: Berichte Deutscher Wetterdienst, Nr 16.

Weihe, W.H. (1992) 'Adaptation to weather fluctuations', in A.R. Maarouf (ed.) *Proceedings of the Weather and Health Workshop, 19–20 November, Ottawa, Canada*, Downsview: AES, 1–12.

Weihe, W.H. and Mertens, R. (1991) 'Human well being, diseases and climate', in J. Jager, and H.L. Ferguson (eds) *Climate Change: Science, Impacts and Policy, Proceedings of the Second World Climate Conference*, Cambridge: Cambridge University Press, pp. 345–60.

Winslow, A.-E. A., Herrington, L.P. and Gagge, A.P. (1937) 'Physiological reactions of the human body to varying environmental temperature', *American Journal of Physiology* 120: 1–20.

Wohwill, J.F. (1974) 'Human adaptation to levels of environmental stimulation', *Human Ecology* 2: 127–47.

Wulsin, F.R. (1949) 'Adaptation to climate among non-European peoples', in L.H. Newburgh (ed.) *Physiology of Heat Regulation and the Science of Clothing*, New York: Hafner, Ch.1.

Wyndham, C.H. (1970) 'Adaptation to heat and cold', in D.H.K. Lee, and D. Minard (eds) *Physiology, Environment and Man*, New York: Academic Press, pp. 177–204.

Wyon, D.P., Kok, R., Lewis, M.I. and Meese, G.B. (1982) 'Effects of moderate cold and heat stress on the performance of factory workers in Southern Africa 1: introduction to a series of full scale simulation studies', *South African Journal of Science* 78: 184–9.

13

TOWN PLANNING, ARCHITECTURE AND BUILDING

Anthony Brazel and Jonathan Martin

INTRODUCTION: CLIMATE AND THE BUILT ENVIRONMENT

Climate can influence the planning of towns, buildings and settlement designs and can evoke strategies to promote the efficiency of thermal comfort (Chapter 12), both outdoors and indoors (Givoni, 1989; Landsberg, 1981; Mather, 1974; Olygay, 1963; Smith, 1975). In turn, the built environment, in whatever form it takes, affects local and regional climate change (Chapter 21), which thus can be a feedback influence on comfort and health. It can subsequently stimulate further design and planning considerations and strategies to alleviate unwanted climate impacts (Garbesi *et al.*, 1989).

A review of urban climates, air pollution, clothing, comfort and human health are topics of other chapters, namely Chapters 21, 22 and 12. In this chapter, we focus on a description of design, planning and climate conditions, with emphasis on the passive responses to climate constraints as evidenced in various kinds of environments. The lessons learned from the past, on how we can take advantage of climate conditions to provide high-quality living in urban and rural environments, may assist with our efficient sustainability in the future. Action plans for the future, to abate the harsh effects of climate change, have been researched. Increased interaction and co-operation among scientists and local constituencies in cities, towns and neighbourhoods will aid the enactment of changes (e.g. ordinances) to cope with climate. This will lead to harmony of the built environment with the forces of climate and nature.

The end result of building structures, and their arrangements on the landscape in various settlements, have been associated with a combination of mystical, religious, military, social and empirical factors, as well as including environmental considerations (Parry, 1979). The focus here is on the nature of climate relations to the built environment and some evolved design and planning ideas that tend to shed light on the influence of climate. Climate can be viewed as making possible certain decisions, not a priori forcing or determining precise outcomes worldwide. However, many new towns (and, of course, buildings) do, in fact, reflect some considerations of climate impacts on them and include some aspects of Landsberg's (1973) 'meteorologically utopian city' (e.g. the preservation of trees, reduced use of cars and adequate spacing of buildings). If we do not continue to consider the climatic factor in our future, particularly in areas such as tropical cities, the consequences may include (Chapter 21) possible excess human deaths from heatwaves, decreased comfort, increased costs for cooling buildings and major flooding problems, among other deleterious outcomes (Oke *et al.*, 1990/1991).

HOT, HUMID CLIMATES

Hot, humid regions are found primarily in a band up to latitudes 10–15° north and south of the equator and include mainly developing or Third World countries. It is estimated that more than 40 per cent of the world's population lives in this equatorial belt (de Blij and Muller, 1994). In these countries, the majority of the population cannot afford modern solutions to climatic stress, such as air conditioning (see Chapter 21, The consequences of greenhouse warming); therefore, it is appropriate to solve climatic problems for the built environment as inexpensively as possible by optimizing climatic building response. Although the hot, humid region can be subdivided into numerous zones, there are common climatic characteristics that pertain to creating a comfortable built environment. Most significant are the persistently high ambient air temperatures which average about 27°C annually and vary only 1–3°C within 1 month (Givoni, 1989). Additionally, a narrow diurnal temperature range of 6–10° and maximum temperatures averaging some 30°C offer minimal relief to inhabitants of these regions. Rainfall can average between 200 and 600 mm per month and cause erosional flooding, which damages infrastructures and greatly impedes mobility within towns and cities. Solar radiation can be intense on clear days and, on days with high diffuse radiation, it is characterized by sharp hourly variations (Givoni, 1989).

The most advantageous strategies to minimize climatic stress on the inhabitants in hot, humid regions are to reduce solar radiation gain, provide ventilation or maximize available ventilation and minimize potential hazards from flooding. One can look to the typical towns, cities and buildings in these regions for appropriate applications of relevant climatic building strategy. Open, widely separated buildings are typical of the region, with buildings placed on stilts. The emphasis is on good roof construction and protection from the elements (high rainfall). Villages in hot, humid regions tend to be organized differently than settlements in hot, arid regions. Building types and arrangements geographically focus on natural ventilation for interior climatic comfort. Buildings are loosely dispersed and built on elevated platforms with wind-permeable walls and overhanging roofs for shade and water shedding. The loosely dispersed arrangement allows maximum ventilation between the structures that are constructed of materials that have little or no heat storage capacity. Often, the siting of a native village will be strategically located to utilize whatever breeze might be perceived to be available.

For example, the tropical rain-forest natives of northeastern Ecuador build their dwellings entirely with wood and thatch and create their living quarters on a platform within the structure at several feet above the forest floor (Plate 13.1). This pavilion-style architecture assists any wind to ventilate below the structure and raises the inhabited space above the wind-sheltering vegetation that surrounds the house and village. Having the main living platform above the ground serves also to protect the inhabitants from frequent flooding of a river that geographically might connect many villages to a local or regional economy. Additionally, building walls are porous enough to allow any breeze to permeate through but still provide privacy. This wall-form can be found throughout hot, humid regions and often serves climatically pragmatic ideals as well as cultural customs.

Possibly, the most extreme example of a well-ventilated dwelling with a minimal heat storage capacity can be found in Colombia. Here, a minimum dwelling consists of nothing more than a grass roof on a wooden frame that also supports a hammock and baskets for storage.

Plate 13.1 Dwelling in northeastern lowland tropical Ecuador, near the equator (taken by A. Brazel)

Walls are not necessary, as the builder's main objective is to maximize ventilation and minimize shade (Rapoport, 1969). Sometimes, native dwellings must rise high above the ground surface to utilize available sea breezes, which is typical of island architecture in hot, humid regions. Examples are the native dwellings of Tumaco, Colombia, an island off the southwest coast of Colombia. The native houses, similar in construction to those in Ecuador in as much as they have thatched roofs and ventilative screen walls, are built on stilts high above the lush jungle floor. This climatic response influenced architects to adopt this native building strategy when plans were being drawn to rebuild the old town that was destroyed by fire in 1947 (Aronin, 1953). Architects designing current typologies (such as single-family detached houses, row buildings, multistorey apartment buildings and tall high-rise buildings in hot, humid areas) can employ these concepts and indeed should do so, given the fact that developing countries currently have limited economic capacity for air-conditioning technology.

HOT, ARID CLIMATES

The hot, dry regions on the earth are found primarily in the subtropical latitudes in central and western Asia, Africa, North and South America, and on the Australian continent and present a difficult challenge to climatologists, urban planners and architects alike. In these regions, the summer daytime temperature reaches in excess of 40°C and, owing to low relative humidity, plummets to perhaps 10–15°C overnight. This diurnal swing in ambient air temperature responds to the arid land surface's diurnally variable energy budget, which can elevate temperature to 50°C during the day and rapidly drop the temperature to 15°C or lower overnight. Some hot, dry regions experience mild winters while others (Astrakhan, 46°N and 48°E) can reach well into the subfreezing temperatures in the coolest winter months (January and February).

The most advantageous strategy to minimize climatic stress on the inhabitants of these regions involves a few basics, namely to minimize solar gain in the summer, maximize solar gain in the winter, minimize wind exposure during both seasons and build within the smallest possible building envelope. The towns, cities, and buildings of the Middle East (and other hot, arid regions) provide excellent examples of the application of these general climatic building strategies. In valleys which are parallel to the prevailing winds, an ideal strategy includes a settlement layout of dense and narrow streets, shaded sidewalks, short walking distances, compact geometry, courtyards with greenery and the use of high-albedo materials as building surfaces. Delaying heat transfer into the building during the day, but providing warming at night, would be advantageous.

In the cities of the Middle East, such as Amman, Jerusalem, Kuwait City and Tehran, and in many other cities and towns, people have chosen to create a comfortable built environment, both indoors and out. To relieve the burning intensity of the sun in outdoor spaces, they have built their narrow streets between clustered groups of buildings which are tall enough to shade the street completely for most of the day, and the materials utilized are for the most part light in colour (Plate 13.2). These buildings, clustered together to reduce exposed building surface area, were usually constructed of high-mass materials that are able to delay the impact of the hot mid-day sun by storing the energy in the mass of the wall and releasing it at night through long-wave radiation emittance. This moderating effect of the diurnal temperature swing, called 'thermal mass', takes advantage of the heat storage capacity. By morning, the mass has released large amounts of its stored energy and is thus cool enough to keep the interior of the building comfortable during the day. Because the material is so dense and usually of substantial thickness, there is a time lag between when the sun's energy heats one side of the material and when heat conducts through to heat the inside of the building. As materials vary in mass so does the time lag for materials of equal thickness, and often the thickness and nature of the material of a wall will be determined by the time lag desired.

COLD CLIMATES – POLAR REGIONS

In cold regions, design objectives should maximize solar gain and minimize wind exposure (see Chapter 12, The human energy balance). Although the design objectives are different from those employed in a hot, arid region, the principles for keeping warm in a cold region are similar to those used to keep cool in hot, arid regions. The primary differences are that the heat source now emanates from the building's interior and solar radiation gain is desired (Rapoport, 1969). Designs, whether at an aggregate town level or individual structure configuration level, should seek to minimize heat loss with minimal external surface areas, because small, compact configurations are understandably more efficient to heat and insulate. For example, high-mass materials should be used, with a good insulating capacity, in order to maximize solar radiation and minimize wind effects. Outer surfaces should be darkened (to reduce the albedo) in order to absorb more energy, and all windows should be located on walls facing the sun.

The most investigated and expert builders of

Plate 13.2 Building design in the arid tropics, Egypt: (a) ancient practices on the West Bank, Luxor; (b) same principles but modern architecture in Aswan (taken by R. Thompson)

the polar regions are the Central Inuit or Eskimo (Rapoport, 1969), since their igloo provides an excellent example of the design objectives mentioned above. For example, its hemispherical shape minimizes heat loss by providing the maximum volume with a minimum of surface area and effectively resists the strong arctic winds (Figure 13.1). The snow used to build the igloo is a readily available material and offers excellent insulating properties. Often several igloos are grouped closely together and interior, protected passages connect one structure to another with only one main entrance for that particular grouping of igloos. The one main entrance is protected by a snow wall that is perpendicular to the wind direction and is usually oriented on the windward side of the structure, parallel to the wind direction to avoid snowdrifts found on the leeward of the structure. The entrance is purposely lower than the surface of the snowpack and provides access to a series of transitional spaces, called the natiq, separated with hanging skins that lead to the hemispherical living quarters called the iglik (Nabokov and Easton, 1989). As one enters the igloo, through the buried and curved tunnel, the floor level rises until the iglik is reached through a low bulkhead. The gradual rising floor elevation allows a pragmatic stratification of the air mass within the igloo that collects the warmer indoor air within the iglik and confines the cooler air to the natiq. Even with the outside air temperature as low as −50°C, the air temperature at the top of the dome can be as high as 15°C and close to zero in the natiq (Stein, 1977). The hanging skins and the final low bulkhead, separating the main space, help further to confine and stratify the air mass and block draughts from the exterior. During the winter, the Inuit capture available solar radiation with a small window of ice that is cut from an available ice pack and sealed within the hemispherical envelope facing the sun. During the long hours of daylight during the summer, skins are hung on the interior of the dome to create a relieving dark environment. The Inuit take purposeful measures of siting the igloo camp, usually siting it where there is a plentiful supply of suitable snow (normally from an unlayered single storm) and at the base of a leeward cliff on a beach (Rapoport, 1969).

OTHER CLIMATES

Regions with more than one stressful climatic season present a special challenge to climatologists, planners and architects because one season can often require conflicting design solutions than that of the other seasons. Therefore, a sound design in such a climatically dichotomous environment aims for a year-round climatic balance indoors and outdoors within the built environment and often requires a compro-

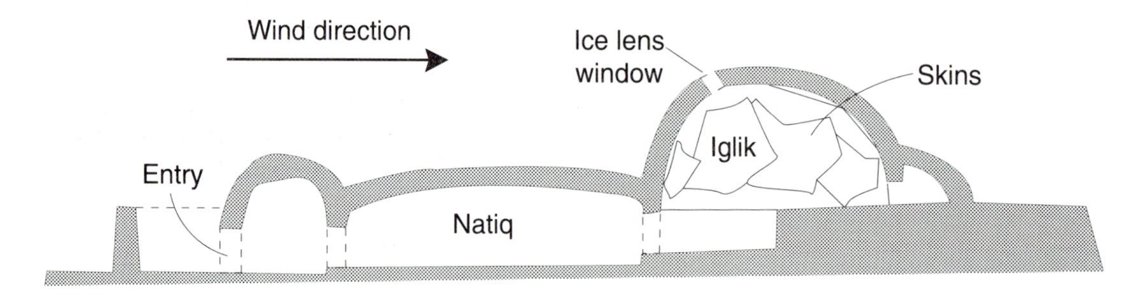

Figure 13.1 Sketch cross-section of an Inuit igloo

mise between the design solutions already discussed.

A 'two-season' challenge calls for different design strategies, and often it is the shift in the wind that allows an architect or planner to accomplish a successful design that encompasses both seasonal strategies. As the summers are both hot and humid, it would be logical that some designs, whether a building or urban layout, provide adequate ventilation through a loosely dispersed open plan as a necessary means of cooling. During the cold winters, a compact plan with accordingly smaller surface area would be both efficient and reduce undesirable ventilation. Finding a balance between these two extremes presents a difficult challenge for the designer. Akbari *et al.* (1992) provide an example of a strategy to shield homes from hot summer sun and cold winter winds through the optimal planting of trees (Figure 13.2).

APPLICATIONS IN URBAN PLANNING

Urban design must be concerned primarily with those challenges that stress the population (see Chapter 12, The human energy balance) when they engage in outdoor activities, while architectural design or building design must address the comfort of those indoors (see Chapter 12, Atmospheric impacts on morbidity and mortality). Urban design should also provide an advantageous platform for architects to design structures that remain comfortable with minimal energy use. Urban planners can provide

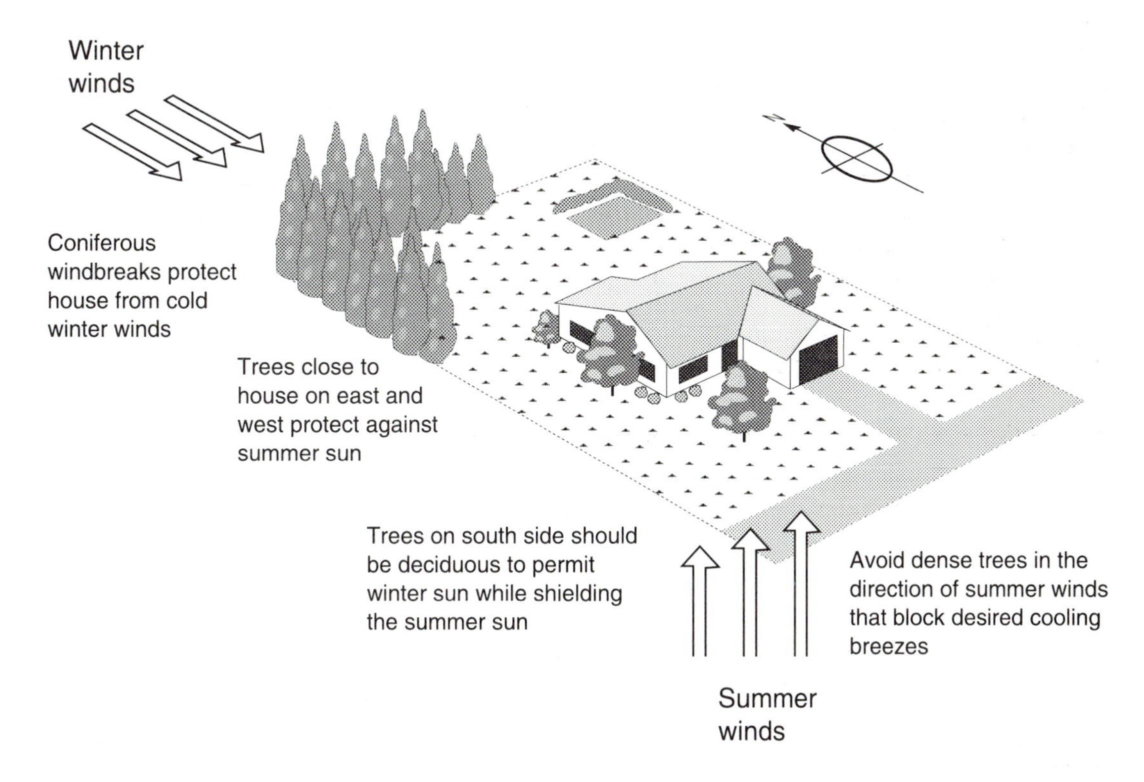

Figure 13.2 Two-season challenge to ameliorate winter and summer advection and to protect dwelling (after Akbari *et al.*, 1992)

solutions to these challenges by reducing exposure to high summer daytime temperatures from intense solar radiation, alleviating high glare from direct and indirect solar radiation throughout the year, providing protection from seasonal afternoon dust storms characteristic of arid regions and providing sufficient protection from the cold winter wind. Urban climate change effects (Chapter 21) have been taken into account in the planning of various cities. Examples are given by Holford (1971), Horbert *et al.* (1985), Noack *et al.* (1986) and Wise (1971). The use of urban parks, greenbelts, shelterbelts and terrain drainage ways for ventilation, for example, all have important useful effects on optimizing the climate factor. As described below, a so-called 'shadowbelt' would be appropriate as well.

Urban planning in hot, arid regions should select a site with as little climatic stress as possible, avoiding floodplains prone to flash floods during the few but often substantial rainfalls. The site should avoid temperature elevation due to the urban heat island (see Chapter 21, The nature of urban climates) and provide shaded areas within the urban fabric and in open spaces during the summer, yet provide sunny areas in both the urban fabric and open spaces during the winter. Strategies that minimize blowing dust and protect public areas from the wind during the winter months prove to be beneficial. Public spaces can be further enhanced by reducing the glare from the environment through vegetative strategies and judicious use of low-albedo materials.

To relieve the heat stress on a hot summer day, walking distances in a town or urban area should be short. Areas utilized by the population should be shaded as much as possible, especially sidewalks, playgrounds for children and parking areas for automobiles. Automobile storage usually requires large expanses of hardscape which contribute to the urban heat island (Chapter 21) and shade is particularly desirable in areas where people tend to congregate socially, namely shopping areas, open spaces and schoolyards. Distribution will vary greatly between cultures and societies and while climatic optimization is a goal, indeed town planning must respect cultural differences.

Urban density and the arrangement of elements within a landscape should be considered carefully to reduce the effect of solar radiation on the indoor climate. Sun angles and landscaping can play a pivotal role in minimizing solar penetration into public and private spaces and reduce available loose dirt that creates dust. Consider, for example, an arrangement of dwellings that orients the building's main axes north/south (Figure 13.3). This arrangement accomplishes a number of objectives which are sympathetic to the main design criteria for a hot, arid region. First, by maximizing the majority of the wall surface of the structure facing either east or west, the exposure to the south or north is minimized. In the northern hemisphere, the glaring sun from the south will have little wall surfaces to charge thermally. One can expect, however, that most of the solar radiation will be absorbed from the east and west in the summer and consideration for the east and west axis should be great. For example, in Phoenix, Arizona, at a latitude of about 32°N, the summer sun rises and sets 30° north of east or west at the summer solstice. Therefore, the spacing between adjacent dwellings east–west of one another should create mutual shading of the east–west walls and reduce the effects of the intense morning and afternoon solar radiation on the indoor climate (Figure

Figure 13.3 (a) Orientation and spacing of buildings and the production of a 'shadowbelt'. Notice the height/width in sections AA and BB. (b) Development of a courtyard with shade, sloping roof and fountain. (c) Method of wind tower and air flow provision for cooling

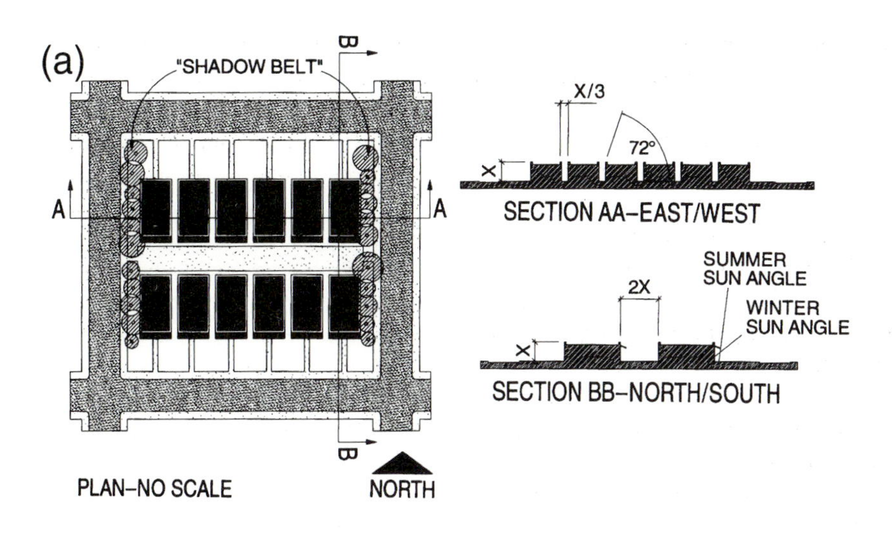
(a)
"SHADOW BELT"
B
A
A
B
X/3
72°
X
SECTION AA–EAST/WEST
SUMMER
SUN ANGLE
2X
WINTER
SUN ANGLE
X
SECTION BB–NORTH/SOUTH
PLAN–NO SCALE
NORTH

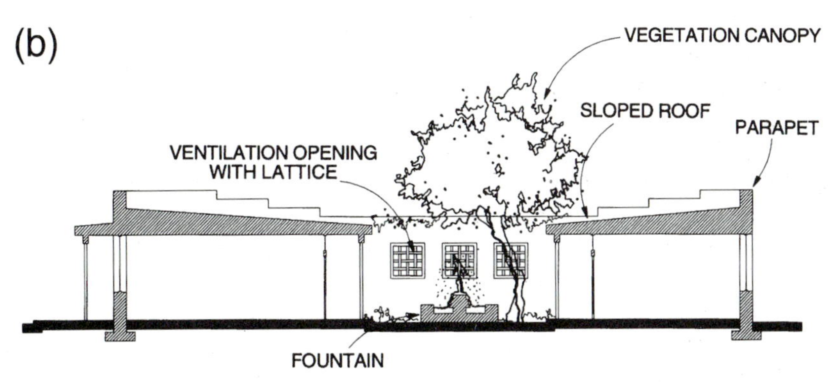
(b)
VEGETATION CANOPY
SLOPED ROOF
PARAPET
VENTILATION OPENING
WITH LATTICE
FOUNTAIN

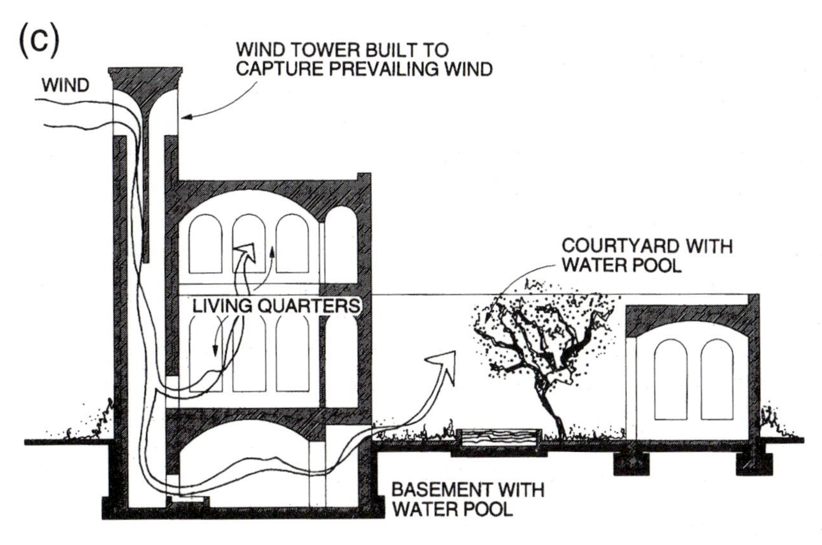
(c)
WIND
WIND TOWER BUILT TO
CAPTURE PREVAILING WIND
COURTYARD WITH
WATER POOL
LIVING QUARTERS
BASEMENT WITH
WATER POOL

13.3(a)). Generally, this spacing should be about three times the height of the buildings thus allowing the sun to strike the east- and west-facing walls between 11.00 and 13.00 hours (Givoni, 1989). Solar protection of the walls at either end of the row of adjacent dwellings can be achieved by employing what we would call a 'shadowbelt'. This shadowbelt could consist of vegetation such as trees or could use other means of creating a specific shaded zone. The shadowbelt will serve as a shading device for the end walls and can also function as a noise and pollution barrier from the adjacent street. Additionally, reducing the space between the structures reduces expensive landscaping required to reduce dust in hot, arid regions.

In many developing countries, much of the population cannot afford planting and maintaining large open spaces between dwellings, and arrangements that reduce these spaces also reduce the population's use of costly imported or processed water. Distances between dwellings along the north/south axis should be carefully calculated so as not to eliminate potential thermal gain in the winter. The distance will be greater than the east–west distance between the buildings, owing to the lower altitude of the sun in the winter. It varies depending upon latitude, but generally a distance of 1.5–2 times the height of the buildings is recommended (Givoni, 1989). Mutual shading along this axis proves detrimental to the indoor climate and shading of the northern or southern elevations of structures should be addressed with shading schemes connected to the building itself.

Urban development schemes such as this increase densities and present difficulties in ventilation; therefore reasonable attention should be given to available prevailing wind patterns. Ventilation is undesirable indoors during the daytime and is secondary to solar protection in the outside built environment. At night, residents will find more comfort outdoors even with light winds. Developed blocks of buildings should be spaced apart from one another so as to allow ventilation between and over the blocks, and tall buildings should not be sited so as to function as 'walls' that inhibit ventilation of an adjacent block. East–west street directions will facilitate a north–south building axis and therefore produce favourable solar orientation.

Building typology in hot, dry climates has, for centuries, traditionally employed courtyards (Figure 13.3(b)). Courtyards can perform a valuable cooling effect when appropriately taking the environment into account. One must consider that internal courtyards enlarge the wall surface area of the house and therefore potentially increase daytime heating of the building's interior. It is important to keep the daytime temperature of the courtyard to a minimum, through means of shade and evaporation. For example, tall and wide canopy trees can offer shade to the courtyard and create a thermal transition zone insulating the courtyard climate from the ambient air above it. However, this insulating effect can diminish desired nocturnal long-wave emittance from the courtyard. To relieve this, one can facilitate cool-air drainage from the roofs into the courtyard by building a parapet on the building's perimeter and sloping the roofs inward to the courtyard. The periphery parapet will reduce wind speed over the roof and thus facilitate radiant heat loss and induce cooling of the boundary-layer air next to the roof surface. The slope of the roof will allow the cooler air to drain towards the courtyard. Effective evaporative cooling of the courtyard requires a large surface area of water. Typically a fountain that sprays a fine mist into a pool works well as does sprinkling the large vegetative canopy, or water sheeting down the surface of a large wall. The evaporative process has one other benefit – namely, psychological. The sound of trickling water and the smell of moisture in the air can go a long way to make one feel cooler. One technique used in the southwestern

USA and elsewhere is the mechanical evaporative cooling system. This is a method employed to produce central cooling in the house, based on the cooling effect of evaporated water to a dry atmosphere.

Houses in hot, arid climates can also be cooled in the interior by employing wind catchers or towers to harness the commonly strong daytime winds (Figure 13.3(c)). Common vernacular structures in Iran and Pakistan employ these vertical tower-like devices to capture wind currents present above roof height. Openings at the top of the tower are directed parallel to the wind, which is funnelled downward into the building and exits across a pool of evaporating water. The net effect is an evaporative cooling process that introduces cooler air into the building at the lowest level and allows warm air to rise and exit through a ventilation opening on the roof (this is known as the 'stack effect').

Any successful urban design should have elements that reduce flooding from runoff outside the urban area, cause disposal of rainwater from the urban 'hardscape', provide rain and sun protection on urban streets, provide sufficient ventilation of urban and open spaces and of the buildings that surround them and minimize the heat island effect (Chapter 21). Perhaps, this latter phenomenon can be affected by using more shade trees and light-colour materials, for example (Akbari *et al.*, 1992).

There are current demands to create a more proactive process in joining together knowledge on microclimate, planning and design concepts and to implement these concepts at the community and city level (see Chapter 21, Conclusions). Such efforts will have an impact on actual applications in the future. One such example is the output of reports by the federal government of the USA through the US Environmental Protection Agency's Office of Policy Analysis and Climate Change Division. These reports were produced in concert with the Department of Energy's Lawrence Berkeley Laboratory (Garbesi *et al.*, 1989; Akbari *et al.*, 1992). Two major concepts put forward are the use of light-colour surfaces and the judicious planting of trees in the urban environment (Akbari *et al.*, 1992). The main role is to reduce the build-up of the urban heat island phenomenon (Chapter 21). These efforts attempt to reach constituencies made up of elected officials, policy-makers, foresters, landscapers, architects, urban planners, utility managers, commercial interests, citizens/community groups, professional schools and programmes, the media and public affairs staff at the local level. What might seem like ideal concepts to employ, in order to ameliorate local and regional climate change impacts in settlements and their neighbourhoods, can only be made successful with a wide range of support from these constituencies which is manifest at the local planning scale. Applications for designing and altering the environment bring together experts from differing disciplines, researchers with different perspectives and consumers of disciplinary and interdisciplinary knowledge. They must not only have an appreciation for the science of the issues related to climate and planning, but also recognize the social acceptance of any planning and design schemes that are derived.

CONCLUSIONS

Climatologists, together with urban and rural planners and architectural designers, hopefully will form more of an interdisciplinary bond and work more closely to study the ways in which relatively simple climate principles can be employed to the populace's benefit, to improve and sustain the quality of life in settlements on the earth and to buffer against future climate change. Several methods have been suggested recently to reach these ends as follows:

1 Governmental global treaties on climate.
2 International decisions to mitigate natural hazards.
3 The establishment of local climate and environmental monitoring networks in settlement areas.
4 Short-range forecasting of such events as heatwaves, flash flooding, and severe storms.
5 Considerations of alternative energy use for transportation and home heating and cooling.

Climate change induced by settlements and cities, together with the global effects of any change, must be considered in the planning of our future. Together with academia, communities and the private and public sectors need to form more effective partnerships at the local level to co-ordinate the technology and knowledge that will result in a better planned environment (see Chapter 21, Conclusions).

REFERENCES

Akbari, H., Davis, S., Dorsano, S., Huang, J. and Winnett, S. (1992) *Cooling Our Communities: A Guidebook on Tree Planting and Light-Colored Surfacing*, US Environmental Protection Agency, Office of Policy Analysis, Climate Change Division, Washington, DC.

Aronin, J. E. (1953) *Climate and Architecture*, New York: Rheinhold.

de Blij, H. J. and Muller, P. O. (1994) *Geography Realms, Regions, and Concepts*, New York: John Wiley.

Garbesi, E., Akbari, H. and Martien, P. (eds) (1989) *Controlling Summer Heat Islands*, Berkeley, CA: Lawrence Berkeley Laboratory, Energy Analysis Program.

Givoni, B. (1989) 'Urban design in different climates', *World Meteorological Organization/TD*, No. 366, Geneva.

Holford, L. (1971) 'Problems for the architect and town planner caused by air in motion', *Philosophical Transactions of the Royal Society of London, Series A*. 269: 335–41.

Horbert, M., Kirchgeorg, A. and Stulpnagel, A. (1985) *Umweltatlas Berlin*, Technical Universität Berlin, Institut für Ökologie, Fachbereich Bioklimatologie, Berlin.

Landsberg, H. E. (1973) 'The meteorologically utopian city', *Bulletin of the American Meteorological Society* 51: 86–9.

Landsberg, H. E. (1981) *The Urban Climate*, New York: Academic Press.

Mather, J. R. (1974) *Climatology: Fundamentals and Applications*, New York: McGraw-Hill.

Nabokov, P. and Easton, R. (1989) *Native American Architecture*, Oxford: Oxford University Press.

Noack, E.-M., Mayer, H. and Baumgartner, A. (1986) *Quantifizierung der Einflusse von Bebauung und Bewuchs auf das Klima in der Urbanen Biosphare, Schlussbericht*, Final Report, Lehrstuhl für Bioklimatologie und Angewandte Meteorologie der Universität München.

Oke, T. R., Taesler, R. and Olsson, L. E. (1990/1991) 'The Tropical Urban Climate Experiment (TRUCE)', *Energy and Buildings, 15/16: 67–73.*

Olygay, V. (1963) *Design with Climate: Bioclimatic Approach to Architectural Regionalism*, Princeton, NJ: Princeton University Press.

Parry, M. (1979) 'Climate and Town Planning', in B. Goodall and A. Kirby (eds) *Resources and Planning*, Oxford: Pergamon Press.

Rapoport, A. (1969) *House Form and Culture*, Englewood Cliffs, NJ: Prentice Hall.

Smith, K. (1975) *Principles of Applied Climatology*, London and New York: John Wiley.

Stein, R. G. (1977) *Architecture and Energy*, New York: Anchor Press.

Wise, A. F. E. (1971) 'Effects due to groups of buildings', *Philosophical Transactions of the Royal Society of London, Series A* 269: 469–85.

14

INDUSTRY AND COMMERCE

Sven Lindqvist

INTRODUCTION: CLIMATE, INDUSTRIAL OUTPUT AND COMMERCIAL ACTIVITIES

There have been few studies on the specific relationship between the variations in climate and weather and their impact on industrial output and commercial efficiency. Compared with agriculture, the number of studies completed on the effect of weather and climate on manufacturing is minimal. The interest of industrialists, retailers and atmospheric scientists, on the effect of weather and climate on manufacturing, is also limited. Nevertheless, a number of general associations are known. Climatology clearly plays a significant role in many aspects of commerce and industry (Landsberg, 1960) but the effect of climate on many industrial or engineering operations is often not recognized. It concerns the effect of climate on the following:

1 Heating and cooling requirements
2 Water supplies
3 Air and water pollution control activities
4 Warehousing, storage and transportation of raw or finished products
5 Weathering or deterioration of stockpiled items such as fuel
6 Health, efficiency and morale of workers
7 All outside activities.

A new approach to this field of applied climatology is not easy to find in scientific papers and consequently today's information differs very little from that given in earlier works (Landsberg, 1960; Griffiths, 1966; Maunder, 1970; 1989; Mather, 1974; Hobbs, 1980). The intention with the design of this chapter has been first to give a more general description of applied climatology in connection with industry and commerce. It then concerns the location and operation of industries and the construction of large structures affected by weather and climate. A third section deals with the influence of weather on commerce relevant to the planning and advertising of products. The chapter concludes with reference to specific applications in Sweden.

LOCATION OF INDUSTRIES

What planning considerations need to be addressed, from the climate point of view, in the location of industries and commerce? Historically, as industries have developed without regard to (for example) topoclimate, some bad decisions have been made. Perhaps the location of industry has been decided upon totally without regard to climate and weather. However, there should be a role for climate analysis in the planning of new industrial sites.

When the need for a new industry or commercial venture is realized and accepted, there are numerous decisions to be made before the plan comes to fruition and many of these decisions depend on certain climatic variables for their solution. Decisions concerning raw materials, buildings and transport may need to rely at some stage on climate data. Other areas which depend upon meteorological factors are water supply, air pollution and space heating and cooling. Will there be a great need at any time of the year to supply an exorbitant amount of heating and/or cooling for the comfort of employees or the safe storage and keeping of products? The storage, warehousing and transportation of the raw, intermediate and end products naturally raises problems dependent upon the climate conditions of the region. Outdoor industries are, of course, even more at the mercy of the climate. After the estimation of the extreme climatic conditions under which operations cannot proceed, it is essential to find out for what percentage of the time these conditions may exist at the chosen site. Optimal conditions have also been worked out for some indoor operations.

However, although many oppose any suggestion of climatic control, it is nevertheless evident that the location of industry, and the migration of labour, is at least partly related to climatic conditions. Some individuals migrate in order to live in a more pleasant environment. One result of this migration is, of course, the accumulation in these areas of a growing labour supply, much of it skilled or professional, and it is this potential labour supply (there in part because of the more favourable climate) that is attractive to new industry. In view of these findings, the study by Bickert and Browne (1966) on the perception of the effects of weather on five eastern Colorado manufacturing firms is significant in assessing the value of weather for manufacturing industry. In their conclusion, Bickert and Browne indicate that although the initial awareness among the five manufacturers of the effects of weather on their operations was found to be minimal, the actual effects of weather variables (upon examination) were found to be considerable, particularly in terms of costs to the companies concerned.

One problem is of course atmospheric pollution, which is discussed fully in Chapter 22. Within this context, there may be problems with industrial gaseous emissions in areas with strong and frequent temperature inversions, occurring typically in large valley environments. Potential problems may be tackled at the planning stage before new industries are established. In Tasmania, Australia, the state government has targeted the Tamar Valley on the north coast for expansion as an industrial zone. Part of the environmental impact study involved the development of diffusion models of airborne contaminants. Emissions from new proposed industries may be readily incorporated in the model and ambient contaminant concentrations can be established.

Planning for extreme climatic conditions may be difficult. Permafrost or permanently frozen ground, resulting from constantly low temperatures, can create significant engineering poblems. Construction of houses, roads, towers, bridges, factories, or other structures requiring foundations, may result in heat conduction into the ground. There are stability problems when the permafrost melts and thermokarst develops. There is no real cure for this situation except to re-establish the permafrost by the removal of all forms of construction.

In a new promotion for the Balearic Islands, climate is claimed to be the main advantage when marketing the islands as a business resort. It is emphasized that the Balearic Islands have one of the best climates in Europe! Furthermore, the urban structure and design are planned to give a good and attractive environment in and outside the buildings. When establishing the theme park and village of Sophia Antipolis at Antibes on the French Riviera, locational factors other than the classical ones were taken into consideration. Climate and site characteristics, bordering mountains and sea provide a wide choice of possible recreational activities. This area, popularly called 'the sun belt', stretches from the northeast of Spain, through the south of France to the north of Italy. Important cities located here are Barcelona, Toulouse, Lyons and Milan.

INDUSTRY OPERATION

The diverse impact of weather and climate on industry may be categorized into two broad divisions, namely those influencing the location of industry and those affecting the operation of the industrial plant, once an industry has been established. Associated with these factors are the design of the plant, and the planning of its operation. Initially, climatic factors rather than weather problems need to be considered. Climatological conditions which influence storage, warehousing and the operation of transport facilities are very important. Furthermore, almost all industries with outdoor activities are to a greater or lesser degree subjected to climatic variations. Such activities as transportation, construction, aircraft manufacturing, shipbuilding and strip mining are often affected by frost, fog, snow, ice, lightning and gales.

Weather is also reflected in almost every phase of manufacturing operation. The experience of management and workers in responding to any climatic challenge is important in limiting its adverse impact. One factor is extreme temperature. However, in some cases the impact may result not from the extreme temperature, but from the declining efficiency of labour and industrial operations with decreasing temperatures. For example, painting is usually restricted by temperature and a temperature below 4°C will usually terminate all outside painting operations. Labour efficiency declines below 16°C. When temperatures drop below freezing, a new set of temperature problems appear, and below −18°C practically all industrial operations are affected.

Snow, sleet and ice will cause vehicles to stall on highways and industrial operations suffer because of the immobility of both workers and materials. In modern logistic systems, very few materials are kept in stock and therefore a delay in delivery, for only a short period of time, will lead to major problems and a break in production. On the other hand, in regions like Sweden and Canada where regular winter snowfalls often reach 100 to 200 mm at a time, there may be hardly a pause in the work schedule because of a moderate snow storm. Gales have a great and obvious influence on many industrial and commercial operations. Secondary wind effects, perhaps increasing the amount of dust or snow transported or making the effects of low temperatures even more severe (see Chapter 12, The human energy balance), can place limitations on many industrial operations. The offshore industry is a major user of weather forecasts (Grant, 1990) since the key parameters for production are wind and waves.

Rainfall in general, and heavy rainfall in particular, play important direct and indirect roles in industrial operations. Heavy rainstorms can cause immediate flooding of low-lying or flat areas where drainage is inadequate. A secondary and more indirect effect concerns the water supply or effluent removal operations that are necessary to keep the industrial enterprise functioning. Depending on the location of the factory and its industrial processes, it may be necessary to rely on the annual rainfall to supply water for processing or to provide adequate dilution of factory effluents.

The application of climatic information is of importance for specific industries. In both the petroleum and chemical industries there is the common low-temperature problem of possible freezing of materials flowing in pipelines or of water in cooling towers. For electricity utilities, climate may influence maintenance equipment, transmission lines, power sources and buildings and the efficiency of workers. Significant effects are the freezing of water supplies and water intake systems, the freezing of coal and the difficulty of starting diesel motors. Low temperatures are often associated with snow or ice conditions, which in turn may affect pylons, towers and transmission lines. Optimal conditions for some selected industrial operations have been suggested by Maunder (1970). Table 14.1 shows that the strict controls of temperature and humidity constitute a major part of the costs of operating the production process, through heating, cooling and dehumidifying demands. The motion picture industry, with more stringent and specific weather requirements, also serves to illustrate the optimal value of weather conditions, notably in this case sunshine, light and visibility. Both weather forecasts and climatic studies are therefore essential in determining the best time and place to photograph outdoors, if filming costs are to be minimized.

Palutikof (1983) studied the impact of weather and climate on industrial operations in the UK, using an index of industrial production (IIP). This incorporated the productivity of the mining, quarrying, manufacturing, construction and utility industries and was seasonally adjusted to take into account the number of weekends/bank holidays and seasonal climates. National productivity anomalies were related to industrial disputes (strikes) and severe weather events, although it must be remembered that regional variations in impact and response are also important considerations. Figure 14.1 illustrates the IIP for the period 1958 to 1979 and it is apparent that the low winter indices in 1962/3 and 1978/9 were due to the severe 'big freeze' conditions experienced at that time. Conversely, the low indices in the winters of 1971/2 and 1973/4 were due more to industrial disputes, with the former season recording a record loss of working days through strikes. The 1973/4 winter low anomaly was due to the infamous coal-miners' strike at that time and the creation of the 3 day week (between January and March 1974) and massive industrial disruption.

Table 14.1 Optimum indoor temperatures and humidity for selected industrial operations

Industrial operation	*Temperature (°C)*	*Relative humidity (%)*
Food processing	18–20	60–80
Milling	15	50–60
Flour storage	25–27	60–75
Baking	18–20	40–50
Process cheese manufacturing	15	90
Textile manufacturing		
Cotton	20–25	60
Wool	20–25	70
Silk	22–25	75
Synthetics	21–29	55–60
Miscellaneous		
Cosmetics manufacturing	20	55–60
Cosmetics storage	10–15	50
Drug manufacturing	20–24	60–70
Electrical equip. manufacturing	21	60–65
Paper manufacturing	20–24	65
Paper storage	15–21	40–50
Photo film manufacturing	20	60
Printing	20	50
Rubber manufacturing	21–24	50–70

Source: Maunder (1970).

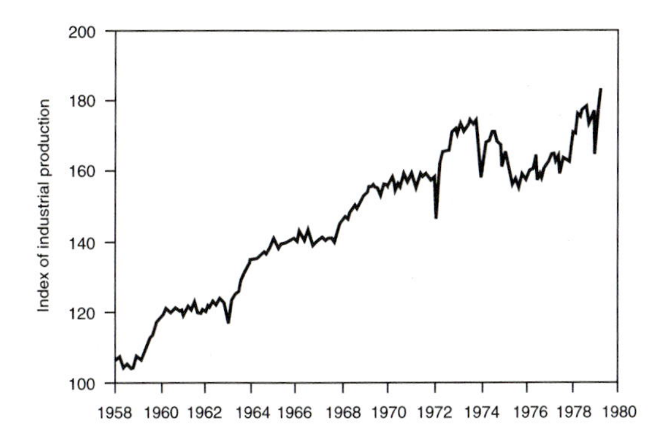

Figure 14.1 Index of industrial production (IIP) for the UK, 1958–80 (after Palutikof, 1983)

Investigating the effect of hot, dry summers on IIP is more difficult since the drought of 1975/6 was accompanied by a time of economic depression and a marked general decline in industrial output. However, Table 14.2 indicates the performances of individual industries at that time and it is apparent that the operations with the greatest industrial decline were those with very high water requirements (e.g. ferrous metals and utilities). The productivity of the drinks industry increased owing to the hot, dry weather as did the (summer) clothing/footwear industry. Table 14.3 shows the industrial consequences of the severe winter of 1962/3 where the most seriously affected operations were those where production took place out of doors, or where the finished product is used out of doors (e.g. timber, bricks/cement, mining/quarrying and shipbuilding). The large fall in output from the clothing and footwear industries was probably due to the related decline in retail sales when transportation systems became disrupted and workforce sickness decimated the labour force. As expected with 'big freeze' conditions, demand and output increased markedley for utilities and the associated coke and oil refining industries.

Industry has for quite some time been engaged in in-depth discussions of the problems relating to the possible future destruc-

Table 14.2 Performance of individual industries in the 1975/6 drought

Industry	*Average deviation per month from mean of preceding season*
Ferrous metals	−10.0
Utilities	−6.4
Metal goods (not elsewhere specified)	−4.6
Non-ferrous metals	−4.3
Pottery and glass	−3.9
Mining and quarrying	−2.3
Paper, printing and publishing	−2.2
Engineering and electrical goods	−2.1
Timber, furniture	−1.8
Vehicles	−1.3
Bricks, cement, etc.	−0.9
Shipbuilding	−0.9
Textiles	−0.6
Coke ovens, oil refining, etc.	−0.3
Leather goods	+0.6
Food	+0.7
Chemicals	+2.1
Clothing and footwear	+3.0
Drink and tobacco	+4.4

Source: Palutikof (1983).

tion of the stratospheric ozone layer by chlorine compounds from chlorofluorocarbons (CFCs) (Chapter 22). It is also concerned with the separate problem of an enhanced greenhouse effect (Chapter 22) attributed to the observed increase in the concentration of carbon dioxide and other trace gases in the atmosphere (Karpe *et al.*, 1990). Industry perceives the problems of projected global climate change as a challenge which requires solutions using the technologies already available or to be developed. It has already made noticeable contributions in this context (see Chapters 5 and 22) and will step up its efforts. In some of the highly industrialized countries, appropriate technologies introduced mainly to improve economic efficiency have already led to a reduction in carbon dioxide emissions. There are technical possibilities for

Table 14.3 Performance of individual industries in the 1962/3 winter

Industry	*Average deviation per winter month from autumn 1962 mean*
Bricks, cement, etc.	−14.4
Timber, furniture	−14.3
Clothing and footwear	−9.6
Paper, printing and publishing	−5.6
Mining and quarrying	−4.7
Shipbuilding	−4.3
Engineering and electrical goods	−3.7
Non-ferrous metals	−2.7
Drinks and tobacco	−2.4
Ferrous metals	−2.3
Food	−2.0
Metal goods (not elsewhere specified)	−2.0
Chemicals	−1.6
Leather goods	−0.6
Pottery and glass	−0.3
Textiles	0.0
Vehicles	0.0
Coke ovens, oil refining, etc.	+3.3
Utilities	+17.5

Source: Palutikof (1983).

reducing the trace gases affecting the global climate. In industry's view, it will be essential at each stage to support the developing nations and to co-operate with them in order to achieve adequate development through suitable alternative technologies.

CONSTRUCTION OPERATIONS

Construction activities may be related to weather and climate in three broad areas, as follows:

1 Their influence on structural and architectural design (Chapter 13).
2 The effect, in economic terms, of the impact of weather on day-to-day operations of the construction industry.
3 The influence that weather and climate has, or more accurately should have, on urban and industrial planning.

For example, a given structure must be designed to withstand stresses caused by climatic factors such as temperature fluctuations, wind forces, humidity changes and snow/ice loads. Of these, temperature is one of the most important climatic elements in any engineering or construction design because of the expansion and contraction of materials with changing temperature. The maximum expected wind speed is also among the stress factors considered in designing bridges, towers and buildings. Suspension bridges are particularly vulnerable to wind, for they tend to sway and to develop destructive oscillations. A notable example of this was the destruction by wind of the first Tacoma Narrows bridge at Tacoma, Washington, on 7 November 1940. Two other important atmospheric factors are high humidity and excessive rainfall. Extremes of precipitation must also be taken into account since dams, canals, pipes and bridges all require their construction to withstand the maximum flood stages likely within a certain 'return period' of the event.

A large percentage of construction operations are sensitive to adverse weather conditions. For example, building constructions are often vunerable to ordinary weather occurrences and building materials and systems are susceptible, to a greater or lesser extent, to degradation by the natural weathering processes. However, many critical weather conditions, when they occur, prevent any useful work being done and large losses in income may result. The ability to forecast more accurately the critical weather conditions is therefore important. The list of construction problems is oriented primarily to short-period weather phenomena although it is possible to utilize it climatologically as well. Important elements are the changing climatic conditions known to occur at different times of the year and the need to adjust cost figures accordingly. Knowledge of the influence of low or high temperatures, winds, or other weather factors, may suggest a range of possible locations for the construction site. The location may be climatically more favourable, or less costly in weather delays, than at other nearby sites.

Weather processes may have a great economic impact on day-to-day operations in the construction industry since enormous sums of money are involved, employing large numbers of workers. Possible weather-related losses have to be considered in economic terms of decreased profits and intermittent unemployment, accompanied by reduced spending power. Precipitation, temperature extremes and high wind speeds are the major causes of disruption to the construction industry, either singly or in combination. The combination of precipitation and low temperatures which produces snow increases construction problems. Steel reinforcing rods, incorporated in most concrete structures, have the ability to conduct heat much more readily than snow, so that attempts to melt snow often result in an impenetrable slab of steel-reinforced ice. Wind presents obvious dangers for working on tall structures or when using tall and slender cranes.

Weather conditions, forecasted as well as actual phenomena, are clearly of considerable value to the construction industry. Even as early as the initial tendering stage, when costs are being estimated, it is useful to know the likely weather interference over (perhaps) the next 12 months. While the job is underway, accurate daily (and perhaps even 3 hourly) forecasts would greatly assist efficient completion. The very localized scale at which forecasts would be required complicates the forecasting problems and emphasizes the possible roles for the consulting meteorologist and applied climatologist.

COMMERCE

Recently, Doyle (1996) has shown that extreme unseasonal weather can play a key role in the overall activity of the economy. The relationship between temperature anomalies and gross national product (GDP) in the UK is shown in Figure 14.2. Below-average temperatures boost economic activity especially in winter and spring, but hot summers have a negative impact on the retail trade and on housing transactions. Weather not only affects the kind of activities we pursue but it also has a significant effect on the overall level of activity.

Some examples have appeared in the literature on the relationship between the retail sales of certain products, such as raincoats, cold drinks, ice cream, overcoats and bathing suits, and weather factors. Year-to-year changes result from the occurrence of favourable or unfavourable weather events such as heatwaves and widespread blizzards. Warm summers in Sweden (1994 and 1995) and in the UK (1995) have affected retailing. People made fewer shopping trips, preferring to go on leisure trips to the beach. Sales, especially in large city-centre shops, were badly affected during the extended spells of fine weather. Important factors are as follows:

1 The weather might physically prevent people from getting to stores.
2 The weather might be disagreeable enough to discourage some people from shopping.
3 The weather might produce certain physiological effects that would change shopping habits.
4 The weather might be such as to make certain items more saleable.

The overall regional climate may detemine the demand for goods such as heaters, air-conditioning equipment and different types of clothing. Much of the published material relating to commerce concerns retailing or insurance. A greater frequency and intensity of global

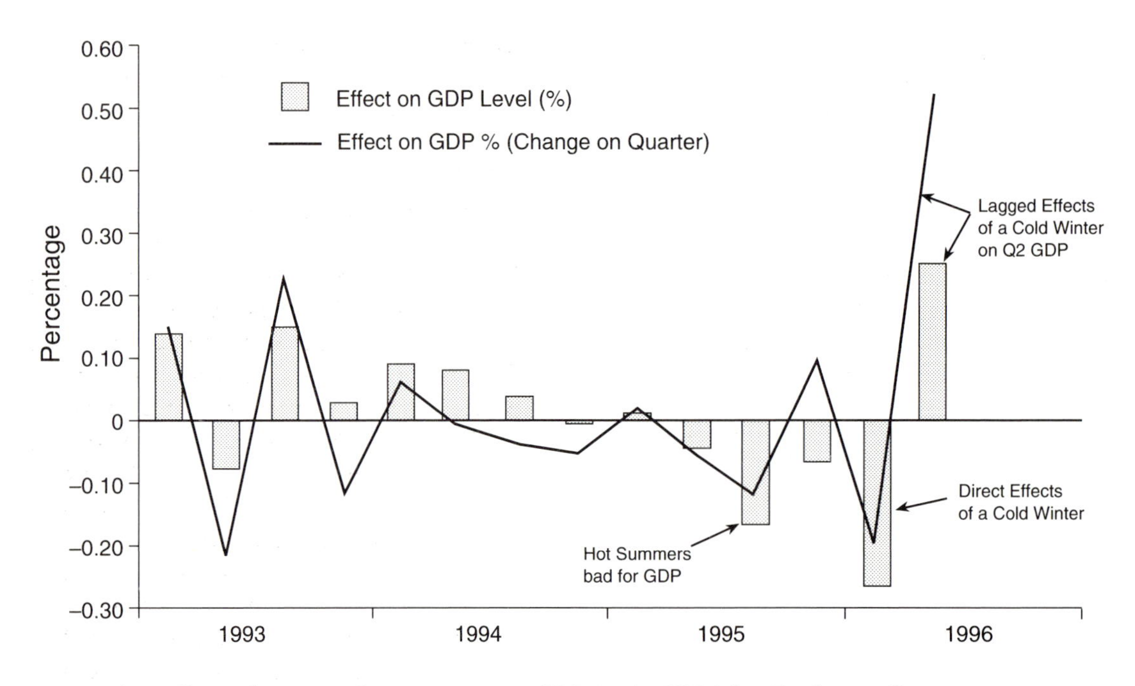

Figure 14.2 Effects of unseasonal temperature on GDP in the UK (after Doyle, 1996)

weather extremes are recognized as symptoms of predicted global warming (see Chapters 18 and 23). These will increase insurance claims in the developed world in order to pay for the more widespread structural damage, with a huge rise in premiums to cover the costs involved. Perhaps the most obvious impact of weather on consumer behaviour concerns decisions such as whether or not to go shopping. It is arguable whether manufacturers, wholesalers and retailers actually make adequate (or any) use of weather information. Linden (1962) and Maunder (1973) have argued for greater and more effective use of weather information to improve economic returns and service to customers. Knowledge of the weather 'reflex' of a particular product could be extremely useful in improving marketing tactics. Distributors, especially those marketing weather-sensitive items such as soft drinks, might try to co-ordinate regional distribution with a long-term weather forecast, and advertising and promotional activities also could be more closely geared to weather forecasts. Early-warning advice is an important service for private weather forecasting, which is increasingly given on a commercial (profit-making) basis.

Agnew and Thornes (1995) focus attention on the weather sensitivity of the food and drinks sector within the UK retail and distribution industry. The potential benefits to accrue from the application of meteorological knowledge are evaluated in light of current advances in retail information systems and management. When the products are ready and available, it is necessary to study the finer points of marketing, such as advertising. This operation also needs careful advance planning. For example, climate statistics may help the producer to decide on the optimal time to advertise antifreeze in an area. Hotels and holiday resorts, of course, are also affected by adverse weather phenomena. The resorts emphasizing winter sports rely upon good snow coverage for a certain period of the year and they are also subject to variable weather conditions, which can severely curtail the skiing season (see Chapter 18, Climate change, tourism and recreation).

CONCLUSIONS

This section analyses some applications of the Swedish Meteorological Institute (SMHI) in the field of applied climatology, with an emphasis on industry and commerce. Work completed by the Department of Physical Geography at Göteborg University, and a related consulting firm, will also be discussed. The applied climatologists in Göteborg utilize the existing road weather information system, including topoclimatological surveys, thermal mapping and investigation of climatological factors, which are relevant in the planning and construction of new roads. Topics in urban climate and air pollution are also being researched.

Planning studies, which investigate the optimum location of industries, can benefit from climatological information (Lindqvist and Mattsson, 1989). The city of Göteborg is situated close to the coast, featuring valleys some of which may have a high frequency of temperature inversions. When establishing new industries, opinions have been given on the risks of increasing a problematic air pollution situation. Even problems associated with bad smells have been taken into consideration. The importance of climatological and meteorological factors is obvious when considering outdoor construction in harsh environments. Bridges, for example, are well-exposed objects and the impact of wind on bridges is well known throughout history. To avoid failure in bridge construction, it is important to estimate accurately extreme conditions at specific locations well in advance.

An example of such an investigation is the project at 'Höga-Kusten' suspension bridge, which crosses the Ångerman river, near Härnösand in central Sweden. A field programme, carried out by climatologists from Göteborg, analysed wind and temperature data from a tower located on the small island of Fyrholmen in the Ångerman river. The recordings from the site provide data from early 1990 with wind speed and wind direction measured at two different levels, 30 m and 60 m, above the ground. The analysis of the data makes it possible to calculate maximum wind gusts occurring during a 100 year return period. Studies of terrain roughness and long-term meteorological data from a nearby synoptic observation site (Sundsvall airport) complement the analysis for the estimation of wind loads that are likely to occur. These field measurements were complemented by wind tunnel tests where a model of the landscape, of scale 1: 1000, was used to simulate the conditions at the bridge's location.

SMHI has a Division for Environment and Energy which, along with other activities, examines the applications of climatology to construction. One important programme concerns the dimensioning requirements for technical systems and materials. Typical tasks involve the estimation of wind loads, snow loads, temperature, heavy rain and air humidity data for applications to projects in building and construction. The Division also supplies dimensioning requirements for electrical materials, components and plants. Climatic data are also needed for

CONCLUSIONS *continued*

the evaluation and standardization of corrosion risks in different climatic environments, namely urban, maritime and continental locations. In some cases, climate data are used to evaluate production results; other applications may involve the insurance assessment of damages and costs caused by various weather-related hazards (see the previous section). The main task of the Division concerns energy and environmental studies in built-up areas, although it also provides data on the properties of building materials, including exposure to weather for long periods.

SMHI also has a Division of Commerce with a business section offering weather-based information to trade and industry. The objectives are to increase benefits and reduce costs in commercial activities. Typical activities include:

1 Analysis of weather sensitivity; the relationships between weather/climate and (for example) retail sales; frequency of customers' visits, established by means of regression analysis.
2 Weather/climate-based prognosis on demand (7–10 days) by means of regression relations.
3 Weather prognosis for months, with respect to the deviation of the mean temperature from the monthly normals. This is produced by means of a statistical method and a 10 day forecast using a dynamic model.

The customers are companies/organizations which need basic data for decisions concerning direct production, both for products and services, direct advertising of products, both content and timetable, and refining their products and services.

Some examples can be given of the first customer group. The food industry, dairies and breweries need weather and climate information for an effective flow of products so that supply is related to demand. The same situation exists when an evening paper requests data to optimize the number of copies printed and to regulate the distribution for different markets. Among service companies and organizations, multiple stores groups require weather and climate information for the optimal staffing of cash-desks. Other examples are zoological and amusement parks who need to regulate staff numbers. In a way, the police authorities and insurance companies have the same demands. Evidently, film and photographic companies need weather and climate information to plan outdoor activities. Multiple stores and other shops specializing in sport and recreation, clothing, footwear, paint services and supplies all need climate information when advertising, in order to maximize their sales in accordance with the weather. Advertisements from travel agencies often include information on the climate and expected weather at home (bad) and in the holiday destination (fair) (see Chapter 18, Weather and climate information for the tourist industry). Of course there are many more examples to support the specific relationship between variations in climate and weather and industrial output and commercial efficiency.

REFERENCES

Agnew, M.D. and Thornes, J.E. (1995) 'The weather sensitivity of the UK food retail and distribution industry', *Meteorological Applications* 2: 137–47.

Bickert, C. von E. and Browne, T.D. (1966) 'Perception of the effect of weather on manufacturing', in W. Sewell (ed.) *Human Dimensions of Weather Modification*, University of Chicago, Department of Geography, Research Paper No. 105: 307–22.

Doyle, L. (1996) *Window on the Economy*, Kleinwort Benson Research, Second Quarter Report, 56pp.

Grant, K. (1990) 'The offshore industry and weather forecasts', *Weather* 45: 404–7.

Griffiths, J.F. (1966) *Applied Climatology – an Introduction*, London: Oxford University Press.

Hobbs, J.E. (1980) *Applied Climatology*, London: Butterworths.

Karpe, H.-J., Otten, D. and Trindade, S.C. (eds) (1990) *Climate and Development*, Berlin: Springer-Verlag.

Landsberg, H.E. (1960) *Physical Climatology*, Du Bois, PA: Gray Printing Company.

Linden, F. (1962) 'Merchandising weather', *The Conference Board Business Record* 19: 15–16.

Lindqvist, S. and Mattsson, J.O. (1989) 'Topoclimatic maps for different planning levels: some Swedish examples', *Building Research and Practice* 5: 299–304.

Maunder, W.J. (1970) *The Value of Weather*, London: Methuen.

Maunder, W.J. (1973) 'Weekly weather and economic activities on a national scale: an example using United States retail trade data', *Weather* 28: 2–12.

Maunder, W.J. (1989) *The Human Impact of Climate Uncertainty: Weather Information, Economic Planning and Business Management*, London: Routledge.

Mather, J.R. (1974) *Climatology – Fundamentals and Applications*, New York: McGraw-Hill.

Palutikof, J. (1983) 'The impact of weather and climate on industrial production in Great Britain', *Journal of Climatology* 3: 65–79.

15

TRANSPORT SYSTEMS

John E. Thornes

INTRODUCTION: CLIMATE AND TRANSPORT SERVICES

An efficient, safe and cost-effective transport system is a vital goal for every country or region in the world. All sectors of a country's economy rely on transport even if it is only to get employees to work and to deliver goods. Table 15.1 shows a comparison of transport statistics for a number of developed countries in 1991. A staggering number of people are killed on the roads each year, with a total of more than 90,000 killed in just the seven countries with figures listed. It can be seen that the USA has more road, rail and air transport than any other country but that Japan has a larger shipping fleet. The UK and Japan have a large number of air kilometres flown, compared with the other countries, but both have a much smaller number than the USA. Table 15.2 shows the amount of freight transport in the same selected countries. Road carries the bulk of the freight in all countries listed apart from the former USSR where rail transport dominates. The small proportion of freight carried by rail transport in the UK stands out as does the huge amount of freight moved in the USA by all modes of transport.

Transport underpins all national economies as well as being a sector in its own right. The 1992 Standard Industrial Classification (SIC(92)) divides national economies into 17 different sections as shown in Table 15.3 which gives the UK gross domestic product (GDP) by industrial sector for 1992 (Central Statistical Office, 1994). Section 'I' covers 'Transport, storage and communications' but it is obvious that every sector has transport requirements. Agriculture, for example, requires delivery of fertilizers, fuel for farm machinery and distribution of products. Health services rely on drug deliveries and ambulances whereas tourism relies entirely on transport.

All transport services are dependent upon the weather and climate. The atmosphere acts as both a resource and a hazard to virtually all forms of transport. From the resource point of view, vehicle combustion engines (in cars, lorries, buses, tractors, aircraft, trains, ships, barges, etc.) require air to mix with the fuel (diesel, petrol, gas) in order for combustion to take place; aircraft rely on the air density to fly and yachts use the wind to drive their sails. Even electric vehicles require the electricity to be generated and that usually involves a combustion process. Nuclear power stations, which may provide the electricity for electric vehicles, rely on the atmosphere to absorb the excess heat generated. The hazards associated with the climate are more obvious and involve effects on the vehicles of transport as well as on the transport infrastructure (roads, railways, canals) and nodes such as ports and airports. Snow, ice, frost, fog, wind, heat, lightning, heavy rain and flooding can all render transport systems inoperable.

Table 15.1 Variety of international transport statistics, 1991

Country	*km of roads (000's)*	*km of rail (000's)*	*Road deaths*	*Shipping fleets (Mt)*	*Aircraft km flown (million)*
United Kingdom	384	16.9	4,753	5.3	511
France	810	33.5	10,483	3.8	380
Germany	550	44.1	11,046	5.5	434
Italy	304	19.6	7,084	7.7	201
Sweden	136	11.0	772	3.0	109
Japan	1,116	n.a.	14,595	24.8	520
USA	6,259	187.7	44,599	17.6	7,037
Former USSR	1,409	147.5	n.a.	18.4	n.a.

Table 15.2 Freight transport by various countries in billion tonne kilometres

Country	*Road*	*Rail*	*Inland waterway*	*Sea-going*	*Air*
United Kingdom	127	15	0.2	54	2.6
France	112	54	8	n.a.	3.7
Germany	174	133	57	0.6	4.1
Italy	168	22	0.1	35	1.2
Sweden	29	19	n.a.	8.5	0.2
Japan	274	27	n.a.	245	5.2
USA	1,183	1,510	550	900	14.5
Former USSR	527	3,827	233	n.a.	2.4

Table 15.3 The UK gross domestic product (GDP) by industrial sector for 1992

	Industrial sector	*£ million*	*%*
A & B	Agriculture, hunting forestry and fishing	9,309	1.7
C	Mining and quarrying	9,842	1.8
D	Manufacturing	114,698	21.4
E	Electricity, gas, water	13,717	2.6
F	Construction	32,002	6.0
G & H	Wholesale and retail	72,549	13.5
I	Transport, storage and communications	41,613	7.7
J & K	Financial and business	121,704	22.6
L	Public administration	36,605	6.8
M & N	Education, health and social Work	52,509	9.8
O, P, Q	Other services	32,892	6.1

As travel has become global, so transport agencies have gained experience in all weather and climate types. People have begun to expect to be able to travel to any part of the globe, on any day of the year, in the same elapsed time, regardless of the weather and climate. This has meant that transport agencies have had to try to become less weather sensitive in order to compete, at a time when the volume of traffic is rapidly increasing. Road and rail systems are more regionalized, whereas air and water transport is often global. Aircraft can take off in freezing temperatures in Alaska, cruise at even lower temperatures in the upper troposphere and then land on a hot Caribbean island. Ships sail half way around the globe to deliver cars from Japan. Thus aircraft and ships have to be designed to cope with anything that the global atmosphere can throw at them. Cars and trains, however, are often more parochial in their climatic capabilities, although the global market for cars has led to recent improvements in their design (Thornes, 1992).

Weather and climate impact upon the means of transport (e.g. car or train) in a different way to the surface or space on or in which the vehicles operate. Anticipated climate extremes such as 'dry' snow may immobilize a train engine whilst not affecting the track. On the other hand 'wet' snow might not affect the train engine but might bring down overhead electric cables. The climatic design and maintenance of roads and railway lines, therefore, present a very different problem to the climatic design of vehicles and trains. Roads and railway lines need to be routed, if possible, to avoid hazards such as frost hollows, preferential sites for fog formation and areas prone to flooding. Also they need to be designed to cope with extremes of temperature, for instance to avoid rails buckling and roads 'bleeding' in hot weather. A knowledge of likely climatic extremes is vital therefore before roads and railway lines are built in different countries (Chapter 23).

Extremes of weather and climate are not always the most costly or dangerous. For example, zero degrees Celsius is a critical temperature because water freezes, and is most slippery at that temperature (Moore, 1975). Hence roads are likely to be more hazardous in climates where the zero degrees Celsius threshold is crossed most frequently (Thornes, 1991). From an economic and safety point of view, it is often the variability of the climate from year to year, and the unpredictable nature of that variability, that is important. The climate presents a range of problems and solutions to global and regional transport. In order to manage effectively the atmospheric impact (atmospheric management) on transport, there needs to be an improvement in the dialogue between transport managers and meteorologists/climatologists. There is still considerable scope for improvement. All transport modes could be made more efficient, safer and cost effective by better atmospheric resource management (Chapter 5).

Effective atmospheric resource management also has to bear in mind whether or not the decision to travel rests with the individual or the transport companies. The decision whether or not a train, bus, boat or aircraft will travel in severe weather depends upon the management. However, the individual driver can decide whether or not to use a road. It is unusual for roads to be closed in severe weather, although it is common for some mountain roads to be closed in winter. It is therefore potentially easier to control rail, bus, water and air transport as long as the public are kept well informed.

It is not easy to distinguish between the impact of weather and the impact of climate on transport, as most managers have to consider both. In general, the climate of a region should determine the design and quantity of equipment required and labour resources that need to be allocated to function efficiently and safely, whereas the weather will determine day-to-day operational expenditure and safety. The operational managers responsible for day-to-day activities are often not the same level of staff as the administrative managers who make the strategic decisions and plan for the future on an annual timescale. There may well be conflicts of interest concerning, for example, the introduction of new weather-related technology to decide when to salt roads for winter maintenance (e.g. the introduction of road weather information systems (Thornes, 1991)), or relating to the weather routing of ships (Mackie, 1982). New equipment may enable the performance of operational activities to be more carefully monitored and therefore consultation between administrative and operational staff is important to avoid resistance to the introduction of the new technology.

The management of the impact of weather and climate also depends upon whether or not the impact can be reduced. Snow and ice on a road or runway or aircraft can be treated cost-effectively and even prevented by using de-icing chemicals, whereas fog, heavy rain and high winds cannot be controlled directly. Thus fog on roads, for instance, can only be managed by reducing speed limits, which unfortunately many motorists will ignore. Ultimately roads may have to be closed, which is preferable to a road having to be closed because of a weather-related accident. The sensitivity of air, rail, road and water transport to weather and climate will now be discussed in turn.

AIR TRANSPORT

Modern aircraft use the atmosphere more than they suffer from it; hence the resource usage of the atmosphere far outweighs the hazard. Aircraft do still have problems in severe weather, mostly when taking off and landing and, because the aircraft use the atmosphere continuously, virtually all weather parameters are important. Beckwith (1985) gave a very useful summary of the links between the atmosphere and aviation particularly in the USA. Bromley (1977) showed that in the USA in 1975, 85 per cent of total delays to aircraft for periods of at least 30 minutes were weather related, and that the weather was responsible for 36 per cent of the accidents. Thunderstorms accounted for 25.6 per cent of delays, snow/ice for 15.5 per cent, wind 18.9 per cent and poor visibility 18.7 per cent. Most modern international airports attempt to operate every day of the year whatever the weather. In 1991, there were 388,000 take-offs and landings and 40 million passengers handled at Heathrow in the UK and 814,000 take-offs and landings at O'Hare in Chicago, which handled 60 million passengers.

Climatic Design and Levels of Maintenance

Aircraft

Today, aircraft and helicopters are designed to operate virtually anywhere in the world and, as such, the climatic design is advanced. Also aircraft fly at height through the atmosphere where the temperature may be as low as −65°C and the winds as strong as 400 knots. Most flights are 'above the weather' and although these conditions are more severe in absolute terms than at the earth's surface, the air density is much less and flying is safer. The climatic design is apparently good in so far as the airlines have an excellent safety record. In the UK in 1992, only 36 passengers and crew were killed in accidents involving UK-registered aircraft in UK airspace. Pilot briefing and on-board radar for weather routing are now computerized, which means that severe weather can usually be predicted and therefore avoided in flight. Significant fuel savings are possible with effective weather routing.

Atkins (1982), when discussing routing for Concorde, stated that

> The two main meteorological aspects which affect the performance of jet aircraft are wind and temperature, efficiency increasing with a

decrease of ambient temperature. At the flight level of most jets the effect of wind is generally more significant but for Concorde, flying above 50,000 feet, the temperature is equally important. For a typical supersonic Atlantic crossing between London and New York, the effect of a 1 °C increase in temperature throughout is to increase fuel consumption by about 100 kg. If the ambient temperature is above −51.5 °C the effect is even greater, each 1 °C change making a difference of 350 kg of fuel. This is because the aircraft surface temperature becomes limiting at −51.5 °C owing to friction, and therefore as the ambient temperature rises above this value, a corresponding reduction in aircraft speed has to be made with a consequent loss of efficiency and increase in fuel consumption.

Clear-air turbulence is still a problem, but quick reporting allows rerouting to avoid the problem. Helicopters are designed to fly at low levels and therefore they are more vulnerable to severe weather. Nevertheless, they are capable of operation in severe conditions and have a reasonable safety record. Standards of vehicle maintenance are high. Most of the stress on the vehicle is encountered during take-off and landing and severe weather adds to the 'wear and tear'. The cost of maintenance is high, but it is difficult to estimate the added cost of severe weather.

Airports and runways

Airports are designed to be kept open every day of the year in all weather conditions. International airports are huge complexes with problems in severe weather of access by road as well as keeping the runways open. The main runway is normally orientated in the direction of the prevailing wind to assist take-off and landing. Obviously the climate will determine the frequency of wind direction, and some locations will have difficulty if there is no identifiable prevailing wind. Cross-winds are always a problem on take-off and landing.

All airports in the UK have provision for 'round the clock' snow and ice control, both for clearing runways and aircraft. De-icing chemicals, applied in advance of snow or ice formation if possible, are expensive and can cause corrosion of the runway surface. If snow is allowed to bond to the surface, damage to the runway can be caused by snow ploughs, and hence brushes are preferred. Access roads will be ploughed, however, and damage may be caused. Also the de-icing chemicals, such as urea and glycols, can get into streams and rivers nearby and cause damage to fish. The urea breaks down into ammonia and then nitrates which encourage the growth of algal blooms that deplete the dissolved oxygen. Glycols also combine with oxygen reducing the amount available for aquatic life (O'Connor and Douglas, 1993). Up to 50 million litres of chemicals are sprayed onto aircraft and runways around the world each year to prevent the build-up of ice on wings and to keep the runway ice-free. This pollution is now being controlled at some airports such as Copenhagen and Stockholm by spraying aircraft on rubber mats that catch the chemicals and recycle them. Some airports have experimented with the use of helicopters or jet engines to try and disperse warm fogs. Some success may be achieved with thin layers of fog, but the method is rather costly and unreliable.

Weather Sensitivity and Thresholds

Aircraft

The thresholds are as follows:

1 Temperature: Aircraft are likely to have to operate in temperatures ranging from about −70°C up to about 50°C. At least four thresholds are present: (a) below about −51.5°C as discussed above for Concorde; (b) below −30°C de-icing chemicals will not be able to keep the aircraft free of ice for take-off; (c) below 0°C de-icing chemicals

may have to be used to prevent ice building up on the aircraft which may hinder take-off; (d) above 25°C payloads may have to be reduced for take-off owing to the lower air density. Thus the problems mostly relate to take-off.

2 Snow: Any falling snow during take-off and landing reduces visibility.
3 Wind: Lack of wind can be a problem for take-off and landing as some wind resistance is useful. Light aircraft are affected by cross-winds as low as 15 knots and, at 25 knots, even heavy aircraft can be troubled, and alternative runways may have to be used. Above about 35 knots, an aircraft will have problems with any runway orientation (Beckwith, 1985).
4 Rain: Heavy rain can reduce visibility which may be a problem on take-off and landing.
5 Visibility: Any visibility below 800 m is a problem for take-off and landing.
6 Low cloud: A low cloud base below 60 m is a hazard to take-off and landing.
7 Humidity: High humidity above about 98 per cent is a problem if condensation or sublimation takes place, leading to possible engine-starting problems or wing icing.

Airports and runways

The thresholds are as follows:

1. Temperature: If the runway temperature falls below 0°C and there is moisture around, then ice is a possibility on the runway. For temperatures down to about −5°C, urea can be used to lower the eutectic point of the moisture but below that temperature, then CMA (Calcium Magnesium Acetate), Konsin or some other glycol-based chemical has to be used. These chemicals are very expensive and therefore it is desirable to spread them only when necessary. Less chemical application (about a quarter) is needed to prevent ice formation than to melt ice and hence accurate predictions are a must. If the chemical is spread too soon, it may be washed off the runway by rain. If the runway surface temperature gets too warm, above 45°C, then the melting of asphalt may be a problem.
2. Snow: Normally anything more than a light covering of snow will be brushed off the runways but continuous brushing may be required if snow persists.
3. Wind: Strong cross-winds of greater than 30 knots may put emergency services on alert.
4. Rain: More than 50 mm in an hour is likely to flood runways at Heathrow but each airport will have a different threshold according to drainage and topography. Heavy rain can be brushed off to avoid aquaplaning on landings.
5. Visibility: Maintenance is affected below about 200 m.
6. Humidity: If de-icing chemicals are hygroscopic then runways or approach roads may be made wet rather than dry, which will reduce friction on the runways or approach roads.

Weather Forecasting and New Technology

Improved weather forecasting is important for pilots, air traffic control and maintenance authorities. Weather routing is still being improved as predictions of upper-level winds become more accurate. The prediction of runway surface temperature could lead to substantial savings in the use of runway de-icing chemicals. The installation of surface sensors to measure runway surface temperatures and residual de-icing chemical is becoming more widespread. Linking this information to weather centres has yet to be achieved, but it is soon planned to happen at Gatwick. Instrument landing systems to enable landings in poor visibility are entering a new era with the use of microwaves rather than radio.

RAIL TRANSPORT

This section is mainly concerned with the situation in the UK but elsewhere there are similar problems encountered. Railways, in having their own special track, could be thought to be less weather sensitive than other modes of transport. Fog, for instance, should not be a hazard if signal personnel do their job properly. However, heavy rain or snow can cause track blockages, particularly in cuttings, and cold weather affects many activities, particularly when accompanied by snow (Ewens, 1994). Smith (1990) showed how sensitive the timing of trains is to low temperatures and snow, from a study of train arrivals in the Eastern Region of the UK, as follows:

> Although major disruptions to rail services were confined to a few days per year, mid-winter on-time punctuality was 6–7 per cent below that achieved for the rest of the year . . . the presence of snow on the ground was the most important single factor in creating delays. Punctuality started to deteriorate at screen minimum temperatures around +2 °C, a threshold which occurs on average every other day during the winter. Overall, for about 15 per cent of the year, the weather was responsible for about half of the delay and disruption experienced.

Strong winds are a hazard and can bring down overhead cables, particularly if trees are blown onto them.

Climatic Design and Levels of Maintenance

Locomotives

Railway engines are powered primarily these days by either electricity or diesel. Both are designed in the UK to operate in UK conditions, and exports are no longer significant. Hence it could be argued that trains designed in more severe climates might have more climatic design features. The cold weather of January 1987 caused significant revenue losses of the order of £15 million and several design faults were identified, such as air-cooled engines freezing, diesel fuel waxing, sliding doors freezing, passenger heating failing, suspension freezing and brakes locking. One wonders if continental trains suffer the same problems. The lessons learned from the snow in January 1987 were not sufficient to prevent chaos during the snow of February 1991. The air intake of locomotive engines caused considerable problems as discussed below.

Track

Climatic design should ensure that the track will not buckle in hot weather, at the one extreme, and that points will not freeze at the other. Considerable investment in point heaters has gone on over the last 10 years, but problems were still rife in January 1987. The effective design of cuttings to limit drifting snow is important, along with the use of snow fences to reduce drifting across the track. However, snow fences are expensive, and the whole railway network cannot be fenced! Bridges are a problem as there are so many of them, some of which are quite old. Even if a track is flooded in places, it is difficult to know if a bridge is still safe with a swollen river beneath. With regard to track maintenance, there are something like 50 pairs of dedicated snow ploughs on British Rail, and 170 small ploughs for diesel engines. In 1981, a snow blower was bought, to be based in Inverness, at a cost of about £250,000. During the snow of January 1987, it was moved to Kent, a journey that took more than 24 hours. There were problems of turning the snow blower on electric tracks and therefore another snow blower has been purchased for southern England.

Weather Sensitivity and Thresholds

For locomotives and track, these include the following:

1 Temperature: At low temperatures (below −20 °C), electric trains may suffer power loss due to icing of overhead cables, or the third rail. Diesel oil may wax at temperatures below about −18°C. Below −3°C, it is dangerous to load coal (Crawford, 1989), and below 0°C, points may freeze. Rails will buckle if they get too hot, and above 32°C, speed restrictions may apply (Allardice, 1989).
2 Snow: Above a depth of 5 cm, problems may be encountered and the effective limit is 15 cm above track height. In Kent in January 1987, 40 cm of snow was deposited in places with drifts up to 7 m deep. Whilst such conditions might be considered more frequent in Scotland, nevertheless Kent does seem to record significant falls every 10 years or so.
3 Wind: If the mean wind exceeds 34 knots or gusts in excess of 60 knots are experienced, then overhead cables may be disrupted if trees are blown onto them, or if trees are blown onto the line.
4 Rain: Crawford (1989) stated that:

   > Normal movement of all rail traction should cease when the water level reaches a point 0.05 m below the top of the running rail. However, locomotives may run on flooded sections at walking pace, provided the level of water is no more than 0.1 m above rail level.

   Heavy rain, as well as causing flooding of the track, may cause landslips, or bridge damage. The thresholds depend upon previous falls as well as local geography.
5 Visibility: This is not really a problem unless people are unable to get to the stations, or signals cannot be seen.
6 Humidity: Problems are associated with starting engines affected by condensation or sublimation in cold weather.

Weather Forecasting and New Technology

A new service (called 'Open Rail') is being provided to British Rail by the Meteorological Office, to provide better weather forecasts and more site-specific weather forecasts. For example, better forecasts of freezing conditions will allow the more economical use of de-icing chemicals. New technology should enable point heaters to be controlled and linked automatically to weather forecasts. Design improvements should enable locomotives to be less weather sensitive in cold conditions. Also, there is a much-improved forecast service for leaf fall in the autumn.

Better contingency planning is essential and better training should be a priority. The lessons learned from the snow of January 1987 did not enable British Rail to be any more effective during the snow and low temperatures of February 1991. In a report following the 1987 snow, British Rail identified 50 action points which included looking at the problem of dry snow getting into traction motors. In three days in February 1991, 455 electric train motors were put out of action. Such an investment programme is to be welcomed but it does not guarantee weather proofing, as the lack of success of similar actions following the 1987 snow has shown. Also it appears that little, if any, new money is to be put into weatherproofing, meaning that money is to be diverted from other identified improvements.

Over the last few years, the number of train accidents caused by snow and landslides are as follows:

1982:	100	1985:	113	1988:	124	1991/2:	107
1983:	98	1986:	120	1989:	170		
1984:	132	1987:	129	1990:	182	Average =	128

This amounts to, on average, nearly one accident per day in winter. In 1991/2 there were 39 failures (1990, 57 failures) of overhead lines, of which more than half were attributed to high winds. The summer of 1991 only experienced 13 track buckle incidents as opposed to 73 in 1990. No trains were derailed as a result of track buckle in 1991. In early 1996 British Rail is the only major customer for weather forecast services, although when parts of British Rail are privatized, this should provide scope for additional customers.

The 'Open Rail' service is provided via telex to the BR NTN telecommunication system. 'Open Rail' has five modules on offer at different times of the year, as follows:

1 'Open Rail' itself, which is primarily a winter service from October to March. The forecasts are issued in the early morning for the day ahead for each rail region. Forecasts of air and ground temperature, windchill, wind direction and speed and snow are made, including an outlook for the next 4 days. The type of snow, a leaf fall index, and other hazards (e.g. floods) are also forecast.
2 Hotrail service, which operates in the summer months and attempts to predict the likelihood of rail buckling. Each report covers a 3 day period and includes air temperature, cloud cover, rainfall, wind speed and wind direction. Forecasts of actual rail temperatures are also being attempted.
3 Railice service, which specifically predicts ice formation on the third rail on the southeast network and London Underground. The forecast is in the form of a 24 hour probability of hoar frost, clear ice, freezing slush, freezing rain, etc.
4 Wind Warning service, when high winds are likely in the next 24 hours and a national wind warning is sent to BR control centres around the UK, which is updated every 3 hours. The forecasts are primarily for the engineers who are responsible for overhead electric lines and who can put maintenance teams on standby.
5 Heavy Rainfall Warnings represent forecasts of heavy rain, which are issued if flooding or landslides are likely.

The 'Open Rail' service is relatively new and no study of its cost-effectiveness has yet been attempted.

A Case Study: British Rail – Network Southeast

Network Southeast is split into nine divisions that carry out normal maintenance activities plus having to de-ice the third rail. For this purpose, the 'Open Rail' weather forecasts are supplemented by rail temperature and icing-risk forecasts from Southampton Weather Centre. No rail temperatures are automatically monitored so there is no feedback as to the accuracy of the forecasts. The de-icing activities are carried out by specially equipped trains which spread a de-icing fluid onto the third rail along preordained routes. Eight of the nine divisions use a chemical supplied by Kilfrost called Electice (made by Shell), whereas South Central is using a newer chemical produced by Texaco which apparently is less messy and less corrosive. Network Southeast as a whole has about 20 de-icing trains which will operate every night during cold weather. Few figures are available on how much de-icing fluid is used but South Central estimates that it uses about 10,000 litres of de-icing fluid in an average winter at a cost of about £1/litre. South Central estimates that a de-icing train costs about £150,000 per annum to maintain and run, including labour costs. South Central has four de-icing trains, which therefore means that each train on average uses 2500 litres per annum. If these numbers are multiplied up for the 20 de-icing trains used, the total costs are approximately £50,000 for de-icing fluid and £3 million to run the trains.

The weather forecasts are also used for many other purposes, the chief one being to decide when to put sandite on the rails to prevent leaves making the lines slippery. Sandite is used throughout the country but Network Southeast has the worst problems because of the loss of power from the third rail if it is covered by leaves. The same trains as used for de-icing are used in Network Southeast for spreading sandite. This creates a big problem, however, because the sandite 'season' overlaps with the de-icing 'season' and it can take up to 2 days to convert the trains. Hence a good weather forecast is essential. The sandite 'season' runs from early October to the middle of December when the leaves are falling. Obviously there can be clashes of need in November and December.

The weather forecasts are used for a host of other activities by BR including the following:

1 If air temperatures below about −2°C are forecast, then water tanks are emptied from the toilets to prevent damage.
2 If low temperatures are forecast, then platforms, car parks and approach roads are treated with de-icers to prevent ice formation.
3 Signal lines are coated with a substance to prevent them freezing, which has to be topped up and checked in cold weather.
4 Weekend maintenance relies heavily on the forecast and if staff cannot work then it can cost up to £500,000 in lost wages, equipment hire, etc., for one weekend. If work starts and then cannot be finished, a line could be closed all day Monday. This does happen even with existing forecasts!
5 If rail temperatures are expected to be close to 32°C in summer, then the whole of BR has speed restrictions applied (50 mph, 80 kph) in case of rails buckling and causing a derailment. Accurate forecasts are therefore critical.
6 Wind forecasts are used where overhead cables may be damaged so that staff are placed on standby.
7 Flood forecasts are used to put staff on standby with an alert for landslides as well.
8 Forty-eight hours' notice is required to cancel weekend working or else wages have to be paid anyway.

ROAD TRANSPORT

A useful review of the relationships between the atmosphere and road transport is contained in *Highway Meteorology*, edited by Perry and Symons (1991). Some 88 per cent of inland freight transport is carried by road transport in the UK, and 93 out of every 100 passenger kilometres are travelled by road (British Road Federation, 1991). Hence, road transport is the most important form of transport for the UK economy and weather and climate interference is of great consequence.

Climatic Design and Levels of Maintenance

Vehicles

Cars, vans, lorries and buses are expected to perform in all weather conditions with a minimum of fuss. The main areas of concern for drivers relate to the starting of the vehicle on a cold damp morning, and the grip that the tyres have on the road surface. Automatic chokes and fuel injection to give instant starting are now quite common, but certainly not universal, and of course vehicle maintenance is the responsibility of individual drivers. Radial tyres and advanced braking systems offer a considerable improvement to road safety, by reducing stopping distances on wet roads. Codling (1974) has shown how weather and climate affect road accidents, although he does not acknowledge that the number of accidents on icy roads would obviously be much higher if there were no winter maintenance. Icy roads are still a problem though despite winter maintenance since not all roads are treated and weather forecasts are never going to be 100 per cent accurate. Studded tyres and snow chains are still used in snowier climates but have been banned by many countries owing to the damage caused to the road surface. It is also important that engines do not overheat in warm summer weather, and that batteries do not go flat in winter.

One might think that these features are obvious and that all vehicles are designed to meet climate extremes, but how long do the vehicles last? There is strong evidence that some vehicles last much longer than others, for instance Volvos have an average 'lifespan' in the UK significantly longer than most UK-manufactured cars. In view of intense competition, and the fact that in the UK most new cars are company cars which are exchanged in 2 or 3 years, there is little incentive for the vehicle manufacturers to design long-lasting vehicles (Valery, 1973).

Roads and Bridges

In recent years, the climatic design of motorways to avoid fog-prone areas has been discussed (Musk, 1988). The special design of crash barriers on the M62 across the Pennines to avoid trapping snow, and the topographic design of cuttings to avoid snowdrifts, are also recent examples. Layton (1985) outlines the more common climatic design features built into North American roads, as outlined by the American Association of State Highway and Transportation Officials. These include the use of a 'fog stripe', which is a white paint stripe at the edge of the road in fog-prone areas, and a limit on the grades approaching intersections in cold climates. There have been considerable problems with bridge maintenance, especially when rock salt is used. In the UK, the elevated section of the M6 in the Midlands is now treated with urea, but there were problems in the cold weather of January 1987 as urea is only effective in preventing ice formation down to about −5°C. Steel-deck bridges like the Severn bridge in England and the Kessock bridge in Scotland use Konsin which is non-corrosive. The winter maintenance of roads is carried out by county councils and district authorities. Priority 1 salting routes include motorways and trunk roads which are paid for by the Department of Transport (DTp), as well as county roads and some district roads which are paid for locally. Priority 2 routes are normally only treated when there is time, for instance clearing snow from housing estates. Each county is normally divided into divisions which have maintenance depots spread across the region. Each depot has the required number of 'gritters' and snow ploughs to treat the priority 1 roads in less than 3 hours but the designation of priority 1 routes is left to the county. On average, a pre-salt of priority 1 routes costs about £20,000 per night per county, including district roads. Hereford and Worcester (Ponting, 1989) estimate that in heavy snow (more than 6 cm), the county spends about £150,000 per day. Thus one night's pre-salting across the country could cost 60 times £20,000, which equals £1.2 million per night, whereas if snow covered the country, it could cost of the order of 60 times £150,000, which totals £9.0 million per day. The total annual bill for snow and ice control in the UK is on average about £140 million, with extremes ranging from only £50 million up to £200 million in a severe winter (Thornes, 1994). Better design of salting routes could reduce the time taken to treat the roads, and recent developments in the thermal mapping of routes may enable selective salting of routes to take place (Thornes, 1985).

The warm weather of July 1983 caused asphalt roads to 'bleed' and stone dust had to be spread to prevent the surface breaking up, although the warm and dry weather of 1975/6 was very beneficial for construction activities owing to a lack of weather interference. Hereford and Worcester estimated that they saved in excess of £100,000, whereas the nation as a whole saved over £100 million. This evidence suggests that these are the likely costs of adverse weather to construction and maintenance in more normal years. The climatic design of winter maintenance activities should take into account the likely variability of the number of days with snow and ice. Should engineers gear up for another winter as severe as 1978/9 or 1962/3, or should they plan for an average winter with reserve resources when required?

Weather Sensitivity and Thresholds

Vehicles and roads will be treated together in this section as the conditions suffered are similar. As with runways, it only requires about a quarter of the amount of de-icing chemical to prevent ice formation than to melt ice, owing to the extra energy required to melt ice. Hence an accurate forecast of the likely formation and timing of ice will enable pre-salting to be carried out (i.e. application of de-icing chemical before ice formation). There are several other thresholds that the engineer is interested in apart from 0°C and their probability can also be relayed to the engineer. Recently, the importance of measurements of road surface temperature has been realized by the installation of road surface sensors. Previously, engineers and the Meteorological Office relied on air temperature information but it has been shown that there can be considerable differences between air and road temperatures.

For roads, the critical thresholds are as follows:

1 Air/Road temperature: Diesel waxing might be a problem below −15°C. The British Standard is −15°C but a recent survey found that of 11 oil companies whose diesel fuel was tested, all functioned down to −16°C and three operated below −20°C. Until the cold weather of the winter of 1982/3, legislation only required diesel fuel to be operative down to −9°C. Below about −10°C, rock salt is not very effective and below about −5°C urea is no longer an effective de-icer, whereas at 0°C ice is at its slipperiest. Below 10°C asphalting and concreting are a problem if there is also a wind above about 25 knots. The wind-chill cools the surface too quickly and the surface may well have a short lifespan. Above 35°C, surface dressing of roads must be suspended as the asphalt will not cool sufficiently quickly.
2 Snow: More than 6 cm of snow will require snow ploughs to be engaged; less than that and the traffic can normally disperse it, or a pre-salt will be sufficient to melt it.
3 Wind: Apart from the wind-chill effect for the surface dressing of asphalt roads with winds greater than 25 knots, winds above 30 knots cause problems for suspension bridges, high-sided vehicles and construction cranes and falling trees may block roads.
4 Rain: Rainfall prevents road repairs and flooding often requires expensive cleaning up afterwards.
5 Visibility: All road repairs have to stop in poor visibility as drivers obviously do not want to run into roadworks. It is always argued that drivers should reduce speed in poor visibility but evidence suggests that this is rarely the case unless visibility gets down below about 100 m.
6 Humidity: The salt used on roads is hygroscopic above 80 per cent humidity. In the UK, the average humidity in winter is usually above 80 per cent, and therefore salt often remains in solution making the roads wet and therefore more slippery than if they were dry. Hoar frost and condensation on the inside and outside of vehicles can considerably reduce visibility for a time, and rock salt splashed onto windscreens can be a problem. Visibility can also be reduced

by low sun angles around dawn and sunset if there is little cloud around.

Weather Forecasting and New Technology

There has been a considerable dialogue over the last 5 years between meteorologists and highway engineers leading to the 'Open Road' Service being offered in the UK by the Meteorological Office. A complete 'turn-key' package is available to highway engineers including thermal mapping of road surface temperatures, the installation of road surface sensors and the link to the Meteorological Office. This enables a forecast of which roads require salting on a given night. A typical system costs about £150,000 and Cheshire County Council have been monitoring the cost-effectiveness of their system. The development of the National Ice Prediction Network has taken about 20 years to achieve, but it should considerably reduce the annual costs of snow and ice control, and enable highway engineers to be fully prepared for the worst that the winter can produce. Even so, weather forecasts are not perfect and engineers and the general public will still be caught out. Improvements to road surface temperature prediction models are still being made (Thornes and Shao, 1991).

The consequences of an incorrect weather forecast can be very different. If ice is forecast and roads are salted, but road temperatures stay above freezing, then money has been wasted, and of course salt is corrosive (Type 1 Error). If the forecast states that road temperatures will stay above freezing, and in fact they fall below, then a number of accidents may occur (Type 2 Error). The aim of a good forecasting system is to reduce the probability of either error occurring (Thornes and Fairmaner, 1989). In the past, forecasts were on the pessimistic side and far too many warnings were given so that Type 2 Errors rarely occurred. Overall, the National Ice Prediction Network has been a success with more than 650 road weather outstations and its computerized links between automatic roadside weather stations, weather centres and highway offices. Similar links are being developed for other transport services, but they are by no means as sophisticated. The highway engineers can actually compare through the day, on the computer screens in front of them, the actual and forecast road surface temperatures and wetness. This enables them to build up, over a few winters, a picture of the accuracy of the weather forecasts, and to understand some of the difficulties as well.

WATER TRANSPORT

Freeman (1985) stated that:

> Any work at sea is extremely weather-sensitive, ranging from the most common activity – the operation of a ship going as efficiently as possible from one point to another – to the most esoteric – such as underwater construction and the operation of oil drilling, production, and transportation facilities.

Weather forecasts, as well as including normal predictions of wind, temperature and precipitation, also have to include forecasts of 'the routine and violent action of waves, tides and currents' (Freeman, 1985).

Climatic Design and Levels of Maintenance

Large vessels

These are normally designed to operate globally and, as such, they must be designed to withstand most conditions, including periods of violent weather. For this reason, a large percentage of ships tend to be overengineered with beneficial safety margins offset by sometimes undesirable weight penalties and maintenance problems. The safety and life expectancy of a vessel is very much dependent on the initial design. This has a fundamental bearing on both the capacity to withstand severe weather and the 'ongoing' maintenance procedures. Maintenance schedules may depend to a certain extent on the climatological regime under which a vessel normally operates. The effect of significant anomalies is likely to be proportional to any non-routine cumulative weather stress. In any environment where competition is fierce and profit margins traditionally slender, there may be pressure to produce more cost-effective designs in the future. A good understanding of the dynamics and variation of weather-related stress will therefore be essential in producing effective and safe designs.

Smaller vessels

These normally operate in a restricted environment and therefore their design can be influenced by a number of factors. These include:

1 Climatological domain
2 Usage
3 Capacity
4 Speed
5 Endurance.

Small vessels, including a number of specialized craft, may only operate under certain conditions but also include some which may have to withstand extreme conditions that are taken for granted by larger vessels.

Examples of vessels which normally operate in a restricted environment are:

1 Hovercraft
2 Jetfoils
3 Catamarans
4 Small charter vessels.

The design of hovercraft, jetfoils and catamarans is usually the best possible compromise between speed, capacity and weather tolerance. The mere fact that such vessels are designed for speed tends to make them more weather sensitive than the larger and more traditional monohull versions.

Harbours

Land-based infrastructure will provide an initial logical basis for choosing potential harbour sites, with exposure and related climatology also having an important influence. Strong winds, whilst possibly favouring certain directions, can, in most cases, come from any direction. The controlling factor (when designing the harbour itself) therefore tends to be providing shelter from the predominantly worst sea conditions and, in particular, heavy swells when relevant.

Weather Sensitivity and Thresholds

There is not sufficient information to list individual weather thresholds but certainly all seagoing vessels are weather sensitive, with a large variation of capability across the spectrum of vessels.

Large vessels

Ultimately severe weather can (possibly in combination with other circumstances) produce catastrophic loss of a vessel no matter how large or well equipped it may be. Thankfully, such occasions are relatively rare and, in conjunction with improvements in weather forecasting technology, can usually be avoided. Apart from the obvious sensitivity to violent events, even large vessels are sensitive to normal weather variations. There can be a profound effect on stability, journey time, safety of cargo and fuel efficiency. Prior knowledge of the weather along a route can therefore become a powerful management tool, contributing towards a safer, more efficient and cost-effective operation.

Small vessels

Limitations of size and price mean that the majority of small vessels are more weather sensitive than their larger cousins. Winds as low as 22 knots can cause problems for certain craft and therefore weather information is very important. Around the UK, competition from the Channel Tunnel has created a resurgence of interest in fast, large-capacity craft. These offer minimal time penalties over the tunnel traffic, potentially lower cost and more point-to-point flexibility. The quest for speed does, however, make such craft weather sensitive, with limits depending on the size of the type of craft.

Apart from the well-accepted hovercraft and jetfoils, effort continues into perfecting the more recent wave-piercing catamaran concept. Generally such vessels become inoperative with wind speeds in the range 35–45 knots and significant sea heights in the range 2.5–4 metres. The likely conditions in a particular area can also have a very important influence on the exacting design technology used on these highly specialized craft. Subtleties such as wave period and direction become just as important as the more obvious wind speed and wave height parameters.

Harbours

The weather can affect the day-to-day running of a harbour with wind and rain critical to loading and docking strategy. On certain occasions severe weather can close a port completely with high seas making it dangerous for vessels to negotiate the harbour entrance. Although rare, advance notice of the likely onset and cessation of these events can facilitate beneficial operational decisions from both the operator and client points of view.

Weather Forecasting and New Technology

New technology has in general made vessels faster, safer and more reliable with better all-weather capabilities than their predecessors. These advances have occurred over a broad front, from basic vessel design and propulsion system through to the more esoteric advances in automation, navigation, and communication. Even so, modern designs can still benefit greatly from the considered application of weather intelligence. This can be from the use of climatology in the early design stage through to assisting safe, efficient and cost-effective running in the operational phase.

Marine forecasting organizations have the capability to monitor and forecast weather in great detail across the entire globe. As well as forecasting wind conditions using the atmospheric models, the wind fields are used to drive a global sea/swell model. For example, the 'Metroute' service takes full advantage of this powerful facility, combining the virtues of machine, scientist and mariner to provide 'state-of-the-art' ship-routing advice. Forecasts can be tuned to provide different types of routing from minimum time tracks, to the most fuel-efficient route and occasionally routes to protect sensitive cargo. As well as forecasting advice, detailed post-voyage assessments are also available and these provide a useful management tool. A large range of supplementary forecasts are also available from specialist meteorological centres providing advice to ferries, fishing boats and other smaller types of vessel. Mackie (1982) outlines the substantial savings that can be made by accurate wind and wave forecasts, and outlines the major aims of ship-routing services. Warning and hence avoidance of severe events such as hurricanes and typhoons are also necessary, and the UK Meteorological Office global model has scored highly in recent years in detecting the development of these systems and tracking their courses.

CONCLUSIONS

There is considerable overlap between the four transport sectors discussed above, for instance road transport is fundamental to get the public and staff to airports, railway stations and ports. In difficult weather conditions, one form of transport may have to substitute for another; for example, in deep snow, air transport may provide the only access into an area. Parry and Read (1988) noted also that if transport is disrupted, then there is a considerable 'knock-on' effect for other industrial sectors. There are clear 'knock-on' effects with those industrial sectors that rely on transport for the movement of labour and materials, and some recent innovations in this area may increase weather sensitivity in manufacturing industry. For example, 'just-in-time' production, where stockpiles of material are kept to a minimum, reduces any margin of flexibility if transport is disrupted (see Chapter 14, Industry operation).

The cost of delays is enormous, and delays due to roadworks have been estimated by the Confederation of British Industry (CBI) to cost UK industry up to £20 billion per annum. Delays due to severe weather might also be measured in billions of pounds in a severe winter (Thornes, 1994). These figures suggest that it should be extremely cost-effective to improve both weather forecasts and the supply of more up-to-date weather information to all transport agencies. Specialist services such as 'Open Road', 'Open Rail', 'Metroute' and 'Airmen' already exist but they are not easily available to the general public. There is therefore a need to get weather information to the transport users as well as the transport managers. Research involving, for instance, in-car information systems means that drivers should be able to obtain 'real-time' weather information for roads around the UK by the year 2000.

Most transport services require daily forecast information but 2 to 5 day planning forecasts are important to aid planning for emergencies. However, agencies should be prepared for false alarms as the new technology is not infallible. New technology may push back thresholds but changes may make the transport system vulnerable in different ways if, for example, a new system relies on computers that may occasionally fail. Staff training is vital for emergency procedures and in the use of new technology. Contingency plans are also very important and need to be continuously updated. The spatial variability of extreme weather events (Chapter 23) needs to be carefully monitored and the information passed effectively to the public so that they know which transport services are operable, and how timetables have been affected. The use of local radio, presto, viewdata, recorded telephone messages and television is to be encouraged, so that the public are effectively informed at home before they commence a journey, during a journey if possible and at the transport nodes.

The issue of climate change, and its likely impact on the transport sector in the UK, has been reviewed by the Climate Change Impacts Review Group (Department of Environment, 1996). They concluded that the importance of climate change on transport should not be exaggerated since the likely impacts will be small. This report is based on the latest regional scenarios of climate change which assume that winters will generally become milder in high latitudes,

CONCLUSIONS *continued*

although much more evidence is required to confirm this. The recent winter of 1995/6 in northern Europe has shown that blocking anticyclones can cause more ice, snow and fog whilst hardly reducing the mean air temperature. The impact of weather and climate on transport in the UK is therefore very significant and further research and development is required before transport managers and users are likely to be fully satisfied with the weather and climate information that they receive.

REFERENCES

Allardice, J.G. (1989) 'Meteorological Office services to transport', in S.J. Harrison and K. Smith (eds) *Weather Sensitivity and Services in Scotland*, Edinburgh: Scottish Academic Press, pp. 65–74.

Atkins, N.J. (1982) '100mb temperature forecasts for Concorde', *Meteorological Magazine* 111: 225–32.

Beckwith, W.B. (1985) 'Aviation', in D.D. Houghton (ed.) *Handbook of Applied Meteorology*, New York: John Wiley, pp. 945–77.

British Road Federation (1991) *Basic Road Statistics 1991*, London: British Road Federation.

Bromley, E. (1977) 'Aeronautical meteorology', *Bulletin of the American Meteorological Society* 58: 1156–60.

Central Statistical Office (1994) *Annual Abstract of Statistics*, London: HMSO.

Codling, P.J. (1974) 'Weather and road accidents', in J.A. Taylor, (ed.) *Climatic Resources and Economic Activity*, Newton Abbot: David and Charles.

Crawford, C.L. (1989) 'The effect of severe weather on the Scottish rail system', in S.J. Harrison and K. Smith (eds) *Weather Sensitivity and Services in Scotland*, Edinburgh: Scottish Academic Press, pp. 83–9.

Department of Environment (1996) 'Transport', *The Potential Effects of Climatic Change in the UK*, London: HMSO.

Ewens, G. (1994) 'When the trains don't come', *New Scientist* 1913: 38–40.

Freeman, J.C. (1985) 'Marine transportation and weather-sensitive operations', in D.D. Houghton (ed.) *Handbook of Applied Meteorology*, New York: John Wiley, pp. 978–97.

Layton, R. (1985) 'Surface transportation', in D.D. Houghton (ed.) *Handbook of Applied Meteorology*, New York: John Wiley, pp. 998–1030.

Mackie, G.V. (1982) 'The Meteorological Office Ship Routing Service', *Meteorological Magazine* 111: 218–24.

Moore, D.F. (1975) *The Friction of Pneumatic Tyres*, Oxford: Elsevier Scientific.

Musk, L. (1988) 'The assessment of local fog climatology for new motorway and major road schemes', *Proceedings of the IVth International Conference on Weather and Road Safety, Florence*, pp. 777–97.

O'Connor, R and Douglas, K. (1993) 'Cleaning up after the big chill', *New Scientist* 1856: 22–3.

Parry, M.L. and Read, N.J. (1988) *The Impact of Climatic Variability on UK Industry*, AIR Report No. 1, School of Geography, University of Birmingham.

Perry, A. and Symons, L. (1991) *Highway Meteorology*, London: E & FN Spon.

Ponting, M. (1989) 'Highway weather forecasting during winter', unpublished MSc thesis, University of Birmingham.

Smith, K. (1990) 'Weather sensitivity of rail transport', *Proceedings of the WMO Technical Conference on Economic and Social Benefits of Meteorological and Hydrological Services, Geneva*, pp. 236–44.

Thornes, J.E. (1985) 'The prediction of ice formation on roads', *Highways and Transportation* 32(8): 3–12.

Thornes, J.E. (1991) 'Thermal mapping and road-weather information systems for highway engineers', in A.H. Perry and L.J. Symons (eds) *Highway Meteorology*, London: E & FN Spon, pp. 39–67.

Thornes, J.E. (1992) 'The impact of weather and climate on transport in the UK', *Progress in Physical Geography* 16: 187–208.

Thornes, J.E. (1994) 'Salt of the Earth', *Surveyor* 8 December: 16–18.

Thornes, J.E. and Fairmaner, B. (1989) 'Making the correct predictions', *Surveyor* 172: 22–4.

Thornes, J.E. and Shao, J.M. (1991) 'A comparison of road ice prediction models', *Meteorological Magazine* 120: 51–7.

Valery, N. (1973) 'The rotten cars we deserve', *New Scientist* 60: 611–14.

16

AGRICULTURE AND FISHERIES

David W. Lawlor

INTRODUCTION: NATURE AND RATE OF ENVIRONMENTAL CHANGE

Rapidly increasing human numbers are altering land use and are massively exploiting ecosystems to provide space for agriculture. This is decreasing biodiversity and is having major ecological effects. Agriculture has removed natural forests and grasslands and competing plants and animals (Chapter 10), replacing them with a very much restricted range of crop plants and domestic animals. Food acquisition by hunter–gatherer methods, once of great importance, has reduced populations of wild plants and animals to relict numbers. Exploitation of unmanaged animal populations is now mainly restricted to fish, mammals and invertebrates from the shallow seas. Consumption of fossil fuels is increasing as a consequence of industrial economic activity and is resulting in pollution, including accumulation of carbon dioxide (CO_2) in the atmosphere (Chapter 22) owing to production exceeding CO_2 sequestration. Concentrations of methane and other trace gases are also increasing (Rozema *et al.*, 1993).

These gases absorb terrestrial infrared radiation and may be causing climate change (see Chapter 22, Greenhouse warming), often equated with increasing global temperatures (global or greenhouse warming). Other aspects of the environment may also change, namely rain- and snowfall, cloud cover and solar radiation, winds and storms and sea level (see Chapter 7, Conclusions). Such global environmental change (GEC) is expected to affect terrestrial, aquatic and atmospheric environments. Agriculture contributes importantly to these changes but it will also be directly affected by them. Crop and animal production is likely to be directly affected by combinations of changing conditions. Fisheries will be affected by factors altering the aquatic environment in spawning and feeding grounds and main fishing areas may shift. Altered conditions will also affect weeds and pests, parasites and diseases of plants and animals. All organisms may be affected and pose a potential threat or opportunity to agriculture (FAO Working Group, 1994).

How will these broad GEC conditions be expressed on the small scale in particular locations? What will the consequences be for organisms (crop plants and animals and species of fish etc.) required by human beings? How will pests, diseases and weeds respond to changes in particular environmental conditions? How important will individual physico-chemical and biological factors be, alone and in interaction? The potential effects can be judged at the agricultural scale from what is already established in much scientific work. This chapter examines what is known of the effects of global environmental change on crops, animals and fisheries and what is likely to happen over the next century or so, and suggests possible remedies.

CLIMATIC FACTORS REGULATING CROP AND ANIMAL PRODUCTION

To assess the effects of changes in atmospheric composition, temperature, cloud and rainfall on the production of crops and animals is difficult: the great range of current conditions under which they are grown means that complex interactions between the changing environmental variables may occur. For many crops grown in current environments, the main causes of poor or variable yield are known, namely harmful temperatures; inadequate water supply (e.g. limited sporadic rain, limited soil water-holding capacity); poor nutrition (e.g. inadequate nitrogen, phosphorus, etc.); pests (e.g. insects such as aphids, borers); diseases (e.g. rusts, mildews); and competition from weeds (including parasitic plants such as *Striga*). Table 16.1 indicates the temperature and precipitation requirements for selected commercial crops. It is apparent that climate plays a significant role in determining the crops grown in any region. However, although the climatic limitations are clearly defined, human activities (especially manipulation of the rainfall requirements through irrigation schemes) can extend crop production into more unfavourable areas (Oliver, 1973). Temperature may affect shoot production, reproductive development and formation of yield and extremes of heat and cold, if occurring at the same time as important processes in growth and development, may greatly reduce yields. Table 16.2 indicates the resistance of selected crops to frost occurrences in different stages of their development. It is apparent that in the case of spring wheat, oats and beans (for example), the germination stage is least susceptible to below-freezing temperatures. Conversely, cabbages display a most effective resistance to frost in the fruiting stage whereas tobacco plants show equal resistance at all stages of their growth development. Environmental factors interact (e.g. high temperatures and low atmospheric humidity) to inhibit pollination in cereals (Lawlor, 1995). For domestic animals, the main limitations to production are known (inadequate nutrition, pests, diseases, etc.): it is likely that changed climate will alter nutrition and increase the incidence and severity of some diseases. The effects may be wide scale (poor fodder quality) or local (increased tick-borne disease).

Economic viability of agriculture, be it highly technical or subsistence, will depend subtly on the changes of the environment and ecosystems. Small changes can have substantial long-term effects, particularly where agriculture is marginal. In much of world agriculture, especially in the less technologically developed areas, it is the lack of application of knowledge and techniques, as well as limited resources, which prevent output reaching potential. Mathematical–simulation models of the processes are sufficiently developed to forecast the major effects of environmental change (Rozenzweig and Parry, 1994), but not the more subtle combinations, helping to assess future climate effects and some interactions (Mitchell *et al.*, 1995). Together, models and experimentation provide a basis for assessing the likely effects of GEC on plant growth and the consequences for yield. Responses are unlikely to be linear over the range, with substantial interactions. We may expect large effects, both desirable and undesirable from a human viewpoint, on plants and animals in agricultural as well as natural (managed and unmanaged) ecosystems (see Chapter 10, Climate and vegetation in the future).

Global environmental changes which may be experienced by agriculture and fisheries are discussed in Chapter 22. In addition to increased atmospheric concentrations of CO_2 from the pre-industrial value of about 280 ppm to 360 ppm now and probably to 700 ppm by the end of the next century, a large range of pollutants, including particulates, may increase (Table 22.1). The current annual rate of increase of CO_2 is probably far faster, longer term and

Table 16.1 Temperature and precipitation requirements for selected commercial crops

Crop type	*Temperature*	*Precipitation*	*Notes*
Cocoa	Since temperatures of between 10 and 15.6°C may be harmful, the crop cannot be profitably grown in regions where mean maximum of coldest month falls to 13.9°C or where absolute minimum of less than 10°C occurs	Tree not resistant to dry weather so generally restricted to areas where dry season does not exceed 4 months	A tree of humid tropical lowlands – mostly grown within 20° of equator below 457 m.
Citrus fruits	Little or no growth where temperature is below 15.6°C. Dormancy in cooler months of subtropical climates. Temperatures slightly below freezing are highly damaging	Requires high soil moisture content. Orchards often irrigated even in fairly moist areas	Can be grown on a variety of soils with high humus content
Coffee	A tropical crop whose temperature requirement varies with species. Generally, optimum temperatures are between 15.6° and 25.6°C	Depending upon temperature, optimum amounts vary from 127 to 229 cm per year. The distribution is important with an ideal minimum in the flowering season. Too much water can promote tree disease	Generally does best on a well-drained loam soil. Thus some species are highland varieties. Others do well in lowlands
Cotton	Needs a frost-free growing season of 180 to 200 days. Does not grow below 15.6°C and optimum temperatures are from 21.1 to 22.2°C during the growing season. Four to five months of uniformly high temperatures are beneficial	Can tolerate a wide range in annual precipitation; the distribution during the growing season is of critical importance. Frequent, but light, showers immediately following planting are an attribute	Needs sunshine to ripen the boll in full maturity
Rubber	For optimum growth and yield, a mean maximum of over 24°C. The maximum should not, however, exceed 35°C	Evenly distributed rainfall of more than 178 cm per year. Lengthy dry periods inhibit growth. Good drainage essential	A tree of the equatorial rain forest (Amazonia) transferred to plantations in southeast Asia
Sugar cane	Susceptible to low temperatures. Little or no growth below 10°C while optimum is appreciably higher. Frost very dangerous to young cane	During vegetative growth requires a considerable amount of moisture and is sensitive to drought. In ripening period should be relatively dry to maintain high sucrose level	Often grown in cleared areas formerly occupied by tropical forests
Tea	Optimum, temperatures should not fall below 12.8°C nor exceed 32.2°C	Can tolerate high amounts of rainfall (254–381cm per year) if rain fairly evenly distributed throughout year	Often grown as an upland crop in tropical areas

Source: Oliver (1973).

Table 16.2 Resistance of selected crops to frost in different growth stages

Crop type	*Temperature (°C) harmful to plant in stages of:*		
	Germination	*Flowering*	*Fruiting*
(a) **Highest resistance**			
Spring wheat	−9,−10	−1,−2	−2,−4
Oats	−8,−9	−1,−2	−2,−4
Barley	−7,−8	−1,−2	−2,−4
Peas	−7,−8	−2,−3	−3,−4
(b) **Resistance**			
Beans	−5,−6	−2,−3	−3,−4
Sunflower	−5,−6	−2,−3	−2,−4
Flax	−5,−7	−2,−3	−2,−4
Sugar beet	−6,−7	−2,−3	–
(c) **Medium resistance**			
Cabbage	−5,−7	−2,−3	−6,−9
Soy beans	−3,−4	−2,−3	−2,−3
Italian millet	−3,−4	−1,−2	−2,−3
(d) **Low resistance**			
Corn	−2,−3	−1,−2	−2,−3
Millet	−2,−3	−1,−2	−2,−3
Sorghum	−2,−3	−1,−2	−2,−3
Potatoes	−2,−3	−1,−2	−1,−2
(e) **No resistance**			
Buckwheat	−1,−2	−1,−2	−0.5,−2
Cotton	−1,−2	−1,−2	−2,−3
Rice	−0.5,−1	−0.5,−1	−0.5,−1
Peanuts	−0.5,−1	–	–
Tobacco	0,−1	0,−1	0,−1

Source: Oliver (1973).

greater than experienced by organisms in ecosystems over many millennia. Global temperatures may increase by up to 4°C, with 2–3°C more likely; indeed, they may have risen by 0.5°C since the early 1900s and particularly strongly in the last 30 years. The incidence of extreme temperatures may be increasing (see Chapter 23, Conclusions). Global rainfall may increase by 15 per cent but relatively more so in high than low latitudes, with some areas such as the Mediterranean suffering severe deficiencies. Snowfall amounts, frequency and duration of cover will decrease, related to alterations in temperature. Cloud cover may increase and solar radiation decrease (Lawlor, 1991). Water pollution, with greater silt, nutrient and pesticide concentrations, will result from agricultural and industrial/urban sources and rising sea levels are expected (see Chapter 7, Conclusions). Ultraviolet B (UVB) radiation incident on crops and animals is increasing, especially at high elevations and latitudes.

RESPONSES OF AGRICULTURAL CROPS TO GLOBAL ENVIRONMENTAL CHANGES

Generic Responses

Many of the responses expected of crop plants will be common to particular groups, allowing some generalizations. Plant characteristics are important, such as the growth cycle in relation to weather and climate and photosynthesis/respiration. Clearly, species and varieties are adapted to particular means and ranges of conditions and are most productive (by definition!) under optimum conditions. Relatively small changes in rain, snow, temperature, solar radiation, nutrient supply and pests and diseases can, and do, greatly affect production and quality. So we may expect changes to intensify as GEC accelerates. Given the great range of crops worldwide and the conditions under which they are grown, it is not possible to predict the consequences at the local level. However, farmers need information to judge agronomic practices better. Consumers may be less affected (in the short term) by local problems if they can purchase foods from elsewhere. As many of the GEC conditions will shift slowly (relative to the annual crop cycle of much of agriculture) and agricultural technology, economics and politics continue to have profound effects on farming in most societies, the responses to changing environment will be included in the overall adaptation of farming.

CO_2, Photosynthesis and Production

The significance of increasing CO_2 for plants, and therefore the biosphere, is substantial. Terrestrial plants use atmospheric CO_2 as the source of carbon for photosynthetic (i.e. light-driven) synthesis of, ultimately, all organic matter; mechanisms are complex but are well understood. The reaction of CO_2 with the acceptor ribulose bisphosphate (RuBP), which is generated using light energy captured by chlorophylls in the chloroplasts of leaf cells, is most significant. This carboxylation reaction is catalysed by the enzyme RuBP carboxylase (Rubisco) in chloroplasts: a three-carbon compound is formed; hence the term C3 photosynthesis. Many crop plants (wheat, barley, sugar beet, oil seed rape, potatoes, rye grass, for example) of cooler regions are C3 but so are rice, yams, cassava and sweet potato of the tropics and subtropics. Rubisco also allows RuBP to react with oxygen, resulting in synthesis of substances which are later metabolized with loss of CO_2 in photorespiration. This is inefficient because it consumes RuBP and wastes already fixed carbon and energy. A large oxygen concentration in the atmosphere competes effectively with CO_2, so photorespiration may be 50 per cent of net photosynthesis under hot, dry conditions. Photorespiration reduces the production and yields of C3 plants. Increasing CO_2 to 700 ppm or more greatly reduces photorespiration and increases net photosynthesis (Lawlor, 1993; Lawlor and Keys, 1993).

An additional form of photosynthesis, called C4, is advantageous in a potentially warmer world. Basically, like C3, it also has a CO_2-accumulating mechanism which initially forms a four-carbon; hence the name. The CO_2 concentration in the chloroplast, which saturates Rubisco and photorespiration, is all but eliminated. C4 plants include maize (corn), sugar cane, millet and sorghum, which are all crops of very sunny, hot climates, with a very large potential for producing biomass and yields.

Crop Types

The world's main food and fodder grain production is from annual, short-growing season graminaceous crops with very clearly defined vegetative and reproductive phases. Grain yield is determined by the number of grains formed and the degree to which they are filled. The former depends on number of shoots (tillers) and the number of ears or panicles formed and how many potential grain sites are made. Then, the production of viable ovules and pollen is important as is fertilization. Grain filling depends on the rate and duration of accumulation of carbohydrates and proteins, together with proper ripening, and largely determines the size and quality and storage characteristics of grain. Conditions during each of these stages are important: temperature affects the rate of germination, tillering, ear and grain development, filling and ripening; radiation affects the assimilation during potential grain formation and filling. Water supply, during early development, affects numbers of tillers and ears and, during grain filling, the grain size. Straw production is important for animal fodder in many parts of the world. Protein (legumes) and oil crops are often very similar in growth and development to cereal grain crops. Storage organs, such as root tubers, are important staples of diets worldwide.

Carbohydrate grain crops are both C3 and C4. C3 cereals, namely wheats (*Triticum* species), are the most widely traded agricultural commodity, with very great genetic variability. Wheat is grown worldwide (except in the humid tropics) and by extensive methods, for example in South America, Australia, USA, although it may be irrigated (India). Rice (*Oriza sativa*) has wide variability, including dryland, irrigated and deep-water types, and is adapted to warm conditions, predominantly tropical and subtropical. A smaller volume enters world trade than wheat and it is particularly produced by small farmers. The effects of climate change on the production of these two commodities will determine the world's food status to a large degree. Other C3 cereals, (e.g. barley, oats and rye) are more locally grown, often in poorer environments of subsistence agriculture than wheat or rice; in developed agriculture, they are used for animal feed and brewing. Broadly, these crops respond similarly to environmental factors although the ranges (particularly of temperature) differ.

C4 cereals include maize (corn) (*Zea mays*) which is a widely grown and traded food and fodder grain in warm to hot, but not very dry, conditions; it is widely irrigated. Maize genetics are very well understood and there are many varieties. Millets and sorghums (*Sorghum vulgare*) are largely grown in warmer and, particularly, drier climates of the tropics and subtropics, under rain-fed conditions and limited fertilization, and are particularly important in subsistence agriculture. Sugar cane (*Saccharum officinarum*), although not a grain crop, is a tropical C4 grass of great economic importance for sugar production which is determined by total biomass production and its sugar content.

Protein crops include legumes which are a very varied group of C3 plants widely grown in industrial-scale and subsistence agriculture, providing a major protein source for human food and animal fodder. Soya bean and other pulses are very widely traded. Crop development and yield determination are akin to cereals although less strictly phasic. The capacity for denitrogen fixation by bacteria in the nodules imposes a large demand for energy as carbohydrates. Denitrogen fixation is important in the nitrogen economy of cropping systems but is of decreasing importance than artificial N fertilizers, except in subsistence agriculture and extensive pastures. Their response to environmental change may be rather different to that of cereals.

Oil crops are derived from many biologically diverse plants grown in many environments: for example, the C3 oil seed rape (often of temperate climates) and oil seed mustard, soya bean and sunflower and C4 maize of hot conditions. Oil production is energy demanding and is sensitive to the environment, so it results in differences in response to environmental change compared with other crops. Carbohydrate root crops include mainly C3 species such as potato and sugar beet of temperate latitudes and sweet potato, yams and cassava of the tropics. Their yields are determined by environmental factors which regulate the development of the plant, including the storage organs and the assimilate supply for storage. They will probably respond to elevated CO_2 and temperature in a similar way to the C3 cereals and legumes.

Weeds affect crops so profoundly that they

must be considered with as much care as the crops themselves, for they threaten production everywhere. Weeds are so diverse (both C3 and C4) that generalization is very difficult: often, weedy characteristics depend on the crop infested and on agronomy. Measures of chemical control frequently induce the evolution of resistance. A gross approximation is that they respond to the environment in the same way as others of their photosynthetic group and are very like the crops they infest.

CROP PRODUCTION AND CLIMATE CHANGE

Elevated CO_2

As C4 plants are so efficient in current CO_2 concentrations, their photosynthesis increases little with increasing CO_2 (compared with C3 plants) and their yields are hardly affected. However, in C3 plants, photosynthesis per unit leaf area may rise by 30–50 per cent in the short term if atmospheric CO_2 is raised to 700 ppm or so, largely due to reduced photorespiration. Consequently C3 plants, grown in elevated CO_2, accumulate more carbon assimilates and grow more biomass, with generally higher yields (Baker and Allen, 1992; Mitchell *et al.*, 1995). The effects can be large, with up to 40 per cent increases, but more generally these are in the region of 25 per cent. Figure 16.1 shows the response of winter wheat to elevated CO_2 and temperatures under simulated UK conditions. One consequence of long-term exposure to elevated CO_2 is that the plants become replete with sugars and starch because, probably, their biosynthetic (growth) mechanisms are regulated to deal with smaller carbon assimilate supply. If the supply of other essential materials (e.g. nutrients) or conditions (e.g. low temperatures) limit growth, then accumulation of carbohydrates is greater. Growth of root systems is enhanced and

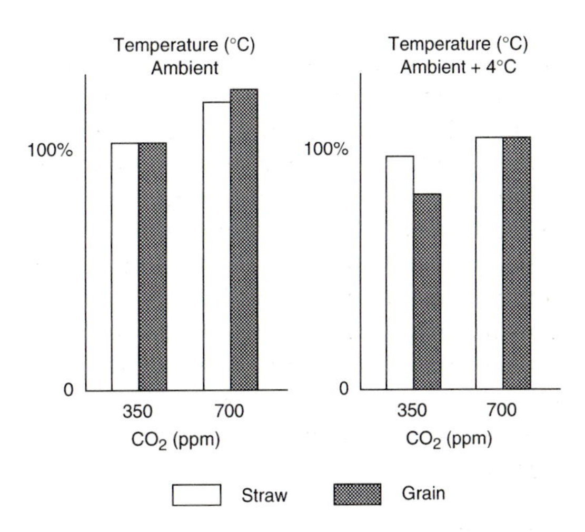

Figure 16.1 Straw and grain production by winter wheat grown under current ambient CO_2 (about 360 ppm) and temperatures and elevated CO_2 (700 ppm) and temperatures (+ 4°C above ambient) expected under climate change conditions. Average of six experiments done at IACR-Rothamsted, UK

more organic matter accumulates in soils from root decomposition and also exudation of organic substances. These will affect the type and activity of the microbial flora and the fauna of the rhizosphere. Increased accumulation of carbon in the soil may be a substantial sink for a part of the anthropogenic CO_2 (see Chapter 9, Soils and the greenhouse gases).

Accumulation of carbohydrates may inhibit photosynthesis. The enzymatic machinery is 'downregulated' (i.e. partially switched off) so the rate of CO_2 assimilation falls, allowing the balance between production and consumption of carbohydrates to be maintained. *In extremis*, synthesis of components of the photosynthetic apparatus may be inhibited (i.e. 'acclimation' occurs). Again, this allows the system to balance (Lawlor and Keys, 1993). The importance of such adjustments to the fine tuning and response of C3 crop plants to environmental change in the field is not yet clear (Delgado *et al.*, 1994). There are other consequences for plants. For example, the demand for nutrients may be greater as growth is stimulated, although frequently there is a decrease in the concentration of nutrients (e.g. N per unit dry matter, the so-called 'dilution effect'). Reduction in nutrient content decreases the nutritional value of herbage for animal (ruminant) fodder and grain for human consumption.

Stomatal responses to CO_2 are very important. The supply of CO_2 to Rubisco is by diffusion from the atmosphere through the leaf surface and water is lost from the leaf to the atmosphere by the same route. Stomata provide a 'valve' which regulates water loss and CO_2 assimilation by the leaf in C3 and C4 plants. In C3 plants, the CO_2 concentration in the chloroplast is substantially smaller than in the atmosphere because of the stomata. This exacerbates competition of O_2 with CO_2 and increases the ratio of photorespiration to photosynthesis. If plants are water-stressed, the stomata close more and so worsen the effect. C4 plants, by virtue of their CO_2-concentrating mechanism, are able to maintain larger rates of photosynthesis and low photorespiration with smaller stomatal conductances than C3 plants, even under water stress. Stomatal conductance is decreased by elevated CO_2 so that rates of transpiration decrease and consequently the water-use efficiency rises substantially owing to that, and increased photosynthesis (Baker and Allen, 1992; Lawlor and Keys, 1993). This is likely to be of major importance to crop production. However, some caution is needed. In the field, absolute rates of water loss may not decrease as much as in controlled conditions, owing to the interaction between evaporation rate and temperature. The increased leaf and crop temperatures, which result from slower evaporation, may be 1 to 2°C which stimulates water loss. Another factor is the long term acclimatization of stomatal response so that the effect is decreased over time. Pests and diseases will probably not be affected directly by CO_2 increases to any large extent. Much greater direct effects will arise from changes in temperature and rainfall and indirect effects will be associated with changes in plants.

Temperature Increases

Temperature has the most profound effects on plants. One intimately related to the photosynthetic system is the temperature dependence of Rubisco. Higher temperatures increase the oxygenation relative to carboxylation and hence photorespiration compared with photosynthesis. C3 plants are particularly affected since C4 plants are protected by the CO_2 concentration mechanism and thrive in warmer (indeed hotter) conditions (Lawlor and Keys, 1993). C3 plants are predominant in cool climes. The development (rate of formation of organs) of crops is temperature dependent and the required number of degrees of heat accumulated over time (the temperature sum or day degrees) is characteristic for species and varieties (Mitchell *et al.*, 1995). Those with a large heat requirement are only successful in warm conditions and those with a low requirement are disadvantaged by warmth (Carter *et al.*, 1992). Crop plants are adapted to a range of conditions. However, at the extremes of their environmental tolerance, crops become very sensitive to conditions and, thus, potentially to climate change. Even within the current environments to which crops are adapted, the range of temperatures may be large. The extremes cause loss of production with inevitable social and economic difficulties (i.e. decreased availability and increased cost of foods).

Extreme temperatures for short periods, coinciding with particular phases of crop growth, can greatly and adversely affect the production of crops which are apparently well adapted to that environment. Hot periods during germination may damage the establishment of sorghum in hot climates but warmth may aid sugar beet in colder areas. A specific process in yield formation may be involved in susceptibility and tolerance: cereals generally are sensitive to heat during the formation of pollen, its release and pollination. Thus crops, although generally adapted and selected for particular environments, are functioning close to an optimum. Consequently, relatively small (on a human scale) changes in temperature, rain, etc., may have large effects on yield. The consequences of global warming and extreme events of many types on agricultural crops must be of major concern in ensuring continuity of food supplies.

A response to a temperature increase may be a shift in the types of crops grown in different areas. Many simple models of the distribution of crops have been used to assess likely changes (United Kingdom Climate Change Impacts Review Group, 1991; Rosenzweig and Parry, 1994). For example, maize, with its poor low-temperature tolerance, is uncompetitive in the cool environment of the UK. However, warmer early springs would foster good germination and establishment whereas hotter summers would encourage yield and warmer autumns would accelerate ripening. Whether the light would be adequate in early spring and autumn is another question. Higher temperatures may stimulate many diseases and pests. Fungal and bacterial plant pathogens could be stimulated. Insect pests will probably be more problematic: for example, aphid populations in the UK would increase dramatically as a consequence of warmer winters (Harrington and Stork, 1995).

Rain, Snow, Cloud and Radiation Changes

Assessment of the potential changes in rain, snow, cloud and radiation amounts, intensities and durations is inadequate. Higher latitudes may receive more rain than low latitudes with a 15 per cent increase in total global precipitation, but desertification of the Mediterranean basin is possible. Snow may decrease with warming. Clearly, water supply is crucial for plant growth as currently large areas with abundant radiation could support active agriculture if water was available. Irrigated agriculture is successful in many hot environments yet may be economically unsustainable owing to cost. Surface and groundwater supplies are finite and agriculture must compete with growing urban/industrial demands; irrigation often leads to salinization. Small changes in rainfall have large, rapid effects and even in humid environments like the UK, droughts decrease yields and quality of most crops. Some crops are more sensitive than others: for example, potatoes are largely grown with irrigation in the UK when rain is inadequate but not cereals. Any reduction in rainfall, for example, in the wheat growing areas of the USA, Australia, South America or Russia, would rapidly reduce production. All crops will be affected with C3 types more sensitive than C4 types. If conditions were to change rapidly, then geographical transfer of crops would probably help sustain production. Breeding for a multigene trait like drought tolerance has proved difficult and slow.

Pests and diseases could respond very markedly to changes in rainfall as they do currently. Many rust and mildew diseases and soil-borne pathogens are greatly stimulated by warm, rainy, humid conditions. Some insects (e.g. tsetse flies), ticks and mites may increase with increased rainfall. The subtlety of the responses of different groups of potentially damaging organisms requires careful and specific analysis which is not possible here.

UVB Radiation Increase

Concern over the increase in UVB radiation relates to both plant and animal responses. Many plants are damaged (e.g. leaf scorching) and animals are also affected (e.g. blindness in sheep) but the effects are not well evaluated for realistic conditions, as early studies applied excessive UVB values. Varietal differences are apparent and adaptation is important: plants may produce more protective compounds which deter pests and diseases. The effects on micro-organisms may be considerable but the long-term effects on the stability of the genetic material of higher organisms cannot be ignored (Krupka and Kickert, 1993).

Nutrient Availability

GEC will not directly affect nutrients but interaction between the supply of nitrogen, phosphorus, potassium, etc., is very likely. Elevated CO_2 and higher temperatures could affect the availability of fixed N for crops from legumes, for example, which play an important role in the N economies of much global agriculture. There may be opportunities to exploit this to reduce the fertilizer dependence of non-legume crops and also of crops of legumes which currently receive N fertilizers.

CLIMATE AND ANIMAL PRODUCTION

Factors regulating domestic animal production are generally well understood including diseases, pests, etc., and the needs for nutrition and husbandry. Cattle, sheep and goats are distributed worldwide and graze pastures. They are fed high-nutrition diets in the developed world but survive on low-quality natural grazing in less developed countries. Cattle and, less so, sheep and goats provide much of the milk consumed as well as meat and hides. There is considerable variation in genetic adaptation to the environment. For example, zebu cattle, and especially goats, are heat tolerant whereas sheep are better adapted to high altitudes and infertile soils. Cattle are susceptible to fly-borne diseases (trypanosomes transmitted by tsetse flies) and sheep suffer from foot rots, flukes and worms.

Other sources of animal proteins are pigs and chickens. Both provide meat/skins and eggs respectively, which have commercial value. They are often restricted in range and are fed relatively high-value energy products under rather protected conditions, even under subsistence agriculture, but particularly with industrialization. Therefore, environmental change may affect free-ranging animals directly (changing temperature, UVB, rainfall, fodder amounts and quality, etc.) but housed-animal production will be most dependent on the economics of housing and climate control and fodder supplies.

Effects of Climate Change on Domestic Animals

Elevated CO_2 over the next century will have little direct effect on domestic animals or their pests and diseases. Higher temperatures, especially extreme values at critical times such as reproduction, will be much more important in determining the efficiency of production of particular sorts and varieties. Rain will also be vital, both directly (drinking water) and indirectly, as it influences plant growth and diseases. This is probably the most critical environmental parameter, especially in existing marginal areas such as the Sahel. Quality of the fodder grown under GEC conditions could have long-term effects on production. Pests and diseases may be very important in determining the future of animals in particular regions and, as with plants, require specifics which must be considered.

Effects of Climate Change on Pests and Diseases

Agriculture worldwide is subject to so many different bacterial, fungal, protozoan, invertebrate and vertebrate organisms which adversely affect animal production that analysis is impossible. Generally, elevated CO_2 will have limited or indirect effects. Higher temperatures will stimulate reproduction and numbers of organisms (e.g. aphid pests of crops) and altered rainfall may have substantial consequences (e.g. wetter conditions will increase mosquitoes and malarial transmission (Harrington and Stork, 1995)). Specific pest–host combinations must be considered for meaningful analysis.

FISHERIES AND GLOBAL ENVIRONMENTAL CHANGE

Fisheries are currently overexploiting stocks and this is a far greater threat to them, and the species on which they depend, than GEC. However, there are grounds for concern at the massive alterations to fresh and marine waters from pollution (e.g. acid rain in Scandinavian fresh waters and pesticides) and changes such as reduced river flow due to water extraction. Often, spawning and nursery grounds are damaged. As reproduction is generally sensitive to changing conditions, these effects may substantially affect population dynamics. Of the GEC conditions, CO_2 may have only very small effects on the pH of waters. Temperature will be much more important, as the ranges of many species are related to water temperature. Also, coastal fisheries and spawning grounds may suffer from sea-level rise (see Chapter 7, Conclusions). However, if phytoplankton production is increased owing to elevated CO_2 and more nutrients (possibly from agricultural pollution), productivity could increase. Probably, geographical distribution would change if temperatures increase. Precipitation will have large effects on fresh waters. Currently, low rainfall and large rates of extraction, for irrigation and urban/industrial use, are greatly affecting lakes and rivers worldwide with effects on fisheries.

Diseases and pests will change but the extent is difficult to assess, as with other systems mentioned where specific organism–host relationships must be considered. Increasing fish farming, with intensive control of nutrition, diseases, etc., may allow food from this sector to increase and be less sensitive to the environment. However, this will depend on the economics concerned and on international agreements to conserve wild stocks.

PLANT AND ANIMAL BREEDING

A substantial degree of genetic variation exists for biochemical and physiological systems, conferring a wide spectrum of responses of organisms to the environment. As conditions change so the better adapted may increase in 'fitness' and exclude the less well adapted. This Darwinian natural selection applies to naturally and human-selected organisms. Plants and animals have been selected and bred to give the current crops, herds and flocks the ability to survive extreme conditions. However, many crops and domestic animals have been bred for large production but they need large inputs of water, nutrients and extensive (often expensive and environmentally damaging) pest, disease and weed control. Such genotypes may be poorly suited to future climates. A substantial part of the world's agriculture is based on relatively low productivity and technology. It is subsistence farming based on self-saved and reared seed and animals, with a good ability to survive extreme conditions.

A major concern is the very narrow species and genetic base upon which much of the food resource of the human population depends, despite the existing variation which is exploited in breeding. There are grounds for optimism that sufficient genetic variation exists to allow adaptation of crops and animals to novel environmental conditions. Both have short breeding cycles relative to the likely rate of GEC. Furthermore, they have already been selected for a wide range of characteristics (quality as well as yield) to stabilize yield in extreme environments and increase the resistance to diseases and pests. By identifying and incorporating genetic material from the wide range available,

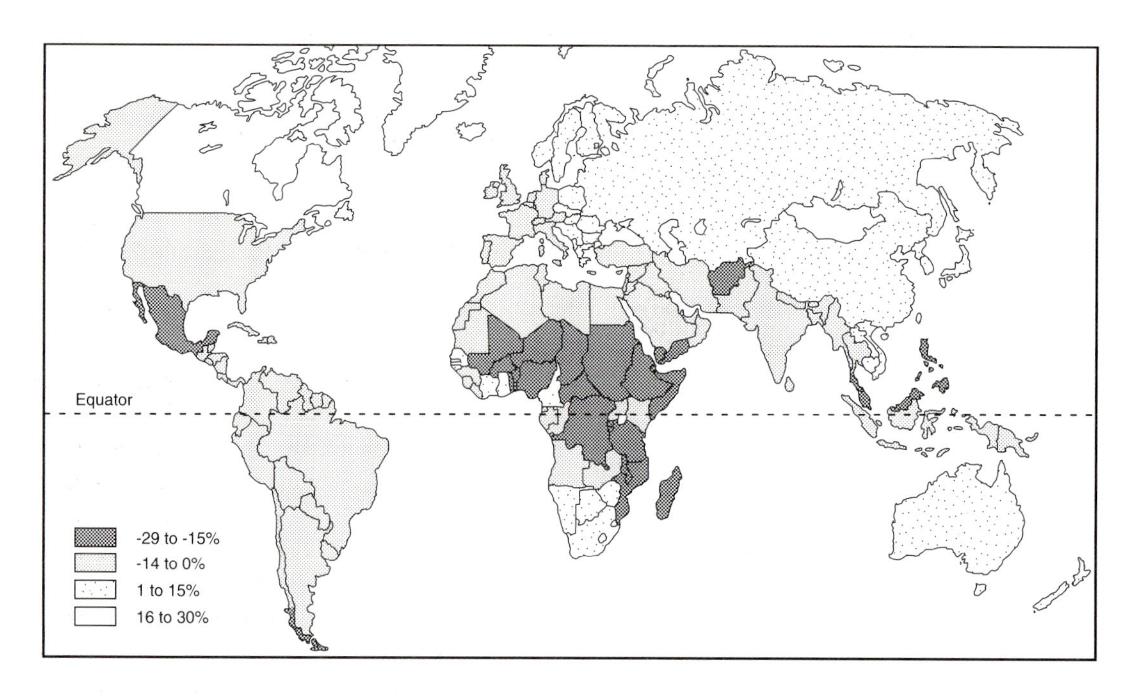

Figure 16.2 Calculated percentage response of average grain yields across the world to a doubling of CO_2 to 555 ppm, 4°C increase in temperature and 8% decrease in precipitation Rosenzweig *et al.*, 1993)

it should be possible to introduce favourable characteristics. Adaptation to CO_2 and temperature may involve general, perhaps unconscious, selection.

Small-scale agriculture in poorer parts of the world may suffer most from lack of access to genetic material for selection. International and national programmes of breeding will be important for improvements but the rate of introduction may be inadequate. Costs of breeding are not trivial although compared with the costs (social as well as directly financial) of not adapting, they could be incalculable and indeed catastrophic. Commercial plant breeding can only provide the adaptation to GEC conditions, if the needs are clearly defined and if it is economical.

CONCLUSIONS

Four main conclusions are evident from the above discussions:

1 Elevated CO_2 levels will generally stimulate C3 plant growth/biomass and yield production provided water, temperature and nutrition are adequate (Figure 16.2). C4 plants will be little affected. Protein and oil crops may benefit more from increased CO_2 than other C3 plants. Responses to elevated CO_2 may be smaller where other factors limit growth. Weeds will respond rather similarly to crops but small changes, relative to the crop response, may have large agricultural impor-

CONCLUSIONS *continued*

tance. CO_2 will not directly affect domestic animals or pests and diseases.

2 Higher temperatures will stimulate growth etc. if crops are limited by low temperatures in current conditions but they will decrease productivity if supraoptimal. Thus, tropical and subtropical regions may not benefit or suffer net loss of production whilst temperate regions gain or remain constant, depending on the relative changes in CO_2 and temperature. Generally, higher temperatures will adversely affect current crops and changes in the types of crops grown may be the principal response. Domestic livestock will not be so affected except in marginal conditions. Higher temperatures will stimulate pests and diseases of plants and animals substantially.

3 Water availability will be a crucial aspect of global environmental change: crop water use may decrease somewhat and water-use efficiency will increase for most C3 crops, with some effects on C4 crops. Animals will be affected by water supplies for drinking and indirectly via fodder supplies. The effects of changing rainfall and humidity on pests and diseases of animals and plants are likely to be large and varied.

4 Increasing UVB radiation will decrease crop and animal production, especially at high latitudes and elevation, but may decrease some diseases.

REFERENCES

Baker, J.T. and Allen, J.R. (1992) 'Response of rice to carbon dioxide and temperature', *Agricultural and Forest Meteorology* 60: 153–66.

Carter, T.R., Porter, J.H. and Parry, M.L. (1992) 'Some implications of climatic change for agriculture in Europe', *Journal of Experimental Botany* 43: 1159–67.

Delgado, E., Mitchell, R.A.C., Parry, M.A.J., Driscoll, S.P., Mitchell, V.J. and Lawlor, D.W. (1994) 'Interacting effects of CO_2 concentration, temperature and nitrogen supply on the photosynthesis and composition of winter wheat leaves', *Plant, Cell and Environment* 17: 1205–13.

FAO Working Group on Climate Change (1994) *Global Climatic Change and Agricultural Production*, Rome: FAO.

Harrington, R. and Stork, N.E. (eds) (1995) *Insects in a Changing Environment*, London: Academic Press.

Krupka, S.V. and Kickert, R.N. (1993) 'The greenhouse effect: the impacts of carbon dioxide (CO_2), ultraviolet-B (UV-B) radiation and ozone (O_3) on vegetation (crops)'. *Vegetatio* 104/105: 223–38.

Lawlor, D.W. (1991) 'Responses of plants to elevated carbon dioxide: the role of photosynthesis, sink demand and environmental stresses', in Y.P. Abrol *et al* (eds) *Impact of Climatic Changes on Photosynthesis and Plant Productivity*, Delhi: Oxford & IBH Publishing.

Lawlor, D.W. (1993) *Photosynthesis: Molecular, Physiological and Environmental Processes*, 2nd edn, Harlow: Longman.

Lawlor, D.W. (1995) 'Photosynthesis, productivity and environment', *Journal of Experimental Botany* 46 (Special Issue): 1449–61.

Lawlor, D.W. and Keys, A.J. (1993) 'Understanding photosynthetic adaptation to changing climate', in L. Fowden, J. Stoddart and T.A. Mansfield (eds) *Plant Adaptation to Environmental Stress*, London: Chapman and Hall, pp. 85–106.

Mitchell, R.A.C., Lawlor, D.W., Mitchell, V.J., Gibbard, C.L., White, E.M. and Porter, J.R. (1995) 'Effects of elevated CO_2 concentration and increased temperature on winter wheat: test of ARCWHEAT1 simulation model', *Plant, Cell and Environment* 18: 736–48.

Oliver, J.E. (1973) *Climate and Man's Environment*, New York: John Wiley, pp. 271–2.

Rosenzweig, C. and Parry M.L. (1994) 'Potential impact of climate change on world food supply', *Nature* 367: 133–8.

Rosenzweig, C., Parry, M.L., Fischer, G. and Frohberg, K. (1993) *Climate Change and World Food Supply*, Oxford: Environmental Change Unit.

Rozema, J., Lambers, H., van de Geijn, S.C. and Cambridge, M.L. (1993) *CO_2 and Biosphere*, Dordrecht: Kluwer Academic.

United Kingdom Climate Change Impacts Review Group (1991) *The Potential Effects of Climate Change in the United Kingdom*, London: HMSO.

17

FORESTRY

Alexander Robertson

INTRODUCTION: CLIMATE AND FORESTS

Climatic shifts are a common occurrence. Over geological timescales (100 million years), the uplift of plateaux and mountains changed global circulation patterns, which invoked changes on major biomes. For example, the mean winter position of the jet stream (i.e. the boundary between polar and temperate air masses) affects the movement of low latitudinal tree lines. Shifts in the mean summer position of the jet stream change the boundary between forest and grassland in the plains of central Asia and North America. Forests undoubtedly have a predominant role in the seasonal cycle of planetary energy exchange and transport. The seasonal heating over Borneo, which is covered mainly by rain forest, is thought to influence the onset of northern monsoons. It has been suggested that large-scale deforestation in Borneo and elsewhere in southeast Asia is changing the 'equilibrium' in energy processes in that region. Similarly, the Amazonian tropical forest recycles rains brought onto the continent by easterly trade winds. A tropical forest is a net absorber of carbon dioxide (CO_2) (Grace *et al.*, 1995). However, burning and decay of tropical rain forest inject as much CO_2 into the atmosphere as the burning of fossil fuels in the industrialized nations of the northern hemisphere. Justifiably, there is concern that large-scale expansion of non-sustainable cultivated environments at the expense of forests will change the current energy balance governing the planetary atmospheric circulation system.

There is an old adage that 'civilizations lead to aridity'. In fact, a popular notion states that regional climate change, caused by deforestation, was at the root of periodic cultural recessions during the Graeco-Roman–Byzantium period. This prompted von Humbolt to comment that: 'By felling trees, which are adapted to the slopes and summits of mountains, men of every climate prepare for the future ages at once two calamities: want of wood and scarcity of water.' However, since the last sub-Pluvial (4000–5000 BP), the climate has been essentially unchanged apart from short-term variations. Also, although climate shifts cause episodic human suffering, the demise of the ancient civilizations had more to do with economic and social instability and much less with environmental influences. Nevertheless, there is often a tenuous relationship between climatic shifts, deforestation and human population. In the late 1960s and early 1970s, the southward shift of rainbelts was coupled with earlier deforestation in the Congo, and was followed by non-sustainable land usage. This triggered the expansion of the Sahara desert and desertification over much of equatorial Africa. In the Sahel region alone, more than 100,000 people starved and one-third of the cattle died.

CLIMATE AND SUSTAINABLE FORESTS

Sustainable forestry is usually defined in terms of the maintenance or enhancement of productive capacity and biodiversity of forests, taking into account the biological capabilities of the land. The standard approach is to treat these relationships as linear systems. In parametric statistics, it is considered that variances are normally distributed and that most events are normally distributed. The problem with this approach is that it cannot deal with the myriad of interscale relationships and cannot deal with the fact that new phenomena are encountered as one scales up or down between, for example, plot, forest type, landscape mosaic and biome.

When linking models by scaling up to progressively larger scales, new phenomena are encountered that are not present at smaller scales. Luxmoore *et al.* (1991) devised a stochastic approach based on a Monte Carlo procedure (called Latin hypercube sampling) to link physiological, succession and forest management models (Figure 17.1). Part of the procedure involves a technique called extended-range modelling which, in simulation models, means that when spatiotemporal domains of soil–forest–atmosphere are extended, new phenomena are added. The products of these simulation models are frequency distributions which can be used to determine confidence intervals for statistical comparison. With this technique, scaling-up spatially from the plot level to the landscape level, and temporally from hourly to annually, makes it possible to generalize linear changes in forest productivity in response to various climatic regimes.

However, natural and cultivated environments are non-linear dynamical systems that are inherently historical, evolutionary and irreversible. Consequently, it is not possible to address the question of sustainability from the standard approach (White and Engelen, 1993). In the cultivated environments, sustainable management implies that forests can be genetically and technologically engineered for a specific regional climate but rarely, if ever, for the vagaries of weather or, rather, weather extremes. However, the ability to describe the coupling between forests and climate depends on how well we model wind, small-scale turbulence within a forest canopy, the passage of frontal systems and shifts in the mean seasonal position of the jet stream.

Forests respond differently to abiotic factors at climate and weather scales. For example, the extent of wind throw in a forest is primarily a function of strong winds associated with the passage of frontal systems. However, dieback, which makes the forest prone to wind throw from less severe storm winds, is a function of seasonal climate. At longer timescales, changes in the trajectory of frontal systems associated with shifts in the mean position of the jet stream, as a result of climate change, have a considerable influence on forestry. These would include changes in wind regimes, regional rainfall patterns and rates of evapotranspiration which affect forest stability, tree growth, biodiversity – wood quality, regeneration – fire hazard, and human activities.

DYNAMICAL PERCEPTIONS

Forest climate is usually thought of as average weather, unchanging over the long term, but comprised of many short-term physical processes that are largely beyond human control. In cultivating forests, silviculturists tend, intuitively at least, to treat climate and weather as separate entities. Different emphasis on climate depends also on the different philosophical viewpoint of the dynamical nature of forests. Basically, there are two basic schools of thought: the 'climax' or 'equilibrium' theorists and 'mosaic-cycle' or 'dynamical' theorists, respectively.

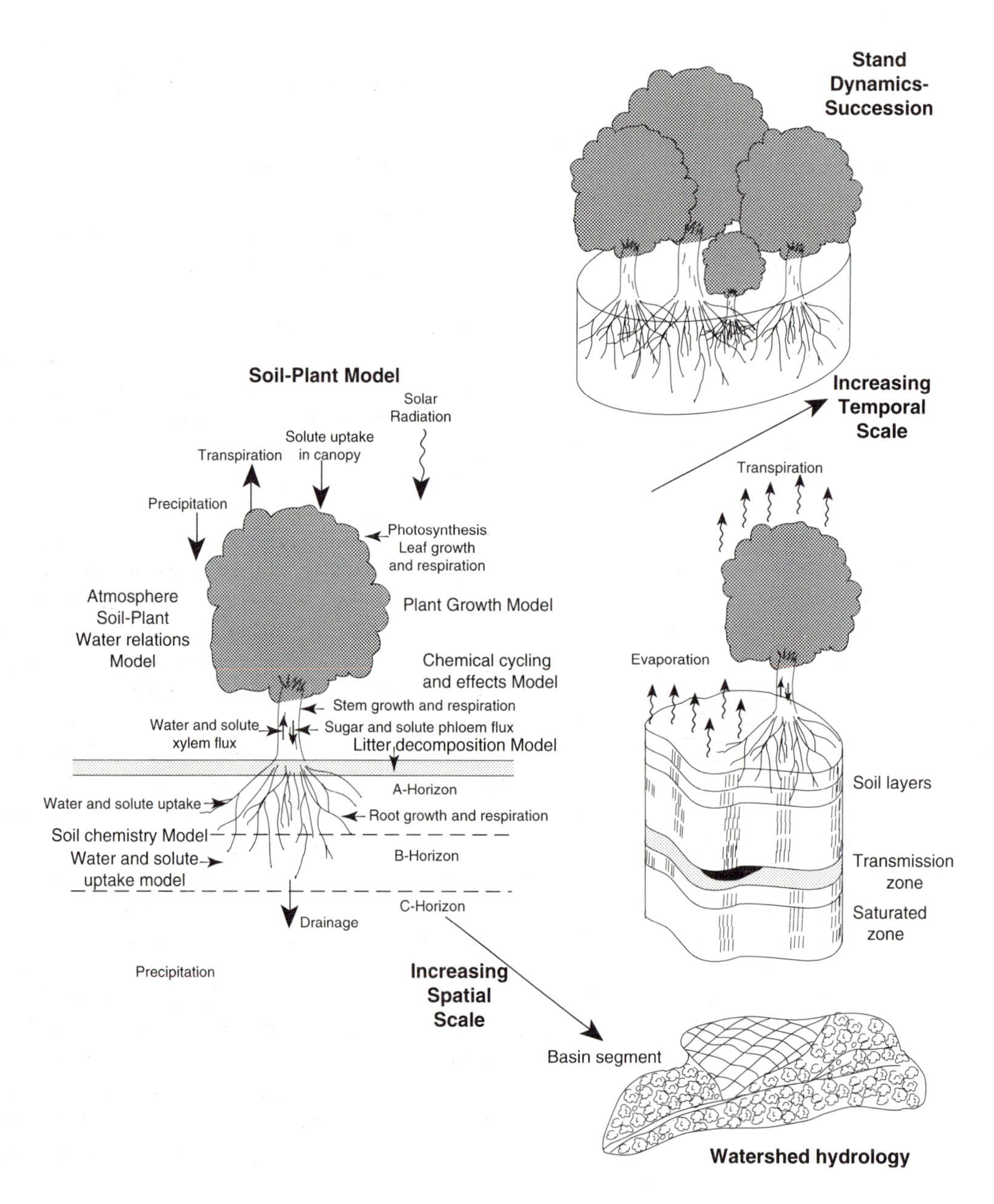

Figure 17.1 The scaling up of a soil–plant model in space and time. This involves the incorporation of additional phenomena not represented at the small scale. Examples of such additional phenomena include topographic effects in watersheds and successional processes of forest communities (after Luxmoore *et al.*, 1991)

Climax Theorists

According to this school of thought, forests are generally long-lived and have a slow response to climate change. That is, forests develop within a unique set of predictable climatic limits such that climate and forest are in equilibrium. In other words, forest succession trends towards a final climax state determined by the local climatic and biophysical environment. This concept of a natural ecosystem is, in a dynamical system's parlance, a fixed-point system whereby its long-term trajectory trends towards a fixed point (climax). This is useful in forest policies with an immediate or relatively short-term interest in the resource – in plantations and managed forests, for example.

Climax theorists tend to stress the biological aspects so that when shifts in climate occur, then forest succession is comprised of a series of definable units of ecosystem which trend towards a new climax. However, the variability of climate and weather constantly changes the forest dynamics so that forests are rarely, if ever, in equilibrium with the climate. This means that it is virtually impossible to define and predict the impact of an 'average' climate on a particular forest type beyond the crudest circumstances. For example, when climax theorists discuss the role of disturbance in gap formation, they focus primarily on biotic factors and rarely ask how gaps are formed and expanded by abiotic factors. One reason for this, of course, is the paucity of reliable abiotic data due to the difficulty in measuring environmental variables over any but the shortest spatiotemporal scales. Also, from an applied climatological perspective, there is the paucity of standard meteorological stations located within forested environments. One reason is that few people live within large, forested regions. Another has to do with the different criteria used for standard meteorological stations compared with those used for a specialized forest climatological station. Standard weather stations are designed to monitor basic weather and climate usually as part of a national or global network.

Mosaic-Cycle Theory

The corner-stone of mosaic-cycle theory is that ecosystems consist of patches of 'mosaic stones' which cycle continuously through a set of states, with adjacent patches cycling asynchronously (out of step). The dynamical system's approach, embodied in the 'shifting mosaic steady state' or simply the 'mosaic-cycle' theory was first proposed by Aubreville (1936) and was revived and expanded by Remmert *et al.* (1991). The concept of a shifting mosaic cycle is an apt description of patterning in managed forests or, stated another way, patterning in managed forests is a function of shifting forestry priorities to reflect market and social demands plus climate and weather effects. From a forest climatologist's and silviculturist's standpoint, the dynamical systems approach is a more enlightened concept for two reasons. First, it tends to stress the role of climate and weather more so than classical climax theorists do. Second, it has a more direct link with other scientific disciplines, notably atmospherics, geophysics and mathematical modelling, which are firmly rooted in dynamical systems theory. The Amazonian forest is a typical example of mosaic cycling. Fifty per cent of Amazonia is covered by typical rain forest on terra firma whereas the rest of the region is composed of a vast number of much more dynamic vegetation types, resulting from much more frequent natural and human disturbances. In addition, there is an increasing amount of secondary growth (capoeira) arising from shifting agriculture.

Mueller-Dombois (1991) emphasized the abiotic factors, namely wind, which cause gaps and many forms of forest dieback, notably the peculiar mosaic cycle of wave forests or 'Shimagare' phenomenon which is characterized by dead tree strips moving across the landscape.

From a different perspective, Wissel (1992) proposed a cellular automaton model of the dynamics of patterning which emphasized the importance of strong solar radiation causing beech dieback at crucial times in the mosaic cycle. Most of the forests in Europe, which have been manipulated by humans through all or part of their history, are examples of 'economic' mosaic cycling. Most studies of climate–forest interactions, including those by mosaic-cycle theorists, are based on conventional climatological, statistical and physical models that are necessary to provide a basic understanding of 'average' relationships between climate and forests and weather and forestry. Conventional approaches necessarily involve reliance on and improvement of a formidable array of classical measurement and modelling of forest climatological techniques. However, before we can define, describe and utilize the relationship between climate and forests, we must improve our understanding of their interactive spatiotemporal characteristics. More precisely, we should focus on initial (deterministic chaotic) events which trigger particular pattern formations and explain how these evolve through disruptive processes throughout the course of the interrelationship between climate, weather, forests and forestry.

SPATIOTEMPORAL CHAOS

Spatiotemporal chaos is variously referred to as 'self-organizing complexity', 'non-linear determinism', 'patterned but unpredictable', 'patterned instabilities' and other seemingly contradictory terms. From a forest–climatological viewpoint, our interest in this phenomenon is aimed at describing the deterministic processes such as the pattern-forming instabilities which govern the dynamics of vegetation mosaics in forested and deforested landscapes. Chaos is the initial condition that results in a patterned but only partially predictable instability, such as a phase change from rain to freezing rain. Various dynamical states evolve from both natural and human non-equilibrium pattern-forming processes with a large number of deterministic chaotic elements and localized random shocks. In some respects, the relationship between silvicultural operations and extreme climatic events is a prime example of a non-equilibrium pattern-forming process. On the one hand, silviculture (such as thinning) is representative of deterministic chaos that equates with stabilizing (exploitive) processes whereas localized random shocks, embedded within the deterministic chaotic system (e.g. wind throw or frost damage), are responsible for the noisy (disruptive) transient states. In natural forests, much the same thing happens through spatiotemporal adaptations (deterministic chaos) in response to disruptive changes in climate and weather (random shocks).

The interaction between forests and climate is a system of interlocking stable and unstable states with a large number of chaotic (non-linear) elements that characterizes atmosphere–forest–soil interactions. These interlocking states include homogeneity (stable), periodicity, quasi-periodicity and spatiotemporal chaos. Spatiotemporal chaos arises from a phenomenon known as intermittency. This is an abrupt change from one dynamical state to another, such as damage to parts of a forest caused by an episode of freezing rain or wind throw caused by moderate winds from an unusual direction. More often than not, the impact of these small, brief weather events is the catalyst for major transformations in terms of the forest architecture. Furthermore, spatiotemporal chaos is both the bane and a product of forest policymakers. This is because the coupling between climate change and changes in forestry, trade and land-use priorities is viewed as a complex interactive process which has a huge number of chaotic elements that are impossible to describe and predict. However, comparatively few degrees of freedom are involved in extreme variability over a wide range of timescales. In fact, many bioclimatic models attempt to take advantage of this property by focusing on 'effective' parameters. White and Engelen (1993) pointed out that landscape patterns are fractal (self-similar) in nature and that only a small number of dynamical systems exist at a phase transition between stable and unstable regimes and that these systems generate elaborate fractal structures.

Fractal geometry has been called the 'calculus of heterogeneity'. Fractals are not defined in a formal, legalistic statement but are expressed by mathematical and graphical representations. A fractal is quantified by its fractal dimension which essentially measures how efficiently an object fills space. For example, a patch created by cutting is likely to have smoother edges and lower fractal dimension than a patch created by climate and weather events. Apparently, fractals are products of dynamical (spatiotemporal chaotic) systems and, as such, may provide a link between the abiotic (i.e. mechanistic) processes of climate and weather and the resultant biotic (architecture) forest. For example, on a poor, exposed site, trees have heterogeneous (asymmetric) profiles with a high fractal dimension whereas, on a rich and sheltered site, trees tend to have homogeneous (symmetrical) profiles with a low fractal dimension. In the case of forest mosaics, the magnitude of a fractal dimension of patches that make up a forest mosaic reflects the geometrical and dynamical complexity of the forest. The geometric complexity can be described by a simple fractal diversity index (Olsen *et al.*, 1993) which relates the roughness of a particular forest patch to that of its neighbours. Fractal diversity indicates the number of degrees of freedom (dynamical systems) that are needed to model forest dynamics. These include predicting the susceptibility, intensity and rate of spread (percolation) of disturbances, particularly weather-related disturbances, across a heterogeneous landscape (Turner *et al.*, 1989).

From a forest–climatology standpoint, the principles and applications of the topologically related paradigms of fractal geometry and chaotic dynamics suggest that overly complex models or analytical systems are not necessary to arrive at a description of spatiotemporal chaos. This principle was demonstrated by cellula automaton models simulating the potential impact of climate change on possible land-use changes on the island of St Lucia (White and Engelen, 1993; Engelen *et al.*, 1995). The model included six cultivated land-use categories (namely, mixed agriculture, agriculture, rural residential, urban, tourism and airport) which tend to dominate the natural land use, including secondary forest growth. While only short-term prediction (10–20 years) of the state of St Lucia is possible, their cellular automaton modelling (Figure 17.2) shows how models of human activities and natural systems can be integrated to explore the future dynamics on the island. Since forest and forestry responses to climate and weather generate fractal structures (i.e. self-similar systems of interlocking states), then their spatiotemporal description may not require complex, intractable, climatological and statistical models. Parsimonious spatiotemporal models, for example, were used by Wissel (1992) and Hendry and McGlade (1995) in a study of regional dieback of beech forests in Europe.

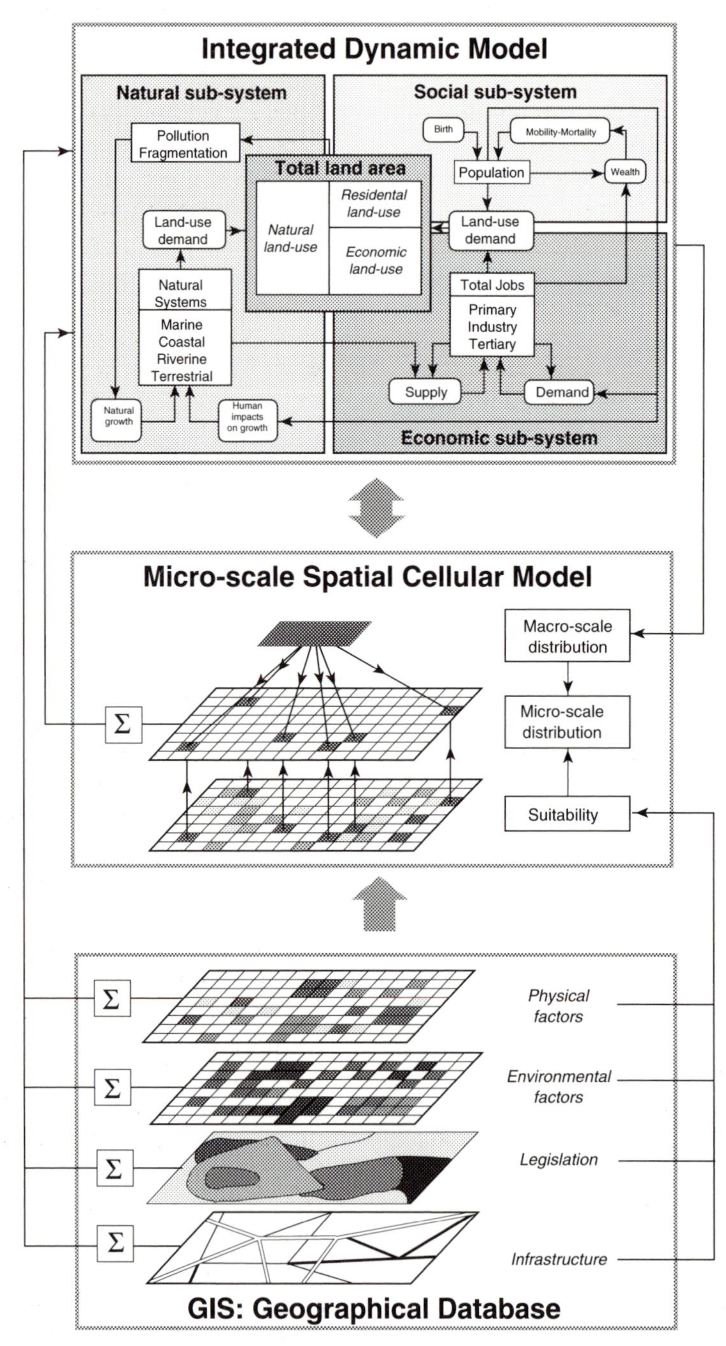

Figure 17.2 A two-level model. On the macro level, long-range interactions are modelled by means of an integrated model. The macrogrowth coefficients are fed into a microlevel cellular model to perform the detailed allocation based on short-range mechanisms. Both levels of the model will retrieve data from the same GIS (after Engelen *et al.*, 1995)

CONCLUSIONS

In many parts of the world, forestry operations have destabilized forests simply by changing surface wind patterns, even in regions with a comparatively benign wind regime (Franklin and Forman, 1987). Changes in surface climate caused by human activities have often led to environmental and social degradation over many timescales. These include deforestation of the Mediterranean basin and Iceland, desertification of the Sahel, reduction to scrub forests (krummholz or elfinwood) in Lapland, eastern Canada, New Zealand and conversion of forests to grasslands throughout most of the semi-arid regions of the world. While there are various topoclimatology classifications, weather events clearly affect silviculture more than climate change *per se*. For example, in terms of climatic regimes, topographical differences in light intensity and periodicity of direct sunlight and precipitation tend to occur as a gradient which determines potential forest growth. However, in sheltered valleys and on mid-slopes with deep soils that are soaked with heavy precipitation (as rain or freezing rain) or subjected to infrequent strong winds (particularly from an unusual direction), forests are prone to intense disturbances. Conversely, forests on upland sites with shallow soils tend to have trees that are resistant to strong winds but, during a few days of warm and windy weather, are prone to wild fires.

The Role of Global Afforestation in Climate Change

There has been much discussion and considerable justification for a global afforestation programme for sequestration of CO_2 to slow or lessen the build-up of atmospheric CO_2. Nilsson (1995) outlined some of the defective costs and benefits of such a programme. There is a reasonable expectation that a global afforestation programme for environmental reasons is wholly justified. In fact, by direct observation and proxy data, the influence of natural forests on climate (such as flood frequencies, energy fluxes, wind flow, etc.) is moderately understood, or at least within the realm of assessment. At societal timescales, causes for abrupt regional climatic and environmental events by human activities (including urbanization, forestry, agriculture and tourism) are also reasonably well understood. However, the role of forests and forestry as a forcing function in global climate change is poorly understood. Presently, baseline forestry–climatological data are scarce and are likely to remain so in the foreseeable future. As noted earlier, this is partly because forests are in sparsely populated areas, where few standard climatological stations are located, and mainly because the criteria for standard climatological stations are designed to monitor and predict weather. A weather network has stringent criteria for instrumentation configurations and the physical site, neither of which should change for as long as possible.

The criteria for forest–climatological stations are quite the opposite because forests are always changing. After all, a forest is a spatiotemporal chaotic system involving changing patterns of biodiversity, dispersal, composition, structure, growth longevity and human activities. Thus, a 'standard' forest–climatological station would require constant revision of instrumentation configurations in order to monitor new phenomena that emerge

CONCLUSIONS *continued*

as a forest stand matures or declines or is changed by episodic natural abiotic/biotic effects and human activities. From the standpoint of monitoring forest response to climate change, the concept of a 'standard' forest–climatological station is neither technically nor economically feasible and, perhaps, is not even realistic. Hence, it is unlikely that the relationship between forests and climate at societal timescales can be modelled by GCMs (Chapter 4) to any reasonable level of confidence.

However, temporary or short-term stations produce baseline data to establish many effects of climate and weather on forests, and vice versa. Aradottir *et al.* (1995) demonstrated the utility of circular statistics to show the qualitative and directional aspects of wind regimes on patterning of birch-forest regeneration and canopy. Conversely, there is an increasing use of the trees and forests as climatic indicators. Yoshino (1987) proposed an international scale of wind-shaped trees to estimate wind conditions. Robertson (1994) indicated how the principles of circular statistics (analyses of directional data), fractal geometry and chaos theory can be applied to studies of the impact of climate and weather on the spatiotemporal dynamics of wind-shaped forests. Gustafsson (1995) combined climatological and biogeochemical parameters to link the decline of forests in western Sweden to the influence of westerly winds and dry acid deposition (Chapter 22). Wind throw hazard classifications were devised prior to the establishment of large-scale industrial forest plantations, in areas where there was no previous forest history or climatological data. These distinguish catastrophic wind throw caused by extreme gale-force winds (> 40 m s^{-1}) and endemic wind throw caused by strong winds from 'normal' fall and winter storms of about 30 m s^{-1} (Quine, 1995). Naturally, the susceptibility of plantations to wind throw depends on the spatiotemporal interrelationship between many biotic and abiotic factors.

However, usually only a few dynamical systems govern the relative stability of a plantation such as tree form, stand density, species composition, soil type and, above all, relative exposure to a particular wind regime. Most of these can be manipulated through management practices (Quine *et al.*, 1995). However, 'storm-proof' designed plantations, purely for industrial purposes, are much costlier and provide no guarantee against the uncertainties of climatic shifts and extreme weather events. The criteria for environmental (i.e. multipurpose) afforestation programmes, on the other hand, include climate as an integral part of the biophysical resource. In many ways, such forests simulate natural forest ecosystems inasmuch as the forest as a whole is inured to windiness and that endemic wind throw is a natural consequence of ecosystem dynamics. That they can also be profitable, more productive and sustainable is demonstrated by the transformation of natural forest to managed forests in Finland and by the transformation of monoculture plantations (established for purely industrial purposes) to mosaics of mixed wood for multiple use, such as The Queen Elizabeth National Forest Park in central Scotland.

CONCLUSIONS *continued*

Public concern about the role of forests in regional and global climate change has put the forestry agenda under the 'banner' of sustainable environmentalism. Emphasis on forests and forestry has shifted from a purely industrial perspective to one with greater emphasis on public and environmental values. A major issue in global environmental change is the potential of a global afforestation programme for the sole purpose of sequestering atmospheric carbon. Prior to the industrial era, 46 per cent of the earth's terrestrial ecosystems were forest whereas today forests account for less than 28 per cent. Forests cover about 3417 million ha (26 per cent) of total land area with 38 per cent tropical forest and 40 per cent temperate/boreal forest, respectively. In addition, 10–20 per cent open woodland and scrub (krummholz or elfinwood) are generated by forest clearing. There are roughly 130 million ha of plantations (Winjum and Schroeder, 1995), about 65 per cent (40 million ha) of which are purely industrial plantations (Pöyry, 1992), which are expected to increase to about 50 million ha at the close of this century. Sixty per cent of the world's forests are considered exploitable whereas 90 per cent of American and European forests are exploitable. Tropical rain forest is disappearing at the rate 2 per cent (17 million ha) annually.

Forests release CO_2 and other greenhouse gases (Chapter 22) into the atmosphere through respiration, decomposition of forest humus and burning, while photosynthesis sequesters CO_2. Burning and decomposition of the tropical rain forest, alone, emits as much CO_2 into the atmosphere as burning fossil fuels in the industrial nations of the northern hemisphere. In 1980, wildland fires were 28 per cent higher than in 1960 (another 'dry' year) owing to greater accessibility into the hinterlands as a result of access roads for exploiting natural resources, all-terrain vehicles and a growing market in ecotourism. For climate change to be an urgent problem, 1500 Gt of CO_2 would be required by the year 2050. Forests account for 80–90 per cent of plant carbon and 30–40 per cent of soil carbon (i.e. 500–800 Gt above-ground and 200–300 Gt below-ground carbon). During the past 100 years, only 75 Gt of CO_2 have been released into the atmosphere by forest clearing and 140 Gt have been emitted by fossil fuel burning. There are 540 Gt of economically available fossil fuel stocks plus 240 Gt of CO_2 from biomass, and together these make up 61 per cent of the 'dosage' required to induce global warming. Hasenkamp (1992) estimated that 500 million ha of new plantations would prevent global warming. However, Nilsson and Schopfhauser (1995) calculated that a total of 340 million ha of land are available for new plantations and agroforestry (270 million ha and 70 million ha, respectively). Furthermore, they also noted that the effect of a massive global plantation programme would not have any significant effect on the carbon balance for at least half a century. Besides, there are many insurmountable impediments facing a global plantation programme. These include innumerable land-use problems involving social, political, organizational and economic impediments such as land tenure, customs and laws, industry and trade (Grainger, 1991).

CONCLUSIONS *continued*

Another major problem facing modellers and policy-makers in general is the lack of appropriate classification techniques for bioclimatic zonation which, for natural forest ecosystems, has very different criteria from those of managed forests. In Finland, the boundaries of various classes of forest productivity and potential productivity have shifted, with most regions showing a dramatic increase due to advances in silviculture. In New Zealand, for example, exotic northern hemispheric 'Gondwana' tree species are cultivated well above the natural altitudinal tree line, because of their vitality and virtually unlimited tolerance to cold compared with southern hemispheric 'Pangaean' tree species. Conversely, because of specialized adaptation to a specific climatic regime, many native tree species are poorly adapted to the quasi-artificial bioclimatic conditions of a plantation either in their native locality or in exotic regions. However, the major difficulty in developing criteria for estimating the potential production of various afforestation and reforestation scenarios in specific regions and landscapes is that large-scale plantation forestry is a unique or relatively new experience for most parts of the world, even for highly industrialized and densely populated countries. The rational for a global plantation programme is generally based on a reductionist philosophy, with the tacit assumption that the link between global climate change and sequestration of carbon by forests is basically a linear process. It is obvious from almost all the studies relating to this issue that the carbon balance is a non-linear process, whereby most of the processes, if not beyond our understanding, are at least beyond our control. However, forest climatology needs to be promoted in order to provide the necessary baseline data and input data for the dynamical system's models, based on a more efficient and effective process for rationalizing various levels of predictability.

REFERENCES

Aradottir, A., Robertson, A. and Moore, E. (1995) 'Circular statistical analysis of the directional growth response of birch colonization and a polar plantation in south Iceland', *Proceedings of the 20th World Congress, International Union of Forestry Research Organizations, Tampere.*

Aubreville, A. (1936) 'La foret coloniale: les forets de l'Afrique occidentale française', *Annales Academie Science Colon Paris* 9: 1–245.

Engelen, G., White, R., Uljee, I. and Drazan, P. (1995) 'Using cellular automata for integrated modelling of socio-environmental systems', *Environmental Monitoring and Assessment* 34: 203–14.

Franklin, J. and Forman, R. (1987) 'Creating landscape patterns by forest cutting: ecological consequences and principles', *Landscape Ecology* 1: 5–18.

Grace, J., Lloyd, J., McIntyre, J., Mirandaa, A., Meir, P., Miraandaa, H., Nobre, C., Moncreiff, J., Massherder, J., Mahli, Y., Wright, I. and Gash, J. (1995) 'Carbon dioxide uptake by an undisturbed tropical rain forest in southwest Amazonia, 1992 to 1993', *Science* 270: 778–80.

Grainger, A. (1991) 'Constraints on increasing tropical forest areas to combat global climatic change', in D. Howlett and C. Sargent (eds) *Proceedings of the Technical Workshop to Explore Options for Global Forestry Management, 24–30 April, Bangkok, Thailand*, London: International Institute for Environment and Development.

Gustafsson, M. (1995) 'Raised levels of marine aerosol deposition due to increased storm frequency cause forest decline in southern Sweden', *Proceedings of the 20th World Congress, International Union of Forestry Research Organizations, Tampere.*

Hasenkamp, K. (1992) 'Global reforestation to solve the problem of CO_2: or mankind will be burning the wrong tree until it finds the right one', *Yearbook of Renewable Energies*, Vol. 1, Bochum: Ponte Press.

Hendry, R. and McGlade, J. (1995) 'The role of memory

in ecological systems', *Proceedings of the Royal Society of London. B. Series* 259: 153–9.

Luxmoore, R., King, A. and Tharp, M. (1991) 'Approaches to scaling up physiologically-based soil-plant models in space and time', *Tree Physiology* 9: 281–92.

Mueller-Dombois, D. (1991) 'The mosaic theory and spatial dynamics of natural dieback and regeneration in Pacific forests', in H. Remmert (ed.) *The Mosaic-Cycle Concept of Ecosystems*', Berlin: Springer-Verlag, pp. 46–60.

Nilsson, S. (1995) 'Valuation of global afforestation programs for carbon mitigation. An Editorial essay', *Climatic Change* 30: 249–57.

Nilsson, S. and Schopfhauser, W. (1995) 'The carbon sequestration potential of a global afforestation program', *Climatic Change* 30: 267–93.

Olsen, E., Ramsey, R. and Winn, D. (1993) 'A modified fractal dimension as a measure of landscape diversity', *Photogrammetric Engineering and Remote Sensing* 59: 1517–20.

Pöyry, J. (1992) *J.P. Data Bank*, Vantaa, Finland: Jaakkoo Pöyry.

Quine, C. (1995) 'Assessing the risk of wind damage to forests: practice and pitfalls', in M. Couttes and J. Grace (eds) *Wind and Trees*, Cambridge: Cambridge University Press, pp. 379–403.

Quine, C., Couttes, M., Gardiner, B. and Pyatt, G. (1995) 'Forests and wind; management to minimize damage', *Bulletin 114*, London: HMSO, 27pp.

Remmert, H. (ed.) (1991) *The Mosaic-Cycle Concept of Ecosystems*, Berlin: Springer-Verlag.

Robertson, A. (1994) 'Directionality, fractals and chaos in wind-shaped forests', *Agricultural and Forest Meteorology* 72: 133–66.

Turner, M., Gardiner, R., Dale, V. and O'Neill, R. (1989) 'Predicting the spread of disturbance in heterogeneous landscapes', *Oikos* 55: 121–9.

White, R. and Engelen, G. (1993) 'Cellular automata and fractal urban form: a cellular modelling approach to the evolution of urban land-use patterns', *Environment and Planning A* 25: 1175–99.

Winjum, J. and Schroeder, P. (1995) 'Forest plantations of the world: their contribution to carbon storage and terrestrial ecology', *Proceedings of the 20th World Congress, International Union of Forestry Research Organizations, Tampere.*

Wissel, G. (1992) 'Modelling the mosaic cycle of a middle European beech forest', *Ecological Modelling* 65: 29–43.

Yoshino, M. (1987) 'A proposal for an international scale of wind-shaped trees as climatic indicators', *Proceedings of the 11th ISB-Congress, West Lafayette, USA*, pp. 173–80.

18

RECREATION AND TOURISM

Allen Perry

INTRODUCTION: CLIMATE–LEISURE INTERACTIONS

Tourism is the world's fastest growing industry and, by early in the twenty-first century, it is likely to be the largest industry in the world. Travel and tourism provide one in every nine jobs worldwide and will add one new job every 2.5 seconds between now and 2005. Within the UK, tourism employs more than 1.5 million workers, more people than in the National Health Service, and generates some £33 billion in annual revenues. Although Europe is the world's leading tourist region, with 64 per cent of world international tourist arrivals, a further 20 per cent of all international tourists visit developing countries, many of which are highly dependent on the industry for their economic well-being. Tourism accounts for 5 per cent of the UK's national income but in Barbados it accounts for 30 per cent (Spink, 1994) and in the Bahamas and Cayman Islands, almost 80 per cent.

Climate constitutes an important part of the environmental context in which recreation and tourism takes place and because tourism is a voluntary and discretionary activity, participation will often depend on favourable climate conditions. The betterment of health has been a common motive for travel since the taking of waters at mineral and hot springs was fashionable in the seventeenth and eighteenth centuries (Kevan, 1993). By the latter half of the eighteenth century, a belief that drinking and bathing in sea water, and taking the sea air, had curative power led to the establishment of the first seaside resorts, notably at Brighton and Bognor Regis on the south coast of England. As it became more fashionable to be seen at the seaside than at spas, it became the vogue to seek out the sun, rather than cultivate pale consumptive complexions. In the twentieth century, a holiday in the sun was for long perceived as vital to well-being and the acquisition of a sun tan as important as the owning of consumer durables. Only very recently has concern about the link between skin cancer and UVB radiation caused some reappraisal.

Nevertheless, a recent UK survey suggested that for over 80 per cent of the more than 9 million people who holiday abroad, the near certainty of warmer, sunnier and more settled weather (than can be normally found during the English summer) was given as the primary reasons for the journey. As Pigram (1983) noted 'for many tourists wanderlust appears to take second-place to sun-lust'. It is then rather surprising that recreation geographers have not investigated more fully the links between recreation and climate. Coppock (1982), writing an introductory article to the journal *Leisure Studies*, noted that 'climate remains a curiously neglected theme in Western Countries, despite its acknowledged importance in all outdoor pursuits'. Boniface and Cooper (1994) also noted that 'climate is one of the key factors influencing tourist development'.

Over the last decade, climate–recreation interaction studies have increased in number and have become organized into three main emerging areas that will form the basis of this chapter. These are:

1 Forecasting how weather and climate affects the participation rates for different types of leisure activities and the levels of personal safety, comfort and satisfaction that ensue.
2 Improving the range and provision of weather and climate information for the leisure industry.
3 Investigating the likely impacts of climate change, particularly projected global warming, on tourism and recreational activities and enterprises.

WEATHER AND LEISURE PARTICIPATION

Smith (1993) has drawn a distinction between weather-sensitive tourism, where the climate is insufficiently reliable to attract mass leisure participation, and climate-dependent tourism, where travel to the holiday destination is generated by the perceived attractiveness and reliability of the basic climatic conditions. In the case of weather-sensitive tourism, the potential recreationist will probably consider the following four circumstances before deciding on whether to embark on an outing:

1 On-site weather
2 Conditions at the trip origin
3 The weather forecast
4 Conditions as anticipated by the recreationist.

For a short afternoon trip in the car, factors 1 and 2 may be essentially the same, while forecasts may be suspected since the general public have less confidence in their accuracy than is really justified (Perry, 1972). This may lead to a greater weight being placed on 'weather expectations'.

For many activities there are critical threshold levels beyond which participation and enjoyment levels fall and safety may be endangered (Figure 18.1). Sometimes, as Taylor (1979) has suggested, these margins are known, measured and may be advertised, especially where there is risk of injury or death. Quite often, however, they are merely the subject of value judgements by recreationists and a sort of recreational 'Highway Code' is needed, spelling out the spectrum of risks involved. In the case of outdoor sports, Thornes (1983) has recognized that the effect of weather can fall into one of three categories, as follows:

1 Specialized weather sports which require certain weather conditions to take place at all, such as skiing, gliding or surfing.
2 Weather interference sports. These may be ball games like soccer or tennis, where adverse factors like wind or rain may affect all the players, or events like horse racing where not only weather, but also the surface conditions, can be all important.
3 Weather advantage sports where participants may gain by playing first, for example golf,

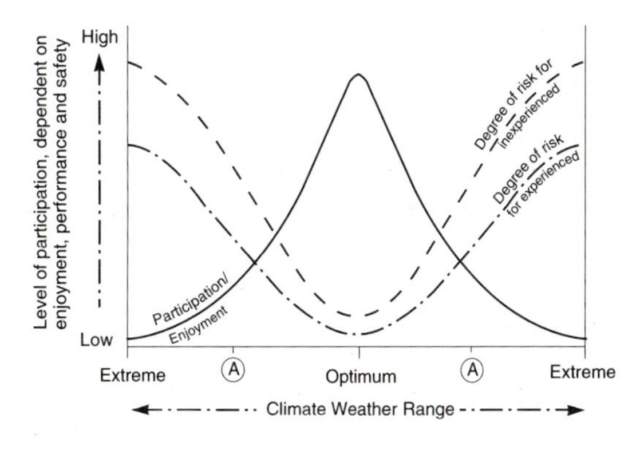

Figure 18.1 The likely impact of weather and climate on enjoyment and safety of outdoor activities

or by choosing a certain position, such as a particular side of a river on which to row.

Many individual sports have attracted a considerable literature; examples include climbing and hill walking (Pedgley, 1979; George, 1993), sailing (Houghton, 1993), parachuting (D'Allenger, 1970), hot-air ballooning (Samuel, 1972) and marathon running (Spellman, 1996).

Predicting spectator numbers has been investigated so that staff can be deployed as necessary. Multiple regression techniques have been used by Illingworth (1977) to predict attendance at premier division football matches and by Thornes (1977) for an open-air swimming pool. More ambitious was the analysis by Shaw (1983) to produce a real-time forecasting model for a mix of recreational sites in northwest England. This suggested that air temperature was the biggest influence on attendance in summer, but that the day of the week is more important in winter. In the summer, participation in outdoor activities (especially visits to the country and seaside) increase in importance relative to visits to town and cities, although local towns prosper on wet days. In very fine summers like 1995, retail sales frequently fall as consumers reduce visits to shops to a minimum (see Chapter 14, Commerce). Avoidance of weather stresses can be economic in the case of large organizations: for example, several football clubs in the UK have installed pitch protection systems to help to maintain their fixture list. Also, in the USA, structures like the Houston Astrodome provide almost perfect weather insulation and can increase attendance numbers threefold compared with open, weather-prone facilities. In Europe the Center Parcs, with their large enclosed 'climatic domes', represent a successful attempt to provide an attractive year-round holiday environment free from the imposed stresses of severe weather.

When the trip being planned is of a major kind, such as an annual holiday, the individual's own 'weather memory' may be influential. Experiences, such as being 'washed out', might result in a complete change of holiday orientation in future years. For some holiday destinations, climate is a clear deterrent: for example, Hay (1989) reported that the weather emerged as the most important feature that visitors found unattractive about Scotland.

The Scottish ski industry is a good example of a tourist enterprise where attendance and profitability is highly dependent on suitable weather conditions (Perry, 1971). The industry is worth about £17 million to the Scottish economy each year, but poor snow seasons like those of 1988–92 meant that equipment manufacturers and tour companies, as well as the chairlift companies, made poor financial returns. These poor snow-cover conditions also occurred at many European winter sports centres and because skiing is a sector of the tourist industry that operates on relatively small financial margins, there was widespread dismay. The future of this industry will be discussed below.

WEATHER AND CLIMATE INFORMATION FOR THE TOURIST INDUSTRY

Tourism and leisure activities have been shown by Smith (1981) to generate a large percentage of the total demand for weather information, especially in summer. Telephone weather information services have been shown by Farrow (1993) to offer good value taken in relation to the potential costs of going out on a day when there might be inclement weather. For a consumer who is purchasing a package holiday or inclusive tour, Perry (1993) has shown that misleading and incomplete weather information in holiday brochures can create a false impression of a resort's weather. Since there are strong vested commercial interests in selling holidays, companies are anxious to portray destinations in as desirable a way as possible. Thus, in the case of winter brochures of Mediterranean destinations, the higher mid-winter temperatures are stressed but not the high rainfall figures.

Summer brochures are no less selective in their presentation of weather information, using adjectives and phrases such as 'ideal', 'delightful' or 'one of the best climates in the world' and not mentioning negative aspects such as the mistral, meltemi or the excessive heatwaves which ruined holidays in Greece in at least two years in the 1980s. Many holiday regions have themselves attempted to create a favourable image by incorporating a climatic factor into their name; thus in Spain we have the Costa del Sol and on the Black Sea the resort of Sunny Beach. Companies specializing in long-haul destinations in general devote more information to weather data, commonly highlighting the three average wettest months of the year. However, the growth of 'off-season' (i.e. summer) packages to the Caribbean in recent years, at prices that are comparable with the eastern end of the Mediterranean, illustrates the need for detailed data, if the holiday-maker is not to be misled. In July, wetter islands in the Caribbean record twice the average precipitation of London, but rain falls for only about half the duration, and hence at four times the intensity.

Many holiday-makers would seem to have no obvious source of information available to consult other than the brochure from which they book their holiday. However, there are a number of sources independent of the tour operator that can provide useful information. These include:

1 Specialized publications. Two examples illustrate the range of information available at the holiday-planning stage. *Holiday Which* is a subscription journal that has dealt with weather matters on occasions. To get around the disadvantages of quoting average temperatures it has given information on the typical spread of temperatures from day to day at a range of resorts (*Holiday Which*, 1974). Some airlines have also done a good job in showing climate data, notably Lufthansa which produces a year-round guide to climate in more than 50 cities and resorts. Some general idea of humidity is given, together with day and night temperature, hours of sunshine, number of rainy days and sea temperature. Information for each month is regularly included in the in-flight magazine for the appropriate month.
2 Holiday weather books. Perhaps the first brave attempt at producing an 'ordinary person's guide to holiday weather' was produced by that well-known television weatherman of the 1970s, Bert Foord (1973). The more popular holiday areas in Europe, together with north Africa and the West Indies, are included and, as well as a useful text, 12 maps give a diagrammatic view of the month's weather. These rather rough maps have outlasted the rest of the book and, tidied up and extended, they have been reissued recently in booklet form (Meteorological Office, 1992). By the 1980s, the respected *World Weather Guide* (Pearce and Smith, 1984) had appeared, claiming to be the 'first and only guide to the world's weather for the serious traveller'. Packed with information, much of it tabular and rather dull, it had some competition for a while from *Travellers' Weather* (B. Ahlstrom and J. Pohlman, 1976) which included pictures and a few simplified weather maps; but the print run of this translated book was small and it soon went out of print. In recent years, two volumes of *The Travellers' Almanac*, covering North America and Europe, have appeared (Bernard, 1991; 1992). As well as giving the usual table of averages, they include some unusual data such as the probability of hurricane force winds in any given year in Florida, and a chapter on bad-weather driving in Europe.
3 Television and telephone. In the UK, the immediate weather outlook across Europe

and the main Mediterranean resorts is given special prominence on the BBC 1 presentation that follows the noon news headlines. Typically, if time permits, attention is drawn to unusual weather, such as extreme temperatures or high rainfall totals. Premium-priced telephone information lines offer forecasts for holiday regions, including information on alpine snow and weather conditions in winter for skiers.

Portraying the suitability of the climatic environment for different types of leisure activity, and for different groups of potential customers, is a difficult task. People's requirements in respect of weather conditions are different according to age, fitness, state of health and what holiday activities are planned. While everyone wants clear factual information in an understandable and unambiguous format, to present the totality of the climatic environment it is often necessary to conceive indices that integrate a number of climatic parameters. While climatologists use various standard measures of comfort and discomfort, these are little known or understood by the general public, many of whom find concepts like wind-chill (already used in TV presentations) confusing. French climatologists (Besancenot *et al.*, 1978) have experimented with a simple summer weather-type classification to show the atmospheric ambience in a holiday region. The clarity with which this information can be portrayed can be seen in Figure 18.2. It does demonstrate that, with a little ingenuity and some cartographic skill, the climatologist can greatly improve the quality and quantity of information that can be put at the holiday-maker's disposal. The examples shown illustrate how, even in high summer, weather varies greatly at Mediterranean destinations. Almeria is much more prone to warm, close weather than Nice or Corfu. Climatologists face a challenge to improve further upon these efforts.

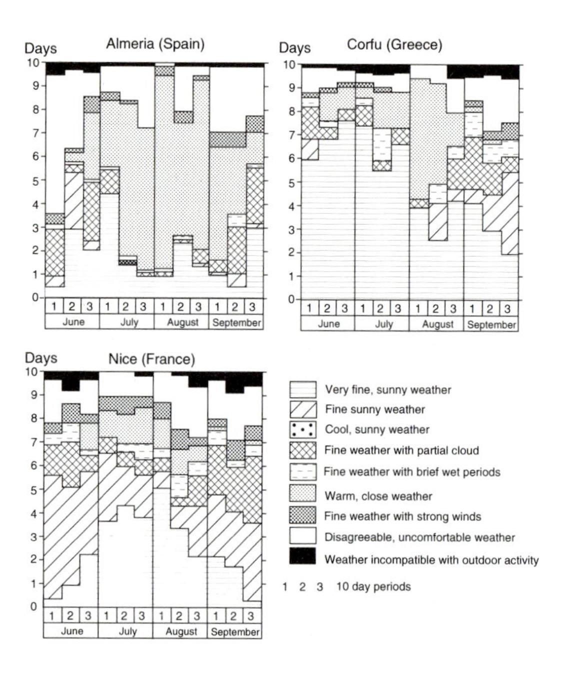

Figure 18.2 Examples of climate information for tourists produced by French climatologists

CLIMATE CHANGE, TOURISM AND RECREATION

Projected climate change, as a result of greenhouse global warming, is likely to provide new opportunities for the tourism industry in some areas, but will restrict both the supply and demand for outdoor recreational facilities in others. Thus as Wall (1992) has noted, changes in global climate are beyond the control of the tourism industry and may have far-reaching consequences for many current tourist destinations as well as for places contemplating involvement in tourism. Because temperatures are expected to increase more markedly in high rather than low latitudes, this may extend the lengths of summer seasons in mid-latitude countries like the UK. Perry and Smith (1996) have noted that an enhanced uptake of many outdoor pursuits ranging from gardening to water-based sports is to be expected. On their own, UK climatic conditions are insufficient to generate either mass travel from overseas or guarantee planned outdoor leisure at any season. Indeed, there has been a sustained erosion of the domestic summer market for many years by holidays taken abroad in 'sunshine' destinations. Mintel (1991) claimed that 73 per cent of respondents to a survey cited 'good weather' as the main reason to go abroad. Such 'import substitution' has been only partially offset by the emergence of short-holiday markets and the growth of the day-visitor trade, stimulated by factors like progressive improvements in road transport.

Despite this, UK tourism displays marked seasonality and weather sensitivity. In Scotland, ferry services to the west coast and the islands carry over half of all passengers during 12 peak summer weeks and 40 per cent of all bed-nights booked into youth hostels are registered in July. Within seasons, the prevailing weather conditions become progressively more influential in affecting tourism as the amount of atmospheric contact necessary for each activity increases (Wall, 1992). This is especially true as tourists adapt to weather conditions by short-term, opportunistic decision-making. For example, during wet weather conditions unfavourable for walking or camping, these activities may be exchanged for visits to nearby towns. The switch may generate additional income for the local economy through visits to shops, restaurants and other indoor facilities. This demonstrates that 'poor weather' is a relative term and that weather-sensitive 'winners' and 'losers' already exist.

The Mediterranean is likely to become less attractive for health reasons in the summer. Apart from the dangers increasingly associated with skin cancer, many Mediterranean beach resorts may simply be too hot for comfort in the peak season, with a much higher frequency of severe heat waves (Perry, 1987; Giles and Balafoutis, 1990). A few destinations (e.g. Cyprus and Corsica) offer the potential to commute from hot beaches to cooler mountains but the Mediterranean area is likely to face other climate-related problems, such as increased insect-borne diseases, marine water pollution and the scarcity of freshwater supplies.

Longer-haul flights to tropical destinations may not offer a solution. Several destinations in the Caribbean, which presently attract charter flights predominantly in the summer (Perry and Ashton, 1994), and the Far East are already prone to severe storms and hurricanes. As marine temperatures are likely to rise further, there is a possibility that the existing hurricane zones may experience greater activity in the future and that such storms may spread to other coastal areas. Even the perception of increased hazards (see Chapter 23, Conclusion) might damage developing tourism in countries such as Malaysia or northern Australia (which have few facilities to offer the holiday-maker apart from climate), as well as more-established tourist areas like Florida. The tourist industry is very vulnerable to natural disasters as 70 per cent of holidays are coast orientated and tourists tend to be more vulnerable that local residents, because of their lack of familiarity with the places they are visiting. In general, destinations with a great reliance on their natural resource base to attract tourists may be at most risk, although some high-latitude destinations may become rather more attractive.

Periods of summer heat stress and associated poor air quality in urban areas may make large cities less desirable places in which to spend leisure time. This could lead to a weekend flight to the countryside and to the coast, where land temperatures will be moderated by sea breezes during the day. Such a pattern could stimulate a rise in the ownership of second homes along the coast and in rural areas by urban residents. Winter holidays are now an important part of tourism. Harlfinger (1991) showed that 86 per cent of Germans visiting Majorca during the winter found the climatic experience to be positive and over half regarded the return to Germany as a potential burden on their health. Milder winter conditions could well provide less cold stress to older people in the UK and mean that vacations currently taken in the Mediterranean, or wider afield, by UK residents could become less compelling.

Generally milder conditions, especially in the winter and shoulder months, could entice more people into the uplands for hill walking, creating opportunities and threats for agriculture and nature conservation. Mountaineering may provide some partial compensation for reduced skiing opportunities in Scotland but there may be a greater risk of snow avalanches in the warmer conditions. In the 1994/5 winter season, an estimated 25 climbers died on the Scottish hills, including ten in avalanches. Although such deaths represent less than 0.1 per cent of those taking part in the sport, more initiatives are likely to be needed to improve people's education about upland hazards.

Warmer winter conditions will bring less snow and snow confidence (the probability of snow lying at particular times during the season) will decline. Higher wind speeds will lead to more snow drifting and less access to the slopes as a result of the closure of ski lifts and tows in conditions of gale-force winds, low cloud and reduced visibility. At present, the ski season in Scotland continues from January to April. However, as the season contracts, suitable snow conditions at New Year and Easter will become less likely and a complete snow cover will become less certain at the critical February mid-term school holiday period. Estimates for the facility at Glenshee suggest that 60,000 skier days per season are required to maintain viability, compared with the 180,000 skier days achievable in a good year. As an indication of the effect of warmer winters, in 1991/2 the number of skier days fell to 12,500.

Poor snow conditions have several consequences, including the inability to use expensive facilities like piste grooming machines. It is important to note that, in Switzerland, the economic effects of recent snow-deficient winters have been greatest on the resorts at low altitudes, impacting adversely on both transport and accommodation (Abegg and Froesch, 1994). Given the prospect of an aperiodic series of snow-deficient winters in Scotland, it is likely that ski clients will either fail to book or at least delay booking until a few days before arrival in order to secure snow. Reduced sales and hire of ski equipment will follow, together with the laying off of staff in hotels and other service sectors.

Rising sea levels (see Chapter 7, Conclusions) will produce increased inundation and erosion along the coast. These processes will affect all the natural and cultural features along marine shorelines, many of which (beaches, golf courses, promenades, swimming pools, piers, marinas and seafront hotels) are tourism related (Boorman *et al.*, 1989). For example, there are an estimated 450 designated 'bathing beaches' in the UK together with 370 coastal marinas, many in the south and east of England. Some of the most important links golf courses (Muirfield, St Andrews, Royal Lytham St Annes) are at risk from a combination of rising sea levels and coastal erosion. Studies of consumer attitudes suggest that the public value the coastal environment highly as a recreational resource and that there is a willingness to pay for its protection. For example, when asked to value their enjoyment of a beach visit in monetary terms, visitors to sites on the east coast of England produced an average figure exceeding £7.00 (Coker *et al.*, 1989).

For a beach backed by a sea wall, as in many resort towns, the likelihood is that increased erosion would lead to a loss of beach material just at the time when the demand for beach-based holidays is likely to be growing. Depending on the supply of sand, the beach facility could be totally lost with subsequent undermining of the stability of the sea wall. Other UK coastal habitats used for recreation, such as sand dunes, shingle banks, marshlands and soft-earth cliffs, would also be affected. Attempts to develop a coastal vulnerability index for Wales suggest that some currently attractive dunes and bays (e.g. Borth Bay) could be the first type of 'beauty spot' to disappear. More flooding of estuaries could reduce the scope for specialist pursuits, such as bird watching, and lead to ecological changes at Sites of Special Scientific Interest.

The cost of shoreline protection for recreational facilities built along the shore, plus any beach preservation and replenishment activities, will be high. For example, Barkham *et al.* (1992) highlighted the £165,000 spent in 1990 on improving sea defences to protect the Royal West Norfolk golf course whilst at Sidmouth, in Devon, the lowering of beach levels by over 4 metres has necessitated the expenditure of over £6 million to provide new breakwaters and enhanced beach nourishment. Once again, these effects may not be worse than impacts elsewhere. The low-gradient beaches characteristic of much of the Atlantic and Gulf coasts of the USA are very vulnerable to erosion (Leatherman, 1989). In other countries, important historic sites are already under threat from rising sea levels and storm surge.

Tourism is a continuously adapting industry, responding to changing demographic and economic conditions as well as to new demands and technologies. In view of the fragmented structure of the industry, climate change adaptation is likely to be gradual with new investment in tune with other strategic decisions. Recreation and tourism impinge on many other sectors. For example, it is estimated that, globally, tourism accounts for at least 60 per cent of all present-day air travel. Within the UK, transport improvements will be particularly important for ensuring accessibility to popular, short-term destinations.

CONCLUSIONS

New research initiatives are urgently needed into the effects of climate, observed and perceived, on tourism which will require more collaboration between applied climatologists and tourist specialists. The challenge will be to draw direct links between weather and climate conditions and the behaviour of tourists (De Freitas, 1990). Recently, the World Meteorological Organization convened a meeting of experts on climate, tourism and human health (WMO, 1995) and, whilst initiatives of this type are to be welcomed, they need extending and broadening.

REFERENCES

Abegg, B. and Froesch, R. (1994) 'Climate change and winter tourism in Beniston', *Mountain Environments in Changing Climates*, London: Routledge, pp. 328–48.

Barkham, J.P., Macguire, F. and Jones, S. (1992) *Sea level rise and the UK*, London: Friends of the Earth.

Bernard, H.W. (1991) *The Travellers' Almanac – North America*, Madison, WI: Riverdale.

Bernard H.W. (1992) *The Travellers' Almanac – Europe*, Madison, WI: Riverdale, 177pp.

Besancenot, J.P., Mouiner, J. and De Lavenne, F. (1978) 'Les conditions climatiques du tourisme litteral', *Norois* 99: 357–82.

Boniface, B.G. and Cooper, C. (1994) *The Geography of Travel and Tourism*, London: Butterworth–Heinemann.

Boorman, L.A., Goss-Gusland, J.D. and McGrorty, S. (1989) *Climatic Change, Rising Sea Level and the British coast*, Institute of Terrestrial Ecology Research Publication No. 1, London: HMSO.

Coker, A.M., Thompson, P.M., Smith, D.I. and Penning-Rowsell, E.C. (1989) *The Impact of Climate Change on Coastal Zone Management in Britain, Conference on Climate and Water, University of Helsinki, Academy of Finland*, Vol. 2, pp. 148–60.

Coppock, J.T. (1982) 'Geographical contribution to leisure', *Leisure Studies* 1: 1–27.

D'Allenger, P.K. (1970) 'Parachuting and the weather', *Weather* 25: 188–92.

De Freitas, C.R. (1990) 'Recreation climate assessment', *International Journal of Climatology* 10: 89–103.
Farrow, R. (1993) 'Weather information for the leisure industry – distribution methods', *Weather* 48: 419–20.
Foord, H.V. (1973) *Holiday Weather*, London: William Kimber, 240pp.
George, D.J. (1993) 'Weather and mountain activities', *Weather* 48: 404–10.
Giles, B.D. and Balafoutis, C. (1990) 'The Greek heatwaves of 1987 and 1988', *International Journal of Climatology* 10: 505–17.
Harlfinger O. (1991) 'Holiday bioclimatology: a study of Palma de Majorca, Spain', *Geojournal* 25: 377–81.
Hay, B. (1989) 'Tourism and Scottish weather', in J. Harrison and K. Smith (eds) '*Weather Sensitivity and Services in Scotland*, Edinburgh: Scottish Academy Press, pp. 162–66.
Holiday Which (1974) London: Consumers' Association.
Houghton, D. (1993) 'Winds for sailors', *Weather* 48: 414–19.
Illingworth, J. (1977) 'Whether to weather the weather or not', unpublished undergraduate dissertation, Department of Geography, University College London.
Kevan, S.M. (1993) 'Quests for cures – a history of tourism for climate and health', *International Journal of Biometeorology* 37: 113–24.
Leatherman, S.P. (1989) 'Beach response strategies to accelerated sea-level rise', in J.C. Topping (ed.) *Coping with Climate Change*, Washington, DC: Climate Institute.
McBoyle, G.R., Wall, G., Harrison, R., Kinnard, V. and Quinland, C. (1986) 'Recreation and climate change: a Canadian case study', *Ontario Geography* 28: 51–68.
Meteorological Office (1992) *Your Holiday Weather*, Bracknell: Meteorological Office.
Mintel (1991) *Special Report – Holidays*, London.
Pearce, E.A. and Smith, C.G. (1984) *The World Weather Guide*, London: Hutchinson, 480pp.
Pedgley, D. (1979) *Mountain Weather. A Practical Guide for Hillwalkers and Climbers in the British Isles*, Milnthorpe, Cumbria: Cicerane Press.
Perry, A.H. (1971) 'Climatic influences on the development of the Scottish skiing industry', *Scottish Geographical Magazine* 87: 25–9.
Perry, A.H. (1972) 'Weather, climate and tourism', *Weather* 27: 199–203.
Perry, A.H. (1986) 'A theme for tourism', *Geographical Magazine* 58: 2–3.
Perry, A.H. (1987) 'Why Greece melted', *Geographical Magazine* 59: 430–1.
Perry, A.H. (1993) 'Climate and weather information for the package holiday-maker', *Weather* 48: 410–14.
Perry, A.H. and Ashton, S. (1994) 'Recent developments in the UK's outbound package tourism market', *Geography* 79: 313–21.
Perry, A.H. and Smith, K. (1996) 'Recreation and tourism' in Department of the Environment *Climatic Change Impact Research Group 2nd Report*, London: HMSO.
Pigram, J. (1983) *Outdoor Recreation and Resource Management*, London: Croom Helm.
Samuel, G.A. (1972) 'Some meteorological aspects of hot-air ballooning', *Meteorological Magazine* 101: 25–9.
Shaw, J.M. (1983) 'The effect of weather on recreation activity in the north west of England', unpublished undergraduate dissertation, Department of Geography, University of Birmingham.
Smith, K. (1981) 'The effect of weather conditions on the public demand for meteorological information', *Journal of Climatology* 1: 381–93.
Smith, K. (1990) 'Tourism and climate change', *Land Use Policy* 7(2): 176–80.
Smith, K. (1993) 'The influence of weather and climate on recreation and tourism', *Weather* 48: 398–404.
Spellman, G. (1996) 'Marathon running – an all-weather sport?', *Weather* 51: 118–25.
Spink, J. (1994) *Leisure and the environment*, London: Butterworth–Heinemann.
Taylor, J.A. (1979) 'Recreation, weather and climate', SSRC-S.C, *Joint Panel Review on Leisure and Recreation Research*, Sports Council, 42pp.
Thornes, J.E. (1977) 'The effect of weather on sport', *Weather* 32: 258–68.
Thornes, J.E. (1983) 'The effect of weather on attendance at sports events', in J. Bole and C. Jenkins (eds) *Geographical Perspectives on Sport*, University of Birmingham, pp. 201–10.
Wall, G. (1992) 'Tourism alternatives in an era of global climate change', in V.C. Smith and V.R. Eadington (eds) *Tourism Alternatives*, Chichester: John Wiley.
WMO (1995) *Report From the Meeting of Experts on Climate, Tourism and Human Health*, World Climate Programme WCASP 33, Geneva: WMO.

19

POLITICAL, SOCIAL AND LEGAL ASPECTS OF CLIMATE

Paul J. Croft

INTRODUCTION: CLIMATE AND POLITICAL, SOCIAL AND LEGAL IMPACTS

Climate and weather affect human activities worldwide, including agricultural, industrial and commercial activities and many other economies (Chapters 14 and 16). Climate has to some extent determined human development, governance and socialization, and it is not unexpected then that a variety of associated political, social and legal impacts may occur because of climate change. These impacts can be quite different according to the realm (i.e. political, social and legal) affected and may or may not be obvious or direct since they are often interconnected. Each impact is related to, and partially responsible for, the very foundation of local, regional and national socio-economic structures affected. Furthermore, the subsequent development and continuous flux of these structures is determined, to some extent, by climate change. When considering regional variations, modification or changes in climate then these impacts can be, or may become, overwhelming.

It is therefore important to consider the broad impact of climatic conditions and climate change on human activities and one way in which this may be accomplished is through the use of climate data and information. For example, climate information plays an important role in the establishment of emergency and disaster response programmes and it is often used as a basis for risk assessments and as a basis for judgements in the insurance market. Climate information is also used in the assessment or prediction of impacts due to climate change. However, this can be quite difficult as climate represents an integration of periodic and chaotic atmospheric conditions over time. Therefore the impact of climate is typically non-linear and non-static and determining the true climate impact is further complicated by human responses to the impact itself.

In addition, natural climatic variability blurs the identification and study of climate change impacts. Short- and long-term variations which may be considered 'natural' produce discontinuous impacts which are not readily nor easily mitigated. These discontinuities, whether record-breaking conditions of hot or cold, or dry or wet, often lead to misinterpretations of climate change. For example, they may lead to poor decision-making by planners unfamiliar with natural climatic variability. These misinterpretations can in turn result in the inappropriate design of infrastructure and other systems, as well as poor resource planning. Commodities traders, construction crews and environmental managers may all be adversely affected when their climate-related decisions, based on expectations, do not match reality. Therefore significant economic and other losses, as well as incorrect decision-making, can occur for both long- and short-term climate change. This ensures that climatic conditions and climate change impacts will be relevant politically, socially and legally.

The existing political framework, and those of the past, incorporate the impact of climate and climate change. The outcome of armed conflicts, colonization and trade, and national policies for drought and climate change, are just a few examples of these impacts. The resulting political structure sets a framework within which local social structures evolve. Within social structures, climate impacts are realized according to the nature of human response. Beyond clothing, health and adaptation (Chapter 12), human responses include personal and group rights, environmental impacts, economic loss or gain, and social compromise. These are inherent to social organizations and structures and are often embellished through complex legal codes. Since personal and group rights may be affected, the ultimate impact of climate and climate change can be evidenced by following legislation and legal judgments. Such legal documentation provides a measure of nature's physical limitations on human activities. It serves to identify and quantify those characteristics associated with a location's climate which limit or prohibit certain activities. It also serves as a standard by which we judge our ability to overcome climate's physical limitations. This facilitates a legal determination of liability together with an assessment of environmental reactions and human responses to climate.

Each of the political, social and legal aspects described above is closely related and one often impacts upon another, causing a cascade of 'reactions' to follow. The human response to such environmental impacts is often unpredictable and attempts to mitigate any environmental changes typically complicate the impact and the response. As all political, social and legal aspects involve the human element, they are not independent and therefore not easily rectified with one another. Thus the impacts to the human political, social and legal structure are somewhat beyond human control and although mitigation measures may be helpful, environmental impacts due to climate change are quickly complicated by human interaction and intervention. These complications may negate any successful mitigation and/or lead to greater impacts and, in spite of these complex interactions, the political, social and legal fabrics tend to coalesce onto common 'climatic' ground. In the process, they provide a basis for structured reaction and response to local and regional climate change.

POLITICAL CONSIDERATIONS

Political structures have traditionally been based in part upon the existing physical environment. Thus, they represent a net integration of regional and secular climate variation. For example, initial human settlement along river valleys led to a natural political structure that varied from that in hilly terrain. Each location offered different natural resources that were politically expedient to exploit, and each location encountered a distinct climate subject to natural variability and real change. The nature of the political response to the climatic environment was therefore iteratively developed according to variations in environmental conditions.

Historical accounts and legends offer specific examples of the impact of climate on the political framework (Lamb, 1977). The politic of taxation accomplished through the establishment of a raingauge network, the passage of ship traders to establish settlements, the use and export of natural resources (e.g. fuel, Chapters 14 and 20) and air pollution control (Chapter 22) are only a few examples (Oliver, 1973). Colonization and trade, expansion, determination of borders, infrastructure (Chapter 15) and local policies (Chapter 13) are dependent upon climate and regional variations in climate. Even the outcome of armed conflicts and individual campaigns or battles have, at times, been influenced and/or decided by regional climatic conditions (Chapter 24).

Climatic conditions may also be considered as determinants in the development of national policies (e.g. on drought or other extremes). Internationally, the World Weather Watch (Rasmussen, 1992) must consider the political repercussions when calling on scientists, governing officials and the public for comment on climate-related issues. For example, the need for regular observations and the exchange of these observations is necessary for global climate study. However, questions of significance, logistics and cost must be considered. In some instances political disputes can lead to censorship of relevant climate information and the reduction of international data exchange (White, 1994). This has often been avoided through the intervention of the United Nations and other political or scientific bodies. However, disagreements often exist with regard to the type of observations necessary, their proper collection, the protocol for evaluation, and the appropriate analysis and interpretation of collected data. These issues become critical when dealing with a global phenomenon (e.g. El Niño), an environmental threat (e.g. ozone depletion) or a perceived environmental threat (e.g. climate modification and global warming).

Politically, such issues may result in significant disputes that are difficult to resolve. For example, a global climate phenomenon demands concerted observation, analysis and response. The phenomenon's pervasiveness, and interconnected physical mechanisms, demand complete international co-operation. The phenomenon is thus usually considered on a scientific basis with little regard to the political and economic implications (Anderson, 1985). Therefore, initial solutions are scientifically based as well. However, when a nation's resources or livelihood are at stake, the economic cost of making changes to implement these solutions may be considered too high. Thus in practice a purely scientific response becomes muddled, or overtaken, by political realities and pressures against which it must be weighed. The final solution will then be an integrated response based on the phenomenon and political structures. This situation is similar to that encountered for an existing environmental threat. However, in this case evidence of damage and/or danger must be acted upon. Whereas a global phenomenon could be evaluated and responses weighed, a real environmental threat often requires swift and decisive action. It also often requires scientific opinion which may be unclear, unformulated or lacking in consensus. This leads to difficulties in the development of response strategies and solutions. An environmental threat also requires that political motivations be suppressed. In the case of a perceived environmental threat, although quite similar to both situations described, political and social pressures can slow response and lead to inaction. The need for 'environmental diplomacy' is quite obvious (Susskind, 1994).

SOCIAL ASPECTS

Human social structures, which are based to some extent on environmental conditions, may be examined according to the nature of human response to climate. Clothing, health, housing and adaptation provide a measure of climate impacts and they are determined, to a large extent, by regional climate (Chapters 12 and 13). They are also affected by the broader community issues (including personal and group rights), environmental impacts, economic loss or gain and social compromise. Therefore, climate change issues are related to quality of life, ownership, hazards, costs and social contracts and each of these requires co-ordinated individual and group responses which are subject to political pressures. Because of these pressures, responses are often diverse and chaotic with little or no scientific basis. For example, policy reversals, based on incomplete or incorrect information or understanding, and social confusion are common to the global warming issue.

Social reactions to climate impacts are also significant and may 'headline' social structures. When weather conditions represent an extreme variation from what is considered natural (or normal), or if certain conditions persist across a region, a rush to judgement and response often occurs. For example, a series of cold winters or wet summers may lead to temporary changes in the nature of local social structures. The changing response of a community may be directed towards mitigation of the impacts of heavier snowfalls. In response, the community may allow more time (and money) for snow removal and more days for business delays. This is because it may be believed or perceived that 'real' climate change is taking place. On shorter timescales a tropical cyclone, snow storm or wind storm may result in delays in travel and may prohibit some modes of travel. This could be considered an immediate threat to the existing social environment requiring an immediate personal response.

The social 'expectations' discussed above also incorporate personal and group rights. However, they typically involve the impact of climate or climate change and modification as it affects and determines the quality of life. In this sense climate impacts may range from aesthetically unpleasant or uncomfortable conditions (e.g. dry and dusty conditions), to interference (e.g. weather modification for rain enhancement) and to hazardous conditions (e.g. toxic air pollutants). However, it is the greater social and political plans of regional communities of any size that determine the level and nature of impact. For example, the building of a large business district with parking garage facilities requires a transformation of the local natural landscape. This immediately alters the nature of local heating and cooling, runoff and air flow. These may be considered unpleasant (local 'hot-spot'), interfering (local flooding) and hazardous (local icing of a new, unnatural surface). However, some impacts could be considered as improvements. For example, the location may no longer be muddy or dusty (more pleasant), it may reduce soil erosion (less interference) and it may provide a safety shelter (less hazardous). However, landscape modifications, whether natural or human made, more often exacerbate local climate impacts or create new ones.

Other social environmental impacts that are related to climate, especially air pollution (Van Der Hoven, 1985), occur locally and regionally (Chapter 22). Pollutants affect personal and group rights and the level of acceptable pollution is determined in part by the social limitations imposed by the community. However, such limitations may lead to economic loss, gain or compromise (Chapter 24) depending upon the social, political and legal structure of each community involved or affected. For example, the addition of a pollutant source in a community must be considered with regard to jobs and quality of life. Social impacts are also related to, and complicated by, real climate variation or change due to natural causes and/or human intervention. They often have socio-political motivations important in evaluating predicted impacts (Nordhaus, 1994). In some cases, climate variation may be advantageous and lead to commercial applications for various industries and users (Herdan and Otten, 1994). In other cases, the social structure may affect the study of climate variations and impacts in terms of research focus (Yarnal *et al.*, 1987) and the application of results (Chapters 10, 11 and 16) across disciplines (Mather, 1993). Indeed, the economic impacts of climate are often featured by the news media in a variety of written and spoken formats. These impacts often lead the media and the public to a preoccupation with secular climate variability. Even on an interannual basis, short-term variations have been historically sufficient for television broadcasters to highlight local (Henson, 1990), national (Teel, 1992) and international conditions (e.g. weather conditions for Olympic events by Lott and Ross, 1994).

Antisocial behaviour (crime) has also been correlated with extreme climatic events and seasonal climate changes. Perry and Simpson (1987) completed a study in Raleigh, North Carolina, in order to examine the relationships between monthly violent crimes (including rape, murder and aggravated assault) and a range of independent environmental variables (Table 19.1), including monthly mean minimum temperature and monthly precipitation totals. These three crimes occurred more frequently in the summer months (e.g. 14.9 per cent of the murders were recorded in August compared with 2.1 per cent in April). It was assumed that the longer daylight hours and warmer weather (see Chapter 12, Atmospheric impacts on performance and behaviour) at this time result in greater activity levels, criminal opportunities and exposure to risk, compared with the dark and cold winter months.

Stepwise multiple regression analyses were applied to the data to determine the best combination of determinant variables involved in the crimes concerned. The results of these analyses are shown in Table 19.1 and it is apparent that the murder rate is poorly explained in terms of the variables studied, since basic psychological imbalance is the key factor. Nevertheless, climatic factors were found to be significant and important in rape and assault offences, with increases in minimum temperatures positively correlated at the 1 per cent significance level. This supported the assumptions made above but the authors warned about making generalizations for other cities in different climatic regimes. Critics also suggested that atmospheric pressure and humidity are equally important atmospheric variables which need to be studied, along with the crime-enhancing effects of alcohol and cocaine consumption.

LEGAL CONSIDERATIONS

The legal structure, including legislation and judgments, has been moulded over time, to some extent, by local climates and regional variations. Climate imposes real physical limitations on human activities which can be

Table 19.1 Stepwise regression analysis of predictor variables with murder, rape and aggravated assault rates

Environmental variables	*Murder rate*	*Rape rate*	*Aggravated assault rate*
County labour force	−0.16	0.52*	−0.38*
County unemployment rate	–	−0.27*	−0.28*
Average minimum temperature	–	−0.24*	0.28*
Amount of precipitation	–	0.10	0.07
Constant	2.12	−4.94	53.96
R^2	0.03	0.30	0.37
F	3.21	12.60*	18.36*
DF	1,118	4,115	4,115

* $p < 0.01$.
Source: Perry and Simpson (1987).

quantified. In this way, climate may be used to attribute a loss to an unusual circumstance, to determine liability and to assess environmental and human impacts of weather and climate variability. Consequent uses of climate data are typically made when a loss or injury has been experienced by one or more persons or broader groups of persons. Such legal actions question the ability of persons to modify the climatic environment in which they live and determine the party responsible for modification and these are major considerations in forensic meteorology. In some cases, the focus is on property rights (Thomas, 1985). In others, it is the anticipated impact of climate variation and the ability of humans to mitigate climate-imposed impacts. In each case, applied climatology is used to determine the physical constraints and extremes of climate and climate change. For example, Table 19.2 emphasizes the dominant role of climatic extremes in the payment of indemnities in Montana during the period 1947–67. It is apparent that drought and hail events were responsible for 88.7 per cent of the payments made at this time. Indeed, in collective terms, climatic extremes directly caused 93 per cent of these payments, which actually increased to over 98 per cent since these events also controlled the spread of insects and diseases. There is considerable overlap between shorter and longer timescales of weather and climate and thus the development of consensus definitions is essential. It should also be clear that none of this can be accomplished without accounting for the social and political structures present.

Table 19.2 Weather and climate factors responsible for the payment of indemnities in Montana, USA, 1947–67

Factor	*Percentage of indemnities paid*
Drought	61.0
Hail	27.7
Insects	4.1
Wind	2.6
Disease	1.3
Frost, freeze, cold spells	1.2
Excess moisture	0.4
Floods	0.1
Others	1.6

Source: Mather (1974).

The evaluation of climate data (Chapters 2 and 3) as a physical limitation is crucial when assessing climate impacts and in the determination of liability and this is accomplished by first reviewing the relevant climatological data. The base climate of a location or region is established and quantified according to standard and acceptable (and consensus) techniques and measures (Chapters 2 and 3). Then comparisons are made between the long-term climate and meteorological data to assess how typical or unusual are the conditions and their impacts. Unfortunately, this does not allow for consideration of climate variation or fluctuation. Further, climate and climate change must also be evaluated with regard to the expected response to these as determined by the political and social communities. This approach has been used successfully to establish whether conditions were 'wetter than normal' or in some way unusual or unexpected (Shulman, 1992). If unusual, then it is assumed that the conditions represented an 'act of God' and therefore were beyond human control. If it is found that the conditions represented typical climatic elements, or variability, and that these conditions could have been mitigated or prevented, a certain degree of liability could be assigned. Other examples of litigation meant to assess liability include accidents involving ice and snow, wind shear and turbulence encounters by aircraft, and the public or private warning of hazardous conditions by weather services. Huston (1992) has shown how meteorologists are increasingly being used as expert witnesses and a new discipline of forensic meteorology/climatology is developing. Even in homicide cases, a determination of the time of death may be made, given exposure to typical climatic conditions (Isaak, 1983). Such a specialized case often requires the integration of other information and science for conclusions to be drawn, such as decay and insect maturity (Goff, 1991).

It should also be noted that the establishment of what is considered typical or 'normal' climate is first a scientific and statistical issue (Chapter 2). Various meteorological and climatic parameters exhibit different empirical distributions that deviate from the statistical normal distribution. Thus, one measure used to establish 'normal' climatic temperature will not be appropriate for establishing a 'normal' climatic rainfall. Many attempts have been made historically (Conrad and Polk, 1962) and recently (Guttman, 1989) to standardize climate 'normals' according to the parameter in question or according to the significance of the parameter to human or related activities. In the case of short-term overlapping periodic and quasi-periodic variations in climate (Peixoto and Oort, 1992) that may be observed (e.g. the El Niño phenomenon and the sunspot cycle), this procedure is more difficult. However, it is necessary when assessing environmental and human impacts on climate (Chapter 5), weather modification and other issues (e.g. acid rain, Chapter 22). Even the simplest observation of weather and climate can lead to litigation. Such litigation may question the nature or design of placement, the ability of a network to provide adequate coverage and information, the expense involved, and the potential hazards or risks posed by the equipment or lack of equipment (e.g. siting and installation of the Doppler weather radar network in the USA.

REFERENCES

Anderson, J.R. (1985) 'Economic Impacts', in D.D. Houghton (ed.) *Handbook of Applied Meteorology*, New York: John Wiley.

Conrad, V. and Polk, L.W. (1962) *Methods in Climatology*, Cambridge, MA: Harvard University Press.

Goff, M.L. (1991) 'Feast of clues – insects in the service of forensics', *The Sciences* 31(4): 30–5.

Guttman, N.B. (1989) 'Statistical descriptors of climate', *Bulletin of the American Meteorological Society* 70(6): 602–7.

Henson, R. (1990) *Television Weathercasting: A History*, Jefferson, NC: McFarland.

Herdan, B. and Otten, H. (1994) 'Development of commercial applications for weather forecasts', *Meteorological Applications* 1: 23–31.

Huston, J.S. (1992) 'The meteorologist as an expert witness', *Bulletin of the American Meteorological Society*, 73: 1831–4.

Isaak, N.J. (1983) 'Climatology helps in solving crimes', *Weatherwise* 36(6): 301–3.

Kunkel, K.E. and Court, A. (1990) 'Climatic means and normals – a statement of the American Association of State Climatologists', *Bulletin of the American Meteorological Society* 71(2): 201–4.

Lamb, H.H. (1977) *Climatic History and the Future*, Princeton, NJ: Princeton University Press.

Lott, N. and Ross, T. (1994) 'Lillehammer 1994 – an Olympic climatology', *Technical Report 94–01* 20 January, 10pp.

Mather, J.R. (1974) *Climatology Fundamentals and Applications*, New York: McGraw-Hill, pp. 360–1.

Mather, J.R. (1993) 'A shared vision', *Annals of the Association of American Geographers* 83(4): 561–7.

Nordhaus, W.D. (1994) 'Expert opinion on climatic change', *American Scientist* 82: 45–51.

Oliver, J.E. (1973) *Climate and Man's Environment*, New York: John Wiley.

Perry, J.D. and Simpson, M.S. (1987) 'Violent crimes in a city: environmental determinants', *Environment and Behaviour* 19: 77–90.

Rasmussen, J.L. (1992) 'The World Weather Watch – a view of the future', *Bulletin of the American Meteorological Society* 73(4): 477–81.

Shulman, M.D. (1992) 'Weather and litigation – the role of the expert', *National Trial Lawyer* 6(2): 3–5.

Susskind, L.E. (1994) *Environmental Diplomacy: Negotiating More Effective Global Agreements*, New York: Oxford University Press.

Teel, L.R. (1992) 'The Weather Channel turns 10', *Weatherwise* 45(2): 9–15.

Thomas, W.A. (1985) 'Property rights in atmospheric resources', in D.D. Houghton (ed.) *Handbook of Applied Meteorology*, New York: John Wiley.

Van Der Hoven, I. (1985) 'Environmental impacts', in D.D. Houghton (ed.) *Handbook of Applied Meteorology*, New York: John Wiley.

White, R.M. (1994) 'Report of the ad hoc committee on international data exchange', *Bulletin of the American Meteorological Society* 75(4): 549–51.

Yarnal, B., Crane, R.G., Carleton, A.M. and Kalkstein, L.S. (1987) 'A new challenge for climate studies in geography', *The Professional Geographer* 39(4): 465–73.

20

THE ENERGY SECTOR

Jim Skea

INTRODUCTION: CLIMATE AND ENERGY SUPPLY

The energy sector is at the heart of the debate about climate change policy since it accounts for most global carbon dioxide (CO_2) emissions and is also responsible for significant levels of emissions of methane and nitrous oxide. Unsurprisingly, most of the literature assesses the role of the energy sector as the cause of climate change. Policies aimed at reducing emissions of greenhouse gases (GHGs) (such as, for example, a carbon tax) would have significant impacts on energy markets. The promotion of greater energy efficiency and renewable energy sources would inevitably form part of a climate change mitigation strategy. However, the energy sector will also be affected by changes in the world's climate and by associated sea-level rise. This chapter assesses climate impacts in the energy sector, drawing out their significance in relation to other driving forces. The analysis draws heavily on the recent Second Assessment Report of the Intergovernmental Panel on Climate Change (IPCC, 1996).

The energy sector comprises a very diverse range of activities. This chapter begins with an overview of the energy sector worldwide, showing the relative importance of fossil, nuclear and renewable energy sources. The breakdown of energy use between industrialized and developing countries is also shown. This section briefly reviews projections of global energy demand over the coming decades, with and without the assumption that aggressive policies to mitigate climate change are put in place. The remainder of the chapter traces climate impacts down through the energy supply chain, starting with primary energy production and then moving on to energy transportation, conversion of primary energy into delivered products (in petroleum refineries and power stations) and, finally, energy demand. Table 20.1 provides a synopsis of the climate impacts discussed in this chapter and the key point is that energy operations are relatively insensitive to climate impacts. The most important sensitivities lie in the availability of energy resources, especially renewable energy, and demand for energy, much of which is tied up with creating acceptable indoor climates. Some of the impacts of climate change on the energy sector will be beneficial rather than detrimental. The chapter concludes with an assessment of the importance of climate sensitivities in relation to other factors, including shifting demographics, changes in technology and policies to reduce GHG emissions.

WORLD ENERGY SUPPLY

There are many authoritative discussions of current and possible future patterns of world energy supply and demand (IPCC, 1996; World Energy Council/IIASA, 1995) and in

Table 20.1 Synopsis of climate sensitivities in the energy sector

	Temperature	*Precipitation*	*Windiness*	*Frequency of extreme events*	*Water availability*	*Sea-level rise*	*Other*
Renewables	More evaporation from reservoirs, shifting seasonal runoff	Hydroelectric potential	Wave potential, reservoir evaporation, wind potential	Many renewable systems vulnerable, especially wind turbines, solar systems	Hydroelectric potential	Design of tidal, wave systems	Insolation affects solar potential
Biomass	Biomass availability	Biomass availability		Damage to biomass crops	Biomass availability		
Energy demand	Less space heating, more air conditioning		Space heating				More humidity, more air conditioning
Energy extraction	Open-cast coal mining	Open-cast coal mining		Offshore oil and gas		Offshore oil and gas	
Energy conversion	Slightly less efficient thermal generation				Cooling water availability	Coastal power stations, refineries	
Energy transport/ transmission	Lower capacity of power lines	Icing of power lines		Effects on power lines			

Key: [dark shading] = significant impact requiring adaptive response at a strategic level; [light shading] = modest impact requiring adaptive response; [no shading] = minor impact

Note: This table identifies impacts and their degree of significance. The direction of impacts and uncertainties are discussed in the text.
Source: UK Climate Change Impact Review Group (1996).

this section it is possible only to sketch out the key features. Table 20.2 shows that 75 per cent of demand for primary energy (the energy embodied in resources as they exist in nature) is met by fossil fuels. Oil accounts for 45 per cent of fossil energy supply while coal and gas make up 31 per cent and 24 per cent respectively. Biomass (plant and organic waste used as fuel) accounts for 14 per cent of world energy supply. Nuclear power meets 5 per cent of demand while renewable energy resources, currently almost all hydroelectricity, satisfy a further 5 per cent. Industry, households and the service sector ('others' in Table 20.2) can make flexible use of coal, oil, gas or electricity. Transport, on the other hand, is heavily reliant on liquid fuels and it is the fastest growing area of final energy demand.

Table 20.3 shows that, in 1990, almost half of the world's energy supplies were used in industrialized countries. Developing countries used about a third and countries with reforming economies (the former Soviet Union together with Central and Eastern Europe) accounted for around 20 per cent. In industrialized countries, per capita consumption of energy is 4.7 tonnes of oil equivalent compared with 0.7 tonnes of oil equivalent in developing countries. Following the 1973 energy crisis, the view was developed that fossil fuel supplies

Table 20.2 Global energy consumption in 1990 by energy source and by sector (exajoules)

	Coal	*Oil*	*Gas*	*Nuclear*	*Hydro*	*Electricity*	*Heat*	*Biomass*	*Total*
Primary	91	128	71	19	21	–	–	55	385
Final	36	106	41	–	–	35	8	53	279
Industry	25	15	22	–	–	17	4	3	86
Transport	1	59	0	–	–	1	0	0	61
Others	10	18	18	–	–	17	4	50	117
Feedstocks	0	14	1	–	–	–	–	0	15

Source: IPCC (1996).

Table 20.3 Projections of world energy demand and CO_2 emissions by region (energy demand: gigatonnes oil equivalent; CO_2 emissions: gigatonnes carbon)

	1990	*2020*			*2050*		
		A2	*B*	*C1*	*A2*	*B*	*C1*
World							
Energy	9.0	15.4	13.6	11.4	24.8	19.8	14.2
CO_2 emissions	6.0	10.0	8.4	6.3	15.1	10.0	5.4
OECD							
Energy	4.2	5.7	5.2	3.7	6.7	5.6	3.0
CO_2 emissions	2.8	3.8	3.3	2.0	4.3	2.5	0.9
Reforming economies							
Energy	1.7	2.3	1.7	1.7	3.7	2.4	1.7
CO_2 emissions	1.4	1.5	1.1	1.0	2.3	1.3	0.8
Developing countries							
Energy	3.1	7.4	6.6	6.0	14.4	11.8	9.6
CO_2 emissions	1.9	4.7	4.0	3.4	8.5	6.2	3.7

Source: World Energy Council/IIASA (1995).

were resource limited although more recent evidence suggests that there is less resource pressure than previously thought. Reserves of fossil fuel energy currently stand at 50,000 exajoules (EJ), equivalent to 130 years' supply at 1990 levels of consumption (IPCC, 1996). Reserves are occurrences of fossil fuel which have been identified and measured and which are economically recoverable using today's technology and at today's prices. Half of these reserves are in the form of coal. Resources, which are considered recoverable with foreseeable economic and technological developments, stand at 186,000 EJ, equivalent to almost 500 years' supply. The biggest potential constraint on the supply of fossil fuels is now seen to be the environmental consequences of combustion, notably the impact of CO_2 emissions on the world's climate.

Work undertaken by the World Energy Council (WEC) and the International Institute for Applied Systems Analysis (IIASA) shows that world energy demand is likely to grow in the twenty-first century and that fossil fuels will continue to dominate energy supply, unless aggressive measures are taken at the international level to protect the global environment (WEC/IIASA, 1995). Tables 20.3 and 20.4 show the results of various scenarios constructed by WEC and IIASA. Case 'A2' refers to a situation in which economic growth is high and coal plays a major role in the world energy scene. Case 'C1' is 'ecologically driven' and incorporates policies which would reduce global CO_2 emission to 2 gigatonnnes of carbon equivalent by 2100. Case 'B' represents a plausible middle course.

Under every scenario, energy demand and CO_2 emissions grow rapidly in developing countries, overtaking the OECD total by around 2020. However, per capita CO_2 emissions in developing countries remain well below the OECD level, even by 2050. CO_2 emissions rise in every scenario except 'C1' and by 2050, global CO_2 emissions in Case 'A2' are 150 per cent higher than they were in 1990. In the ecologically driven scenario 'C1', renewable energy sources account for 21 per cent of energy supply in 2020, rising to 40 per cent by 2050. This degree of success for renewable energy is not reflected in the other scenarios. Even in scenario 'C1', world demand for fossil fuels in the year 2050 is still 16 per cent higher than it was in 1990.

ENERGY RESOURCES

Overview

The world's energy resources derive either from materials buried in the ground (fossil fuels and uranium) or from renewable resources which

Table 20.4 Projections of world energy demand and CO_2 emissions by source (energy demand: gigatonnes oil equivalent; CO_2 emissions: gigatonnes carbon)

	1990	*2020*			*2050*		
		A2	*B*	*C1*	*A2*	*B*	*C1*
World	9.0	15.4	13.6	11.4	24.8	19.8	14.2
Coal	2.2	4.3	3.4	2.3	7.8	4.1	1.5
Oil	3.1	4.5	3.8	3.0	4.8	4.0	2.7
Gas	1.7	3.4	3.2	3.1	5.5	4.5	3.9
Nuclear	0.5	0.6	0.9	0.7	1.1	2.7	0.5
Renewable	1.6	2.6	2.3	2.4	5.7	4.4	5.6
CO_2 emissions	6.0	10.0	8.4	6.3	15.1	10.0	5.4

Source: World Energy Council/IIASA (1995).

derive directly from solar energy or indirectly through winds, tides, biomass and the water cycle. The availability of fossil fuel/uranium is unaffected by climate change, though the extraction process may be affected to a small degree. The availability of renewable energy resources on the other hand is intimately tied to the earth's climate system.

Fossil Fuels

Most of the world's oil and gas production takes place onshore where climatic impacts are minimal whereas offshore oil production is much more sensitive to extreme weather and sea-level rise. It has been reported, for example, that Shell has added 1 metre to the height of a North Sea oil platform in order to counteract the effect of sea-level rise over its lifetime (National Academy of Sciences, 1992). This would not however have a significant impact on the total cost of the facility. Sea-level rise (see Chapter 7, Conclusions) could conceivably have impacts on onshore support facilities which are inevitably exposed to water levels. Perhaps, more importantly, changed storm activity could have an impact on the transportation operations which support offshore oil exploration and production (see Chapter 15, Water transport).

The offshore oil industry may experience some benefits from climate change and, for example, production conducted at high latitudes could be helped through a longer ice-free season (Lonergan, 1989). A recent Canadian study relating to the Arctic petroleum industry (McGillivray *et al.*, 1993) concluded that, if global temperatures were to rise by 1–4°C over a period of 50 years, the open water (ice-free) season in the Canadian sector of the Beaufort Sea would increase from 60 to 150 days; the maximum extent of open water in summer would increase from the current 150–200 km offshore to 500–800 km offshore and the occurrence of larger wave heights and periods would increase. These factors would tend to reduce costs for the offshore petroleum industry, though increased wave activity would push up design requirements. Changes in temperature and precipitation may have very minor impacts on surface coal mining. In particular, increased winter rain in high latitudes could increase the number of days when it would not be possible to operate. However, relative to other pressures on mining, this effect is almost trivial.

Biomass

Biomass resources

Biomass energy involves the harvesting of trees or crops for conversion to energy. Biomass is a 'renewable' energy source only to the extent that the natural resources on which it relies are managed sustainably. In some of the least-developed countries, there is limited access to electricity grids and 90 per cent of energy needs are met by biomass. The world's poorest populations are therefore susceptible to climate-induced changes in the availability and sustainability of biomass resources. The sensitivity of biomass to climate depends upon the sensitivity of the underlying biological resources. Trees account for 64 per cent of biomass use, with the remainder coming from crop residues and animal dung. Most biomass fuels are used in the countryside, but poorer people in cities may be forced to use biomass fuels as opposed to commercial energy sources. Most wood is used directly as firewood but 11–12 per cent is first converted into charcoal (Smith *et al.*, 1992).

Household biofuels are used mainly for cooking although they are also used for space heating in regions with colder winters, for example Nepal and northern China. In rural Asia, many families and villages burn biogas generated by anaerobic fermentation. More than 4 million family house generators have been installed in China and 1 million in India (Dutt and Ravindranath, 1993) and new generators are being installed rapidly. A small amount of fuelwood is used in industrialized countries, principally for home heating, and in Scandinavia, for example, fuelwood accounts for 13 per cent of primary energy consumption. The use of biomass is not restricted to households since biomass wastes are used extensively for electricity generation or co-generation (the simultaneous generation of heat and electricity) in agro-industries in many parts of the world. Residue from sugar cane (bagasse) is used in the sugar industry in the USA, Cuba and Mauritius (Turnbull, 1993) and Brazilian steel mills use charcoal derived from eucalyptus (Sampaio, 1994). In Brazil, 18 per cent of transportation fuels are biomass based (Goldemberg *et al.*, 1993) and in 1987, ethanol fuel from grain, mainly maize, accounted for 8 per cent of the US gasoline market (Wyman *et al.*, 1993).

Climate impacts

Fuelwood supply depends on the quantity and pattern of rainfall and forest yields will fall if rainfall declines, particularly in low-latitude countries. In dry areas, fuelwood may therefore become scarcer. Climate change impacts will be exacerbated by other pressures since deforestation around cities in developing countries will mean that fuelwood must be transported over ever greater distances. The projected deficit of demand over supply could be most critical in tropical Africa although, on the other hand, the growth of annual and perennial plants will increase as atmospheric concentrations of CO_2 increase and temperatures rise (see Chapter 10, Climate and vegetation in the future).

New biomass conversion and production techniques could be developed to help countries which face fuelwood shortages. Biofuel supply from trees, shrubs, grass or crop residues can be increased or maintained by giving local populations a stake in sustainably grown forests (Chapter 17) (Bertrand, 1993), planting more trees on agricultural land, using higher-yielding species which are better adapted to extreme conditions (Riedacker *et al.*, 1994) and increasing the productivity of agricultural land. In addition, technologies, such as pyrolysis and pressurized charcoal manufacture (Mezerette and Girard, 1990; Antal and Richard, 1992), can improve the efficiency with which biomass is converted into energy.

The rate of conversion of biomass to methane gas in anaerobic digestors increases with temperature. Currently, methane production may be interrupted in colder regions, such as northern China, in winter (Rajabapiah *et al.*, 1993) so global warming could therefore improve yields, reduce periods of interruption and extend biogas production into new regions. Climate change could affect the availability of supplies of annual biomass crops and hence electricity production and co-generation based on agricultural wastes. However, the impact will be no greater than that on the supply of food or agricultural raw materials. Generation plants which use annual or perennial crops will be less vulnerable (Larson, 1993). Energy systems which make use of both biofuels and fossil fuels will be less susceptible to climate change. Biofuels and fossil fuels can be combined by injecting gas derived from biomass into natural gas networks, manufacturing 'gasohol' (a mixture of gasoline and ethanol) or manufacturing methanol which can be derived from biomass, natural gas, or coal.

Hydroelectricity

Hydroelectricity is the most widely developed of the renewable energy sources, though the development of new large-scale schemes is becoming constrained by environmental and social concerns in many countries. The displacement of indigenous peoples is a particular issue at stake. The impacts of climate change on the availability and usefulness of hydroelectric power will be the result of a complex series of interacting factors. The use of water resources (Chapter 6) for electricity production must be balanced against other potential demands including the provision of habitats for wildlife, crop irrigation, navigation and recreational facilities.

The impacts of climate change could be positive or negative depending on the specific location. On the one hand, increased precipitation, likely in high latitudes, will increase water resources whereas, on the other hand, higher temperatures will increase evaporation from reservoirs and reduce power availability. Evaporation losses depend on particular features of the hydroelectric system, such as the proportion of the hydrological basin covered by reservoirs and lakes, the configuration of a basin's reservoirs relative to precipitation sources and vegetation cover (Cohen, 1987a, b). Losses also depend on temperature, wind and humidity. Climate change can affect the timing of water runoff (Chapter 6) and thus the utility of available energy. Higher temperatures will lead to reduced snow accumulation and hence will reduce the tendency for runoff to be concentrated in spring and early summer. This could be advantageous because electricity is of greater value in winter when, in many areas, demand is higher. Where water storage is dominated by snowmelt rather than by reservoirs, river basins are likely to be more sensitive to temperature shifts than to hydrological changes (Lettenmaier and Sheer, 1991).

A number of the individual studies which have been carried out indicate generally positive effects. In the James Bay region of Quebec, an increase in precipitation would outweigh an increase in evaporation, causing generation to increase by 7–20 per cent (Singh, 1987). Climate change could result in increased hydroelectric production in the South Island of New Zealand, with generation shifting to more useful times of the year (Fitzharris and Garr, 1995). A 2–4°C warming in northern California would reduce snow accumulation and shift peak runoff from spring to winter and this would result in a closer match between runoff and peak power demand in the region. However, water supply would become less reliable and flood risks would increase (Lettenmaier *et al.*, 1992). More efficient reservoir management would not be sufficient to deal with these additional risks and new reservoir storage would be needed. Climate change could also affect the operation of hydroelectric systems to a minor degree. For example, changed precipitation patterns and the influence of climate on the vegetation of a watershed could affect rates of siltation and reservoir storage capacity.

Solar Energy

Solar energy encompasses solar thermal energy, used to heat water, and photovoltaic (PV) systems, used to generate electricity. Many major companies are investing heavily in the development of PV and the cost of systems has dropped very rapidly. The viability of both solar thermal and PV energy systems is dependent on local conditions and solar energy has a high value because it is dependably available at times when demand for energy is highest. The impacts of climate change on solar energy production are less than for most other forms of renewable energy. Systems that concentrate sunlight, using either mirrors to reflect it or lenses to refract it, require direct sunlight. Increases in humidity, haze or cloudiness will reduce the effectiveness of concentrating systems, although tracking systems can offset this to some extent (Kelly, 1993). As a result of climate change, there may possibly be more clouds and less direct solar radiation (Enquête Kommission, 1991).

Flat-plate PV systems make use of both direct sunlight and diffuse light scattered by clouds or humidity. Electricity generation in a region with poor conditions can be 40 per cent lower than in a region with excellent conditions, assuming identical systems (Zweibel and Barnett, 1993). PV cells lose about 0.5 per cent of their efficiency per °C above their rated temperature (Kelly, 1993). The performance and reliability of solar energy is somewhat sensitive to climatic factors. When there is little rainfall, dust can soil reflective surfaces and reduce energy capture by up to 8 per cent (Radesovich and Skinrood, 1989). Larger systems are being developed in order to reduce unit costs but this makes a system more susceptible to wind damage. The electronic components of PV systems can be damaged by moisture from storms or dew, lightning strikes, overheating and voltage surges as the sun passes out from behind clouds (Boes and Luque, 1993). The structures which support PV collectors would need to be strengthened if peak wind speeds were to increase or if storms were to become more frequent or severe (Chapter 23) (Kelly, 1993).

Wind

On a global scale, the availability of wind energy is related to the temperature difference between high and low latitudes. It appears that the high latitudes will warm relative to the tropics which, by itself, would reduce temperature differences, pressure gradients and hence global wind resources. However, wind resources at individual sites depend largely on regional climate patterns and local topography, and may be linked only weakly to changes at the global level. Theoretically, wind energy depends on the cube of wind speed so that any changes in wind resources would therefore cause disproportionately large changes in energy capture. Some studies have attempted to demonstrate this effect at specific locations. For example, Baker *et al.* (1990) showed that, at good but undeveloped sites, a 10 per cent change in wind speeds could change wind energy generation by 13–25 per cent, depending on the particular features of the site and seasonal factors. A change in the direction of the prevailing wind could have a significant impact on existing wind installations, depending on the orientation of the array.

Higher temperatures would have a small but negative impact on wind generation because air density decreases with increasing temperature. However, an average increase of several °C at a developed site should reduce output by at most 1–2 per cent (Cavallo *et al.*, 1993). As with other renewable energy sources, climate could have minor impacts on the performance and reliability of wind turbines. Dust, insects or ice can reduce wind energy production by about 8 per cent (Lynette and Associates, 1992) and heavy rain reduces energy output by increasing the turbulence of the wind (Baker *et al.*, 1990). On the other hand, light rain cleans the blades and can increase energy production by up to 3 per cent whereas wind machines can be damaged by tornadoes, hurricanes, snow and ice (Jensen and Van Hulle, 1991).

Other Renewable Energy Sources

Other climate-sensitive renewable energy sources include tidal barrages, wave energy systems, ocean thermal energy conversion (OTEC) and geothermal supplies. Currently, there are few plans to develop ocean energy systems, though their use might grow in the longer term. Projected sea-level rise should be taken into account over the very long design life of tidal barrages (UK Climate Change Impacts Review Group, 1991). Wave energy resources are sensitive to global wind patterns which generate waves and nearshore winds which affect wave size (Cavanagh *et al.*, 1993). Ocean energy systems in general are vulnerable to storm damage. Geothermal resources, where the steam resource is associated with aquifers recharged by surface water or rainfall, may be affected by a change in precipitation patterns. In New Zealand, increased rainfall would increase the recharge of groundwater for most geothermal fields (Mundy, 1990).

TRANSPORTATION OF ENERGY

The transportation of energy products (principally fossil fuels) is subject to the same vulnerabilities as the transportation of other goods (Chapter 15). For example, higher temperatures could ease winter weather conditions, reducing problems associated with moving coal and oil by road or rail. Fossil fuels, especially oil and coal, are widely traded in international markets and shipping operations may be affected by changes in storm activity. The transportation of oil from offshore production facilities in high latitudes, notably the Arctic and the North Sea, may be particularly affected by increasing storm activity (Chapter 15).

There are several specific features of the energy sector which deserve mention although the transmission and distribution of electricity is a special case. Overhead cables are susceptible to icing and storms and while icing is likely to decrease with climate change, the frequency of storms could increase. If storm frequency was to increase, customers would have to accept a less reliable electricity service or pay for the costs of strengthening lines. Storms in northern Europe in October 1987 led to an average loss of 250 minutes of supply to electricity customers in England and Wales (Electricity Council, 1988). The cost of interruptions to power supply in many service sectors is rising owing to the increased use of advanced information and communication technologies.

The capacity of electricity transmission lines drops at higher temperatures. For example, the capacity of the typical line used in the UK falls from 2720 MV A in winter to 2190 MV A in summer (Eunson, 1988). Climate change would therefore have a perceptible effect on capacity but this could be overcome, at a cost, by redispatching plant or, in the longer term, reinforcing the transmission network. Another specific feature of the energy sector is the transportation of oil and gas by pipeline. In Arctic regions, melting permafrost may make it necessary to redesign oil pipelines in order to avoid slumping, breaks and leaks (Brown, 1989). Also, compressors on gas pipelines would have to work slightly harder at higher temperatures, thus increasing the gas industry's 'own use' of energy.

ENERGY CONVERSION

Power Station Operations

Many fossil-fuel-fired power stations make use of steam turbines which rely heavily on the availability of cooling water (Smith and Tirpak, 1989; Solley *et al.*, 1988). The exception is the gas turbine plant which is used to meet peak electricity demand. Nuclear power stations also need large amounts of cooling water. Climate change could cause water availability/quality to become a constraint for power stations situated on rivers. Two types of cooling are available, namely the once-through systems which abstract large quantities of water but return almost all of it to the river, albeit at a higher temperature, and alternatively, with 'evaporative' cooling (cooling towers), less water is abstracted but all of the make-up water is 'consumed' in the process (Miller *et al.*, 1992).

Low levels of river flow can limit the output of power stations if thermal pollution of the river becomes excessive, if water consumption in evaporative systems becomes excessive in relation to river flow, or, in extreme cases, if river levels fall below the level of the water input pipe. Under drought conditions, plant output has been curtailed for all of these reasons in France and the USA. Nuclear plants might need to be shut down to comply with safety regulations under more extreme temperature conditions (Miller *et al.*, 1992). There is a general trend away from once-through cooling towards evaporative cooling and in the USA at least, there is also a movement away from river-based power plants towards coastal siting

(Smith and Tirpak, 1989). New technologies, such as the combined cycle gas turbine plant which makes use of a mixture of steam turbines and gas turbines, are reducing dependence on the availability of cooling water (UK Climate Change Impacts Review Group, 1991). For all of these reasons, the impact of climate-induced changes in water availability on power generation should be relatively minor. The efficiency of electricity generation would be affected negatively, but in a minor way, by higher temperatures (Ball and Breed, 1992).

Sea-level Rise

Many energy sector operations are best located in coastal zones because of ready access to fuel supplies or because of the availability of cooling water. For example, all UK oil refineries, and more than half of the thermal power stations, are located on coastal sites (UK Climate Change Impacts Review Group, 1991). Sea-level rise (see Chapter 7, Conclusions) could result in additional expenditure on protection against storm surges at existing sites (Smith and Tirpak, 1989; UK Climate Change Impacts Review Group, 1991). Existing power stations and oil refineries are unlikely to close as a result of sea-level rise since there is too much investment associated with the infrastructure (transmission lines, pipelines, transportation routes) which serves the facilities.

ENERGY DEMAND

Overview

Some categories of energy demand are strongly sensitive to climate. The most important are space heating, air conditioning and agricultural applications such as irrigation pumping and crop drying. Climate-induced changes in energy demand will in turn affect patterns of energy supply. Although changes in demand will be one of the most important climate impacts in the energy sector, it is worth noting that the demand changes discussed in this section are still likely to be smaller than those caused by economic growth, technological change or even policies designed to reduce GHG emissions. As shown in Table 20.5, many regional and national studies have now assessed the impact of climate change on energy demand. The studies have varied greatly in terms of time-frames, the nature of the climate scenarios used, the focus in terms of fuels/energy end use and the methodologies used. There has been a strong emphasis on the impacts on electricity demand, probably because analytical tools (weather–demand relationships, computer models) are readily available. Few studies address energy demand in the agricultural sector (Chapter 16) or on impacts on energy supply investments.

Space Heating

The mix of energy sources used for space heating varies widely from one region of the world to another. For example, biomass is used heavily in many developing countries, while coal is used extensively in countries such as China and Poland. The use of natural gas has increased greatly in Europe and North America over the last two decades whereas many new homes rely on electric heating. The use of energy for heating buildings is closely related to 'degree-days', which measure the extent to which mean daily temperatures fall below a base temperature, conventionally set at 15.5°C. Degree-day statistics can be manipulated to take account of temperature rise and hence derive an approximate indication of the impact which climate change would have on energy demand. Table 20.6 shows that UK space heating demand could fall by 7 per cent by the 2020s and 11 per cent by the 2050s. The biggest impacts would be felt in the household and service

Table 20.5 Summary of results of studies relating climate change to energy demand

Study	*Country/ region*	*Temperature change (°C)*	*Date*	*Method*	*Coverage*	*Change in annual demand*	*Change in peak demand*
Aittoniemi (1991)	Finland	1.2–4.6			Electricity	7–23% down	
Darmstader (1991)	US MINK	0.81	2030	Degree-days	Agriculture cooling heating	3% up <2% up 7–16% down	
Matsui *et al.* (1993)	Japan		2050		Electricity	5% up	10% up
Rosenthal *et al.* (1995)	USA	1	2010	Degree-days	Space heating and cooling	11% down	
Smith and Tirpak (1989)	USA	3.7	2055	Weather analogue	Electricity	4–6% up	13–20% up
UK CCIR Group (1990)	UK	2.2	2050	Degree-days	All energy Electricity	5–10% down 1–3% down	

Source: IPCC (1996).

Table 20.6 Possible impacts of climate change on UK final energy demand (reductions attributable to space heating)

	Degree-day method		*Statistical method*	
	2020s	*2050s*	*2020s*	*2050s*
Household	7%	11%	5%	7%
Services	7%	11%	7%	9%
Iron and steel	<1%	<1%	–	–
Other industry	2%	3%	2%	2%
Agriculture	–	–	–	–
Transport	–	–	–	–
Total	**3%**	**5%**	**3%**	**4%**

Source: UK Climate Change Impacts Review Group (1996).

sectors where space heating accounts for a high proportion of demand.

In practice, householders in many countries, including the UK, cannot afford to heat their homes to the temperatures which they might desire (Boardman, 1991) and climate change would allow comfort levels in these homes to improve. Statistical relationships, based on historical associations between temperature conditions and energy demand, can take account of this factor. Table 20.6 shows that statistical relationships, applicable in the UK, result in a lower estimate of the impact of climate change on energy demand than does the degree-day method. The UK results are reflected in all other studies of space-heating demand, as Table 20.5 shows. In the MINK (Missouri, Iowa, Nebraska, Kansas) region of the USA, it has been suggested that fossil fuel demand for space heating would decrease by 7–16 per cent as a result of a temperature change of 0.8°C (Darmstadter, 1991).

Electricity

Electricity demand will be affected by changes in the use of air conditioning as well as changes in space-heating patterns. Air conditioning is still relatively rare in temperate regions of the world, but its use is growing because office equipment is adding to internal heat gains while sealed buildings help to isolate occupants from noisy or polluted urban environments (Herring *et al.*, 1988). Air conditioning is also growing rapidly in some developing countries (Schipper and Meyers, 1992). The relative importance of space heating or air conditioning will determine whether electricity demand will rise or fall as a result of climate change (Linder *et al.*, 1989). Climate change will cause electricity demand to increase in areas where there is a high summer load associated with cooling. Conversely, demand will fall where there is a high winter load associated with space heating. In many temperate zones, it is not clear whether the increase in demand for cooling will exceed the reduction in demand for heating (UK Climate Change Impacts Review Group, 1991).

Total Energy Demand

It is equally uncertain whether total energy demand (electricity plus fossil fuels) will rise or fall. Temperature rise will lead to reduced energy demand in Sapporo, Japan (43°N), while in Tokyo (36°N), reduced energy use for heating in winter will balance an increase due to cooling in summer and, at Naha (26°N), energy demand will rise. In the USA, a 1°C global warming would reduce total energy use associated with space heating and air conditioning by 11 per cent by the year 2010 and energy costs would fall by $5.5 billion. A 2.5°C warming would reduce energy costs by $12.2 billion (Rosenthal *et al.*, 1995). Local energy demand patterns will determine whether peak demand for energy increases or decreases as a result of climate change (Linder and Inglis, 1989). If peak demand occurs in winter, maximum demand is likely to fall whereas if there is a summer peak, maximum demand will rise. Climate change may cause some areas to switch from a winter-peaking to a summer-peaking regime

Meeting Changed Patterns of Demand

Energy suppliers, particularly in the electricity and gas sectors, are highly conscious of the link between energy demand and climate because peak capacity needs are determined with reference to extreme weather conditions (Chapter 23). Only a limited amount of work has been carried out on the way in which climate change might affect investments in energy supply. The US national climate impacts study (Smith and Tirpak, 1989), based on large regional temperature rises in the range 3.4–5.0°C by 2055, concluded that:

1 By 2010, peak electricity demand would increase by 4 per cent above the baseline level and by 2055, peak demand would be 13 per cent above the baseline level.

Table 20.7 Asset lifetimes in the energy sector

Conventional light bulb	Weeks up to 3 years
Electric white goods	5–10 years
Central heating boilers/ systems	10–15 years
Motor vehicles	10–15 years
North Sea oil field	10–30 years
Gas supply contract for combined cycle gas turbine	15 years
Life of renewable energy project	20 years
Conventional power plant	40–45 years
Housing stock	50 years, but some very long
Infrastructure (roads/ rail/ ports)	50–100 years
Tidal barrage, dam	100+ years

Source: Skea (1995).

2 By 2010, additional investment attributable to climate change would amount to 13 per cent of new capacity requirements and by 2055, the extra capacity needed could be 16 per cent of new capacity requirements.
3 Generation costs could be 5 per cent higher by 2010 and 7 per cent higher by 2050.

The study also concluded that the generation fuel mix would be altered, that there would be more interregional power interchanges and that construction requirements and greater levels of fuel use would have adverse environmental impacts.

CONCLUSIONS

The energy sector covers a range of activities which are sensitive to climate and the overall effect of climate change will be the aggregation of diverse individual impacts. However, climate sensitivity is generally low relative to that of agriculture and natural ecosystems (Chapters 16 and 10). The technological capacity to adapt to climate change depends partly on the rapidity of climate change set against the rate of replacement of equipment and infrastructure. The lifetimes of most energy equipment is short compared with the projected timescale for climate change, perhaps 60–70 years for a CO_2 doubling. Table 20.7 shows that many short-lived assets (such as consumer goods, motor vehicles and space-heating/cooling systems) will be replaced several times over this time period, offering considerable opportunities for adaptation. Even medium-life assets such as industrial plants, oil and gas pipelines and conventional power stations are likely to be completely replaced, though there will be less opportunity for adaptation. More difficulties could arise with long-lived assets such as dams or tidal barrages which can have a design life of over a century.

The importance of the climate signal in relation to other pressures for change is a good indicator of the capacity for adaptation to climate change. How do the pressures of changing demographics, market conditions, technological innovation or resource depletion compare with those which could arise from climate change? Over periods of half a century or more, many sectors will change beyond recognition, while others may disappear completely. New products, markets and technologies will also emerge and an industry coping with other, more significant changes may well be able to take climate change in its stride. Different regions of the world vary greatly in their capacity to adapt to climate change and the biggest barriers to adaptation will be faced by people in less developed countries who rely on biomass for their basic livelihoods. In other respects, the energy sector has the capacity to adapt to a changed climate. However, the potential for adapting to climate change will be fulfilled only if the necessary information is available, enterprises and organizations have the institutional and financial capacity to manage change and there is an appropriate policy framework within which to operate.

REFERENCES

Aittoniemi, P. (1991) 'Influences of climatic change on the Finnish energy economy', in E. Kainlauri *et al.* (eds) *Energy and Environment 1991*, Atlanta, GA: ASHRAE.

Antal, M.J. and Richard, J.R. (1992) *A New Method for Improving the Yield of Charcoal from Biomass, Proceedings of the 6th European Conference on Biomass*, Athens: Elsevier, pp. 845–9.

Baker, R.W., Walker, S.N. and Wade, J.E. (1990) 'Annual and seasonal variations in mean wind speed and wind turbine energy production', *Solar Energy* 45(5): 285–9.

Ball, R.H. and Breed, W.S. (1992) *Summary of Likely Impacts of Climate Change on the Energy Sector*, Washington, DC: US Department of Energy.

Bertrand, M. (1993) *La Nouvelle Politique Forestière du Niger et les Marchés Ruraux du Bois Énergie: Innovations Institutionelles, Organisationelles et Techniques*, Montpellier: CIRAD.

Boardman, B. (1991) *Fuel Poverty: From Cold Homes to Affordable Warmth*, London: Belhaven Press.

Boes, E.C. and Luque, A. (1993) 'Photovoltaic concentrator technology', in T.B. Johansson *et al.* (eds) *Renewable Energy: Source for Fuels and Energy*, Washington, DC: Island Press, pp. 361–401.

Brown, H.M. (1989) *Planning for Climate Change in the Arctic – the Impact on Energy Resource Development, Symposium on the Arctic and Global Change, Climate Institute, Ottawa.*

Cavallo, A.J., Hock, S.M. and Smith, D.R. (1993) 'Wind energy: technology and economics', in T.B. Johansson *et al.* (eds) *Renewable Energy: Source for Fuels and Energy*, Washington, DC: Island Press, pp. 121–56.

Cavanagh, J.E., Clarke, J.H. and Price, R. (1993) 'Ocean energy systems', in T.B. Johansson *et al.* (eds) *Renewable Energy: Source for Fuels and Energy*, Washington, DC: Island Press, pp. 513–47.

Cohen, S.J. (1987a) 'Projected increases in municipal water-use in the great lakes due to CO_2 induced climate change', *Water Resources Bulletin* 23(1): 91–101.

Cohen, S.J. (1987b) 'Sensitivity of water resources in the Great Lakes region to changes in temperature, precipitation, humidity and wind speed', in S.I. Soloman, M. Beran and W. Hogg (eds) *The Influence of Climate Change and Climate Variability on the Hydrologic Regime and Water Resources*, IAHS Publication No. 168, Wallingford.

Darmstadter, J. (1991) 'Energy', *Report V of Influences of and Responses to Increasing Atmospheric CO_2 and Climate Change: The MINK Project*, Report DOE/RL/01830–H8, Washington, DC: US Department of Energy.

Dutt, G.S. and Ravindranath, N.H. (1993) 'Bioenergy: direct application in cooking', in T.B. Johansson *et al.* (eds) *Renewable Energy Sources for Fuels and Electricity*, Washington, DC: Island Press, pp. 787–815.

Electricity Council (1988) *Annual Report and Accounts 1987/88*, London: Electricity Council.

Enquête Kommission German Bundestag (1991) *Protecting the Earth: A Status Report with Recommendations for a New Energy Policy*, Bonn University, pp. 251–329.

Eunson, E.M. (1988) *Proof of Evidence on System Considerations*, Hinkley Point 'C' Power Station Inquiry, London: Central Electricity Generating Board.

Fitzharris, B. and Garr, C. (1995) 'Climate, water resources and electricity', in *Greenhouse 94, Proceedings of a conference jointly organized by CSIRO, Australia and the National Institute of Water and Atmospheric Research, CSIRO, New Zealand.*

Goldemberg, J., Monaco, L. and Macedo, I. (1993) 'The Brazilian fuel alcohol program', in T.B. Johansson *et al.* (eds) *Renewable Energy: Source for Fuels and Energy*, Washington, DC: Island Press, pp. 841–63.

Herring, H., Hardcastle, R. and Phillipson, R. (1988) *Energy Use and Energy Efficiency in UK Commercial and Public Buildings up to the Year 2000*, Energy Efficiency Series 6, Energy Efficiency Office, London: HMSO.

IPCC (1996) *Climate Change 1995 – Impacts, Adaptation and Mitigation of Climate Change: Scientific Technical Analyses*, Contribution of Working Group II to the Second Assessment Report, Cambridge: Cambridge University Press.

Jensen, P.H. and Van Hulle, F.J.L. (1991) *Recommendations for a European Wind Turbine Standard Load Case*, in E.L. Petersen (ed.) *RISO Contribution from the Department of Meteorology and Wind Energy to the ECWEC'90 Conference in Madrid*, pp. 33–8.

Kelly, H. (1993) 'Introduction to photovoltaic technology', in T.B. Johansson *et al.* (eds) *Renewable Energy: Source for Fuels and Energy*, Washington, DC: Island Press, pp. 297–336.

Larson, E.D. (1993) 'Technology for electricity and fuels from biomass', *Annual Review of Energy and Environment* 18: 567–630.

Lettenmaier, D.P. and Sheer, D.P. (1991) 'Climatic sensitivity of California water resources', *Journal of Water Resources Planning and Management* 117(1): 108–25.

Lettenmaier, D.P. Brettman, K.L., Vait, L.W., Yabusaki, S.B. and Scott, M.J. (1992) 'Sensitivity of Pacific Northwest water resources to global warming', *Northwest Environmental Journal* 8: 265–83.

Linder, K.P. and Inglis, M.R. (1989) 'The potential effects of climate change on regional and national demands for electricity', in J.B. Smith and D.A. Tirpok (eds) *The Potential Effects of Global Climate Change on the United States: Appendix H – Infrastructure*, Washington, DC: US Environmental Protection Agency.

Linder, K.P., Gibbs, M. J. and Inglis, M.R. (1989) *Potential Impacts of Climate Change on Electric Utilities*, Report EN-6249, Palo Alto, CA: Electric Power Research Institute.

Lonergan, S. (1989) *Climate Change and Transportation in the Canadian Arctic, Climate Institute Symposium on the Arctic and Global Change, Ottawa.*

Lynette, R. and Associates (1992) *Assessment of Wind Power Station Performance and Reliability*, Report EPRI TR-100705, Palo Alto, CA: Electric Power Research Institute.

Matsui, S. *et al.* (1993) *Assessments of Impacts of Climate*

Change on Management of Electric Utilities (I): Development of Impacts Assessment Methods, Report Y92000 8, CRIEPI, Tokyo, Japan.

McGillivray, D.G., Agnew, T.A., McKay, G.A., Pilkington, G.R. and Hill, M.C. (1993) *Impacts of Climatic Change on the Beaufort Sea-Ice Regime: Implications for the Arctic Petroleum Industry*, CCD 93–01, Downsview: Environment Canada.

Mezerette, C. and Girard, P. (1990) 'Environmental aspects of gaseous emissions from wood carbonisation and pyrolysis processes', in R.V. Bridgwater and G. Glassi (eds) *Biomass Pyrolysis: Liquid Upgrading and Utilisation*, Amsterdam: Elsevier Applied Science, 263–88.

Miller, B.A. *et al.* (1992) *Impact of Incremental Changes in Meteorology on Thermal Compliance and Power System Operations*, Report WR28–1–680–109, Tennessee Valley Authority Engineering Laboratory, Norris.

Mundy, C. (1990) 'Energy', in *Climate Change: Impacts on New Zealand*, New Zealand Ministry of the Environment.

National Academy of Sciences (1992) *Policy Implications of Greenhouse Warming*, Washington, DC: National Academy Press.

Radesovich, L.G. and Skinrood, A.C. (1989) 'The power production operation of Solar One, the 10 MWe solar thermal central receiver pilot plant', *Journal of Solar Energy Engineering* 111(2): 145–51.

Rajabapiah, P., Jayahumar, S. and Reddy, A.K.N. (1993) 'Biogas electricity: the Pura village case study', in T.B. Johansson *et al.* (eds) *Renewable Energy Sources for Fuels and Electricity*, Washington, DC: Island Press.

Riedacker, A., Dreye, E., Joly, H. and Pafadnam, C. (1994) *Physiologie des Arbres et Arbustes en Zones Arides et Semi-Arides*, Montrouge: John Libbey.

Rosenthal, D.H., Gruenspecht, H.K. and Moran, E.A. (1995) 'Effects of global warming on energy use for space heating and cooling in the United States', *Energy Journal* 16(2): 77–96.

Sampaio, R.S. (1994) *The Production of Steel through the Use of Renewable Energy, 8th European Conference for Energy Environment, Agriculture and Industry, 3–5 October 1993, Vienna, Austria.*

Schipper, L. and Meyers, S. (1992) *Energy Efficiency and Human Activity: Past Trends and Future Prospects*, Cambridge: Cambridge University Press.

Singh, B. (1987) 'Impacts of CO_2-induced climate change on hydro-electric generation potential in the James Bay Territory of Quebec', in S.J. Solomon, M. Beran and W. Hogg (eds) *The Influence of Climate Change and Climate Variability on the Hydrologic Regime and Water Resources*, IAHS Publication No. 168, Wallingford.

Skea, J. (1995) 'Energy', in M. Parry and R. Duncan (eds) *Economic Implications of Climate Change in Britain*, London: Earthscan.

Smith, J.B. and Tirpak, D.A. (eds) (1989) *The Potential Effects of Global Climate Change on the United States*, EPA-230–05–89, Office of Policy, Planning and Evaluation, Washington, DC: US Environmental Protection Agency.

Smith, K.R., Khalil, M.A., Rasmussen, R.A., Thorneloe, S.A., Smanegdeg, F. and Apte, M. (1992) 'Greenhouse gases from biomass and fossil fuel stoves in developing countries: a Manila pilot study', *Chemosphere* 29(1–4): 479–505.

Solley, W.B., Merk, C.F. and Pierce, R.P. (1988) *Estimated Use of Water in the United States in 1985*, Circular 1004, Washington, DC: US Geological Survey.

Turnbull, J. (1993) *Strategies for Achieving a Sustainable, Clean and Cost Effective Biomass Resource*, Palo Alto, CA: Electric Power Research Institute.

UK Climate Change Impacts Review Group (1991) *The Potential Effects of Climate Change in the United Kingdom*, Department of the Environment, London: HMSO.

UK Climate Change Impacts Review Group (1996) *The Potential Effects of Climate Change in the United Kingdom*, Department of the Environment, London: HMSO.

World Energy Council/IIASA (1995) *Global Energy Perspectives to 2050 and Beyond: Report 1995*, London: World Energy Council.

Wyman, C., Bain, R., Hinman, N. and Stevens, D. (1993) 'Ethanol and methanol from cellulosic biomass', in T.B. Johansson *et al.* (eds) *Renewable Energy: Source for Fuels and Energy*, Washington, DC: Island Press, pp. 865–923.

Zweibel, K. and Barnett, A.M. (1993) 'Polycrystalline thin-film photovoltaics', in T.B. Johansson *et al.* (eds) *Renewable Energy: Source for Fuels and Energy*, Washington, DC: Island Press, pp. 437–81.

PART 4:

THE CHANGING CLIMATIC ENVIRONMENTS

These four chapters (21–24) concentrate on the changing climatic environment, due mainly to the key anthropogenic activities of urbanization and air pollution, in terms of their impact on climate change *per se* and vice versa. Extreme climatic events are also considered, in the context of hazards and disasters for society, *vis-à-vis* climate change predicted for future decades. This part concludes with a review of climate and human historical processes and a discussion of climatic adversity in the future, associated with predicted climate change.

21

URBAN CLIMATES AND GLOBAL ENVIRONMENTAL CHANGE

T.R. Oke

THE NATURE OF GLOBAL ENVIRONMENTAL CHANGE

In Part 3 of this book, the influences of climate (including both present-day elements and proposed future changes) were discussed in terms of human responses and a wide range of socio-economic activities. In this first chapter of Part 4, urban climates are considered, particularly in terms of their impacts on, and responses to, global environmental change (GEC), which is the focus of much discussion and some action. The forces of change are no mystery since they are rooted in the growth of the earth's human population, and the increased resource consumption and degradation of natural systems that follow (Simmons, 1991). The extension of such trends into the future has led to dire warnings from respected commentators (Anthes, 1993; Hardoy *et al.*, 1992; McLaren 1993; Union of Concerned Scientists, 1993) regarding the diminished carrying capacity of the planet and the probability of human misery and environmental mutilation. Cities are increasingly the focus of these forces and their environments show many of the clearest examples of anthropogenic change.

GEC is almost synonymous with climate change and is similarly treated here. Nevertheless, we should remind ourselves that climate should be seen in the context of many other facets of global change (Table 21.1) and that urban development is connected to all of the changes. The linkage is complex, but there is no doubt that the flight to cities, especially in the Third World (Hardoy *et al.*, 1992), is driven or exacerbated by the effects of desertification, deforestation and reductions of biodiversity and soil fertility in rural areas. Furthermore, as will be discussed below, cities are intimately linked to questions of climate change. Urban atmospheres have already demonstrated the strongest evidence we have of the potential for human activities to change climate. In addition, cities are the source of most of the pollutants (Chapter 22) that are suspected to alter the state of the global atmosphere, and are likely to be significantly impacted if the global climate changes in the future. As a result, cities represent potentially important 'laboratories' in which to study climate change and could be one of the most effective agents to mitigate unwanted change.

The 'best estimate' of the most recent consensus on future global climate change suggests that compared with 1990, the following changes may happen by the year 2100: the global mean temperature may increase by about 2°C; sea level may rise by 0.5 m (Chapter 7); the mean global hydrological cycle (Chapter 6)

Table 21.1 Commonly recognized foci of global environmental change

Desertification
Deforestation
Reduction of biodiversity
Loss of soil fertility
Climate change (related to urbanization and deforestation)
Ozone depletion
Tropospheric pollution (aerosol, ozone)
'Greenhouse' warming/precipitation change and storm frequency
Sea-level change (related to climate)
Urbanization (related to or impacted by all of the above)

may be enhanced with greater precipitation and soil moisture at high latitudes in winter; droughts and floods (Chapter 23) may shift to different regions and there may be more extreme rainfall events (IPCC, 1995; 1996). Unfortunately, possible changes in storm characteristics are not known with any clarity. All of the changes will show regional variability but present models (Chapter 4) are not able to depict them reliably. On the other hand, it is known with certainty that stratospheric ozone (Chapter 22) has decreased and ultraviolet radiative input at the surface has increased, at middle and high latitudes. This is likely to continue until the turn of the century before the ozone begins a gradual recovery through the first half of the twenty-first century (IPCC, 1996).

THE NATURE OF URBAN CLIMATES

The surface and atmospheric modifications associated with the construction and operation of cities are massive. The introduction of new surface materials, the creation of the urban canopy (layer of buildings) and the emission of heat, moisture and pollutants are the most important factors involved. Together, they create a new set of aerodynamic, radiative, thermal and moisture surface boundary conditions (Table 21.2(a)) and an atmosphere which is laden with aerosols and gases. These in turn perturb the pre-urban fluxes and balances of heat, mass and momentum and lead to changes in every climatic element (Table 21.2(b)). When a new housing site is developed (trees felled, foundations dug, buildings erected, furnace or air conditioners installed, soil compacted, ground paved over and the pathways and landscaping laid), it is apparent that a mosaic of new microclimates is being created. At the scale of a person walking amongst the buildings, the spatial variation of microclimates is huge. That person can be in a sunny spot or in shade, buffeted by swirling winds or becalmed, in relatively hot/dry or cool/moist surroundings, all just by walking a few metres from the street into a building courtyard. However, within a neighbourhood of similar building types, the mix of microclimates tends to be repeated so that an ensemble climate at the local scale is created. In turn, when these are combined with the climates of the other urban land uses, an urban climate at the city scale is produced. The change in climate from that existing before the city was built (known as the 'urban effect') is often approximated by comparing measured values inside the city with those at surrounding rural sites. This surrogate estimate of urban effects is open to error (Lowry, 1977) but, in the absence of pre-urban observations, it is a reasonable compromise. Table 21.2(b) is an approximate consensus of such urban effects for a typical large city.

A built-up area is one of the roughest surfaces anywhere. Therefore, strong winds flowing from the smoother countryside are retarded and thrown into a more turbulent state by the excess drag of the city. This slowing produces two side effects. One is reduced Coriolis turning force, so the wind direction veers cycloni-

Table 21.2 Typical (a) surface and atmospheric properties, and (b) urban climate effects for a mid-latitude city with about 1 million inhabitants. Values for summer unless otherwise noted

(a) *Property*	*Change*	*Typical magnitudes*
Roughness length	Greater	Rural: 0.01–0.5 m Suburban: 0.6–1.0 m: urban; 1.5–2.5 m
Albedo	Lower	Rural: 0.12–0.20 Suburban: 0.15; urban: 0.14
Emissivity	Greater?	Rural: 0.92–0.98; urban: 0.94–0.96
Thermal admittance	Greater	Rural: 600–2000; suburban: 800–1700, Urban: 1200–2100 J m^{-2} $s^{½}$ K^{-1}
Anthropogenic heat	Greater	Rural: absent; suburban: 15–50 W m^{-2} Urban: 50–100 W m^{-2} (winter up to 250 W m^{-2})
Condensation nuclei:		
Aitken	Greater	Rural: 10^2–10^3; urban: 10^4–10^6 cm^{-3}
Cloud	Greater	Rural: 2–5 × 10^2; urban: 10^3–10^4 cm^{-3}

(b) *Variable*	*Change*	*Magnitude of change or comment*
Turbulence intensity	Greater	10–50 %
Wind speed	Decreased Increased	5–30 % at 10 m in strong flow In weak flow with heat island
Wind direction	Altered	1–10°
Tornadoes	Less	
UV radiation	Much less	25–90 %
Solar radiation	Less	1–25 %
Infrared input	Greater	5–40 %
Visibility	Reduced	
Evaporation	Less	About 50 %
Convective heat flux	Greater	About 50 %
Heat storage	Greater	About 200 %
Air temperature	Warmer	1–3°C per 100 years; 1–3°C annual mean up to 12°C hourly mean;
Humidity	Drier More moist	Summer daytime Summer night, all day winter
Cloud	More haze More cloud	In and downwind of city Especially in lee of city
Fog	More or less	Depends on aerosol and surroundings
Precipitation:		
Snow	Less	Some turns to rain
Total	More?	to the lee of rather than in the city
Thunderstorms	More	

cally over the city. It recovers direction in the downstream rural area but the trajectory is offset by a few kilometres. The other effect, due to horizontal convergence, causes the whole airstream to be uplifted over the city and then subside again when it exits. These changes are capable of disrupting mesoscale and synoptic weather fronts, cloud development and perhaps precipitation. The effect of the city on speed and direction is reversed when the large-scale flow is weak, which tends to coincide with times when the city is warmer than its environs.

The resulting relative low-pressure centre over the city accelerates winds into the city centre but opposes them in the downstream suburbs.

The polluted urban atmosphere (Chapter 22) contains anomalously high concentrations of aerosol particles and radiatively active gases (e.g. ozone, carbon dioxide). These interfere with the transmission of solar and infrared radiation (see Chapter 22, Global air pollution problems) and the degree of disruption depends on the abundance of pollutant sources and the local air pollution meteorology. In general, incoming solar radiation is depleted, especially at the shortest wavelengths such as the ultraviolet, although some of the direct beam lost in this way is recovered as an increase in diffuse input. The geometrically convoluted surface gives the city a relatively low albedo so the loss due to pollution is partially compensated by better absorption. Similarly, fortuitous offsetting occurs in the long-wave exchange since much of the solar energy absorbed by the polluted air is reradiated to the surface, which boosts the incoming infrared radiation. However, since the surface temperature of the city is usually higher than the countryside, the emitted long-wave radiation is also greater. Individually these alterations can be quite large but, when taken together as the net all-wave radiation at the surface, urban–rural differences are relatively small (Oke, 1988). What is less well understood is the net radiation budget of the urban atmosphere and whether on balance it contributes to net warming or cooling.

The surface energy balance of the city differs from that of its surrounding countryside in two main ways. First, because urban development removes much of the plant cover and seals much of the surface, and routes water rapidly away, evaporation is usually decreased. This means that absorbed heat cannot be 'hidden' as latent heat so it remains as sensible heat, which raises the temperatures. Second, whilst most of this sensible heat convects and warms the atmosphere, a significant portion is conducted into the building fabric. The capacity of the materials to store the daytime heat input is large and this storage, plus the extra heat released from the combustion of fuels, creates a huge heat reserve. At night, within the deep street canyons which screen off much of the cold sky and restrict air flow, the heat store can only be released slowly. The contrast between this cooling environment, and that of the usually more open colder rural areas, is what creates the nocturnal warmth of cities that has been dubbed the 'heat island' (Oke, 1982; 1995).

The heat island effect is pervasive; it is found in the ground, at the surface, in the near-surface air and in reduced form extends well above the roof level of all cities. It fluctuates in magnitude with the prevailing weather and the time of day and season. In the air, it is usually relatively small or slightly negative by day, growing rapidly after sunset to a maximum around midnight or later in the night. The net effect is to reduce the daily range of temperature in the city and this is mainly due to an increase in the daily minimum. Over a year, these fluctuations are averaged so urban–rural differences are relatively small. Nevertheless, based on observations over the past century, both the magnitude and the rate of the warming is comparable with that considered possible at the global scale by 2100 (Table 21.2(b), Figure 21.1 and the first section of this chapter). On 'ideal' calm and cloudless nights, the difference between the city centre and the countryside may be greater than 10°C (Table 21.2(b)), depending on the size of the settlement. The slope of the relation between city size and heat island magnitude differs in different parts of the world (Figure 21.2). This is partly due to differences in the nature of the cities (i.e. street geometry, greenspace and heat release) but also to differences in the cooling potential of their 'rural' surroundings which can vary from deserts and snow fields to paddy fields and rain forest (Oke *et al.*, 1991).

Lower evaporation and greater turbulent

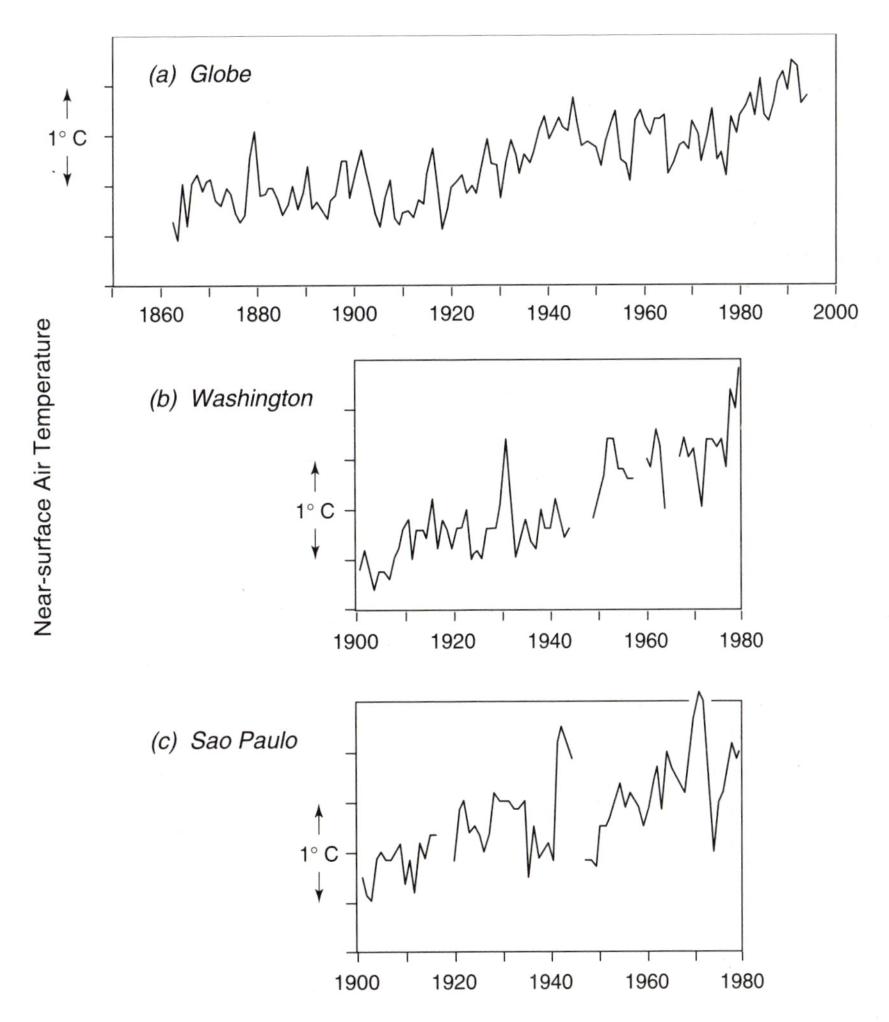

Figure 21.1 Long-term records of average annual near-surface air temperature: (a) for the globe, and station differences between an urban and a nearby less developed site near (b) Washington, DC (Jones *et al.*, 1986a) and (c) Sao Paulo (Jones *et al., 1986b)*

mixing is thought to explain the observation that cities have lower atmospheric moisture than the countryside on summer days. The more surprising finding that cities are more humid may be related to the release of water vapour from combustion and the possibility that cities have less dewfall. However, more research is needed on this topic, along with studies of urban-fog effects. Without doubt, the question of urban effects on precipitation remain the most perplexing. Many of the requirements for precipitation are anomalously present over cities, including enhanced uplift due to wind convergence and heat island instability, and abundant condensation nuclei due to pollution. However, whilst there is considerable evidence to support the idea that a large city modifies cloud and precipitation in its vicinity, especially the massive METROMEX Project in St Louis (Changnon, 1981),

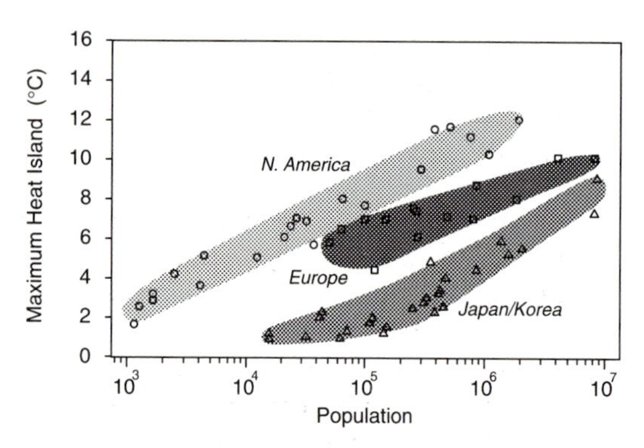

Figure 21.2 The largest heat island magnitudes observed in temperate climate cities under 'ideal' conditions in relation to their size (as given by their population). Based on results in Oke (1973) and Park (1987). Comparable results for tropical cities cover the same range with hot, dry values as large as the North American, and hot, wet values as low as the Asian data (Oke, unpublished)

there remains much to be done to develop a full understanding of the physical mechanisms (Cotton and Pielke, 1995). This brief summary of urban effects on climate should be sufficient to confirm that city dwellers live in climatic environments which have been greatly modified by human activities. For a more complete treatment of this topic, reference should be made to Cleugh (1996), Landsberg (1981) and Oke (1987).

IMPACTS OF URBAN CLIMATES ON GEC

The Cities' Contributions to GEC

The analysis of this relationship has to be considered in two parts. First, surface-generated urban climate (UC) effects in Table 21.2, such as the increase in temperature, can simply be integrated over all the cities on the earth, but the global impact is insignificant. Urban development covers less than 1 per cent of the world's land area (Heilig, 1994) or about 0.25 per cent of the earth's total surface (Sagan *et al.*, 1979). Also, since urban effects on thermal, moisture and kinematic properties only extend downwind for, at most, tens of kilometres, dilution renders the impact on global climate negligible. Cloud effects, however, may be more persistent (Auer, 1981).

On the other hand, cities are major contributors of greenhouse gases (GHGs) such as carbon dioxide (CO_2), methane, ozone and chlorofluorocarbons (CFCs) to the atmosphere (see Chapter 22, Greenhouse warming). Estimates are not available for GHG emissions which identify the urban contribution separately, but it seems reasonable to suggest that at least 85 per cent of the global total of anthropogenic CO_2, CFCs and tropospheric ozone originates in cities. Furthermore, there is ample evidence to show that the urban plumes of these substances extend for hundreds and, in the case of megalopolitan plumes, even thousands of kilometres downstream (Figure 22.5). Alternatively, Charlson *et al.* (1991) and Mitchell *et al.* (1995) showed that the greatest cooling effects of aerosols are located in the vicinity of the urban–industrial regions of North America, Europe and China. It is therefore reasonable to state that cities are the primary source of many of the radiatively active GHGs, especially the aerosols that have both direct and indirect cloud radiative effects (see Chapter 22, Global air pollution problems), which cause or ameliorate changes in the earth's climate. Finally it should be noted that cities are the nerve centres of global society. They drive the quest for economic growth and, as such, are at the core of the consumption ethic. Hence, they are central contributors to the syndrome which leads to GEC.

Cities and the 'Contamination' of the GEC Temperature Signal

This is less of a concern about the impact of UC on GEC, but is more one of whether the UC effects (such as the heat island or cloud enhancement) are making it difficult to detect the influence of human activities on GEC. The simple response is a positive one since the surroundings of many of the long-term climatological stations, which provide the data for global datasets, have changed over time. The simple fact is that a site, deemed interesting enough to attract people and observers many years ago, is likely to have seen more and not less development since. Common sites such as airports and universities have seen intense development in the last 50 years. Armed with a basic understanding of micro- and urban climates, a visit to a local climate station is likely to raise questions as to the ability of the observations to represent a planet, only 0.25 per cent of which is urbanized, and to discriminate changes as small as tenths of a degree over periods as long as a century. Temporal trends of the global and urban temperature signals are visually similar although the rate of change in cities is usually much larger (Figure 21.1). The two are, however, responses to very different scales of climate change and an accurate indicator of global change should not include a disproportionate number of stations with urban effects.

Therefore, there have been attempts to remove urban effects without overly reducing the total number of stations. The approach is either to 'correct' the temperature record of urban-affected stations using empirical relations between city size and heat island magnitude (using paired urban–rural sites) such as those in Figure 21.2 (Karl *et al.*, 1988), or to 'cull' suspect stations from the database (e.g. both urban stations in Figure 21.1(b),(c) were excluded from Jones *et al.*, 1986b; Jones *et al.*, 1986a). There are problems with both approaches. 'Corrections', using population-based relationships and heat islands estimated from urban–rural station pairs, can be large, especially in the record for the USA (Karl *et al.*, 1988). For example, in the period 1901–84, an urban bias of between +0.1°C and +0.4°C was detected; this is larger than the suggested overall temperature trend (+0.16°C) in the USA in the same period (Karl and Jones, 1989). Problems with this approach include the following:

1 The areal basis of reporting population statistics is often at an inappropriate scale.
2 The station may not be located at a site representative of the city's size.
3 Population is a poor surrogate for what is really required, which is the physical character of the area surrounding the site.
4 It is common to fail to estimate the effects of development and topography at the rural site.

'Culling' relies upon the application of criteria, such as the population in the vicinity of the site, to accept or reject a station. Problems arise from the following:

1 The lack of detailed information about the site.
2 The limitations of population statistics (noted above).
3 The fact that the threshold to reject a station, as being affected by urban development, is often surprisingly large (e.g. 10,000 to as high as 160,000 persons, Jones *et al.*, 1990; Karl *et al.*, 1993). It seems questionable whether the methods currently used to correct and cull records are sufficiently precise enough to be used with confidence. Therefore, subtle bias probably remains owing to both urban and rural development around sites.

IMPACTS OF GEC ON URBAN CLIMATE

Except inasmuch as the base climate and boundary conditions are altered for a given city, there is little reason to expect that GEC change will impact UC effects in an easily predictable fashion. For example, contrary to what is stated in the report of the Intergovernmental Panel on Climate Change (IPCC, 1990), if the earth warms significantly in the next century, it does not necessarily follow that urban heat islands will intensify. What it would do is to raise the base temperature on which the heat island effect is superimposed, which would certainly raise the total thermal burden on the city, but not necessarily alter the urban–rural difference. In fact, it is difficult to say what, if anything, will happen to the heat island. Cities in hot climates are likely to be even warmer which may well lead to increased heat emission due to extra energy use for air conditioning. Conversely, in cities with a cold climate, a decrease in space-heating demand may release less heat than at present.

The magnitude of heat islands is modulated by the prevailing synoptic weather. If the climate of the earth changes to a warmer state, it is expected that the synoptic circulation will alter the frequency of weather types and precipitation patterns. If the shift in a region is towards more anticyclonic conditions (perhaps with less ventilation, less cloud and less rainfall), the heat island may be enhanced. However, with increased cyclonic weather, the heat island is likely to be dampened. Given this uncertainty about the relation between urban effects and climate change, it is more fruitful to determine that if global climate changes (in line with model (Chapter 4) and other predictions) will the impacts on cities be important? Also, if this is so, will it render them relative beneficiaries ('winners') or victims ('losers') in terms of the economic, health and other costs incurred by the city? Furthermore, what does the future growth of cities mean for their climatic environments?

Global Increases of UVB Radiation

If stratospheric ozone concentrations continue to decrease (Chapter 22), and ultraviolet (UVB) radiation transmittance through the upper atmosphere increases, will the health of plants and people in cities be damaged? It seems unlikely that this is a serious threat in the immediate future because the urban atmosphere is noted for the preponderance of ozone, aerosols and cloud which all lead to the attenuation of UV radiation intensity. Urbach (1994) noted that increases in tropospheric ozone and aerosol in industrialized regions may have decreased UVB radiation sufficiently to have offset the increases estimated to have occurred owing to stratospheric ozone depletion. So, if not exactly 'winners', at least cities are buffered from the potential negative impacts which are likely to affect biological systems in clearer atmospheres (see Chapter 11, Conclusions).

The Consequences of Greenhouse Warming

There are several aspects of cities that are temperature dependent and which have some feedback upon the health of the inhabitants and the costs of running a city (Table 21.3). Biological activity is usually stimulated by an increase in temperature and this is mostly positive in cold climates where growing seasons are lengthened, survival of some species may be improved and migration patterns are altered (Chapter 11). On the negative side, extra warmth may increase the range of infectious-disease-carrying vectors such as the mosquito, water snail, tsetse fly and black fly (Stone, 1995).

In relation to human comfort and stress (Chapter 12), the distinction is clear. Inhabitants of cities in cold climates are likely to benefit from a reduction of wind-chill (see Chapter 12, The human energy balance and Figure 11.1) and cold stress but those in already hot climate cities, and especially those irregularly exposed to high temperatures, will bear an increasing heat burden (Figure 21.3). Thus, the IPCC (1996) consensus that global warming will lead to more extremely hot days and less extremely cold days is exactly the type of

Table 21.3 Impacts of urban heat islands, and/or a warmer base climate, on temperature-sensitive aspects of cities in cold and hot climate regions

	Climate impact Cold	Hot
Biological activity (plant growth, disease)	+	?
Human bioclimate (comfort, wind-chill, heat stress)	+ (W) – (S)	–
Energy use (space heating, air conditioning)	+ (W) – (S)	–
Water use (garden irrigation)	–	–
Ice and snow (transport disruptions)	+	n.a.
Air pollution chemistry (weathering, photochemistry)	–	–

+, beneficial ('winners'); –, undesirable ('losers'). (W), winter; (S), summer; n.a., not applicable.

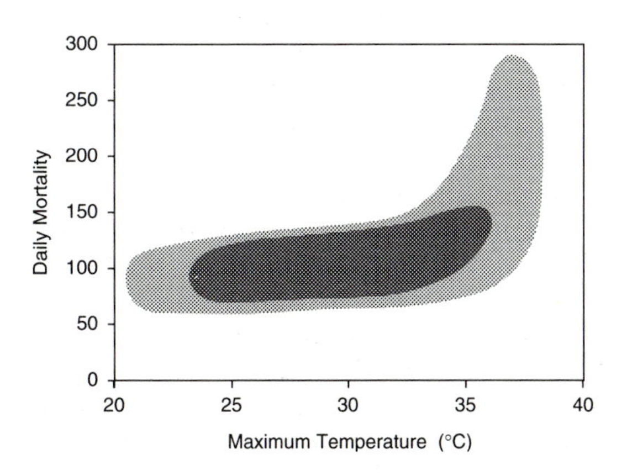

Figure 21.3 Relation between maximum air temperature and mortality in Shanghai from 1980 to 1989. Above a threshold of about 34°C, heat deaths increase sharply. Generalized from Kalkstein and Smoyer (1993). Shading proportional to density of data points

pattern considered most hazardous to humans. Kalkstein and Smoyer (1993) have projected mortality statistics for several cities assuming future warming, as based on projections generated by a global climate model (Chapter 4). After allowing for possible acclimatization, the total summer mortality in Montreal increased from 69 to 218 in response to projected monthly temperature increases of 2.6 to 3.3°C. However, in New York, the number of mortalities increased from 320 to 880 with increases of 2.6 to 3.5°C, in Cairo from 281 to 1125 for 2.1 to 4.0°C and in Shanghai from 418 to 3587 with a rise of 4.2 to 4.7°C. Mortality arising from heatwaves (Chapter 12) often is not perceived to be a major weather hazard, so it often comes as a surprise to learn that in the USA, deaths due to heatwaves are greater than those arising from violent events such as hurricanes (Chapter 12). Auliciems (1994) constructed maps of thermoregulatory 'winners' and 'losers' based on climate warming scenarios projected by the output from a global climate model and what this might mean for human comfort relative to recognized thresholds (Figure 21.4). The clear result is that the populations likely to be worst affected are those in tropical climates most of which belong to the Third World. This has heightened significance when it is appreciated, first, that these are the same areas where urban growth is anticipated to be greatest in the next century, and second, that these estimates do not incorporate the extra burden of urban heat islands.

The relation between energy use and temperature is well defined. If the large-scale climate becomes warmer, cities in cold climates can anticipate fuel savings for space heating. Conversely, cities in hot climates will face extra costs for air conditioning, at least in those places where affluence permits this form of space cooling. For example, in Los Angeles, given typical electricity rates and cooling demand, each extra degree Celsius of extra warmth costs the city about US$20million

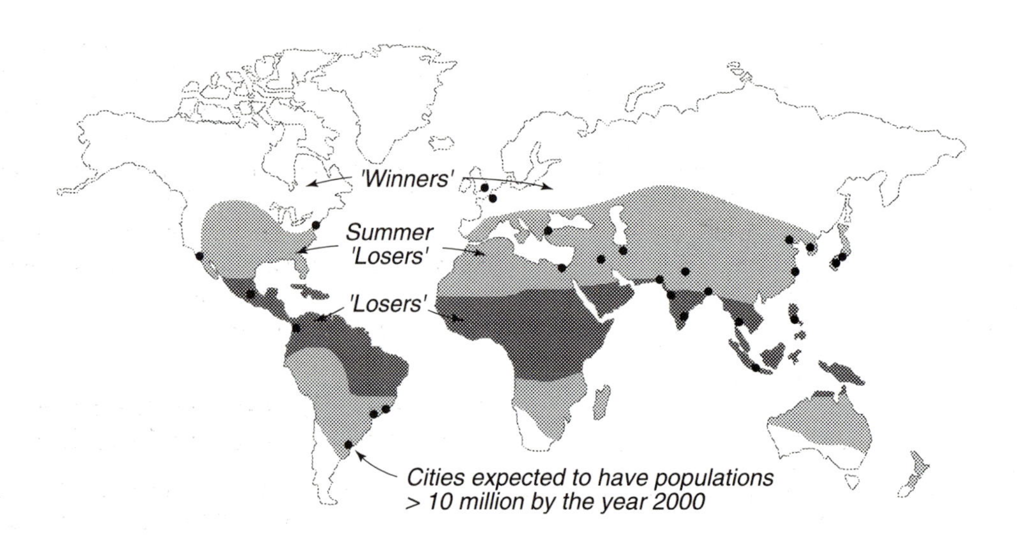

Figure 21.4 Areas projected to become 'winners' and 'losers' in terms of human comfort, if global climate warms in accord with model predictions. 'Winners' benefit and 'losers' suffer in all seasons; in the intermediate category problems will increase in the summer. Adapted from Auliciems (1994). Dots are the locations of cities expected to have populations above 10 million by the year 2000

per year (Figure 21.5). Air-conditioning use sets off significant feedbacks. First, cooling the interiors of buildings results in the expulsion of hot air; this contributes to an extra heat island load which is likely to lead to greater demand for more air conditioning. Second, depending on the source of power for air conditioning, extra fuel use for cooling is likely to increase GHG emissions, which in turn may contribute to global warming. Outdoor water use for garden sprinkling, street and courtyard washing, and some indoor uses including water-based space cooling, are positively correlated with temperature. This places further demand on a scarce and increasingly costly resource (Chapter 6). On the other hand, if temperatures increase in cold-climate cities, costs associated with snow clearance and the disruption of transport are likely to be lessened, because a greater proportion of the precipitation is likely to fall as rain.

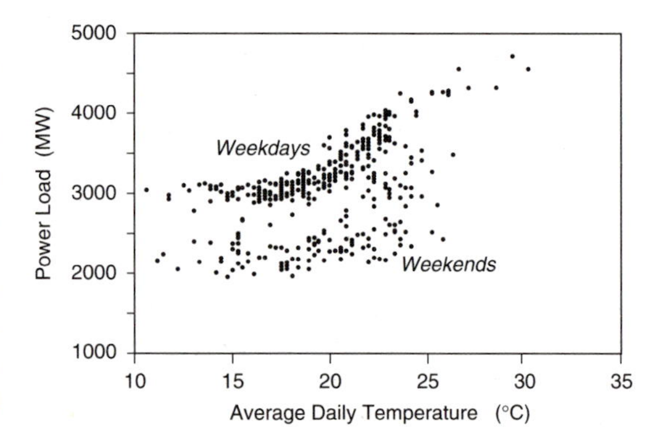

Figure 21.5 The relation between electricity power load at 16.00 hours and average daily temperature in Los Angeles for every day in 1986. The upper, more dense, population of points is for weekdays and the lower set for weekends. After Akbari *et al.* (1989)

Atmospheric chemistry (Chapter 22) is strongly related to temperature and this is especially true for ozone concentrations (Figure 21.6). Both the emission of some of the necessary precursor gases and the subsequent photochemical reaction rates are stimulated by increased temperature. Hydrocarbons (see

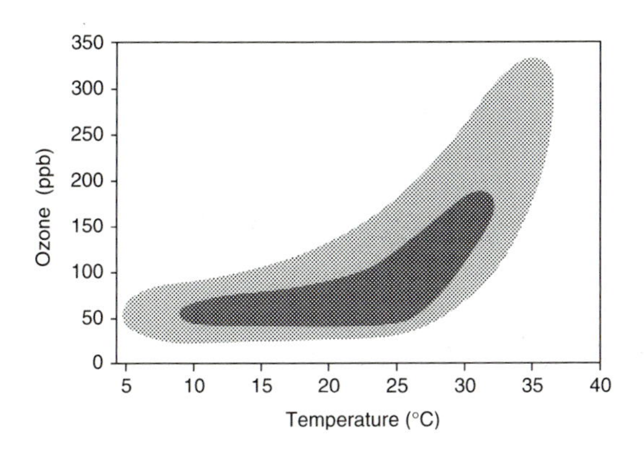

Figure 21.6 Relation between daily maximum air temperature and daily maximum ozone concentration in Connecticut. Shading proportional to density of data points. After Wackter and Bayly (1988)

Chapter 22, Primary pollutants) are important to the production of ozone and these come from the activities of both humans and plants. The primary source of human-related hydrocarbons is automobile use, and hot weather increases both the inefficiency of the engine and the evaporative loss of gasoline during refuelling and from the fuel tank when the automobile is parked in a hot place. Shrubs and trees naturally release a wide range of hydrocarbons such as isoprene and pinene at rates which increases with leaf temperature. When mixed with nitric oxide and nitrogen dioxide, which are always produced by the combustion of fuel and acted upon by ultraviolet solar radiation, ozone and several other deleterious oxidant pollutants are formed (see Chapter 22, Secondary pollutants). These reaction rates are themselves accelerated by increased temperature. Although the exact effects are uncertain, it is thought that increased urban and global temperatures will worsen urban air quality (National Research Council, 1992). The combination of air pollution and heat stress may synergistically create very unhealthy conditions in some cities. One of the relatively few negatives for very high-latitude settlements in a warmer world is the possible destabilization of structures built on permafrost. In summary, as Table 21.3 emphasizes, an increase in temperature is a mixed blessing for cities in cold climates but, except for those with hot summers, it is mostly beneficial. The outlook for the burgeoning populations of cities in hot regions of the Third World is almost wholly negative because it is likely to lead to more stressful, unhealthy and costly conditions than at present.

Associated Changes in Sea Level, Storm Activity and Precipitation

Global sea level has risen by between 0.10 and 0.25 m over the past 100 years and the 'best' scientific consensus is that it will rise a further 0.5 m by 2100, if present model estimates of GHG warming are on target (IPCC, 1996). This is of very real concern to many coastal cities. At present, about 46 million people are at risk of flooding and a rise of 0.5 m, without adaptation measures, would double the number at risk. Incorporating expected population growth further raises the risk substantially (IPCC, 1996). The potential for the inundation of settlements, sewer and drainage disruption, salt intrusion and damage to infrastructure is large. This is especially true in delta lands such as Bangladesh and northern Egypt, where population densities are high and are concentrated on small islands (such as coral atolls) and where the possibility of relocation to higher land is limited.

Current understanding of the nature, or the probability, of future changes in storm activity is insufficiently developed to be of use in planning (IPCC, 1996). This is extremely unfortunate since the threat to coastal communities, open to storm surges associated with tropical cyclones, is already severe and sometimes catastrophic (e.g. 300,000 lives were lost in a single storm in Bangladesh in 1970). If GEC increases the severity or frequency of tropical cyclones, and at the same time sea-level rises and coastal cities expand, the potential for escalating devastation, misery and environmental refugees is obvious. The most vulnerable sites are located in the Third World which lacks the resources to cope. However, any city open to storm damage could witness deteriorating safety and increasing insurance risk.

The recent IPCC view is that global warming would result in a 'more vigorous hydrological cycle' (Chapter 6) with the severity of droughts and floods increasing and decreasing in different places (IPCC, 1996). Several models indicate an increase in more extreme precipitation events (Chapter 23), which provides little basis for the assessment of urban impacts. This is serious because it takes a long time to plan, finance and construct facilities such as dams, sewers and irrigation channels which are required to accommodate increasing floods or drought. In summary, again Third World cities in regions with hot climates are most at risk in a warmer world.

The Future Growth of Cities and Their Climatic Regimes

The trend towards the concentration of people in cities continues. However, whereas it is approaching a steady state in more highly developed areas, the growth of cities in the Third World is booming and will continue to do so in the next century. Urban regions in the Third World will grow by more than 700 million people between 1990 and 2000 but elsewhere, the total increase will only be about one-tenth of that increase (Hardoy *et al.*, 1992).

In the immediate future, unless major urban-design changes are instituted, assumed growth will follow similar patterns to that of today. Even if we neglect the impacts of GEC, Third World cities face a bleak environmental future since increased pollution emissions will decrease air quality further, and increase both haze and respiratory diseases. Pressure on land will increase the density of development in the city *per se* whilst peripheral lands will be settled by squatters in an unorganized way, leading to the removal of more of the natural vegetative and soil cover and its replacement with built-up uses. This will exacerbate blowing dust and vulnerability to floods and landslips after strong rain. At street level, ventilation will be decreased and temperatures will rise, causing greater discomfort and probably higher mortality. In more highly developed parts of the world, regulation of space (and perhaps fuels and the less hectic pace of population growth) will probably see an emphasis on increasing urban densities rather than further sprawl, and attention to climate-ameliorating practices such as fuel conservation and urban forestry.

CONCLUSIONS

The debate about the role of human activities in the global climate, and how we should react to future changes, is hampered by the inability to simulate future atmospheres. Therefore, we should not overlook the fact that cities possess several characteristics which render them interesting analogues or laboratories in which to study questions relevant both to the mechanisms and the impacts of GEC change (Glantz, 1990).

Concentrations of GHGs are especially high in cities; for example, Reid (1995) measured CO_2 concentrations in Vancouver that are more than twice the present global mean value. Similarly, aerosol loadings are often more than an order of magnitude higher than those in background air (Table 21.2(a)). Thus urban atmospheres act as potential analogues of

CONCLUSIONS *continued*

future global radiation environments and there are also similarities between urban climates and projected GEC thermal climates. City dwellers, plants and socio-economic activities, therefore, have already had to cope with thermal stresses equivalent to those we may face in the future. These measurable and ongoing changes might be used as analogues of the way systems can or cannot accommodate GEC changes and they could also be used to assess the effectiveness of mitigation strategies. However, as Changnon (1992) noted, the apparent ability of present urban dwellers to cope with urban climate changes may not apply in the event that GEC threatens food, water and energy supplies. Again, cities in the Third World are especially poorly placed to cope.

Working to mitigate future GEC can be undertaken by most cities where the political will is present. Improvement of local air quality and improved efficiency of energy and water use in cities almost automatically contribute to the mitigation of global concerns. This is no more than doing what we should already be doing to clean up locally – often referred to as a 'no regrets' approach. Examples of measures which would serve this purpose include a number of wide-ranging schemes, namely: reducing emissions; conserving energy and water; switching to fuels with less GHG side effects; greater use of passive heating and cooling technology; more compact city design and greater use of mass transportation; the intelligent use of trees to shelter or shade; and increased use of light-coloured surfaces in hot cities (EPA, 1992). These are measures that can be, and have been, instituted by proactive cities. Given the foregoing discussion, it seems curious that cities are almost absent from GEC research and in the IPCC exercises, since global and regional-scale concerns seem to dominate these activities and the resultant debate. This may be due to the tendency of meteorologists to focus on the products of global climate models (Chapter 4), which are limited to large grid averages which do not mesh with the scale of cities.

This apparent myopia must be rectified. Cities are not only the major source of the problem, they are partial analogues of the processes and consequences of change, and they are a potent force for improvement that should be harnessed. Cities exist at a scale where cohesive action on environmental matters is possible. Often, higher orders of government wallow aimlessly with platitudinous edicts that lack the means for grassroots action or enforcement. On the other hand, a city council can be motivated to set standards of stewardship, curtail unnecessary consumption and waste, enforce the attainment of standards and targets, and form partnerships with other cities, industries and governments. For example, in Canada, the cities of Edmonton, Montreal, Toronto, Vancouver and Victoria have made commitments to reduce GHG emissions by 20 per cent (compared with 1990 levels) by 2005. However, before we get carried away, it has to be said that the larger picture is not so rosy since, whereas in western Europe many cities and nations are seriously addressing such questions, most local governments in North America seem indifferent to questions relating to their role in GEC (Gilbert, 1991). In this vein, Changnon (1992) even remarked

CONCLUSIONS *continued*

that, unlike some European cities, little has ever been done to improve urban climates in the USA. The recent emergence of interest in urban forestry as a means of cooling cities, especially in the southwestern 'sunbelt', may be a start. It almost goes without saying that in most Third World cities, GEC is hardly on the agenda. Greater attention must be paid to fostering local initiatives at the city scale (see Chapter 13, Conclusions), otherwise most of the world's people are destined to live in hot, dirty, dysfunctional and unsustainable settlements, for whom any GHG warming will be a mere extra irritant.

REFERENCES

Akbari, H., Rosenfeld, R. and Taha, H. (1989) 'Recent developments in heat island studies: technical and policy', in K. Garbesi, H. Akbari and P. Martien (eds) *Controlling Summer Heat Islands*, Berkeley, CA: Lawrence Berkeley Laboratory, pp. 14–30.

Anthes, R.A. (1993) 'The global trajectory', *Bulletin of the American Meteorological Society* 74: 1121–30.

Auer, A.H. (1981) 'Urban boundary layer', *Meteorological Monographs* 18(40): 41–62.

Auliciems, A. (1994) 'Thermoregulatory adaptation to global warming – winners and losers', *Proceedings of the 13th International Congress of Biometeorology*, Part 2, Vol. 1, pp. 109–22.

Changnon, S.A., Jr (ed.) (1981) 'METROMEX: a review and summary', *Meteorological Monographs* 18(40), American Meteorological Society.

Changnon, S.A., Jr. (1992) 'Inadvertent weather modification in urban areas: lessons for global climate change', *Bulletin of the American Meteorological Society* 73: 619–27.

Charlson, R.J., Langner, J., Rodhe, H., Leovy, C. and Warren, S. (1991) 'Perturbation of the Northern Hemisphere radiative balance by backscatter from anthropogenic aerosols', Tellus 43: 152.

Cleugh, H.A. (1996) 'Urban climates', in A. Henderson-Sellers (ed.) *World Survey of Climatology, Vol. 16: Future Climates of the World*, Amsterdam: Elsevier.

Cotton, W.R. and Pielke, R.A. (1995) *Human Impacts on Weather and Climate*, Cambridge: Cambridge University Press.

EPA (1992) *Cooling Our Communities*, Washington, DC: US Environmental Protection Agency.

Gilbert, R. (1991) 'Activities related to the prevention of climatic change in some major urban areas of North America', *Working Paper Series* WP-3, Toronto: Canadian Urban Institute.

Glantz, M.H. (1990) *The Use of Analogies in Assessing Physical and Societal Responses to Global Warming*, Boulder, CO: National Center for Atmospheric Research.

Hardoy, J.E., Mitlin, D. and Satterthwaite, D. (1992) *Environmental Problems in Third World Cities*, London: Earthscan.

Heilig, G.K. (1994) 'Neglected dimensions of global land-use change: reflections and data', *Population and Development Review* 20(4): 831–59.

IPCC (1990) *Climate Change: The IPCC Impacts Assessment*, ed. W.G. McTegart, G.W. Sheldon and D.C. Griffiths, Canberra: Australian Government Publishing Service.

IPCC (1995) *Climate Change 1994*, ed. J.T. Houghton, L.G. Meira Filho, J. Bruce, Hoesung Lee, B.A. Callander, E. Haites, N.Harris and K. Maskell, Cambridge: Cambridge University Press.

IPCC (1996) *Climate Change 1995: The IPCC Synthesis*, Geneva: Secretariat, World Meteorological Organization.

Jones, P.D., Groisman, P.Y., Coughlan, M., Plummer, N., Wang, W.-C. and Karl, T.R. (1990) 'Assessment of urbanization affects in time series of surface air temperature over land', *Nature* 347: 169–72.

Jones, P.D., Raper, S.C.B., Bradley, R.S., Diaz, H.F., Kelly, P.M. and Wigley, T.M.L. (1986a) 'Northern Hemisphere surface air temperature variations: 1851–1984', *Journal of Climate and Applied Meteorology* 25: 161–79.

Jones, P.D., Raper, S.C.B. and Wigley, T.M.L. (1986b) 'Southern Hemispherec surface air temperature variations: 1851–1984', *Journal of Climate and Applied Meteorology* 25: 1213–30.

Kalkstein, L.S. and Smoyer, K.E. (1993) 'The impact of climate change on human health: some international implications', *Experientia* 49: 969–79.

Karl, T.R. and Jones, P.D. (1989) 'Urban bias in area-averaged surface air temperature trends', *Bulletin of the American Meteorological Society* 70: 265–70.

Karl, T.R., Diaz, H.F. and Kukla, G. (1988) 'Urbanization: its detection and effect in the United States climate record', *Journal of Climate* 1: 1099–123.

Karl, T.R., Jones, P.D., Knight, R.W., Kukla, G., Plummer, N., Razuvayev, V., Gallo, K.P., Lindseay, J., Charlson, R.J. and Peterson, T.C. (1993) 'Asymmetric trends of daily maximum and minimum temperature', *Bulletin of the American Meteorological Society 74: 1007–23.*

Landsberg, H.E. (1981) *The Urban Climate*, New York: Academic Press.

Lowry, W.P. (1977) 'Empirical estimation of urban effects

on climate: a problem analysis', *Journal of Applied Meteorology* 16: 129–35.

McLaren, D.J. (1993) 'Population and the Utopian myth', *Ecodecision* June: 59–63.

Mitchell, J.F.B., Johns, T.C., Gregory, J.M. and Tett, S.F.B (1995) 'Climate response to increasing levels of greenhouse gases and sulphate aerosols', *Nature* 376: 501–4.

National Research Council (1992) *Rethinking the Ozone Problem in Urban and Regional Air Pollution*, Washington, DC: National Academy Press.

Oke, T.R. (1973) 'City size and the urban heat island', *Atmospheric Environment* 7: 769–79.

Oke, T.R. (1982) 'The energetic basis of the urban heat island', *Quarterly Journal of the Royal Meteorological Society* 108: 1–24.

Oke, T.R. (1987) *Boundary Layer Climates*, London: Routledge.

Oke, T.R. (1988) 'The urban energy balance', *Progress in Physical Geography* 12: 471–508.

Oke, T.R. (1995) 'The heat island of the urban boundary layer: characteristics, causes and effects', in J.E. Cermak, A.G. Davenport, E.J. Plate and D.X. Viegas (eds), *Wind Climate in Cities*, Dordrecht: Kluwer Academic, pp. 81–107.

Oke, T.R., Johnson, G.T., Steyn, D.G. and Watson, I.D. (1991) 'Simulation of nocturnal surface urban heat islands under "ideal" conditions: Part 2. Diagnosis of causation', *Boundary-Layer Meteorology* 56: 339–58.

Park, H.-S. (1987) 'Variation in the urban heat island intensity affected by geographical environments', *Environmental Research Centre Papers* No. 11, University of Tsukuba, Ibaraki.

Reid, K. H. (1995) 'Observation and simulation of atmospheric carbon dioxide in Vancouver', unpublished MSc thesis, University of British Columbia.

Sagan, C., Toon, O.B. and Pollack, J.B. (1979) 'Anthropogenic albedo changes and the Earth's climate', *Science* 206: 1363–8.

Simmons, I.G. (1991) *Earth, Air and Water*, London: Edward Arnold.

Stone, R. (1995) 'If the mercury soars so may health hazards', *Science* 267: 957–8.

Union of Concerned Scientists (1993) *World Scientists' Warning to Humanity*, Union of Concerned Scientists, Cambridge, MA.

Urbach, F. (1994) 'Biologic effects of increased ultraviolet radiation: plankton, plants and people', *Proceedings of the 13th International Congress of Biometeorology*, Part 2, Vol. 1, pp. 88–98.

Wackter, D.J. and Bayly, P.V. (1988) 'The effectiveness of emission controls on reducing ozone levels in Connecticut from 1976 through 1987', in G.T. Wolff, J.L. Hanisch and K. Schere (eds) *The Scientific and Technical Issues Facing Post-1987 Ozone Control Strategies*, Pittsburgh: Air and Waste Management Association, pp. 398–415.

22

AIR POLLUTION

Howard Bridgman

INTRODUCTION: DEFINITIONS AND CATEGORIZATION OF AIR POLLUTION

Natural air pollution originates from volcanoes (Thompson, 1995), surface dust storms and other phenomena which are not affected by human activities. Conversely, anthropogenic air pollution is a result of human activity creating an additional burden on air quality. This chapter will focus on anthropogenic air pollution, with examples from global, continental and urban scales. International publicity about global warming, stratospheric ozone depletion, acid rain and urban photochemical smog has done much to heighten global awareness about anthropogenic air pollution. Solutions are very complex, often costly, and may only show results over long periods of time. Air pollution is divided into two general categories: particulate matter, including aerosols and dust, and gases. Particulate matter is most often described in micrograms per cubic metre of air ($\mu g\ m^{-3}$). Gases are most often described in parts per million parts of air by volume (ppmv), parts per hundred million (pphmv), parts per billion (ppbv) or parts per trillion (pptv). The units may be interchanged for the gases, depending on the molecular weight of the gas and the temperature of the atmosphere.

GLOBAL AIR POLLUTION PROBLEMS

Air pollution affecting the global atmosphere is limited to gases with medium to long lifespans, which are dispersed to relatively even concentrations in atmospheric circulation patterns around the world. Table 22.1 presents examples and typical concentrations and lifetimes. The listed gases have important influences on greenhouse warming, the first of the global air pollution problems described in this section. Two of the gases, chlorofluorocarbons (CFCs) and nitrous oxide (N_2O), also have major influences on the second problem considered here, stratospheric ozone depletion.

Greenhouse Warming

In 1965, measurements of concentrations of carbon dioxide (CO_2) in the background atmosphere began at Mauna Loa in Hawaii. Continuous measurements of CO_2 to date show a regular increase in concentration of about 0.5 per cent per year, as shown in Figure 22.1(a). Analysis of gases in air pockets in ice cores (see Chapter 7, The conversion of snow to glacier ice and the role of climate) has established that, from a base concentration of about 280 ppm, CO_2 began increasing with the Industrial Revolution and the growth in the use of fossil fuels. The measurements at Mauna Loa have been verified at four other background air quality stations on the globe: Point Barrow, Alaska;

Table 22.1 Examples of gases having important influences on global air pollution problems

Item	*Carbon dioxide*	*Methane*	*Nitrous oxide*	*Chlorofluorocarbon**	*Hydrochlorofluorocarbon*†
Pre-1850 concentration	280 ppmv	700 ppbv	275 ppbv	0	0
1992 concentration	355 ppmv	1714 ppbv	311 ppbv	503 pptv	105 pptv
1980s rate of change	1.5 ppmv per year 0.4% per year	13 ppbv per year 0.8% per year	0.8 ppbv per year 0.25% per year	20 pptv per year 4% per year	8 pptv per year 7% per year
Residence time (years)	50–200‡	12–17	120	102	13.3

* Represents at least five CFCs, including the most prevalent, CFC-11 and CFC-12.
† A chlorofluorocarbon substitute.
‡ Unclear, due to incomplete knowledge of the carbon cycle.
Source: IPCC (1994) as presented in CCS (1995).

Western Samoa; Cape Grim, Tasmania; and the South Pole. Increases have also occurred in other greenhouse gases, such as methane (CH_4) (Figure 22.1(b)), chlorofluorocarbons (CFCs) and nitrous oxide (N_2O). Absorption by gases in the atmosphere acts to create a natural heat trap through capture of infrared (IR) energy emitted from the earth's surface (termed radiative forcing), depending on the wavelength of that energy, as shown in Figure 22.2. This keeps the overall global temperature at a 15°C average and without the atmosphere, temperatures would average −18°C. While the atmosphere does not operate the same way as a glass greenhouse (there is no control of convective heat), 'greenhouse warming' is now accepted as the descriptive title for anthropogenic impacts on IR absorption. With the exception of the 8–11.5 μm wavelength band, a large number of gases, dominated by CO_2 and water vapour, absorb most IR wavelengths almost completely. The addition of anthropogenic gases, such as those listed in Table 22.1, increase natural warming processes in two ways. First, there is increased absorption on the edges of wavelengths already 100 per cent absorbed. This process is relatively inefficient. Far more efficient is absorption by gases such as CFCs and N_2O in wavelengths not saturated, such as in the 8.5–11 μm band (Table 22.2). On a global scale, CO_2 contributes about 55 per cent to the anthropogenic component of global warming with other contributions from CFCs (24 per cent), CH_4 (15 per cent) and N_2O (6 per cent) (IPCC, 1994). Anthropogenic emissions add about 5 per cent to the overall absorption activities of the troposphere, an amount considered by many scientists to have a significant influence on global temperatures.

The results from computer-based global circulation models (GCMs) (Chapter 4) suggest that the additional greenhouse gases from anthropogenic emissions are likely to increase global average atmospheric temperatures between 1.5 and 4.5°C by the year 2030, with the largest changes in the polar regions. Figure 22.1(c) shows that the global temperature has increased by 0.3 to 0.5°C over the past century, but direct correlations to increasing greenhouse gas levels have not been established. The possible impacts on climate and the environment are subjects of major controversy. Through the World Meteorological Organization and the United Nations Environment Programme, the International Panel on Climate Change (IPCC) was established to provide relatively uniform international scientific opinion on greenhouse warming. The IPCC reports of

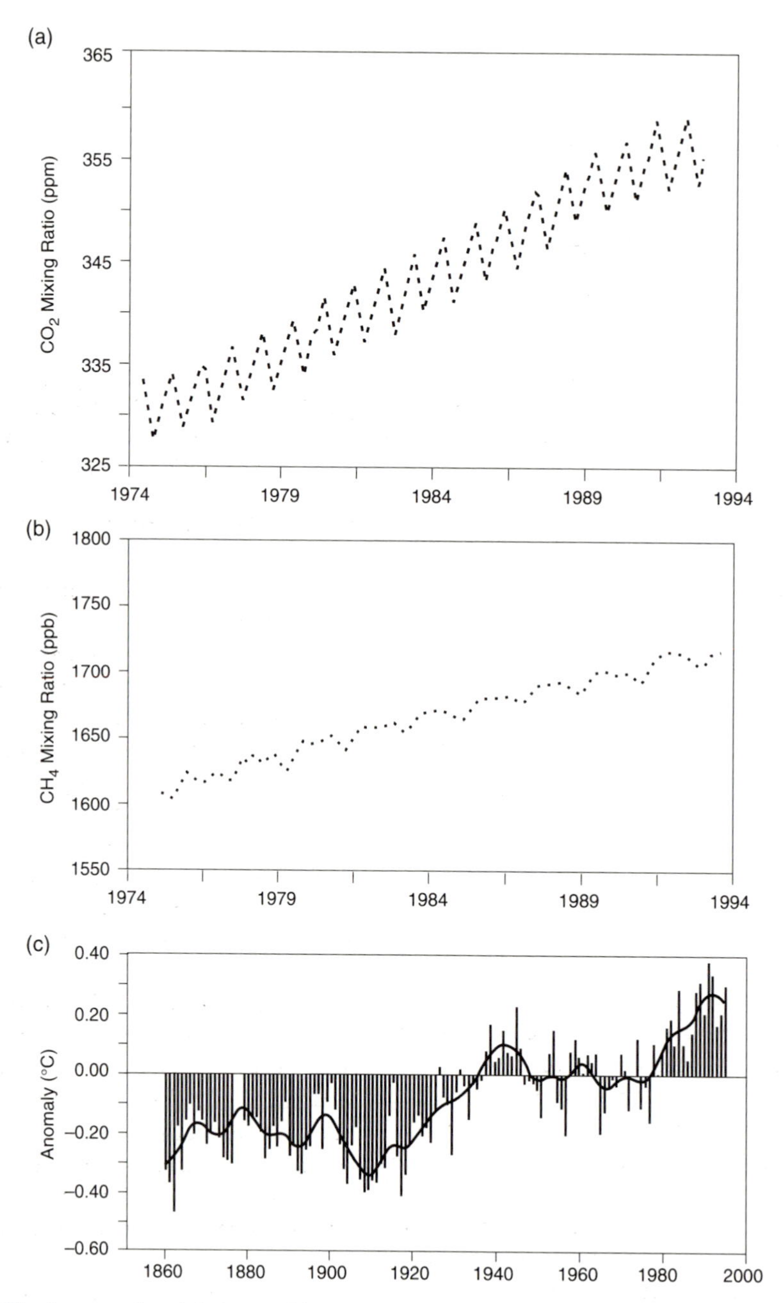

Figure 22.1 (a) The increase in global monthly mean mixing ratios of carbon dioxide (ppm) recorded at Mauna Loa from 1965 to 1993, and (b) methane (ppb) from the National Oceanographic and Atmospheric Administration Network (CDAIC, 1994). The variations are caused by seasons when vegetation is active in the northern hemisphere. (c) The increase in global temperature anomalies between 1860 and 1994 compared with the average for 1951–80, as determined from measurements analysed by the Hadley Centre, UK (Parker and Folland, 1995)

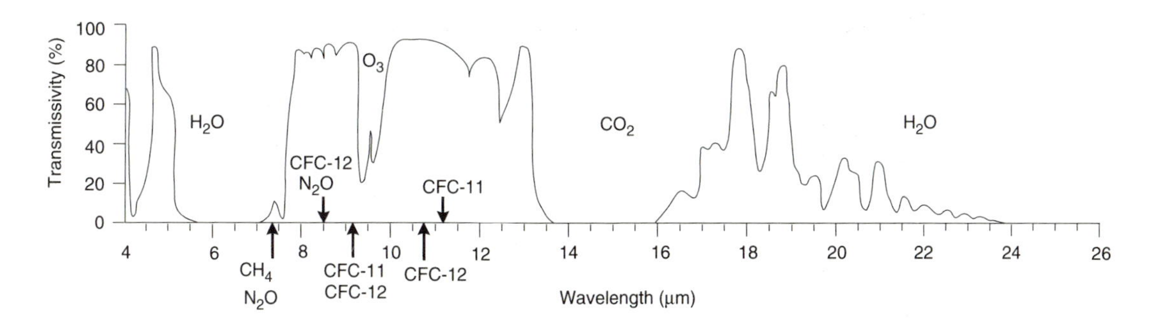

Figure 22.2 The infrared energy spectrum emitted by the earth's surface after absorption by the atmosphere, showing the main wavelengths of absorption by CO_2 and water vapour. In the 'atmospheric window', between 8.5 and 11 μm, stratospheric ozone creates some minor absorption. Anthropogenically emitted gases such as CFCs and N_2O also absorb within this band, creating extra heating of the atmosphere

Table 22.2 Radiative forcing of greenhouse gases compared with CO_2 per unit molecular change, and the reductions required to stabilize concentrations at present-day levels

Gas	*Radiative forcing*	*Reductions for stability*
CO_2	1	<60%
CH_4	21	15–20%
N_2O	206	70–80%
CFC-11	12400	70–75%
CFC-12	15800	75–85%

Source: IPCC (1990).

1990, 1992, 1994 and 1995 provide detailed background to the greenhouse warming problem and also act as a guide to assist decision-making for control measures. One area of major controversy is the accuracy of GCMs in predicting changes in climate associated with greenhouse warming (Chapter 4). Despite great developments in the size and accuracy of GCMs, they still cannot provide the detail needed to model regional changes in climate properly. Smaller, regionally oriented models are being developed that can be nested within the GCMs, to estimate possible regional changes. However, the accuracy of both the GCMs and the regional models is open to question (See Chapter 4, Climate impact models).

Feedbacks and complications

If the atmosphere warms, what feedbacks between the atmosphere, ocean and land will occur? A warmer atmosphere will allow greater evaporation from ocean and land surfaces, and perhaps through CO_2 enhancement, greater transpiration from active plants (Lockwood, 1990). The higher atmospheric moisture content, coupled with greater atmospheric instability (Table 22.3), suggests that clouds will form more often. Water vapour and clouds are very effective absorbers of long-wave radiation. While the greenhouse gases listed in Table 22.1 contribute 40 W m^{-2} (watts per square metre) to heating in the natural atmosphere, cloud and water vapour contribute 110 W m^{-2} (IPCC, 1994). However, clouds are also very effective reflectors of short-wave radiation. More cloud mean less short-wave radiation reaching the earth's surface, less short-wave absorption and surface heating, and lower long-wave radiation emissions. Therefore, there is likely to be cooling due to short-wave radiation reflection to offset, at least in part, extra warming from long-wave radiation absorption. The details of this feedback mechanism have not yet been established.

Another complication is the presence of particulate matter (aerosols) in the atmosphere.

Table 22.3 Potential environmental impacts from greenhouse warming, and estimates of regional changes adopted by IPCC (1990)

Potential environmental impacts

- Sea levels rise between 20 and 40 cm by 2030
- Increases in the frequency of extreme weather
- Shifts in the location of atmospheric circulation zones
- Increased atmospheric instability
- Changes in precipitation and soil moisture
- Changes to agricultural zones and crop growth potential
- Changes to living habitats of native flora and fauna
- Changes to surface water availability

IPCC 1990 estimates of changes in five regions of the world

Region	*Temperature*	*Rainfall*	*Soil moisture*
Central N. America winter	+2–4°C	+0–15%	?
Central N. America summer	+2–3°C	−5–10%	−15–20%
Southern Asia winter	+1–2°C	No change	No change
Southern Asia summer	+1–2°C	+5–15%	+5–10%
Sahel, Africa summer	+1–3°C	Increase	Decrease
Southern Europe winter	+2°C	Increase (?)	?
Southern Europe summer	+2–3°C	−5–15%	−15–25%
Australia winter	+2°C	?	?
Australia summer	+2–3°C	+10%?	

Over 90 per cent of anthropogenic aerosols are emitted in the 30–60° latitude band in the northern hemisphere, from the same sources that emit large amounts of greenhouse gases. Particulate matter may or may not enhance global warming, depending on the thickness and altitude of the particulate layer, the size distribution of the particulates and their chemical make-up. The most abundant anthropogenic particulate is ammonium sulphate, with an average size of <1 μm in diameter and a lifetime of several days in the atmosphere. This type of particulate tends to backscatter short-wave radiation, creating a cooling effect that can offset global warming influences, especially in the mid-northern hemisphere. However, a change in size distribution or an increase in the number of particulates emitted may enhance absorption capabilities, adding to greenhouse warming.

Potential environmental problems

The IPCC has taken a conservative approach to predicting the potential impacts on the global environment. Warming is the global scenario but, owing to changes in atmospheric circulation patterns, some regions may experience cooling. Confidence in regional estimates is low. Table 22.3 presents a list of possible impacts from global warming and the estimates of changes in temperature, precipitation and soil moisture for five regions of the world (IPCC, 1990), assuming no further control over greenhouse gas emissions. These estimates cannot be considered predictions, just likely scenarios, given the present state of knowledge about the worldwide impacts of global warming.

In 1992 the United Nations Conference on Environment and Development (UNCED) was held in Rio de Janeiro and was attended by many of the political leaders and decision-makers. Emerging from UNCED was the Rio Declaration and the Framework Convention on Climate Change, part of which was an agreement to limit greenhouse gas emissions to 1990 levels by the year 2011. As of mid-1995, 135 countries have ratified the convention. However, activation of procedures to minimize greenhouse gas emissions is slow. Details of greenhouse gas emissions from various sources are being completed in most developed countries, to establish the basis for action.

Stratospheric Ozone Depletion

The stratospheric ozone layer exists between 12 and 55 km above the earth's surface, with the highest elevations in the tropics and the lowest at the poles. Ozone (O_3) is produced and destroyed naturally through the interaction of ultraviolet (UV) sunlight with oxygen atoms and molecules and naturally produced nitrogen compounds. Most ozone is created in the tropics, and the stratospheric circulation carries the gas to higher latitudes. The poles in both hemispheres are a sink, where ozone is removed from the stratosphere. The stratospheric ozone layer is essential to life on earth because it protects living beings from harmful UVB wavelengths (280–320 nm). For humans, overexposure to UVB radiation can cause deadly melanoma skin cancer, plus other skin problems. Damage to the eyes and the exacerbation of infectious diseases may also occur. In plants, damage to the photosynthetic activity and growth processes may occur and reproductive capability altered. In water bodies, excess UVB radiation may destroy near-surface phytoplankton, essential as the base of the food chain in the ocean (UNEP, 1991).

Trends and basic chemistry

For long-term monitoring purposes, ozone is measured by a Dobson spectrophotometer which compares UV radiation measurements in a wavelength strongly absorbed by ozone with a wavelength where no absorption takes place. During the last 15 years, satellite-based instruments have also measured the spatial distributions and temporal trends in ozone. The results from these measurements show a general decrease in stratospheric ozone concentration of about 1 per cent per year since 1979 (WMO, 1995), but with considerable variation depending on latitude and altitude. Over all latitudes, important decreases have occurred at altitudes of 15–18 km and 35–55 km. However, an increasing trend in surface UVB radiation has yet to be established. The chemistry of stratospheric ozone loss is complex. The link with tropospheric CFC (Table 22.1) production was first proposed in 1976 by two atmospheric chemists from the USA, F. Sherwood Rowland and Mario Molina. They, along with Paul Crutzen from Germany, were awarded the 1995 Nobel Prize in Chemistry for their research on the formation and destruction of stratospheric ozone. Figure 22.3 provides a schematic view of what happens. CFCs, invented in the 1930s, are widely used as refrigerants, in foam material and as aerosol spray propellants and, as shown in Table 22.1, they have long lifetimes. Chlorine atoms, contained in CFCs, are transported from the troposphere to the stratosphere by convection currents. The CFCs are destroyed by UV radiation, releasing free chlorine (Cl) atoms. Highly reactive Cl interferes with the normal cycle of ozone formation and destruction, capturing an oxygen atom as ClO, and creating a stable oxygen molecule (O_2). ClO then reacts with another oxygen atom, releasing Cl to destroy another ozone molecule. This may occur 100,000 times, until the Cl atom is finally stored in one of two reservoir molecules, either hydrogen chloride (HCl) or chlorine nitrate ($ClONO_2$). Halons and bromine (Br) destroy O_3 in a similar manner.

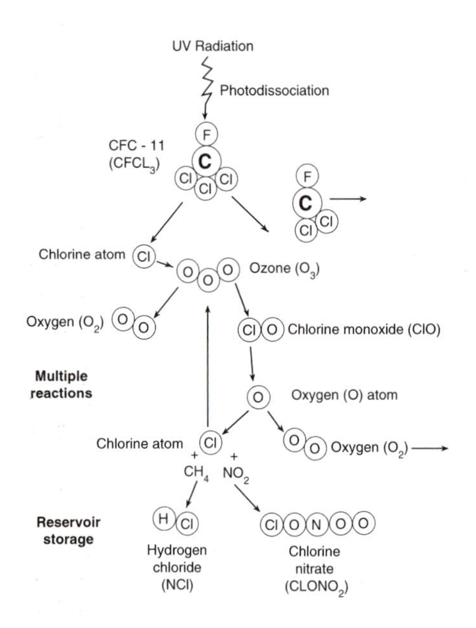

Figure 22.3 A schematic representation of the interaction of chlorine atoms with the stratospheric ozone layer, using CFC-11 as an example. Free chlorine atoms destroy the ozone molecule, capturing an oxygen atom in ClO. Further interaction with an oxygen atom frees the Cl atom to react with ozone again. This may occur many thousands of times until the Cl atom is permanently captured by hydrogen (HCl) or in chlorine nitrate ($ClONO_2$)

The polar 'holes'

A major thinning of the ozone layer over the poles occurs in late winter and early spring. Dubbed the 'ozone hole' by the international media, the seasonal loss of ozone varies by altitude and by year. There are also major differences between the Arctic and Antarctic, with the loss over the Antarctic being much more regular and dominant. Four important physical features determine the strength and variations of ozone loss over the poles in the springtime: the underlying topography; the strength of the atmospheric circulation; the stratospheric temperature; and the chemistry associated with polar stratospheric clouds. Ozone depletion over Antarctica begins with the strengthening of the circumpolar westerlies around the continent in late winter. As the world's highest continent, with an average altitude of over 2000 metres, Antarctica acts as a major topographic anchor for the strong circumpolar westerlies or vortex. In late winter and early spring, the vortex reaches jet stream speeds in the lower stratosphere. Air interchange with lower latitudes is blocked, isolating the atmosphere and its constituents over the continent, allowing a loss of ozone over most of the region inside the vortex (Plates 22.1 and 22.2).

The Arctic stratosphere has no major terrain feature to act as a circulation anchor. The circumpolar westerlies in early spring tend to wander latitudinally to a much greater extent than over Antarctica. Regular interchanges of air with lower latitudes occur, bringing fresh ozone into the Arctic stratosphere. Ozone loss occurs in pockets (Plate 22.3) that develop for short periods and may migrate, but there is no universal 'hole' similar to that existing over the Antarctic. The chemistry of ozone destruction is the same for both areas. As temperatures in the stratosphere fall towards −90°C, very thin polar stratospheric clouds (PSCs) form, made mainly of water or nitric acid ice. PSCs form an ideal surface on which the reservoir species (HCl and $ClONO_2$) can react, releasing free Cl atoms once more into the stratosphere. Aircraft measurements of ozone and ClO over both poles clearly establish the inverse relationship between the two molecules (Figure 22.4(a) and (b)), showing that Cl is the major destroyer of ozone. Assistance is provided by the weak sunlight produced during the polar sunrise. Figure 22.4(c) shows a vertical profile comparing ozone concentration over Antarctica before and during ozone depletion, confirming that major losses occur between 15 and 25 km altitude. With increasing sunlight later in the spring, the polar vortices become weaker, allowing interchange of air with the lower latitudes to be re-established, and the ozone levels over the poles to be restored.

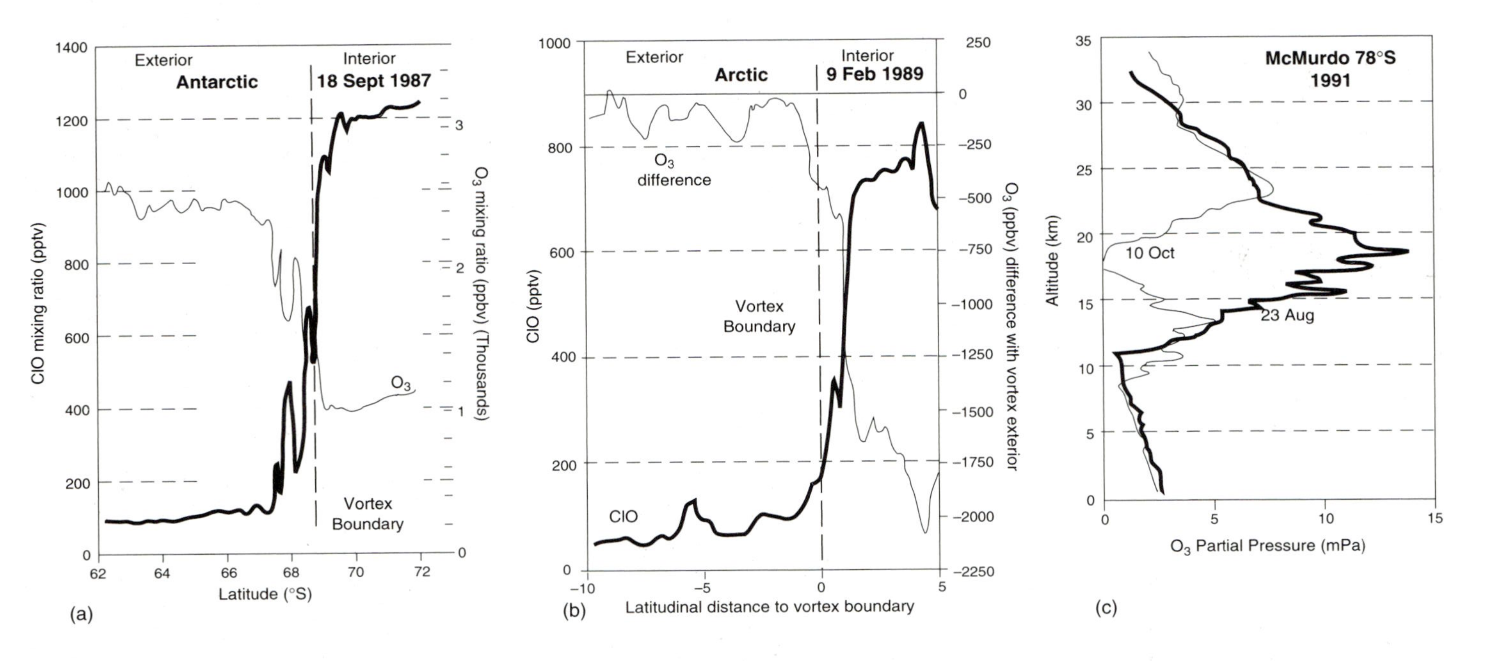

Figure 22.4 Examples of aircraft measurements of ozone and ClO across the circumpolar westerly vortex boundary in (a) the Antarctic and (b) the Arctic, demonstrating the opposite changes of ClO and O_3 concentration. The vertical profile in (c) shows the loss of ozone in early spring over Antarctica compared with early winter (developed from Proffitt *et al.*, 1991; Anderson *et al.*, 1991; Johnson *et al.*, 1992)

Solutions

The lifetimes of the CFC and N_2O molecules listed in Table 22.1 suggest that ozone depletion problems are likely to last several decades. Even if CFC production and use is stopped worldwide immediately, further losses in stratospheric ozone, especially at the poles, will occur until at least the middle of the next century. The potential problems with increased UVB radiation finally galvanized the international decision-making bodies into action. The historic Montreal Protocol limiting use of CFCs (and halons) on a global scale was signed in 1987. Since then, meetings in 1990 and 1992 have updated the Montreal Document and tightened the requirements for phase-out. Table 22.4 presents examples of the control measures recommended in the Copenhagen amendments in 1992. Although significant progress has been made to reduce CFC and halocarbon use, reflected in decreases in background air quality trends in these substances, the 'goals' of the Copenhagen amendments have not been fully met.

Table 22.4 Phase-out periods of CFCs, halocarbons and similar substances agreed to in Copenhagen, 1992

Ozone-depleting substance	*Phase-out period*
CFC-11, CFC-12, CFC-113, CFC-114, CFC-115	Reduce by 75% by 1994 (from 1986 levels) Total phase-out by 1996
Halon-1211, Halon-1301, Halon-2402	Total phase-out by 1994
CC_4 (carbon tetrachloride)	Reduce by 85% by 1995; total phase-out by 1996
CH_3Br (methyl bromide) (fire extinguishers)	Freeze by 1995; further research needed

Substitutes for these substances have been developed and are being tested for potential environmental damage. Some substitutes contain chlorine atoms, such as HCFCs. A slow phase-out of these substitutes is scheduled by 2030, to accommodate the technological developments needed for non-chlorine replacements.

CONTINENTAL-SCALE POLLUTION PROBLEMS

Continental and subcontinental pollution problems are strongly linked to transport from multiple source regions (surface dust sources, industry/power plant agglomerations, cities) with impacts extending from a few hundred to several thousand kilometres from sources. The most well known of these pollution problems is acid deposition or acid rain. Areas of the globe influenced by the continental-scale transport of pollution are depicted in Figure 22.5, with brief descriptions of each problem, and general transport shown by arrows. Continental-scale transport, often termed long-range transport (LRT), is highly dependent on meteorology (Bridgman, 1990). When a large high-pressure system persists for several days over multiple pollution source regions, subsiding air (combined with surface inversions due to long-wave radiation loss and associated cooling at night) creates a highly stable situation. Pollution concentrations build under relatively calm air flow. As the high slowly moves out of the area, increasing air flow on the western, or back, side transports the polluted air mass away from the source region. As the atmosphere becomes more unstable, through convective or frontal activity, the pollution may be deposited several thousands of kilometres downwind.

Acid Deposition

Acid deposition, better known to the public as acid rain, consists of rainwater, fogwater or dry matter which is contaminated by anthropogenic pollution and falls to the earth. As Figure 22.5 shows, the main areas of the globe where acid deposition is a problem are eastern North America, northern Europe, and, more recently, some areas of China. Rainwater acidity generally creates greater problems than acidity from dry deposition, assisted by chemical reactions during LRT from source areas. Acidity in rain-

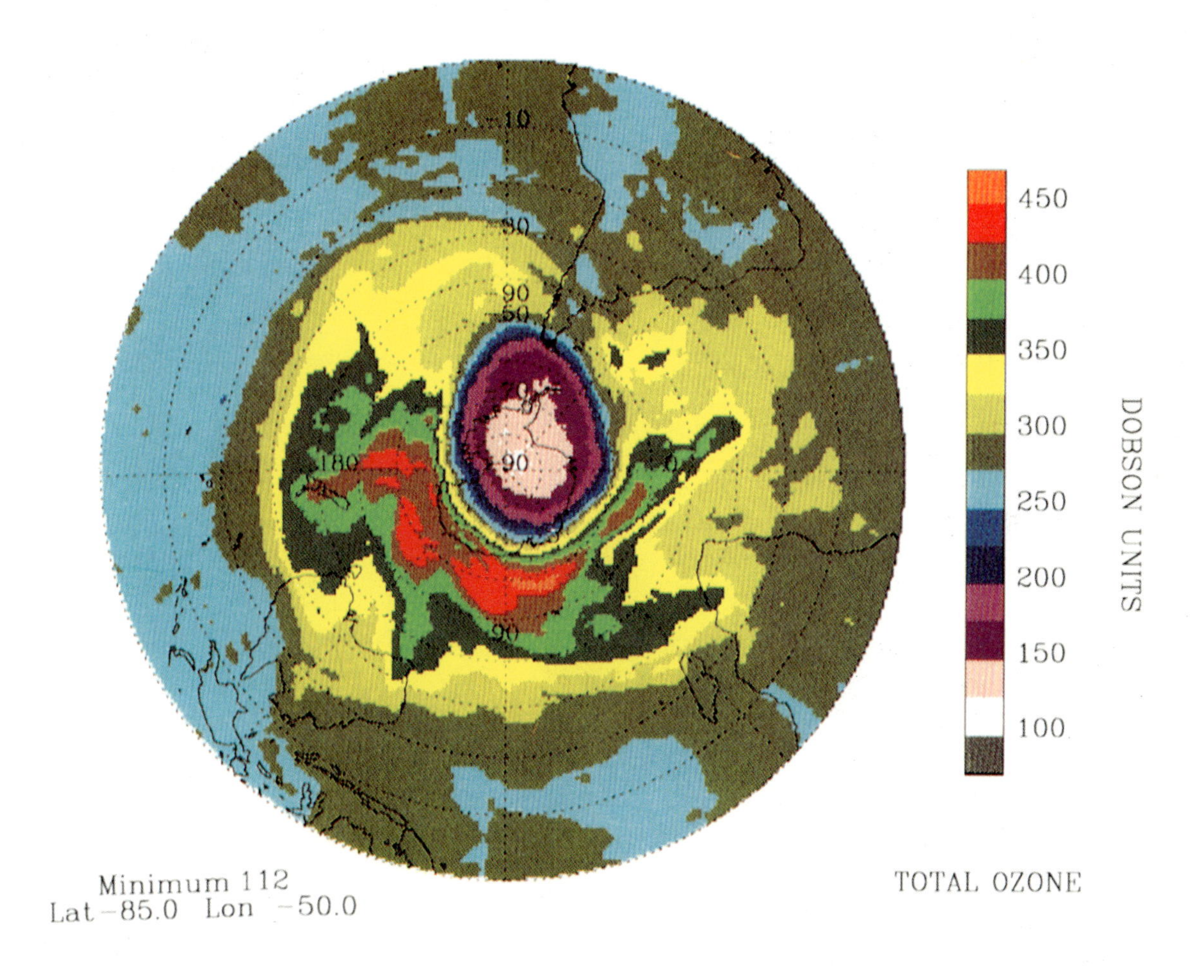

Plate 22.1 Example of false-colour satellite imagery of ozone depletion over Antarctica from the Total Ozone Mapping System (TOMS) for 14 October 1994, depicting a circumpolar shape. The main area of depletion is enclosed within the light-blue band (250 Dobson units or less)

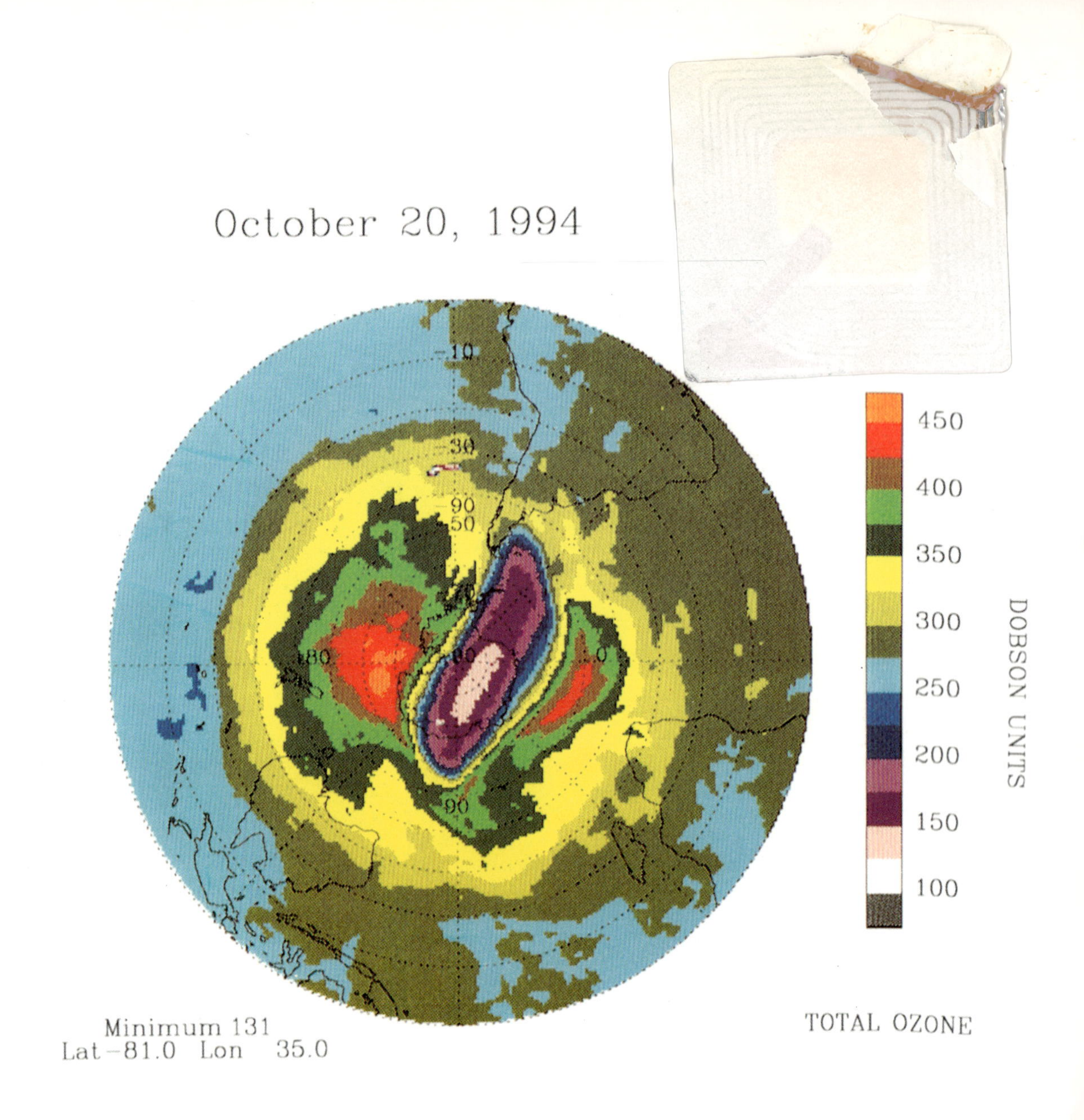

Plate 22.2 As in Plate 22.1 but for 20 October 1994, depicting an elongated shape. The change in shape is caused by day-to-day variations in the circumpolar vortex

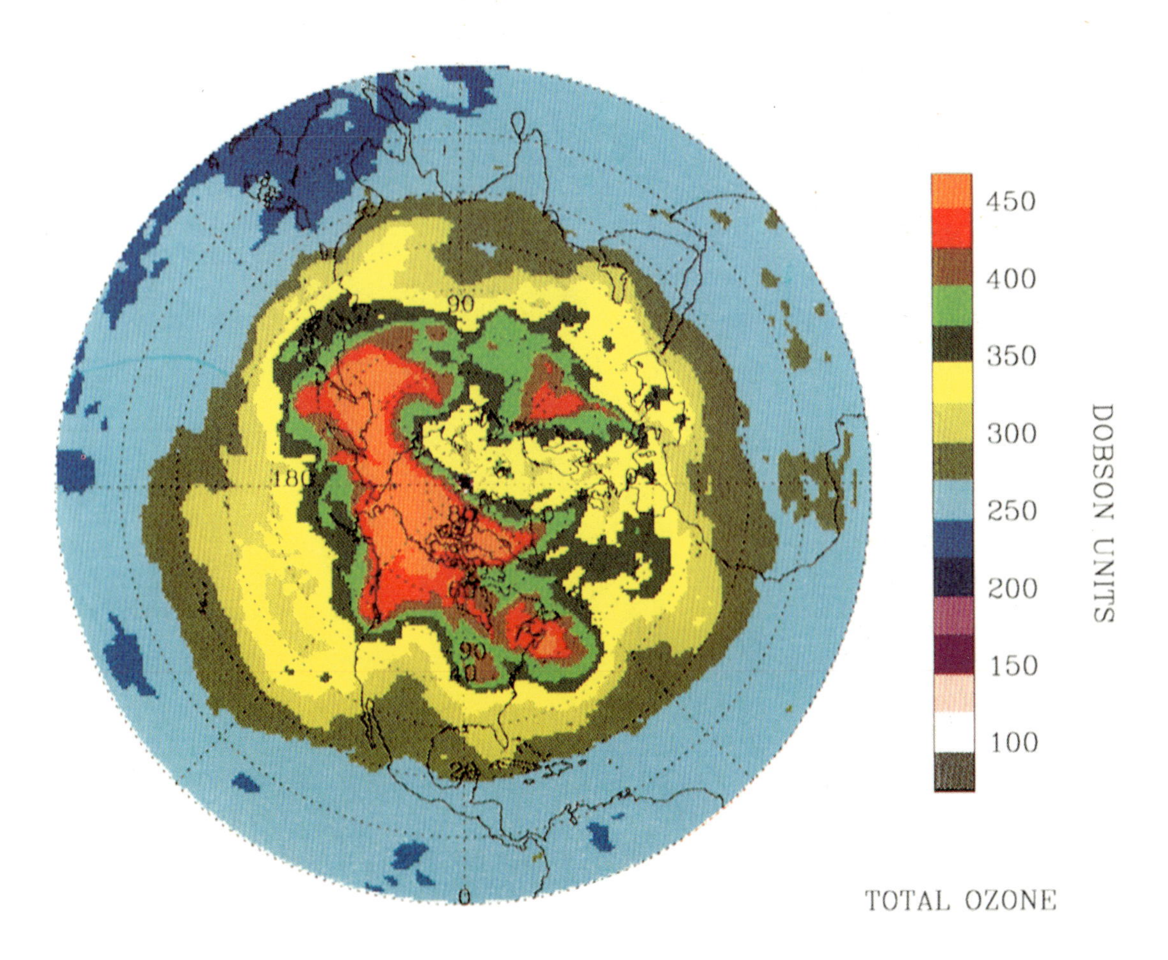

Plate 22.3 Example of false-colour satellite imagery of ozone depletion over the Arctic from the Total Ozone Mapping System (TOMS) for 18 March 1994. In comparison with Plates 22.1 and 22.2, the ozone depletion is neither as strong or as well defined

Source: Photos courtesy of Dr Paul Fraser and Clive Elsom, CSIRO Atmospheric Physics, Melbourne, Australia

Plate 23.1 The Sahel region of Africa experienced more than a quarter of a century of continuous drought conditions, starting in the late 1960s. This is a reminder that the impacts of some climatic extremes are neither rapid in onset nor short-lived in duration
Source: Photo courtesy of the International Federation of the Red Cross and Red Crescent Societies

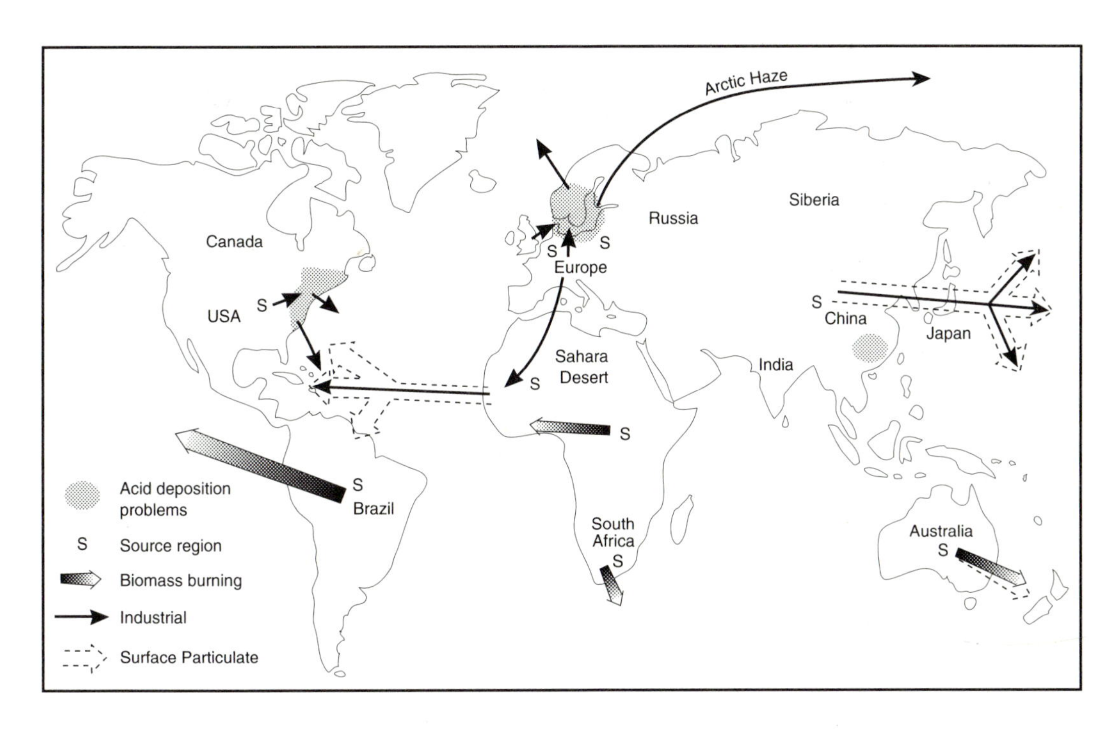

Figure 22.5 World map showing general areas of the globe influenced irregularly by long-range transport of pollutants. Included are areas affected by acid deposition, biomass burning, industrial pollutants and particulate matter. The majority of transport problems occur in the northern hemisphere. Transport varies by season and prevailing winds

water is measured as pH, on a logarithmic scale of 1 (most acid) to 14 (most alkaline). Neutral pH is 7. Natural rainwater is slightly acid, owing to a reaction between carbon dioxide and water, creating carbonic acid. A mixture of other natural influences creates a background pH in rainwater of between 4.8 and 5.6 (Bridgman, 1989). In some areas of the world, such as northern Australia, organic emissions from natural vegetation create pH levels of around 4.4.

Tall industrial and power plant stacks emit pollutants into the middle troposphere, where they react with water vapour and in water droplets to change the quality of the rainwater, dropping the pH to levels below 3 in some rainfall events. The gases sulphur dioxide (SO_2) and nitrogen dioxide (NO_2), emitted from the burning of fossil fuels, react rapidly with water in the atmosphere to create weak sulphuric acid (H_2SO_4) and nitric acid (HNO_3). The main controllers of this reaction are the anions SO_4^{2-} (sulphate) and NO_3^- (nitrate). Acidity depends on the amount of buffering from cations (ammonium, NH_4^+; calcium Ca^{++}, for example). Also important are gases which act as catalysts, to speed up reactions, such as hydrogen peroxide and ozone (Smith, 1991). LRT, over several hours to several days, creates the time for these reactions to happen. The result is acid deposition, often several thousand kilometres from sources, over a wide area. Table 22.5 presents some examples of average rainwater acidity for background and polluted locations around the world. Included also are comparative concentrations of SO_4^{2-} and NO_3^-, the major acid ions, and of NH_4^+, the

Table 22.5 Examples of pH, representing acidity, and concentrations of sulphate (SO_4^{2-}), nitrate (NO_3^-), and ammonium (NH_4^+) in rainwater, at background and polluted sites around the world

Location	*pH*	SO_4^{2-}	NO_3^-	NH_4^+
Background				
Cape Grim, Australia	5.99	4.0	5.0	2.0
New Plymouth, New Zealand	5.57	10.0	3.1	1.5
Amsterdam Island	4.92	8.8	1.7	2.1
Poker Flat, Alaska	4.96	7.1	1.9	1.1
Polluted				
Ithaca, NY, USA	4.16	61.4	29.1	15.7
Sudbury Basin, Canada	4.15	70.8	42.9	25.0
Southern Sweden	4.40	71.0	41.0	42.0
Hradek, Czech Republic	4.40	104.2	198.5	73.6
Beijing, China	6.54	162.5	33.9	160.0

Units for rainwater concentration of ions are microequivalents per litre ($\mu eq\ell^{-1}$).
Sources: reviewed in Bridgman (1990).

major buffer. There is a wide range of ion concentrations at the polluted locations, with SO_4^{2-} and NO_3^- being more than eight times higher than at background sites. The pH for Beijing is unusual, given the high SO_4^{2-} concentrations, but buffering action from high NH_4^+ and calcium (460 $\mu eq\ell^{-1}$) act to neutralize the acidity.

A major international conference held in Sweden in 1995 reviewed the impacts associated with acid deposition (Rodhe *et al.*, 1995). Acid deposition has been a major problem in Europe and North America for several decades. In some areas of the developing world, such as east and southeast Asia, southern Africa and South America, recent industrialization has created major increases in pollutant emissions and acidity. The environmental impacts can be major, including:

- increased acidity in freshwater lake and streams, killing fish populations, and in many cases most other biological life;
- depletion of base cations in soil, leading to nutrient deficiency (especially magnesium and potassium) and increased release of toxic ions such as aluminium;
- reduced root growth in trees, leading to forest decline;
- nitrogen deposition adding fertilization to many ecosystems, creating increased pollution in groundwater and precipitation runoff;
- corrosion damage to buildings, monuments, statues and drinking water supply systems;
- reductions in biodiversity in severely stressed ecosystems, with acid-tolerant species flourishing.

Other Examples of Continental-scale Pollution Problems

Figure 22.5 shows other examples of the long-range transport of pollutants that are not specifically acid deposition. Most of these relate to a mix of industrial and urban emissions, biomass burning emissions and surface particulate matter. For example, biomass burning for forest clearing and 'slash and burn' agriculture during the dry season in the tropics (central Africa, the Amazon basin, and southeast Asia) create a wide range of pollutants, affecting the quality of the local and downwind atmosphere (Harriss *et al.*, 1988). These pollutants include not only greenhouse gases, but smoke components (carbon, soot, etc.), dust and ozone-forming precursors (see the next section). Emissions from Europe and the former Soviet Union are transported into the Arctic basin in the springtime, moving thousands of kilometres as 'Arctic haze' across the polar region to Alaska and northern Canada (Bridgman *et al.*, 1989). Surface dust material, originating in the Sahara desert of north Africa, and occasionally mixed with anthropogenic pollutants from European sources, is blown westward across the tropical Atlantic to resting places in northern South America and the Caribbean Sea (Dickerson *et al.*, 1995). Similar transport occurs from the deserts in China across industrial and urban regions towards the central Pacific.

URBAN AIR POLLUTION PROBLEMS

The growth of cities since the Industrial Revolution has been as exponential as global population growth. The United Nations now recognizes 24 megacities of the world, defined as those with a current or projected population of at least 10 million by the year 2000. Of these 24 cities, only five (London, New York, Los Angeles, Tokyo and Moscow) are in developed countries. Thus, the major urban expansion and growth is occurring in developing countries. Leading the list is Mexico City, with a present population of more than 20 million and a projected population of 24 million by the year 2000 (WHO/UNEP, 1992). The top five include Sao Paulo, Brazil (19 million at present; 23 million projected), Tokyo/Yokohama (20.5 million; 21 million), New York (15 million; no growth) and Calcutta (12.5 million; 15.5 million).

Air pollution over cities is generally confined to two main categories. Primary pollutants are those emitted directly to the atmosphere from either point (industrial plant, power station), line (traffic, freeway) or area (central industrial complex) sources. The major primary pollutants are sulphur dioxide (SO_2); nitrogen oxides (NO_x); carbon monoxide (CO); non-methane hydrocarbons (NMHCs), especially volatile organic carbons (VOCs); and suspended particulate matter (SPM), especially particulate matter smaller than 10 μm in diameter (PM10). Lead (Pb), mainly produced by automobiles, is an important part of SPM. Secondary pollutants are those created in the atmosphere from a chemical reaction between primary pollutants, sunlight and meteorological components, such as temperature and water vapour. An example is tropospheric ozone (O_3). Although each city has its own unique set of air pollution emissions, it is possible to provide a general picture of air quality problems characteristic of urban areas.

High pollution concentrations in urban areas usually occur in a series of episodes, lasting from a few hours to a few days, under stable atmospheric conditions. Those cities located in areas where strong, shallow high-pressure systems dominate the weather for at least one season are most at risk. These include Mexico City, Los Angeles, Athens (Greece) and Sydney (Australia). The polluted air masses are not dispersed until a frontal system brings a clean air mass, or until precipitation occurs. Pollution build-up is enhanced if the urban area is located in a topographic basin, limiting air flow possibilities (Mexico City, Los Angeles). If the city is located on the coast, sea breezes during the day and opposite drainage flows during the night can recirculate the pollutants back over the city (Los Angeles, Athens, Sydney). Medical research has established concentration levels where air pollutants are considered safe for the vast majority of the population. The World Health Organization and environmental agencies in various countries, such as the United States Environmental Protection Agency (USEPA), have established standard levels for the major pollutants, above which a health risk is considered possible (WHO, 1987). Table 22.6 presents examples of these standards, which are then used as a comparison guide with measured concentrations in cities. Where episodes of high pollution occur, strong measures to minimize pollution emissions are

Table 22.6 Examples of World Health Organization and United States Environmental Protection Agency standards for urban pollutants

Pollutant	*Standard (maximum time period)*
Suspended particulate matter	90 μg m^{-3} (annual)
	260 μg m^{-3} (24 h)
PM10	50 μg m^{-3} (annual)
	150 μg m^{-3} (24 h)
Lead	1.5 μg m^{-3} (90 day)
Carbon monoxide	35 ppm (1 h)
	14 ppm (8 h)
Non-methane hydrocarbons	24 pphm (3 h)
Nitrogen dioxide	5 pphm (annual)
Sulphur dioxide	2 pphm (annual)
Ozone	12 pphm (1 h)

Table 22.7 General pollution levels in representative megacities of the world as recorded by the World Health Organization. Data quality problems and representativeness of site locations prevent highly accurate measurements

City	SO_2	*SPM*	*Pb*	*CO*	NO_2	O_3
Beijing	S	S	L	ND	L	H
Bombay	L	H	L	L	L	ND
Cairo	ND	S	S	H	ND	ND
Jakarta	L	S	H	H	L	H
London	L	L	L	H	L	L
Los Angeles	L	H	L	H	H	S
Mexico City	S	S	H	S	H	S
Moscow	ND	H	L	H	H	ND
Sao Paulo	L	H	L	H	H	S
Seoul	S	S	L	L	L	L

Key: S = Serious problem, over 2 times WHO guidelines; H = High to moderate levels, 1–2 times WHO guidelines; L = Low, guidelines usually met; ND = no data.
Source: WHO/UNEP (1992); World Resources (1994).

essential, including the closing of major industries and the minimization of vehicular traffic.

Primary Pollutants

Primary pollutants in the megacities of the world are monitored by the World Health Organization through the Global Environmental Monitoring System (GEMS) network. Table 22.7 lists the relative seriousness of pollution levels in representative megacities for five primary pollutants and one secondary pollutant (WHO/UNEP, 1992). Mexico City consistently exceeds WHO standards for all pollutants listed, with Sao Paulo, Jakarta and Los Angeles having problems with four on the list. Serious and high levels of SO_2 in Table 22.7 show little control over industrial and power sources. SPM problems indicate dry, dusty environments, such as desert regions, and/or industrial and power pollution. High Pb, CO and NO_2 generally represent cities with high traffic levels and limited traffic flow. Agreements between countries with major emissions, signed during the mid-1980s (Convention on Long-Range Transboundary Air Pollution) have shown progress in reducing SO_2 emissions, especially in Europe and North America. Reductions of NO_2 have not been successful. At present, it is the large cities in developing countries that have the highest, and most uncontrollable, pollution levels.

Secondary Pollutants

The last column of Table 22.7 shows the seriousness of levels of the most important secondary pollutant, ozone. Ozone in urban atmospheres should not be confused with stratospheric ozone, discussed previously. While the gas and the chemistry are similar, stratospheric ozone is a natural phenomenon, while tropospheric ozone in urban atmospheres is generally anthropogenic in origin. Urban ozone is formed when the mix of nitrogen oxides and non-methane hydrocarbons, emitted mainly from the traffic fleet, interacts with sunlight in a 'witches brew' of complex photochemical reactions. The result is photochemical smog, of which ozone is a major component. The name was adopted from the famous London smogs (SMoke and fOG) in the early decades of the twentieth century which were created by a mix of SO_2, SPM and fog. However, the chemistry of photochemical smog is very different and the result is the most prevalent air pollution problem in urban areas today.

High levels of ozone occur on clear sunny days with light air flow, under stable atmospheres. Figure 22.6 shows a classical diurnal variation of a typical ozone episode. Increasing traffic levels begin at 06.00, associated with the journey to work with a rapid increase in emissions of nitric oxide (NO) and hydrocarbons (ozone precursors) from exhaust and evaporation. As sunlight also increases during the morning, NO rapidly converts to NO_2. The decrease of NO and increase of NO_2 is crucial to ozone formation since NO destroys ozone. Around 09.00 hydrocarbons peak, and VOCs, NO_2 and sunlight all begin to work together to create ozone. The chemistry is aided by carbon monoxide, which acts as a catalyst. As ozone levels increase, VOCs and NO_2 decrease. Ozone formation ceases with the exhaustion of NO_2. Depending on the city, and the atmospheric situation, ozone levels may remain high during the afternoon and only begin to diminish after dark. In Table 22.7, cities with serious ozone problems also have major traffic fleets. Ozone does not necessarily remain over its city of origin. A series of research projects in Europe and the eastern USA (Comrie, 1994) has established that ozone and its precursors can be subject to LRT. In Europe this means that ozone produced in one country crosses borders to other countries, leading to important transboundary air pollution problems. In the USA, ozone created over cities in the central midwest moves to the east coast, affecting remote rural areas.

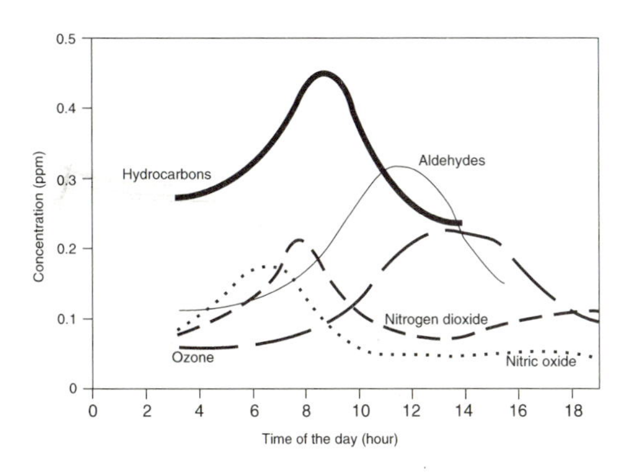

Figure 22.6 Classical development of ozone, representing photochemical smog, during a polluted day in Los Angeles (Brimblecombe, 1986). Variations occur but the development is similar in other cities of the world

The environmental problems created by photochemical oxidants are wide ranging (Rodhe *et al.*, 1995). Ozone creates visible damage to sensitive plants, destroying millions of dollars' worth of agricultural crops each year. Major damage to some forest species, through growth reduction and foliage contamination, has been recorded in the Appalachian mountains in the eastern USA, and in central and eastern Europe. Such damage may occur in conjunction with other pollutants, such as SO_2. Damage to historical monuments created by ozone and other pollutants is widespread throughout Europe. Photochemical smog affects the health of human beings through respiratory distress (bronchial problems, asthma) and eye irritation.

CONCLUSIONS

Over the next few decades, air pollution problems are likely to increase in developing countries, especially southeast Asia and South America, and slowly decrease in the developed countries (Rodhe *et al.*, 1995). The trends in SO_2 are already established whereas NO_x trends are not as clear. The exception to this pattern is photochemical oxidants, which will be a continuing problem in all major cities with large traffic fleets. While the introduction of the catalytic converter on newer automobiles has been successful in reducing tailpipe emissions, the vast numbers of automobiles in cities such as Mexico City, Los Angeles, Sydney and Athens have negated the pollution control benefit.

Controls over point-source emissions, such as SO_2 from a power station, are expensive, but much easier to establish than controls over a traffic fleet. Arguments about whether to reduce traffic nitrogen oxides or hydrocarbons abound, but the chemistry of the urban atmosphere is so complex that the reduction of one may actually lead to an increase in photochemical smog. The only realistic answer is to reduce the number of vehicles travelling in the city. This means convincing people to pool cars, use public transportation or use other vehicles besides the car. Conservationists and planners have been presenting this argument for decades, but a change in a lifestyle presently highly dependent on the automobile is needed. Populations are largely unwilling, despite major education campaigns.

Perhaps somewhat greater progress is being made on global air pollution problems. Although there have been some delays, the phase-out of CFCs and halons is proceeding, as the problems of stratospheric ozone depletion have gained international recognition. Limiting greenhouse gas emissions is another matter, with economic disadvantage a major consideration, and a very large number of complex sources to consider. Despite a lack of specific knowledge in many areas associated with greenhouse warming, scientists must convince decision-makers that the control of CO_2, CH_4 and other gases is essential. There are enough indications that anthropogenic influences will cause important climatic and environmental changes in the near future if the dependence on fossil fuels continues unabated.

REFERENCES

Anderson, J.G. *et al.* (1991) 'Free radicals within the Antarctic vortex: the role of CFCs in Antarctic ozone loss', *Science* 251: 39–45.

Bridgman, H.A. (1989) 'Acid rain studies in Australia and New Zealand', *Archives of Environmental Contamination and Toxicology* 18: 137–46.

Bridgman, H.A. (1990) *Global Air Pollution Problems for the 1990s*, London: Belhaven Press.

Bridgman, H.A., Schnell, R.C., Kahl. J.D., Herbert, G.A. and Joranger, E. (1989) 'A major haze event near Point Barrow, Alaska: analysis and probable source regions and transport pathways', *Atmospheric Environment* 23: 2537–50.

Brimblecombe, P. (1986) *Air Composition and Chemistry*, London: Cambridge University Press.

CCS (1995) *Climate Change Science: Current Understanding and Uncertainties*, Steering Committee of the Climate Change Study, Parkville, Victoria: Australian Academy of Technological Sciences and Engineering.

CDAIC (1994) *Trends '93. A Compendium of Data on Global Change*, ed. T.A. Bolin, D.P. Kaiser, R.J. Sepanski, and F.W. Stoss, ESD Publication 4195, Carbon Dioxide Information Analysis Center, Oak Ridge, T.

Comrie, A. C. (1994) 'Tracking ozone: air-mass trajectories and pollutant source regions influencing ozone in Pennsylvania forests', *Annals of the Association of American Geographers* 84: 635–51.

Dickerson, R.R., Doddridge, B.G., Kelley, P. and Rhoads, K.P. (1995) 'Large-scale pollution of the atmosphere over the remote Atlantic Ocean: evidence from Bermuda', *Journal of Geophysical Research* 100: 8945–52.

Harriss, R.C. *et al.* (1988) 'The Amazon Boundary Layer Experiment (ABLE 2A): dry Season 1985', *Journal of Geophysical Research* 93: 1351–60.

IPCC (1990) *Climate Change, the IPCC Scientific Assessment*, ed. J.T. Houghton, G.J. Jenkins and J.J. Ephraums, International Panel of Climate Change, World Meteorological Organization/United Nations Environment Programme, Cambridge: Blackwells.

IPCC (1994) *Radiative Forcing of Climate Change. The 1994 Report of the Scientific Assessment Group of IPCC: Summary for Policymakers*, Geneva: World Meteorological Organization and United Nations Environment Programme.

Johnson, B.J., Deschler, T. and Thompson, R.A. (1992) 'Vertical profiles of ozone at McMurdo Station, Antarctica: spring 1991', *Geophysical Research Letters*, 19: 1105–08.

Lockwood, J.G. (1990) 'Interactions between the earth's surface, atmospheric greenhouse gases and clouds', *Progress in Physical Geography* 14: 549–56.

Parker, D.E. and Folland, C.K. (1995) 'Comments on the global climate of 1994', *Weather* 50: 283–7.

Proffitt, M.H., Margitan, J.J., Kelly, K.K., Lorewenstein, M., Podolske, J.R. and Chan, K.R. (1990) 'Ozone loss in the Arctic polar vortex inferred from high-altitude aircraft measurements', *Nature* 347: 31–6.

Rodhe, H. *et al.* (1995) 'Acid Reign '95 – conference summary statement', *Proceedings of the 5th International Conference on Acid Deposition, Goteberg, 26–30 June 1995*, pp. 1–15.

Smith, F.B. (1991) 'An overview of the acid rain problem', *Meteorological Magazine* 120: 77–91.

Thompson, R.D. (1995) 'The impact of atmospheric aerosols on global climate: a review', *Progress in Physical Geography* 19(3): 336–50.

UNEP (1991) *Environmental Effects of Ozone Depletion: 1991 Update*, United Nations Environment Programme.

WHO (1987) *Air Quality Guidelines for Europe*, Copenhagen: World Health Organization Regional Publications, European Series 23.

WHO/UNEP (1992) *Urban Air Pollution in Megacities of the World*, Oxford: Blackwells.

WMO (1995) *Scientific Assessment of Ozone Depletion: 1994*, Global Ozone Research and Monitoring Project, Report 37, Geneva: World Meteorological Organization.

World Resources (1994) *A Guide to the Global Environment*, Oxford: Oxford University Press.

23

CLIMATIC EXTREMES AS A HAZARD TO HUMANS

Keith Smith

INTRODUCTION: CLIMATIC EXTREMES, HAZARDS AND DISASTERS

At the present time, there is no reliable physical basis for predicting atmospheric conditions several months or years ahead and planning for climate-sensitive activities has to be based on probability methods. These methods show that atmospheric events are statistically distributed around the long-term mean value and that the majority of events lie within the so-called socio-economic 'band of tolerance' (Figure 23.1). Such events are conveniently interpreted as *resources* because they supply the near-normal conditions on which many human activities are based. Climatic *hazards* are recognized only when more extreme events impose severe negative stress on human systems. In Figure 23.1, near-normal rainfall provides water resources on which domestic consumption, agriculture and industry depend but extreme events, which breach the locally defined threshold of socio-economic tolerance, may well lead to water out of control and the creation of damaging floods and droughts.

Extreme damaging events are not considered to be *disasters* unless they inflict a minimum level of adverse impact on humans and what they value (Smith, 1996). There is no generally agreed definition of 'disaster' but it has traditionally been taken to mean events causing at least 100 deaths or direct economic losses of at least US$1 million. As extreme events deviate further and further beyond the threshold of tolerance (i.e. increase in magnitude and also increase in duration), they create a progressively greater potential for disaster. However, with our present level of understanding, it is rarely possible to define a quantitative relationship between the magnitude of severe events and the exact scale of loss which they will create. This is because the actual losses also depend on many human factors, such as the local population density, the social conditions which affect the vulnerability of people to disaster and the amount of economic wealth exposed to atmospheric processes.

The atmospheric processes which produce disaster potential may last for a few seconds, as with a lightning strike, a few hours in the case of a wind storm, a few weeks in the case of floods on major rivers or for several decades in the case of drought. Extremes of individual atmospheric elements can create important disasters, such as when excessively high temperatures create deaths from physiological heat stress. But the greatest cause of death and economic damage worldwide often arises from the compound-element events and the secondary, or derived, hazards (Table 23.1). For example, wind storms of various types have the capacity

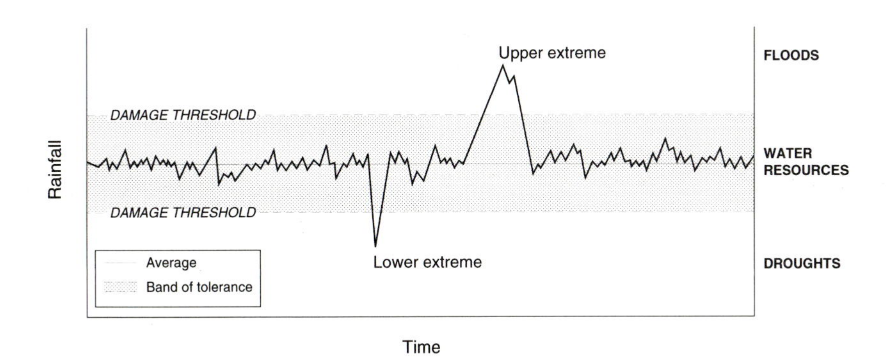

Figure 23.1 Sensitivity to flood and drought hazards as a function of the variability of rainfall around the mean and the degree of socio-economic tolerance

Table 23.1 A classification of climatic hazards

Single-element extremes (common hazards)	**Compound-element events (primary hazards)**
Temperature	Cyclones – wind + rain
Precipitation	Blizzards – wind + snow
Snowfall	Thunderstorms – rain + lightning
Wind speed	Tornadoes – wind + vortex
	Glaze storms – rain + frost
Single-element extremes (less common hazards)	**Secondary hazards (derived from climate elements)**
Lightning	Floods
Hail	Droughts
Fog	Wildfires
	Avalanches
	Landslides
	Epidemics

to inflict massive loss in a short time. The hydroclimatic hazards of floods and droughts, which are mainly derived from extremes of rainfall, are also important and floods are usually regarded as the most common of all environmental hazards.

IMPACT OF ATMOSPHERIC DISASTERS

The impact of atmospheric disasters on people, communities and countries varies greatly according to the state of economic development and the vulnerability of the population. Generally speaking, climatic extremes claim most lives in the less developed countries (LDCs) and cause the greatest economic losses in the more developed countries (MDCs). In recent years, there has been a trend towards establishing databases of so-called 'significant' disasters which reflect the importance of relative, as opposed to absolute, economic losses. For example, the database described by Sapir and Misson (1992) uses the following criteria for individual countries:

1 Number of deaths per event – has to attain 100 or more.
2 Amount of significant damage – has to total 1 per cent or more of annual GNP.
3 Affected people – has to total 1 per cent or more of the national population.

Although the mortality threshold remains an absolute criterion, not related either to total national population or to population density, the 'significant' measures for economic loss and affected people indicate more accurately the impacts of disaster on countries with small populations and weak economies. They are, therefore, a better measure of infrastructure losses in many LDCs. Using this concept of significant disasters, Figure 23.2 shows that tropical cyclones, floods and droughts together accounted for an estimated 48 per cent of all global disaster-related deaths during three recent decades. If secondary, derived atmospheric hazards (epidemics and landslides) are included, the proportion rises to over three-quarters. Tropical cyclones, floods and droughts were also responsible for 84 per cent of all economic damage and 85 per cent of all the persons adversely affected by disaster. Overall, floods are clearly the most important environmental hazard in creating 'significant' disasters.

The same database can be used to identify annual trends since the early 1960s for those disaster types directly responsible for 10 per cent or more of the total loss in each significant category, i.e. excluding drought for mortality. Figure 23.3 indicates a steep increase in flood-related deaths, which may be attributed mainly to the difficulties of forecasting and warning for flash floods. Conversely, the lack of any real trend for mortality from tropical storms reflects the marked success of forecasting, warning and evacuation schemes for Atlantic hurricanes, in particular, over the period in question. These schemes have saved many lives in the Caribbean and the USA despite the growing size of the coastal population at risk. In the cases of significant damage and people affected, flood impacts have increased steeply, whilst drought has also shown some clear upsurges, especially

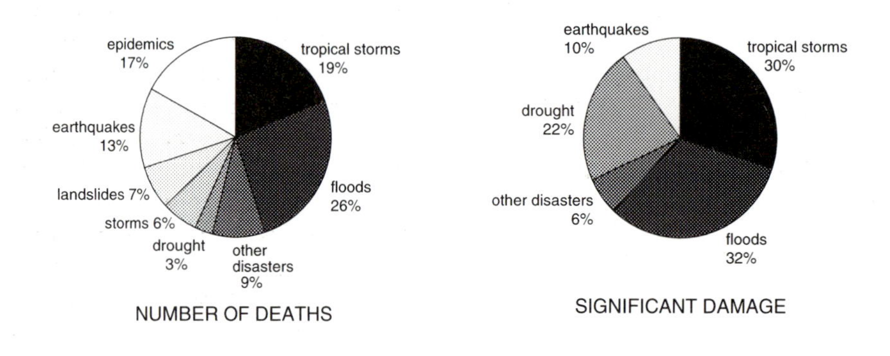

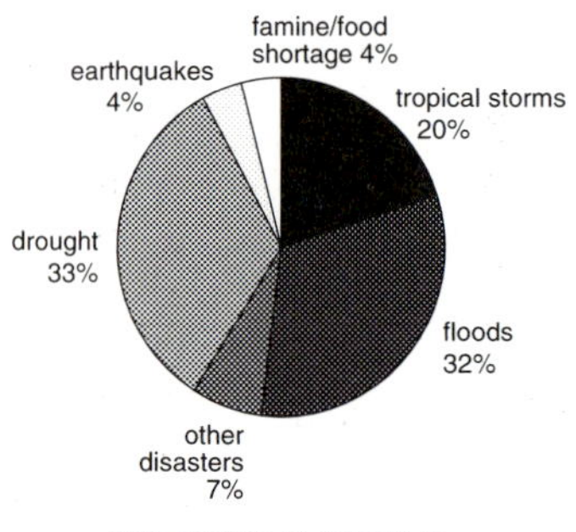

Figure 23.2 The percentage of significant disasters around the world caused by atmospheric hazards 1963–92 (after Department of Humanitarian Affairs, 1994a)

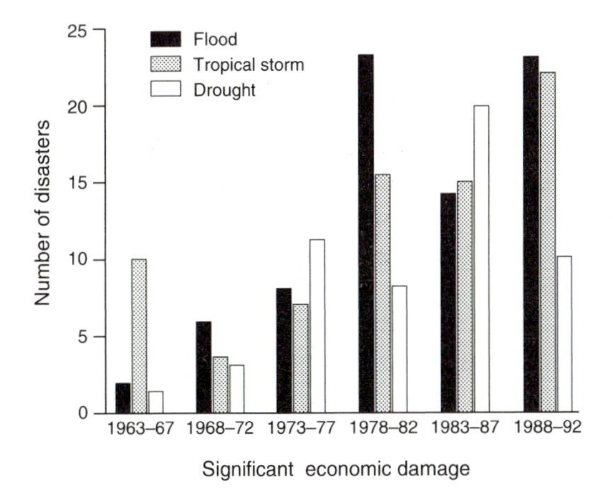

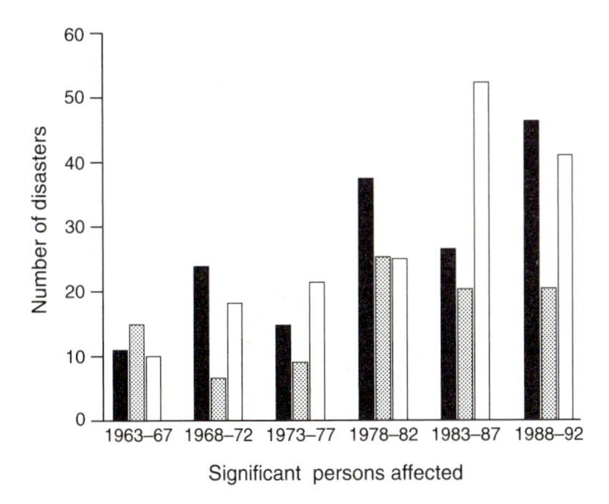

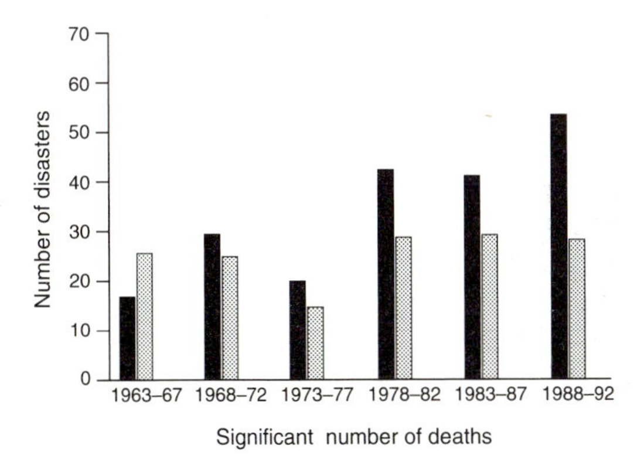

Figure 23.3 The global trend of significant atmospheric disasters by 5 year averages 1963–7 to 1988–92 (after Department of Humanitarian Affairs, 1994a)

during the mid-1980s when drought in the African Sahel helped to create a major famine disaster.

The contrasting pattern of disaster impact between rich and poor countries can be illustrated with reference to losses in the USA compared with some LDCs. One of the worst weather disasters this century in terms of mortality occurred in 1970 when approximately 300,000 people were killed in Bangladesh by the storm surge generated by a cyclone originating in the Bay of Bengal. Once again, in 1991, an estimated 200,000 people were killed and up to 10 million were made homeless as a storm surge washed away houses made of mud, bamboo and straw. The greatest devastation occurred on the many offshore islands created by sediment deposited at the head of the bay. A chronic lack of resources hampered relief efforts after the disaster and disease became widespread, largely because the well heads had not been protected against storm surge with the resultant contamination of water supplies by salt water and sewage. The underlying cause of Bangladesh's vulnerability to tropical cyclones is the extreme poverty that compels more than 10 million people to live in the lowest-lying areas of the country. On the other hand, the relative lack of public investment and of personal wealth means that, at least in absolute terms, the economic loss is modest.

By contrast, hurricane 'Hugo' which hit the Atlantic coast of the USA in 1989 caused only 35 deaths, although it led to the evacuation of hundreds of thousands of people and disrupted the lives of about 2 million persons throughout the entire region. Compared with Third World cyclone-related deaths, which are mainly associated with drowning in the storm surge, only 13 US citizens were drowned or were crushed directly by falling objects. The remaining 22 people were killed in house fires, electrocution accidents or stress-related heart attacks. In addition, adverse psychological effects of weather hazards have been well identified in countries like the USA yet have hardly been explored in the LDCs. However, the main impact of tropical cyclones in the MDCs is the property loss, much of it borne by commercial insurance companies. After hurricane 'Hugo', insurance companies in the USA paid out $2.6 billion for wind-related damage alone and some estimates for the insured losses from hurricane 'Andrew' in 1992, which killed about 65 people, were placed at $15 billion.

Flood losses show the same picture. In China, for example, river floods claim an enormous loss of life. Throughout history, the Huang Ho river has possibly killed more people than any other natural feature on the planet and in 1931 3,500,000 people were killed by floods on the Yangste river. On the other hand, the historic Midwest floods of 1993 on the Mississippi and Missouri rivers, which affected nine states comprising more than 15 per cent of the contiguous USA, killed only 48 people. However, more than 50,000 homes were damaged or destroyed, over 4 million ha of farmland were flooded and estimates of the total losses ranged between $15 and 20 billion (US Department of Commerce, 1994).

In many ways, drought disasters provide the clearest illustration of the differential effect of climatic extremes. Drought affects countries at all levels of economic development but no-one dies today in countries like the USA or Australia, although deaths from famine-related drought remain a problem in many of the poorer countries of the world. It can only be roughly estimated that, each year on average, famine kills nearly 200,000 people and affects about a billion persons worldwide. Most famine-related deaths occur in the semi-arid areas of sub-Saharan Africa but the climate is only one of a large number of factors, including poverty and population growth, responsible for drought disasters in this region (Plate 23.1). In February 1985, the UN Office for Emergency Operations in Africa estimated that 30 million people were seriously affected by drought and in urgent need of food aid. An estimated 10 million of these people abandoned their homes in search of food and water and 100,000 to 250,000 people died. On the other hand, in Australia, the most important consequences of drought are economic. The 1979-83 drought was especially severe and reduced incomes on over half the nation's farms. The ecological significance of such episodes, leading to the removal of vegetative cover and accelerated soil erosion, is also important.

REDUCING THE IMPACT OF ATMOSPHERIC DISASTERS

Given the differential pattern of impact from climatic extremes described above, the two main priorities for disaster reduction are the better protection of life in the LDCs and the better protection of property in the MDCs. These objectives, which are not mutually exclusive, can be achieved by a mixed approach of short-term and long-term measures. These measures incorporate both the physical mitigation of the impact of climatic extremes and the human avoidance of areas prone to such extremes.

Physical Mitigation

Given that climatic extremes such as hurricanes and droughts cannot be suppressed at source by weather modification techniques, physical disaster-reducing measures are mainly concerned with defending property (and lives) by strengthening the built infrastructure on which humans depend. For example, in the case of tropical cyclones, such actions might range from the short-term emergency use of wooden boards to protect the windows on an individual house to the long-term construction of a sea wall designed to protect many thousands of people and their property from storm surge along a low-lying coast.

Hazard-resistant design

In recent years, steeply rising insurance losses in the MDCs from severe storms have highlighted the fact that many properties (especially private residences) have been inadequately designed and constructed to resist attack from wind and water. Figure 23.4 shows an example of a structure predesigned by engineers to cope with a typical degree of wind stress, namely the storm which occurs on average only once in 100 years. As can be seen, gusts which occur up to, and even slightly beyond, the design limits are unlikely to cause any real damage but much larger events, outside the planned 'zone of tolerance', may well cause complete structural collapse. It is important, therefore, that properties are built to perform well, at least up to a specified design standard, if economic losses are to be reduced. Unfortunately, this is not always the case and even advanced countries may lack appropriate building codes or may fail to enforce regulations properly.

Before hurricane 'Hugo' in 1989, South Carolina did not have a state-wide building code in place. As a result, almost all the severe wind damage was caused by wind speeds below what would have been a normal design threshold. Fully engineered buildings, like high-rise hotels and office blocks, proved highly wind resistant and mostly lost some exterior wall panels and cladding only. Similarly, many historic buildings performed well because of their traditional style of construction. On the other hand, small commercial buildings, motels and unreinforced or lightly reinforced concrete-block structures fared worst (Miller, 1990). The wind forces led to the failure of roofs and windows in substandard housing with the subsequent collapse of the walls. Many of these structural failures were due to inadequately sized hurricane clips and poor roof to exterior wall connections. Much of the total damage cost was due to rain penetration. According to Sill (1991), the damage to the interiors of buildings as a result of roof failure was 20–30 times the cost of roof damage.

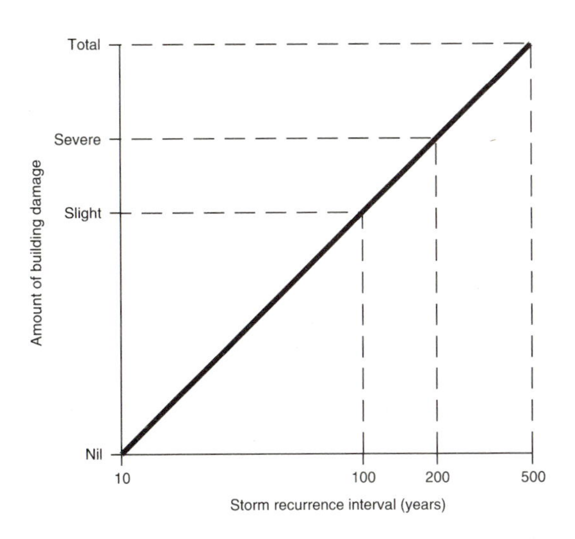

Figure 23.4 The resistance of an engineered building to wind stress from storm return intervals up to and beyond the design standard

Other property losses were associated with the storm surge. Most buildings survived intact where the floor elevation was high enough to stop the structure getting wet or suffering wave impact. For example, homes raised on a piling foundation performed well whereas those built directly on concrete slabs were undermined, and often destroyed, by wave erosion. Improved design and better construction methods, backed by a political will to enforce such standards, are needed for the coastal zones of many countries. The adoption of such measures would probably add less than 10 per cent to the overall capital cost of buildings and enable the structure to withstand hurricane forces with only minor damage, even in an exposed location.

Engineered flood defences

The physical mitigation of coastal storm surges and river floods is based primarily on engineered flood defences, such as sea walls, river levees and dams, designed to keep all but the highest flows away from developed areas. These methods of flood defence have been extensively deployed in both rich and poor countries. For example, for more than a thousand years Vietnam has built a system consisting of approximately 5000 km of river dykes and 3000 km of sea dykes in order to protect rice cultivation and the 70 per cent of the national population estimated to be at risk from flood disasters (Department of Humanitarian Affairs, 1994b). Large-scale engineering methods are not without problems because reservoirs necessitate a sacrifice of land upstream for the protection of floodplain areas whilst levees involve the protection of floodplains at a cost of increasing downstream flood peaks since the excess flows cannot be locally stored in the overbank area protected by the levee (Yevjevich, 1994). In addition, structures may create a false sense of security amongst floodplain residents. This can encourage further economic development in the 'protected' area of the floodplain, thus creating the risk of even greater losses from catastrophic floods in the future.

River engineering has been subjected to much criticism, from environmentalists and others, in recent years. It is important, therefore, to document some of the practical advantages of flood control reservoirs and levees working together, as in the historic floods in the Midwest of the USA in 1993 when the highest-ever flows were recorded at many stations within the Mississippi–Missouri basin. Most of the flood control storage upstream from St Louis is located along the Missouri river, especially within the 150,000 km^2 Kansas river basin. Here 18 reservoirs, with a total flood control capacity of 9112 cubic hectometres, control the discharge from 85 per cent of the drainage area above Kansas City (Perry, 1994). Over the 4 month period April–July 1993, this reservoir system held back some 5548 cubic hectometres of flood water with a beneficial effect on flood discharges on the Kansas river at DeSoto, immediately upstream of Kansas City. The simulation of uncontrolled discharges in the absence of reservoirs (Figure 23.5) shows that the highest daily mean flow would have been 7137 m^3 s^{-1} on 10 July, with a slightly lower peak on 26 July. By comparison, the observed peak flow was limited to 4871 m^3 s^{-1} on 27 July. Generally speaking, flood discharges were reduced by 30–70 per cent.

Although the levees along the upper Mississippi basin confined the flood and caused waters to rise higher than they would otherwise have done, even to the extent of causing localized damage, the net effect was to reduce flood damage. According to the General Accounting Office (1995), 157 of the 193 levees maintained by the Corps of Engineers in the upper Mississippi river basin prevented the flooding of about 1 million acres and additional damage

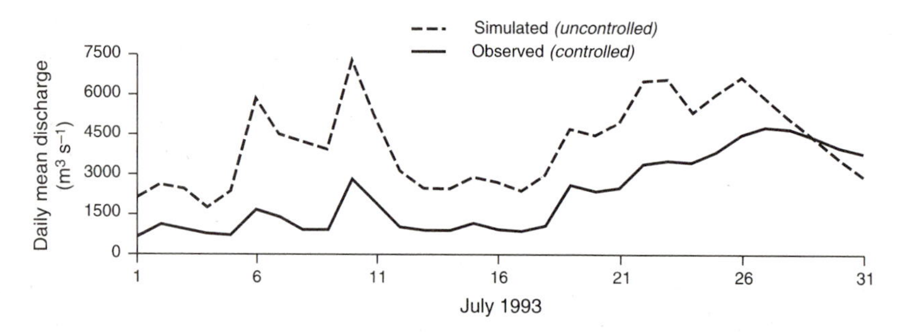

Figure 23.5 The effect of storage reservoirs on the Midwest river floods of 1993 shown by the observed (controlled) and the simulated (uncontrolled) flows on the Kansas river at DeSoto, Kansas (after Perry, 1994)

estimated at $7.4 billion. The remaining 36 levees were either overtopped or beached causing damage estimated at about $450 million. The fact that most federal levees held, many because they had been designed for 1 in 200 year events, protected several cities from inundation, including Topeka, Kansas City and St Louis.

Human avoidance

The human avoidance of climatic extremes, whether on short or long timescales, is the prime requirement for reducing mortality in atmospheric disasters. The most effective short-term response involves the use of forecasting, warning and the temporary evacuation of people from the threatened area, a strategy that has proved effective against severe storms and floods in the MDCs. Where this is not possible, reliance has to be placed on the mobilization of disaster aid. In the longer term, people and hazardous processes are best kept apart by land-use planning.

Forecasting and warning

The improved forecasting of all climatic extremes depends on an appropriate mix of physical theory, data monitoring and conceptual modelling. These developments, many of which are dependent on new technology, are all designed to improve the accuracy of the forecast and to reduce the length of time before warnings are issued. This can be illustrated by the evolution of river flood forecasting procedures in the USA (Clark, 1994):

1930s – introduction of the unit hydrograph runoff concept and the start of rainfall–runoff model development;

1940s – initial establishment of regional River Forecast Centers responsible for issuing forecasts for major river basins;

1950s – early computerization of flood forecasting techniques, for example the graphical antecedent precipitation indices for individual basins converted to mathematical algorithms;

1960s – telemetry linkage of rainfall and river gauges provided rapid telephone access to hydrometeorological conditions at remote sites; the Stanford watershed model gave a better basin-scale representation of rainfall–runoff relations;

1970s – satellite technology, for example the GOES satellite launched in late 1960s, gave fully remote collection and relay of real-time hydrometeorological data for large drainage basins;

1980s – advent of automated local evaluation in real time (ALERT) systems provided some improvement in flash flood forecasting for basins less than 250 km^2 in area;

1990s – new emphasis on quantitative precipitation forecasting (QPF) for less than 24 hours to up to 10 days ahead using expanded computing capability and improved numerical models of storms.

Such developments, replicated in many MDCs, have produced major advances in flood forecasting ability, although problems remain with flash floods, which are now the main source of mortality. As shown by Hall (1981), floods which peak within 6 hours of the causative event are not amenable to conventional flood forecasting techniques. Such floods are becoming more important owing to the spread of urban development, and its effect on the peak flow rates in small streams. Also greater recreational opportunities encourage people to visit remote mountain areas and the narrow canyons where flash floods often occur. Most of these floods result from intense localized thunderstorm activity, or from tropical storms remaining stationary over an area. Forecasting is difficult without better quantitative precipitation forecasts based on remote sensing of local rainfall by radar and satellites.

The path of severe storms is less easy to forecast than the arrival of river floods but hurricane track forecasts are issued routinely by the National Hurricane Center in Miami, USA, in order to post hurricane watches and issue warnings of storm landfall. Preparedness for hurricanes, including large-scale evacuation procedures, is now at a high level along the Gulf and Atlantic coasts of the USA. However, forecasting is still not as accurate as it might be since warnings are generally posted over a 200 km stretch of coast, although storm conditions are typically experienced along only 80 km length of coast (Pielke, 1995). This can result in what are perceived, in some areas, to be false alerts, which bring reduced credibility for future warnings. As with river floods, hurricane forecasting relies on a wide variety of real-time data from satellites, weather radar, aircraft reconnaissance, oceans buoys and shipping, which are then input into computer models of storm track and intensity. Figure 23.6 shows forecasts of the progress of hurricane 'Hugo' compared with the actual track taken. Although the maximum variation in predicted landfall was quite wide, this was, in practice, a well-predicted storm.

Conventional forecasting and warning strategies are less relevant for drought. This is partly because the atmospheric processes leading to drought are not sufficiently well understood to provide reliable precursors of disaster but also because it is less possible to relocate people out of harm's way. According to Glantz and Degefu (1991), drought-warning systems are possibly based on scientific information (such as rainfall measurements, remotely sensed indicators of vegetation stress and rangeland condi-

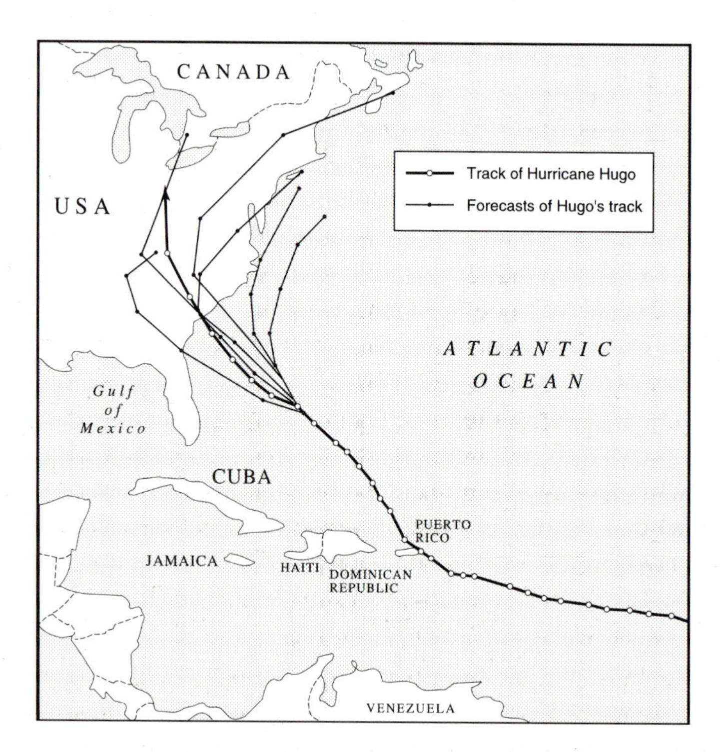

Figure 23.6 The forecast movement of hurricane 'Hugo' over a 72 hour period according to six different hurricane models together with the actual track (after Pielke, 1995)

tions) as well as seasonal forecasts linked to global phenomena, such as ENSO (El Niño and the Southern Oscillation) events. But early indications of drought are just as likely to come from key socio-economic factors (such as rising grain prices, combined with falling livestock prices and wages) as the economic balance starts to shift away from assets and services towards basic food supplies.

Disaster aid

Given the relative lack of success with drought forecasting, greater reliance has to be placed on disaster aid. Food aid can create its own problems, including dependency and 'donor fatigue', but considerable success was achieved in the 1990–2 drought in southern Africa. In this event, harvests were 30–80 per cent below normal and over 80 million people were affected. However, from April 1992 to June 1993, about five times more food and relief goods were shipped into southern Africa than during the 1984–5 famine and a 'Programme to Prevent Malnutrition' ensured a better distribution of available supplies (IFRCRCS, 1994). Aid is sometimes necessary for flood recovery as in Sudan in 1988, when there was damage to agriculture, property and social services totalling around US$1 billion with about 2 million people left homeless. After the emergency phase was over, the World Bank helped the Sudanese government to prepare a US$408 million flood reconstruction programme to present to the donor community (Brown and Muhsin, 1991).

Aid after climatic disasters is provided by many organizations. For example, within the USA, the Federal Emergency Management Agency (FEMA) provides aid under the terms of the Robert T. Stafford Disaster Relief and Emergency Assistance Act of 1974. The Disaster Relief Fund of FEMA is the major source of federal disaster assistance to state and local governments and typically covers 75 per cent of the costs, with the remaining 25 per cent paid by the state. In addition, the Office of US Foreign Disaster Assistance (OFDA) is the main governmental organization responsible for providing humanitarian assistance in response to disasters overseas. During the financial year 1994, OFDA responded to 60 emergencies, of which 22 were caused by climatic extremes (BHR/OFDA, 1995). Aid is also supplied by many nongovernmental organizations (NGOs), such as the Red Cross and charitable bodies.

Land-use planning

The first requirement of successful land-use planning is to halt, and then reverse, the human invasion of hazardous floodplain and coastal areas which was first identified over 30 years ago (Burton and Kates, 1964). Migration to the coastal zone has been a major demographic feature of the twentieth century and it has been estimated that, by the year 2000, more than 75 per cent of the world's population will live within 60 km of the coast (Bernal and Holligan, 1992). In coastal areas it is important to maintain the highest and most continuous sand dunes, as well as the widest beaches, to minimize the property loss from storm surge. For example, of the South Carolina coastal buildings completely destroyed or removed from their foundations by hurricane 'Hugo', 84 per cent were fronted by dune fields less than 15 m wide and 50 per cent were fronted by a disastrous combination of dry beaches less than 3 m wide and dune fields less than 15 m wide (Thieler and Young, 1991). Unfortunately, many sandy barrier islands, such as the exposed North Carolina Outer Banks, are vulnerable because they have already been developed. Restrictions on such near-shore development are a vital tool in reducing the economic losses from hurricanes because much structural damage is caused by the wind-driven waves which expend their energy either on the foreshore dunes or on the first block or two of houses within 100 m of the shore.

In the USA, the National Flood Insurance Program (NFIP) was introduced by the federal government in 1968 to provide financial assistance to flood victims but also to establish better land-use management for coastal and river floods (Arnell, 1984). If the flood hazard is judged to be serious, the local community is obliged to implement controls whereby further development in some areas is prohibited and certain properties must be elevated to at least the 1 in 100 year flood level. Mapping of the hazardous area into risk zones allows the allocation of differential insurance premium ratings to individual properties. The scheme has been criticized by Carper (1990) who claimed that only some 13 per cent of insurable household units in the 100 year floodplain were actually covered. These uninsured homeowners face unaffordable costs, obtain federal disaster relief and deprive the NFIP of much-needed premium income.

The National Flood Insurance Reform Act of 1994 expanded the loss mitigation provisions of the NFIP and also closed some loopholes in the legislation. This Act set up a National Flood Mitigation Fund to provide grants for mitigation projects such as floor elevation, house relocation, land acquisition, flood proofing and beach nourishment schemes. When a community can demonstrate that its policies have reduced flood losses, or have improved the sale of flood insurance beyond certain minimum requirements, homeowners are eligible to receive reductions in their flood insurance premiums. The 1994 Act also made it less easy for residential real estate buyers in flood-prone areas to avoid the purchase of flood insurance, and for flood victims, who should have purchased flood insurance under the scheme, to receive federal disaster assistance.

CONCLUSIONS

Concerted efforts are currently being made to reduce atmospheric disaster impacts through the activities of the United Nations International Decade for Natural Disaster Reduction 1990–1991. Advances in technology will continue to benefit disaster reduction and, for example, Hunt (1995) has provided some estimates of the likely improvements in forecasting some damaging meteorological events by the year 2005 (Table 23.2).

However, other factors may well conspire to increase climatic extremes and their impact. Climate change is the most prominent geophysical change expected in the next few decades. This will have a number of effects on climatic extremes, one of the most obvious being the expected increase in hot weather, and Mearns *et al.* (1984) suggest a two-to six-fold increase in heatwaves. Kalkstein (1989) used a statistical model regressing mortality on climate variables to show that for 15 American cities, without acclimatization, the number of summer deaths would rise by 6246 per annum, corresponding to a further 294 casualties per million inhabitants. By comparison, winter mortality would decrease by only nine deaths per million.

Sea-level rise (SLR) is another of the most certain, and costly, outcomes of global warming (see Chapter 7, Glacier mass balance and climate controls). The consequences of SLR illustrate the fact that most atmospheric hazards create damage mitigation (adaptation) costs, in this case those associated with strengthening coastal defences in addition to the direct costs of damage. According to Fankhauser (1995), damage costs for a 1 metre SLR

CONCLUSIONS *continued*

Table 23.2 Warning times and levels of accuracy for forecasts of atmospheric events with hazard potential in 1993 and 2005

Forecast/Performance	*1993*	*2005*
Flash flood warnings		
Period of warnings (minutes)	15	40
Accuracy of magnitude (%)	55	90
Events with no warning (%)	70	20
Severe thunderstorm warnings		
Period of warning (minutes)	13	25
Accuracy of magnitude	70	95
Temperature		
Percentage of correct forecasts	82	90
Accuracy of the onset of frost	65	90
Heavy snowfall amount		
Forecast accuracy	35	75
Precipitation forecasts		
Period of warning (days) for 25 mm fall (same accuracy as 1 day forecast in 1971)	1.8	3.0
Tropical cyclones		
Accuracy of landfall (km) 24 h ahead	185	150

Source: Hunt (1995).

are likely to range from less than $10 billion for some countries, to over $400 billion in the USA. For most countries the largest costs are associated with the loss of coastal wetlands rather than with the costs of protection. This is because wetlands provide many economic benefits, including commercial fisheries, recreation, flood protection, wildlife habitats and amenity.

A combination of elevated mean sea levels and higher sea surface temperatures could imply more storm surge hazards in the tropics. Coupled with hydroclimatic change, and the possibility of more river floods and droughts, there could be severe effects on countries already prone to water disasters, such as Bangladesh. Little consensus presently exists about the appropriate balance between structural and non-structural measures in coping with such scenarios. However, there is a growing awareness of the need to consider more environmentally sustainable strategies for the future, such as the 'living with floods' approach (Cuny, 1991).

REFERENCES

Arnell, N.W. (1984) 'Flood hazard management in the United States and the National Flood Insurance Program', Geoforum 15: 525–42.

Bernal, P. and Holligan, P.M. (1992) 'Marine and coastal systems', in J.C.I. Dooge, G.T. Goodman, J.W.M. la Riviere., J. Marton-Lefevre., T. O'Riordan and F. Praderie (eds) *An Agenda of Science for Environment and Development into the 21st Century*, Cambridge: Cambridge University Press, pp. 157–71.

BHR/OFDA (1995) *OFDA Annual Report FY 1994*, Washington, DC: US Agency for International Development.

Brown, J. and Muhsin, M. (1991) 'Case study: Sudan emergency flood reconstruction program. in A. Kreimer, and M. Munasinghe (eds) *Managing Natural*

Disasters and the Environment, Washington, DC: Environment Department, The World Bank, pp. 157–62.

Burton, I and Kates, R.W. (1964) 'The floodplain and the seashore: a comparative analysis of hazard-zone occupance', *Geographical Review* 54: 366–85.

Carper, T.R. (1990) 'The National Flood Insurance Program – beating a retreat', *Natural Hazards Observer* 14: 1–2.

Clark, R.A. (1994) 'Evolution of the national flood forecasting system in the USA', in G. Rossi, N. Harmancioglu and V. Yevjevich (eds) *Coping with Floods*, NATO Advanced Study Institute, Vol. 257, Dordrecht: Kluwer Academic, pp. 437–44.

Cuny, F.C. (1991) 'Living with floods: alternatives for riverine flood mitigation', in A. Kreimer and M. Munasinghe (eds) *Managing Natural Disasters and the Environment*, Washington, DC: Environment Department, The World Bank, pp. 62–73.

Department of Humanitarian Affairs (1994a) *Disasters Around the World – A Global and Regional View*, Information Paper DHA/94/132, *World Conference on Natural Disaster Reduction, Yokohama, Japan, 23–27 May 1994.*

Department of Humanitarian Affairs (1994b) *Strategy and Action Plan for Mitigating Water Disasters in Vietnam*, Report No. DHA/94/1, New York and Geneva: United Nations.

Fankhauser, S. (1995) *Valuing Climate Change*, London: Earthscan.

General Accounting Office (1995) *Midwest Flood: Information on the Performance, Effects and Control of Levees*, Report No. GAO/RCED-95-125, Washington, DC: General Accounting Office.

Glantz, M.H. and Degefu, W. (1991) 'Drought issues for the 1990s', in J. Jager and H. L. Ferguson (eds) *Climate Change: Science, Impacts and Policy*, Cambridge: Cambridge University Press, pp. 253–63.

Hall, A.J. (1981) *Flash Flood Forecasting*, Operational Hydrology Report No. 18, Geneva: World Meteorological Organization.

Hunt, J.C.R. (1995) 'Forecasts and warnings of natural disasters: the roles of national and international agencies', *Meteorological Applications* 2: 53–64.

IFRCRCS (1994) *World Disasters Report 1994*, Dordrecht: Martinus Nijhoff.

Kalkstein, L.S. (1989) 'The impact of CO_2 and trace-gas induced climate changes upon human mortality', in J.B. Smith and D.A. Tirpak (eds) *The Potential Effects of Global Climate Change on the United States*, Appendix G, Washington, DC: Environmental Protection Agency.

Mearns, L.O., Katz, R.W. and Schneider, S.H. (1984) 'Extreme high temperature events: changes in their probabilities with changes in mean temperatures', *Journal of Climate and Applied Meteorology* 23: 1601–13.

Miller, H.C. (1990) *Hurricane Hugo: Learning from South Carolina*, Report to the Office of Ocean and Coastal Resources Management, Washington, DC: National Ocean Service, US Department of Commerce.

Perry, C.A. (1994) *Effects of Reservoirs on Flood Discharges in the Kansas and Missouri River Basins, 1993*, Circular 1120E, Denver, CO: US Geological Survey.

Pielke, R. (1995) 'Taming the wind', *The Geographical Magazine* October: 17–20.

Sapir, D.G. and Misson, C. (1992) 'The development of a database on disasters', *Disasters* 16: 74–80.

Sill, B.L. (1991) 'Lessons learned from Hurricane Hugo and future mitigation activities', in B.L. Sill and P.R. Sparks (eds) *Hurricane Hugo One Year Later, Proceedings of a Symposium and Public Forum, American Society of Civil Engineers*, pp. 286–9.

Smith, K. (1996) Environmental Hazards: Assessing Risk and Reducing Disaster, 2nd edn, London and New York: Routledge.

Thieler, E.R. and Young, R.S. (1991) 'Quantitative evaluation of coastal geomorphological changes in South Carolina after hurricane Hugo', *Journal of Coastal Research* (SI) 8: 187–200.

US Department of Commerce (1994) *The Great Flood of 1993*, Silver Spring, MD: Natural Disaster Survey Report.

Yevjevich, V. (1994) 'Classification and description of flood mitigation measures', in G. Rossi, N. Harmancioglu and V. Yevjevich (eds) *Coping with Floods*, Dordrecht: Kluwer Academic, pp. 573–84.

24

CLIMATE CHANGE, HISTORY AND THE FUTURE

Neville Brown

INTRODUCTION: CLIMATE AND HUMAN HISTORICAL PROCESSES

Most of those who have written, during this century, about the past impact of climate change have been geographers. Historians have mainly been dismissive although notable exceptions can be found among the Francophone scholars associated with *Annales*, a journal which dates from 1929 and which has always sought to forge links with other subjects. Thus, quite the most quoted of all the climate and history studies published of late is *Times of Feast, Times of Famine* by Emmanuel Le Roy Ladurie (1971). This work stressed the complexity in the task of judging, say, the influence on crop yield of an alteration of mean temperature by 1°C across several decades. Also, it discerned a disposition to consider things 'both ways' over folk migrations. For example, the Teutons of the first millennium before Christ are supposed to have left their countries of origin because of the cold. The Scandinavians before AD 1000 are supposed to have done the same thing for exactly the opposite reason – the mildness of the climate, stimulating agriculture and thus also population growth (Ladurie, 1971).

It may usually be possible to square this particular circle by distinguishing between migrations born of economic stress (that have involved the wholesale displacement of peoples) and those which have been stimulated by enrichment, in which peoples expand out of what still remain as their core homelands. Nevertheless, one must consider Ladurie's (1971) more fundamental point. A lot of inferences about correlations between climate change and the human historical process have to be tentative until more work has been done on the history of climate. Fortunately, such work is now in progress extensively around the world. Much of it looks back from 1 to 3000 years and seeks a temporal resolution of a decade or better, wherever it may be possible across the time span chosen. The generation of such information will confront historical scholars of every hue, from neo-Marxists to Popperian conservatives. They will make ineluctable judgements about the part climate 'shifts' play in historical change and, of course, how far these 'shifts' may be the result of human activities, namely deforestation, cultivation and, in modern times especially, industrialization. In the meantime, however, it is important to ask what correlations are apparent within our existing knowledge. The provisional answers should inform a debate not only about the historical past but about the near future, not least as regards the climatic roots of social instability and international conflict.

INTO THE MEDIEVAL OPTIMUM

Europe has a singularly intricate and diverse geography and it is also rich in historical materials. To begin a climate-related review of the continent with the fall of the Roman Empire, will be to follow a scholarly tradition going back 200 years. Although the focus has varied in time and locality, a recurrent theme has been the worsening coldness and aridity. Above all, this dual trend has been seen as responsible for stimulating the *Völkerwanderung*, the barbarian migrations that bore so hard upon Rome. Gibbon (1970) laid stress on the progressive worsening of that polity's intrinsic defects, most of which would relate to climate fluctuation very indirectly or not at all. Nevertheless, it was accepted that often a winter freeze afforded 'a vast and solid bridge of ice' over a river, across which tribesmen could deploy for offensive action. It was further noted how reindeer flourished then in the forests of Germany and Poland (Gibbon, 1970). Two generations later, Thomas Arnold, famed not only as a reforming headmaster but also as a classicist, described the Alps that Hannibal crossed in 218 BC as more snowladen than in his own day (Arnold, undated). That particular contention has been challenged of late (Neumann, 1992), but the proposition that the Roman era became 'coldish' around the Mediterranean has been supported by a recent study. This identifies the centuries of most severe cold as the first and the fifth AD (Issar, 1992).

A higher-profile debate, however, has been whether drought cycles deep in Eurasia impelled forward the Huns and other barbarians. It first got under way with Ellsworth Huntington's claim that the Caspian Sea shrank continually from 500 BC to AD 500, with an acceleration of this trend from AD 200. This was recognized as part of a cyclical downward phase over a very wide area (Huntington, 1907). Later on, it was specifically asserted that the Dead Sea, Caspian Sea and Lop Nor were regularly coincident in cyclical phase (Huntington, 1911). The assertion was not based on solid proof and the proposition was, and is, implausible. Lately, consideration of this subject has become more sophisticated. Gumilëv (1968) argued from the fact that the River Volga (quite the biggest source of Caspian Sea water) originates in a stream flowing out from the Valday hills, half way between Moscow and St Petersburg. The levels of the Caspian Sea would be higher whenever rainfall and runoff increased in the upper Volga basin because the storm tracks were displaced to those latitudes (Chappell, 1970). It was concluded that, across the period 2000 BC to 400 AD, such a pattern did prevail, peaking in the third century AD (Gumilëv, 1968). Correspondingly, the Caspian Sea rose even as its own littoral area became prone to drought. More recently, a peat-bog analysis (by what was then the Soviet Academy of Sciences) found that, in a broad zone across the Baltic out as far as Moscow, annual precipitation levels in the fifth century tended to be 50 to 100 mm below the longer-term norms (Chernovskaya, 1990).

Over the years, the literature has very consistently concluded that the fifth century AD, at any rate, was abnormally arid across large areas of Europe, west of the Urals. It is apparent that there is a coincidence between that abnormality and the collapse of the Roman Empire in the west. Allowing for the fact that the rigid division of history into centuries is a human convention not a natural scientific law, the key developments are as follows. In 372, the Huns marched westwards across the lower Volga and in 378, a large Roman army was vanquished by the Goths in the sweltering August heat at Adrianople in northern Thrace, which represented Rome's worst battlefield defeat in nearly 600 years. In 406, the Huns impelled German tribes to surge into Gaul across a frozen Rhine. Four years later Visigoths, also seeking to avoid them, sacked Rome itself and, as the situation deteriorated, the last recognized emperor in the west (Romulus Augustulus) was deposed by the German Goths in 476. It is important not to be too deterministic about all these activities since it has been well appreciated from the writings of Gibbon (originally published between 1776 and 1788) that scores of factors can be adduced as contributive to Rome's decline. However, it should be noted that some of these ideas (e.g. that the barbarians used cavalry more) are poorly supported by modern research (Ferrill, 1986).

Within a century or two, a tendency towards warming had set in, however irregularly. From 900 to 1250, the increase in the mean temperature across England and the adjacent continent was close to 1°C and by the year 1300, the tree line across northern Europe was characteristically at altitudes 80 metres higher than today (Grace, 1989). The literature emphasizes the expansion of the Vikings as this long warming trend got firmly underway. In 789, the first Viking raid broke upon what had become a tranquil Anglo-Saxon England. Around the turn of that century, the Vikings reached Iceland and the frontier society they established there was to lead the medieval Nordic world in parliamentary democracy, the rule of law and inspirational literature. In 985 or thereabouts, they founded what became a colony of 10,000 people in what they described as 'Greenland' and they even made a settlement in Newfoundland. By 865, Swedish Vikings, called the Rus, had established Kiev and Novgorod in a thrust that was to take them into the Caspian Sea and Black Sea.

Climate change has been neglected by historians as a causative factor of this Norse expansion. On the other hand, they seem agreed that land hunger was a factor, manifestly so in Norway but also in Denmark (Logan, 1983). To paraphrase that argument, an underlying cause was population expansion against a background of climatic amelioration. What is clear is that, unlike the Renaissance voyages of discovery, the excursions were not triggered off by breakthroughs in nautical science and weaponry. An alternative version is that the Norsemen also pushed tentatively northwards and in 1194 (so the sagas tell us), they eventually reached Svalbard. However, not long after its discovery, Svalbard was forgotten until it was rather ostentatiously rediscovered by William Barentz, the Dutch navigator, in 1596. Still more to the point, the Norsemen did not actually settle in the territory of Finnmark (the furthermost province of modern Norway) until the thirteenth century. Also, these Norwegian Vikings did not urbanize very far up the coast since only eight settlements in medieval Norway were ever accorded the legal status of townships and the northernmost town established (at latitude 63°20′) was Trondheim (Lunde, 1985). Moreover, analysis of quasi-fossilized trees suggests that the period 1100 to 1300 was anomalously cold in northern Scandinavia. Summer temperature would have been the most critical parameter as regards tree growth but one can also envisage a significant effect from the longer, colder and drier winters that could have resulted from a polewards displacement of the Eurasian anticyclonic belt.

Across western Europe, medieval Christendom reached its zenith between 1100 and 1300. During this period, the upland limits of agriculture typically ascended 60 metres and hundreds of new towns were established. Great monastic orders were founded and scholasticism and Gothic architecture flourished. In 1241, the Hanseatic League was formed, showing that the sombre Baltic was poised to rival the exotic Mediterranean as a commercial thoroughfare. Meanwhile, the infant Rus (alias Russian) nationhood had been encouraged by the longer growing seasons to shift its centre of gravity from Kiev towards Novgorod. That shift was probably crucial to its survival, however traumatized, in the face of the Tatar branch of the Mongol expansion. Moscow was founded in 1156.

Around 1200, rainfall was tending to rise in inner Eurasia, particularly around Lop Nor which was a fluctuating lake situated near the main watershed in Sinkiang. Infantile and foal mortality can be presumed to have fallen and, regionally together, these tendencies will have given the Mongol peoples the extra manpower and horsepower that allowed the upsurge of military adventurism, as far afield as eastern Europe, led by Genghis Khan (1162–1227). Meanwhile, the seas around western Europe were getting rougher and increased volcanic dust in the stratosphere was tending to depress temperatures everywhere. More fundamentally, mean temperatures had begun what was to become five or six centuries of unsteady but insistent decline virtually worldwide, the climate phase we now know as the 'Little Ice Age' (Grove, 1988).

INTO THE LITTLE ICE AGE

By the fourteenth century, the deteriorating climate of Europe had become more prone to extremes of all kinds (Lamb, 1970), paradoxical though this may sound in the light of everything said these days about the positive correlation between warming and weather variability. Tree lines and arable margins began to recede and villages were abandoned. Famine became endemic and the continent lost the relative freedom from rampant disease it had enjoyed for some centuries, with pulmonary tuberculosis markedly increasing. Between 1348 and 1351, perhaps 'a third of the world' that the Europeans knew had died in the 'Black Death', the spread of its bacillus having been assisted by the Mongols and Tatars, inadvertently and otherwise.

Various factors had accentuated the vulnerability of European society to a worsening climate particularly the fact that the population of Europe had trebled between 950 and 1250 to reach about 80 million. Wheat monoculture had accentuated soil erosion and although transportation had improved sufficiently to spread disease, it had not done so enough adequately to correct harvest imbalances. Social tensions found expression in the viciously millenarian visions entertained by fringe movements, like the flagellants in and around the Rhine valley. On the rather more constructive side, the development of a chronic labour shortage hastened the emancipation of the mainstream peasantry. Nor should one forget that in certain spheres perhaps affected less by climate (e.g. in higher education and philosophical thought), progress was well maintained through the fourteenth century (Utley, 1961). Thus the 'razor' enunciated by William Ockham (*c.* 1285 to *c.* 1349), 'What can be done with fewer is done in vain with more', is still a basic axiom of scientific enquiry.

In Iceland, the climate crisis came early as did the political chaos, when a lapse into feudal strife ended in the Icelanders losing their freedom, first to Norway and then (in 1380) to Denmark. By 1480, the Viking community in Greenland had died out, having been isolated from Europe since 1408 and having lacked the resilience that had characterized its early days. Archaeological research on the western settlement (about 62°N, 51°W) confirms that during three centuries of loose association with the Thule-culture Inuit, the Norse never adopted Thule-hunting techniques (Buckland *et al.*, 1996). This lack of adoption was typical of the rigidity of a people who were locked into a territorial enclave. Additionally or alternatively, it can be seen as illustrative of the tenet that people respond positively to environmental change only if that change is not impossibly drastic.

During the period 1565 to 1600, the secular cooling trend resumed in northwest Europe, following some alleviation earlier in the sixteenth century. A feature of the resumption of this cooling will have been a steepening of the thermal gradient that pitches downwards from, say, the latitude of the Azores to that of southernmost Greenland. The armada launched by Spain against England in 1588 was savaged by storms that contemporaries agreed were quite without precedent and two follow-up armadas, in 1596 and 1597, were aborted early because of gales. Moreover, political historians have clearly identified the 1590s as a time of general crisis across Europe with England certainly included. Throughout these 'Little Ice Age' centuries, disastrous European harvests (e.g. in 1313–20, the 1430s, 1594–7, 1788 and 1816) readily led to social unrest. Whether, as with the 'great fear' that swept a hungry France in 1788–9, this culminated in political upheaval, depended on other factors.

The period between 1675 and 1700 recorded a phenomenal cold plunge. Just before the turn of the century, representative temperatures in central Europe were more than 0.5°C lower than any of those recorded over the next 300 years, owing evidently to strong volcanic eruptions (Hekla and Derua in 1693 and Aboina in 1694) and associated dust-veil cooling (Grove, 1988; Thompson, 1989). Across northern Europe, bad harvests were experienced over the next several years, which had demonstrable geopolitical consequences. For example, a weakening of Scotland's economy left the country more prepared to accept the union with England, which was eventually effected in 1707. In the meantime, the loss of a third of the population of Finland in the famine of 1696 had left Sweden strategically enfeebled. In 1700, the Great Northern War duly broke out between Sweden and a Baltic alliance arching from Norway and Denmark to Russia. A particular problem for Stockholm was that a greater spread than before of winter ice made it hard to deploy its navy flexibly. Sweden finally lost the initiative with defeat on land at the battle of Poltava in June 1709, in the wake of the same bitter winter as drove France, similarly encompassed and embattled, to make overtures in the War of the Spanish Succession.

Napoleon's invasion of Russia is, of course, quite the most celebrated example of a campaign turned into a total disaster by an unexpectedly early and bleak winter, once again part of a secular renewal of general cooling. Plans were based on the advice from Marquis Pierre Laplace, a brilliant mathematician and astronomer but also someone with an altogether uncritical faith in the possibilities for scientific prediction. This advice was that Russian winters 'really' begin in January whereas, in 1812–13, winter came in deadly earnest in December.

Some years ago, Lamb (1966) suggested that 1850 might be the best date to take for the end of the Little Ice Age, in Europe at any rate. This was based on accelerated temperature amelioration, as indicated by glacial recession (Lamb, 1966). In Britain, grain prices had risen sharply in the years 1836–41, driven by bad harvests associated with a secular fall in temperatures. During those years, there was much vulcanism worldwide and stratospheric aerosol loading, after which the chains of volcanoes lapsed into comparative somnolence for a good two decades. The 'warmish' and very wet Irish summer of 1846 favoured the explosive spread of a potato blight which was already endemic. This blight aggravated an agrarian economy which was already stagnating in a series (1839–44) of 'extremely bad harvests' across the board (Grado, 1990). The thermal divide which Lamb identified is important in terms of the climatic contribution to political instability. The 1840s, the so-called 'hungry forties', was a decade of ferment. Mass protests in Britain by the Chartists and the Anti-Corn Law League were followed, in 1848, by actual insurrections in various continental cities. That year also saw the publication by Karl Marx and Friedrich Engels of *The Communist Manifesto*. By then, too, Ireland was deep in the grip of a potato famine, which caused her population to decline by a quarter (through starvation or emigration) by 1851.

POST-1850 CLIMATIC AMELIORATION

However consequential this long experience of cooling and coldness had been, mean air temperatures in middle northern latitudes were not more than a couple of degrees Celsius lower in the early nineteenth century than they had been 600 years before. Then, between 1850 and 1940, there was to be a rather unsteady rise of just under 1°C and there is no doubt that this amelioration (i.e. a modest greenhouse warming) contributed to the decline in western Europe of revolutionary pressure (at least until after 1918). In the 1870s, there was a temporary slackening of the temperature rise in Europe and elsewhere, particularly between December 1878 and the harvest months of 1879 when temperatures were well below the seasonal norms across much of Europe. About that time, competition from American food producers was mounting, which was encouraged by improved transportation systems coupled with seven successive years of plentiful rainfall on the High Plains, between 1878 and 1885 (Miller, 1958). Australasia and Argentina were gaining prominence as well although, more generally, there were structural weaknesses in world trade.

These developments lent strength to the respective European lobbies for agricultural protection, a significant element in the burgeoning of nationalisms that contributed so fundamentally to the outbreak and pursuance of the First World War. By much the same token, continual drought on the American prairies and plains through the early 1930s kept the USA locked into isolationism that bit longer, and so helped to precipitate the outbreak of the Second World War in 1939. There is therefore a case for modifying the view that climate need not figure in the many explanations adduced for the outbreak of the Second, and even more particularly the First, World Wars.

From 1940 to 1965, there was a levelling out, and even a slight reversal of the upward temperature trend due to dust-veil cooling overwhelming greenhouse warming (Thompson, 1989). Its consequences were apparent throughout the 1940s in a Europe in the throes of total war, and then of a 'Cold War', and bitter winters, early on, had direct strategic consequences. Monthly mean temperatures were 8–10°C below the mid-1930s averages, which compromised horrifically a clumsy Soviet war of attrition against Finland from November 1939 to March 1940. That episode helped to persuade Berlin that the USSR could readily be overrun and to convince Moscow that more had to be done to thwart that Nazi ambition. Humanity owes a big debt to the Finns and to a savage winter. In 1941, the German meteorological service (in many ways 'the best and the brightest at that time') erred by supporting the view that June was not too late in the season to begin the invasion of Russia. In the event, both Moscow and Leningrad survived by the narrowest of margins. On 8 October, a panzer pincer trapped 650,000 Red Army troops at Vyazma, a mere 150 miles (240 km) from the Soviet capital. However, within 3 days, heavy autumnal rains had turned all roads and tracks into quagmires that persisted until the hard frosts came in early November. Those frosts utterly disorganized the Wehrmacht's logistic back-up, based as that was on the railways (Van Creveld, 1980), which just about saved the situation. Again, humanity can be grateful for dogged national resistance and climate perversity.

As the battle lines of the 'Cold War' were building up, there came the winter of 1946–7 which proved to be the bitterest in Europe since the 1840s, a decade with which comparisons were pointedly drawn. A coal crisis, which hit Britain in February 1947, accentuated the effects of the cold and introduced doubts within the ruling Labour Party about a British military mission being involved in the campaign against Communist activities in Greece. A decision to disengage was therefore taken, thereby obliging the USA either to take on the commitment or else to accept that the 'Red Flag' would soon fly over the Acropolis. The upshot in March was President Truman's enunciation of the doctrine of containment of 'direct and indirect aggression'. This was a geopolitical milestone of great consequence, whether or not it derived from an accurate assessment of the relationship between Moscow and Balkan Communism.

The launch, several months later, of the Marshall Plan was a follow-through intended, above all, to stabilize the three Western zones of occupation in Germany, since the slowness of their post-war recovery had been worsened by the cold. Meanwhile in the USSR, the ideological fervour, especially associated with Andrei Zhdanov (the Leningrad party veteran, who had returned to Stalin's favour in the spring of 1946), intensified as the cold came hard upon a prolonged drought. What the West may have failed to recognize was that Zhdanov's brief ascendancy (he died, ostensibly from natural causes, in August 1948) also ushered in a phase of retrenchment by the USSR in various sectors around its strategic periphery. If this interpretation is correct, then the explanation might be that not only the militancy at home but also the retrenchment abroad were needed to consolidate internally. This was because the adverse weather was making all the more protracted and distressing the material and social devastation the Nazis had left behind them (Watt *et al.*, 1968).

CONCLUSIONS

The message that emerges out of this discussion is that the impact of climate change tends to be at its most pronounced on peoples or polities that are in marginal situations to start with. They could be societies that are on the tribal or colonial fringes of settlement or near upland, polar or desert limits. The allusion can alternatively be to regimes that are losing control internally or under threat of defeat externally. Moreover, the point about marginality is confirmed by everything this narrative can safely leave unsaid. In other words, climate adversity scarcely enters the reckoning when things are stable in other respects. As was intimated above, the middle America droughts of the early to middle 1930s were big events, politically as well as climatically. However, those which occurred during the next period of rainfall deficiency (1950–4), in a quite well-established 20 year rainfall cycle, were to cause only a temporary and localized stir. This was because, in the interim, American agriculture had become more adaptive and resilient and, also, because there was no coincidence this time round with a world 'economic blizzard' and severe depression of the kinds precipitated by the Wall Street crash of 1929. Likewise, the European winter of 1946–7 may have been the worst for a century but that of 1962–3 was the most severe, in England at least, since 1740. However, in the annals of international politics, it was an unremarkable event due to the fact that, in 1963, much of Europe was at a peak of post-war stability and confidence, with abundant fuel supplies.

Looking to the next century, the prospect appears to be a world 'teetering on the brink' more or less perennially because of social and ecological stress. Climate

CONCLUSIONS *continued*

change can be seen as just one extra factor. However, it may *ipso facto* be quite a critical one, particularly in regions like the Middle East where (1) big contradictions are created by imbalances in economic and cultural change and (2), exposure to climatic vagary is acute. This point is underlined by a paradox that suffuses this analysis. It is that, historically speaking, the dangerous trend has been global cooling since it has weakened and foreshortened growing seasons and it has meant tougher winters for societies desperately short of commercial sources of energy. However, today, our concern is with global warming, which is partly because it is a by-product of our own energy richness and mainly because of the high rates of change currently apprehended (Table 22.3) with all their implications for rainfall patterns and, in the ultimate, sea levels (Chapter 7). The consequences of a given rate of temperature change are a subject in respect of which the sceptics about the 'greenhouse effect' can be quite wrong-headed. A 'general consensus' suggests that a temperature rise of, say, 1.2°C in half a century or so 'would present few if any problems' (Lindzen, 1994). With respect, there could never be such a consensus. The evidence from Europe over these last two millennia is that changes (averaged out at 1–2°C over several centuries) have aggravated repeatedly, and often acutely, the difficulties of societies and polities which were every bit as robust as many around the modern world are liable to be in the decades ahead.

REFERENCES

Arnold, T. (undated) 'Hannibal's passage of the Alps', in L. Valentine (ed.) *Half Hours With Standard Authors*, London: The Library Press, pp. 133–64.

Buckland, P.C., Amorosi, T., Barlow, L.K., Dugmore, A.J., Mayewski, P.A., McGovern, T.H., Ogilvie, A.E.J., Sadler, J.P. and Skidmore, P. (1996) 'Bioarchaeological and climatological evidence for the fate of Norse farmers in medieval Greenland', *Antiquity* 70(267): 88–96.

Chappell, J.E. (1970) 'Climate change reconsidered: another look at "the pulse of Asia"', *Geographical Review* LX(3): 347–73.

Chernovskaya, M.M. (1990) 'Moistening changes in the Russian plain during the historical period', in R. Bradzil (ed.) *Climate Change in the Historical and Instrumental Period*, Brno: Masaryk University, pp. 130–3.

Ferrill, A. (1986) *The Fall of the Roman Empire, the Military Explanation*, London: Thames and Hudson, Ch. 3.

Gibbon, E. (1970) *The Decline and Fall of the Roman Empire*, London: Dent, Ch. IX.

Grace, J. (1989) 'Tree lines', in P.G. Travis *et al.* (eds) *Forests, Weather and Climate*, London: The Royal Society, pp. 233–45.

Grado, C.O. (1990) 'Irish agricultural history: recent research, 1800–1850', *The Agricultural History Review* 38(2): 165–73.

Grove, J.M. (1988) *The Little Ice Age*, London: Routledge.

Gumilëv, L.N. (1968) 'Heterochronism in the moisture supply of Eurasia in the Middle Ages', *Soviet Geography. Review and Translation* IX(1): 23–35.

Huntington, E. (1907) *The Pulse of Asia*, Boston: Houghton Mifflin, Ch. XVII.

Huntington, E. (1911) *Palestine and its Transformation*, Boston: Houghton Mifflin, Ch. XIV.

Issar, A.S. (1992) *The Impact of Climate Variations on Water Management Systems*, Sede Boker: Ben Gurion University of the Negev.

Ladurie, LeRoy, E. (1971) *Times of Feast, Times of Famine*, London: George Allen and Unwin, pp. 88–9, 293.

Lamb, H.H. (1966) *The Changing Climate*, London: Methuen, p. 65.

Lamb, H.H. (1970) 'What can historical records tell us about the breakdown of the medieval warm climate in Europe in the fourteenth and fifteenth centuries – an experiment', *Contributions to Atmospheric Physics* 60(2): 131–43.

Lindzen, R.S. (1994) 'On the scientific basis of global warming scenarios', *Environmental Pollution* 83(1 and 2): 125–34.

Logan, F.D. (1983) 'Vikings', in J.R. Strayer (ed.) *Dictionary of the Middle Ages*, New York: Charles Scribner's, pp. 12, 271–2.

Lunde, O. (1985) 'Archaeology and the medieval towns of Norway', *Medieval Archaeology* XXIX: 120–35.

Miller, W. (1958) *A New History of the World*, New York: George Brazillier, Ch. 9.

Neumann, J. (1992) 'Climatic conditions in the Alps in

the years about the year of Hannibal's crossing (218 BC)', *Climate Change* 22: 139–50.
Thompson, R.D. (1989) 'Short-term climatic change: evidence, causes, environmental consequences and strategies for action', *Progress in Physical Geography* 13(3): 315–47.
Utley, F.L. (ed.) (1961) *The Forward Movement of the Fourteenth Century*, Columbus: Ohio State University Press, pp. 106–7, 127.
Van Creveld, M. (1980) *Supplying War*, Cambridge: Cambridge University Press, Ch. 5.
Watt, D.C., Spencer, F. and Brown, N. (1968) *A History of the World in the Twentieth Century*, New York: William Morrow, Part Three, Ch. III.

PART 5:

OVERVIEW

25

CONCLUSIONS AND SYNTHESIS

Allen Perry and Russell D. Thompson

INTRODUCTION: AIMS AND OBJECTIVES REASSESSED

The Preface outlined the aims and objectives of the book and emphasized a 'broad-brush' approach required by all the authors of Chapters 6–24, in terms of the climate–environment responses which are functioning at present. Predictions were also to be made on how these relationships will vary with proposed climate change. At the same time, it was stated that the aim of Part 1 (Chapters 2–5) was to examine and evaluate the current methodologies involved in applied climatological practices.

The complete academic 'freedom' given to the authors, under the 'broad-brush' constraints, is evident in the variability of the content, both in terms of individual coverages, emphases and the respective chapter lengths. Some of the chapters (especially 8, 10, 11, 12, 16, 21, 22 and 23) could easily be expanded into textbooks in their own right. Conversely, other chapters (especially 14 and 19) represent neglected areas of research, where up-to-date material is much more limited. It is hoped that the aims and objectives have been achieved to an acceptable degree and that the cross-referencing, and the two following synthesizing sections, will provide the necessary cohesion and integration of the material presented.

THE SIGNIFICANCE OF CLIMATE IN THE FUNCTIONING OF THE PHYSICAL, BIOLOGICAL AND CULTURAL ENVIRONMENTS

Climate affects all the pathways of the hydrological cycle (Figure 6.1) although the actual location of the drainage basin and the vegetation cover are also key (non-climatic) factors. Likewise, water resources become critical under extreme climate conditions, especially when supply and demand lose their synchronization under drought episodes. Frozen water bodies (i.e. glaciers) are the true 'progeny' of climate which controls all their physical and thermal characteristics. For example, the conversion of snow to glacial ice and the mass balance of glaciers are critically related to climatic conditions (especially atmospheric heat transfers). The influence of climate on geomorphic processes and landforms is not so convincing since, at scales finer than the global or regional, generalizations about the role of climatic inputs in landform evolution have proved more difficult to sustain. The response of a landform to climate is known as landscape sensitivity and low-sensitivity landforms retain their characteristic forms mainly in response to local features and actual morphology, for example the supply of debris to rock glaciers.

Over the past century, the above-ground climate has been accepted as one of the main environmental factors involved in soil formation. Soils and climate are indeed intimately interrelated and they bring together two of the most critical environmental parameters of life on the earth. The most important climatic elements are solar radiation, soil temperatures (which control chemical reactions and biological activity) and the soil moisture budget. However, it is evident from present-day soil classifications that there are important deviations in the relationship between soil patterns and climate data, since climate is only one of some five factors responsible for soil formation (Jenny, 1941). Also, soil zonations, based on present-day climate, do not necessarily reflect the conditions under which soils were originally formed, which could have been so different in the past. However, the FAO classification of eight agro-ecological zones (based on the length of the growing season and temperature) appears to close the gap between soils and the atmosphere.

The relationship between plant species and climate is complex and individualistic since each species has a particular tolerance to the prevailing conditions and they group into biomes accordingly (Figure 10.4). However, there has always been a mutually reinforcing relationship between climate and vegetation and, indeed, vegetation has influenced climate through its role in determining atmospheric composition (e.g. the carbon cycle). It is apparent, though, that non-climatic factors affect this critical relationship, namely plate tectonics, which shifts plant zones into new and foreign climate zones as in Pre-Quaternary times. Also, the astronomical features, which are associated with the earth's orbit around the sun, can change the receipt of solar radiation (namely, the Milankovitch cycles of the Quaternary period).

Temperature is a key element in life and organisms seek constantly to maintain their bodies at temperatures which allow their metabolic processes to work efficiently. Every animal must take into account two significant heat sources, namely from the environment/atmosphere and the body chemistry, and keep them in balance. Apart from temperature, humidity levels and wind speeds/wind-chill also affect animal behaviour. However, the microscale energy exchange between the animal and the atmosphere (the first equation in Chapter 11) is vital to its existence. For example, ectothermy (where atmospheric fluxes are in control) and endothermy (where metabolic heat is the key component) are convincing examples of the influence of these exchange controls.

Homeothermy is controlled by the physiological and behavioural adjustments which are required to balance energy exchanges between the human body and the environment. The course of human evolution and distribution over the earth's surface may be chartered by the responses to the prevailing thermal environment. Physiological adaptation and technological development have gone on 'hand in hand' to permit human settlement well beyond the likely original warm human birthplace. The human energy fluxes (the first equation in Chapter 12) again balance metabolic heat sources/body tissue storage with the atmospheric transfers associated with radiation, convection and evaporation. Thermoregulation mechanisms control the physiological and psychological responses of human beings in a variety of forms, for example thermal stress/comfort and acclimatization (when full attainment results from everyday thermal experiences). However, the clothing 'envelope' provides a 'private' climate which insulates the body from the atmosphere. Thermal stress also impacts on human performance/behaviour with its influence on a wide range of antisocial activities (from moods to street riots) although non-climatic factors (like drugs and alcohol) are also important motivators. Finally, thermal stress also contributes to morbidity and mortality (especially cardiovascular problems) although climate control is only part of the complex biological–environmental interactions.

Climate can be viewed as making possible certain architectural/building decisions, and not a priori forcing or determining precise built environment outcomes worldwide. However, many buildings and towns do, in fact, reflect some consideration of climate impacts on them and include some aspects of Landsberg's (1973) 'meteorological utopian city'. Optimum climatic–building responses are evident in architectural styles in all climatic zones, namely the hot, humid tropics (reducing solar radiation/heat storage again and maximizing available ventilation); the hot, dry tropics (maximizing solar gain in winter and minimizing it in summer and increasing heat storage in outer fabrics); and cold, polar climates (maximizing solar gain and decreasing wind exposure and heat loss). Urban planning should be designed to reduce climatic stress (the so-called shadow-belt).

There have been few studies on the specific relationships between variations in climate/weather and the impact of industrial output, commercial efficiency and manufacturing potential. It is apparent that climate plays a significant role in many aspects of these relationships, although the specific effects are not clearly recognizable. In Colorado, studies of five manufacturing firms revealed that even though their initial atmospheric awareness for the location of industry was minimal, on examination the actual effects were considerable, especially in terms of costs and profits. Air pollution is a real problem which must be tackled at the planning stage with the use of diffusion models to establish the emissions from proposed industrial sitings. Climate affects industrial operations (from warehousing to all outdoor activities), construction programmes (from wind stress on suspension bridges to the deterioration of materials) and commercial activities (weather 'reflex' products/marketing tactics in particular).

All transport services are dependent on the weather and climate, with the atmosphere acting as both a resource and a hazard. The design and operation of aircraft and airports is clearly influenced by critical climate and weather-sensitivity thresholds. Aircraft use the atmosphere continuously so virtually all the weather elements are important and, in the USA in 1975, 85 per cent of delays and 36 per cent of the accidents resulted from adverse weather effects. Rail transport (and the availability of 'Open Rail' forecasting) should be less sensitive to weather elements since locomotives operate on their own tracks. However, in winter, strong winds, heavy rain/snow and frosty weather remain a problem. In the UK, midwinter-train punctuality is some 6–7 per cent below the rest of the year owing to snow. Also, between 1982 and 1992, accidents (due to snow and landslides) averaged 128 per annum. Serious weather/climate interference is evident with road transport since, in the UK, 88 per cent of inland freight is carried by road and 93 out of every 100 passenger kilometres involve road travel. The 'Open Road' forecasts are designed to facilitate winter salting on UK roads, which costs £140 million every winter. 'Metroroute' services represent 'state-of-the-art' ship-routing advice to avoid violent storms and prohibitive sea heights, to increase shipping safety and efficiency in terms of time and fuel.

Agricultural production is at the mercy of weather and climate since the basis of food chains and webs is photosynthesis, relying on solar radiation trapped by green plants. Temperature increases have a very profound effect on crops since they increase rates of photorespiration (at the expense of photosynthesis), which reduces yields substantially. Rainfall is also vital since drought episodes decrease the yield and quality of crops and impose considerable stress on stock, whereas wet spells increase pests and disease. For example, rust, mildew diseases and soil-borne pathogens are greatly stimulated by warm wet weather and insects (i.e. tsetse flies, ticks and mites) also thrive in these conditions (at the expense of crops and domestic animals). The current drought in Texas has decimated the beef cattle industry and has caused the abandonment of many ranches.

Baseline forestry–climate data are rare and are likely to remain so for the foreseeable future since data collection is extremely sparse within forest environments. However, sustainable forest management implies that plantations can be genetically and technologically engineered for a specific regional climate but rarely, if ever, for weather extremes. Also, the ability to couple forests and climate depends on how well modelling can represent, for example, small-scale turbulence within the canopy and the passage of frontal/squall lines. Forests respond differently to abiotic factors at climate and weather scales. For example, the extent of wind throw is primarily a function of strong winds (associated with the passage of fronts) whereas dieback (which makes the forest more vulnerable to wind throw from even less severe winds) is more a function of seasonal climate fluctuations. Over the longer term, changes in the trajectory of storms (related to jet stream meandering) have a considerable impact on forests. Changes in wind regimes and rainfall patterns clearly affect forest stability, growth, biodiversity and regeneration rates.

Climate constitutes an important part of the environmental context in which recreation and tourism take place. Furthermore, because they are voluntary and discretionary activities, actual participation will often depend on favourable climatic conditions. For example, in the UK, in a survey of 9 million people (who holiday abroad), more than 80 per cent gave a warm and sunny climate as the primary reason for travel. 'Sun-lust' has now overtaken 'wanderlust' since 73 per cent of the respondents cited good weather as the main reason for going abroad. Despite these obvious relationships, the links between recreation and climate have been neglected even though they have acknowledged importance in all outdoor pursuits. It is apparent that climate is one of the key factors influencing tourist development in terms of forecasts for safety, comfort and general satisfaction. It also impacts on improved weather information for operating leisure activities and predictions of the likely affect of global warming on recreational/tourist activities and enterprises.

It is evident that climatic conditions are relevant in political, social and legal considerations. Past and current political frameworks have clearly incorporated the impact of climate, for example with the outcomes of armed conflicts, colonization, trade, drought policies and pollution control. Within social structures, climatic impacts are realized according to the nature of the human responses and beyond clothing, health and adaptation, these responses include personal/group rights, environmental impacts, economic loss/gain and social compromise. Social reactions to climatic impacts are also significant and may 'headline' social structure. For example, weather extremes (like hurricanes, snow storms and 'big freezes') cause a rush to judgement and response. It is apparent that the impact of air pollution represents a major social problem in terms of the cost to health and infrastructure. To some extent, the legal structure (including legalization and judgments) has been moulded over time by local climates and regional variations. This control can be quantified and imposes a real physical limit on human activities. Therefore, it can be used to attribute personal losses to an unusual climatic circumstance, to determine liability and to assess the environmental and human impacts of weather/climate variability. Unusual atmospheric extremes represent an 'act of God' if they are beyond human control and were totally unexpected. Conversely, if the events are typical of the 'normal' climate or expected seasonal variability (and indeed could have been mitigated or prevented), then a certain degree of liability could be assigned.

The energy sector is maintained by resources derived either from materials buried in the ground (fossil fuels and uranium) or from renewable resources which derive directly from solar energy or indirectly through winds, tides, biomass and the water cycle. The availability of fossil fuels/uranium is unaffected by climate, although the extraction processes may be affected to a small degree (especially if located offshore). Conversely, the availability of renewable energy resources is intimately tied to the earth's weather and climate systems, especially solar energy and wind power which are the most commonly operational. Hydroelectricity is also at risk from fluctuations in the hydrological cycle although pump–storage schemes can reduce this vulnerability. Energy conversion is also climate dependent (relying mostly on water cooling) and energy demand is quite strongly sensitive to climate (i.e. air-conditioning upsurges in heatwaves).

CLIMATE CHANGE AND THE RESPONSES OF THE PHYSICAL, BIOLOGICAL AND CULTURAL ENVIRONMENTS

It is generally accepted that greenhouse/global warming will be a threat to all the environments considered, if the model predictions are proved to be correct in future decades. As this section is being written (July 1996), the UK Climate Change Impacts Review Group have just suggested that, within 25 years, southern England will experience the climate of the Loire valley and, by 2050, the Bordeaux region (with the so-called 'Garden of England' shifting to Yorkshire). The report also highlights the associated increase in extreme weather events, the loss of rare plant and animal species, damage to existing patterns of agriculture and the increased risk of malaria and water-borne infections.

These environmental threats were considered in most of the chapters in Parts 2 and 3 to a varying degree, depending on the seriousness of the impacts involved. Climate change scenarios are most significant in the hydrological cycle with all the pathways affected by changes in energy and mass inputs/outputs. Water resources will become increasingly vulnerable to the impact of climate change with future increases in the demand for this resource being hindered by the increased evapotranspiration rates in warmer, drier and sunnier conditions. Indeed, the 1996 report of the above Review Group clearly indicates that the south of England will become more prone to drought by 2050 whereas the north will be considerably wetter with flooding a problem. This precipitation disparity highlights the immediate need for a National Water Grid in the UK so that agriculture, industrial activity, public health, water quality and power supplies will become immune to this north–south water 'divide'.

The response of glaciers to climate change remains more confused although the above Review Group report suggests sea-level rises of 19 cm by 2020 and 30 cm by 2050, partly related to increased glacial ablation. A key issue relates to the query of whether global warming would lead to the decay of ice sheets through increased ablation or their growth through increased accumulation. It is apparent that 'best estimates' suggest that by 2036, the projected increase of all trace (greenhouse) gases could increase ablation by some 11 per cent, which could threaten the existence of low-lying temperate glaciers. However, above the firn line and on high, continental ice sheets in particular, a considerable amount of the increased meltwater is absorbed in firnification and ice-sheet enhancement. Global warming and increased atmospheric moisture capacity (coupled with increased storminess/perturbations) can clearly lead to a marked increase in precipitation and ice-sheet thickness. This seems to be the case at present in Greenland where ice-sheet thickening equals 23–28 cm per year, and modelling of ice-sheet ablation indicates that a 3°C warming would generate only a few millimetres of sea-level rise over hundreds of years. The East Antarctic ice sheet appears to have been immune from Pliocene warming, which is analogous to the predicted 3°C global warming by 2050. Worldwide flooding does seem unlikely in the twenty-first century even if the predicted global warming is proved to be correct since there is no need to postulate any contribution to global sea-level rise from the Greenland and Antarctic ice sheets.

Changes in landscape forms, in response to climate change, are also difficult to predict. Compared with plant/animal associations and soil types, they are more complex systems with the individual components (e.g. rock type and weathering) characterized by different response thresholds. Furthermore, it is difficult to separate the climatic response from the tectonic signal in landform evolution and many landform types (especially tors) occur in a wide range of distinctive climatic environments. However, there are obvious uncertainties in current models of future global warming and complexities in the response rate of geomorphic systems. Consequently, the rate of change at the landform scale (excluding, for example, soil erosion processes) is likely to be measured in centuries rather than decades. This delay will be especially apparent in areas remote from low-lying coastal areas, where more potent sea-level changes will be evident on a more immediate timescale.

Soils will be affected by climate change if model predictions are proved correct, with conspicuous alterations of soil moisture budgets, vegetation cover, organic matter and soil stability. Also, climate change itself will be affected by soils emitting radiatively active greenhouse gases (namely, methane, nitrous oxide and carbon dioxide), which will help to reinforce the global warming effect. However, soils are dynamic entities and are the products of multiple factors involved in pedogenesis, as revealed by Jenny's equation. Consequently, they will readily adjust to changing climatic conditions although some responses will be more rapid than others. Vegetation will be affected by hydrological changes, especially increased drought frequencies, by reduced carbon fixation via photosynthesis and with the extinction of species through habitat changes and inadequate management (especially the tundra/boreal biomes). Also, decreases will be apparent in net primary productivity owing to increased respiration rates by primary producers and enhanced water stress.

Animals have developed through periods of chronic environmental stability and their fairly recent ancestors have had to cope with temperature changes that far exceeded those predicted with global warming in both rate and magnitude of change. Only ectotherms (living in a constant marine environment) show little ability to survive small shifts in temperature. Most animals (as individuals and species) would find the postulated increases (even if doubled) well within their zones of tolerance. Changes in ultraviolet radiation are more difficult to quantify and assess since levels are chronically higher than average on high mountains but they have little effect on animal life in these regions. Habitats and environments change constantly and animal life is well adapted to these fluctuations.

Human beings are also affected by climate change and the reconciliation between global warming and homeostasis proves difficult for 20 per cent of the world population over the year. Furthermore, increasing heat stress is becoming a seasonal problem for an additional 65 per cent of the global populace. The solution to heat stress does not imply that air conditioning should be more widely adopted to improve human comfort. Indeed, it is recommended that the active energy use is reduced by 30 per cent and that sensible management strategies should become the accepted way forward (e.g. cybernetic comfort management systems). These strategies should be based on improved acclimatization, clothing variations and a variety of indoor comfort concepts (including improved ventilation, microclimate management and a 'smart-building' approach). These would improve and sustain the quality of life in settlements and would act as a buffer against future climate change (including an increase in extreme events such as cold spells and heatwaves). The importance of climate change in terms of transport systems should not be exaggerated since the likely impacts are small, related mainly to the milder winters which will be experienced in high latitudes, although more evidence is needed. Also, during the winter of 1995–6 in northern Europe, blocking anticyclones caused considerable travel chaos due to increased frequencies of ice, snow and fog, whilst hardly decreasing temperatures. Further research is essential to clarify the relationship between climate and transport systems.

Agriculture will be at particular risk from the effects of climate change, especially since increases in carbon dioxide emissions will cause a substantial increase in photosynthesis, with decreases in photorespiration. Consequently, carbon dioxide increases could increase crop yields by some 25 per cent and will also increase stomatal conduction and water-use efficiency, both of which are of major importance to crop production and yields. Temperature increases in cool environments will foster good germination and the establishment of maize, which is currently uncompetitive owing to poor low-temperature tolerance. Hotter summers would encourage yield formation and warmer autumns would accelerate ripening rates. However, increased temperatures would stimulate the activity of disease and pests (e.g. fungi and bacterial plant pathogens) which would increase dramatically with warmer winters. Ultraviolet radiation (UVB) changes are difficult to assess since the effects are not well evaluated. However, plant damage due to leaf scorching and animal blindness will become increasing problems if ozone depletion continues. Domestic animals will be quite immune from carbon dioxide increases although rising temperatures (and especially extreme thermal stress at critical times) will be more important in determining the efficiency of production. Also, rainfall changes will be the most important factor for stock, especially in terms of available drinking water and disease transmission.

In terms of forest management, weather extremes clearly affect silviculture far more than climate change *per se*. For example, changes in solar radiation/light intensity and precipitation tend to occur as a gradient which determines potential forest growth. However, serious and intense forest disturbances are due more to infrequent, strong winds especially when they are from an unusual direction. Afforestation will increase carbon dioxide sequestration and reduce the atmospheric build-up of this gas. Deforestation and the burning/decomposition of the tropical rain forest emits as much carbon dioxide into the atmosphere as burning fossil fuels in the industrial nations of the northern hemisphere. Conversely, 500 million hectares of new forest plantations would increase carbon dioxide sequestration sufficiently enough to prevent global warming.

The projected climate change will provide new opportunities for tourism in some areas (like the north of England and Scandinavia) but will restrict the supply and demand for outdoor recreational facilities in other more established regions. For example, in the Mediterranean basin, a greater frequency of heatwaves, water shortages and threats of skin cancer will diminish tourist aspirations but areas with cooler mountain 'retreats' (like Corsica and Cyprus) will be less affected. Increased storminess in the humid tropics (e.g. Florida, Malaysia and northern Australia) could make these exotic regions far less attractive. This is especially so since 70 per cent of long-haul holidays are coast-orientated and are therefore more vulnerable to increased hurricane frequency and storm surges. Popular city locations (like London, Paris and Rome) will become increasingly hotter, with more heat stresses, and more polluted, which will decrease their attractiveness. Milder UK winters would cause fewer pensioners to desert the seasonal chill in order to seek the Mediterranean warmth. Warmer winters in Scotland and the Alps would decimate the ski industries since, for example at Glenshee, skier days average 180,000 in a normal season but this collapsed to only 12,500 skier days in the warm winter of 1991–2. Finally, rising sea levels, and increased coastal erosion from more frequent storms, will affect all shoreline attractions like beaches, golf courses, promenades, marinas and seafront hotels. This potential threat will necessitate greater shoreline protection which will prove to be very expensive indeed.

The energy sector is at the heart of the debate about climate change policy since it accounts for most of the global emissions of carbon dioxide, and significant levels of methane/nitrous oxide emissions. Policies aimed at reducing these emissions (e.g. carbon tax) would have significant impacts on energy markets. Climate change will have minimal impacts on fossil fuel exploitation, especially onshore, but offshore oil and gas production is more sensitive to extreme weather events (sea-level rises and increased storminess). Offshore gains will be associated with longer ice-free seasons for Arctic petroleum extraction. In the Beaufort Sea, Canada, for example, a 1–4°C temperature rise over 50 years will increase the ice-free season by 60–150 days and extend the open water from 200 km to 800 km (with, of course, increased wave heights and periods, which are strongly negative impacts). Conversely, the effect of climate change on renewable energy supplies is far more important. Biomass is particularly sensitive to climatic fluctuations, particularly rainfall changes which will decrease fuelwood supplies. For example, increasing drought frequencies will reduce forest yields and increase scarcity problems whereas increased carbon dioxide levels/temperatures will stimulate plant growth and biomass potential. Hydroelectric power is vulnerable to precipitation variations with increases in high latitudes increasing the power potential (by up to 20 per cent), which could then be diminished by increased rates of evaporation in warmer/sunnier environments. Snow accumulation would decrease with obvious reductions in the level of spring-meltwater discharge potential.

Solar energy potential will be affected less than other renewable sources, although increasing haze, cloud and dust will all reduce the effectiveness of concentration systems and increased storminess will damage structures. Increased wind speeds locally will also reduce the effectiveness of this energy source, since a 10 per cent change in wind velocity could change power generation by some 13–25 per cent. Energy transport will be affected by increased storm damage to overhead cables and transmission capacity will have to be reinforced since it declines with increasing temperatures. Energy conversion at fossil-fuel-fired and nuclear stations requires constant supplies of water for cooling purposes, which will be seriously curtailed in drought situations. The preferred coastal location of many of these stations will also be vulnerable to sea-level rises and increased storm surges, with a significant increased expenditure on protection systems. Finally, energy demand will face a critical balance between reduced space heating in winter and increased air conditioning in summer, in order to improve the comfort of world populations.

THE MONETARY IMPACT AND VALUE OF CLIMATE CHANGE

In almost every chapter of this volume, the concern and threat from the impact of projected climate change is clear. The long debate as to whether global warming is a myth or a scourge that will impact on almost every aspect of the life of future generations is finally being resolved. The IPCC in its Second Assessment Report issued in December 1995 includes the following two important conclusions:

1. The balance of evidence suggests a discernible human influence on global climate;
2. The challenge is not to find the best policy to address climate change today for the next 100 years but to select a prudent strategy and adjust it over time in the light of new information.

Considerable advances have been made in distinguishing or 'attributing' human-induced climate change from that occurring naturally. Recently, Santer *et al.* (1996) have provided the most convincing demonstration yet that human actions are producing an 'anthropogenic fingerprint' that can be recognized when the patterns of surface temperature change that are actually occurring, are compared with the expected changes of the model predictions.

While the regional distribution of climate change still cannot be well projected, the broad patterns are becoming clearer and these include:

1 Continents, particularly at high latitudes, will warm more quickly than the oceans.
2 Deserts are likely to become more extreme as they become hotter but not significantly wetter.
3 Developing countries will probably be more seriously affected by climate change than developed countries, since they have fewer adaptation options.

Perry (1995) has noted that resilience to climate change requires flexibility and adaptability and a capacity to bounce back and recover, qualities that are associated with economic well-being, a strong highly developed infrastructure and a young demographic structure. Significantly, ecologists testing the resistance and resilience of vegetation subjected to extreme events (McGillivray and Grime, 1995) find that in a similar way to society at large vegetation tolerance is a function of nutrient stress tolerance. Ecological crisis together with economic conditions have weakened the capacity of many countries to adjust to changes in the climatic environment. A prudent way to face the climate of the future is through a portfolio of actions aimed at mitigation, adaptation and improving knowledge. Significant 'no regrets' opportunities are available in most countries and typically they include implementing energy efficiency measures and phasing out existing policies that increase greenhouse gas emissions, through subsidies and regulations.

Countries like the UK are in a fairly good position to respond to climate change and studies of the impact on the economy suggest a 1 per cent loss of GNP over 50 years, which would be barely detectable to most of the population. Elsewhere, regional conflicts over increasingly scarce resources (like water) could escalate and a refugee crisis on a scale never before seen could instigate further social and economic tensions.

Recognizing climate change as a common concern of humankind leads to an inevitable set of questions which are difficult to answer. How much variation in the climate will constitute a significant climate change? What is the real value of its climate to any nation? Maunder (1994) has agreed that it is extremely important that we regard climate change as a challenge rather than a threat, but to do this we need to know much more about the national and regional impacts of climate variability and change on society. In considering the impacts of projected climate change some countries believe they will be 'losers' in economic terms but in many ways such notions are over simplistic. For example, they take no account of the question of the effect on competitors and thus whether there will be significant changes in competition in the marketplace for crops and raw materials.

People and governments will adapt to

climate change – there is no choice. The need to survive and to develop will ensure this adaptation happens. However, climate change is but one of a large number of constraints which will face decision-makers in the future since there are going to be many other aspects of change that will determine human welfare. To put the expected climate change in the context of other changes remains a challenge for social scientists, at least equal in magnitude or difficulty to the task facing physical scientists in refining and validating their predictions of the climatic future.

CONCLUDING COMMENTS

It is perhaps the breadth, variety and scope of the field of applied climatology that will come as the greatest surprise to many readers of this volume. The editors are well aware that the applied climate community is large and hard to define; indeed, the best working definition of the subject probably has to be that applied climatology is what applied climatologists do. At the outset, when planning an undergraduate text that might be expected to shape and focus courses, it was felt that a review of the current state and status of applied climatology needed to reflect this diversity which is one of the hallmarks of this field of study.

Forty years ago, definitions of the subject tended to emphasize the use of climatic data for planning and operational purposes (Landsberg and Jacobs, 1951), reflecting the era of post-war reconstruction and planning. By the mid-1960s, Griffiths (1966) noted that 'applications of climatology are not discussed in most textbooks. A few texts exist which cover one specific branch of the applied field, for instance agricultural meteorology, but no publication deals with a large number of applications'. During the 1970s, a trend developed to produce a more 'man-oriented type of physical geography' (Smith, 1975) and this resulted in the production of three texts which were to dominate the field of applied climatology for almost two decades. As well as Smith's book, Hobbs (1980) produced a southern-hemisphere-flavoured text that was subtitled 'a study of atmospheric resources'. Most influential were a series of books by Maunder (1970; 1986; 1989) that attempted to value atmospheric resources and give an overview of the economic dimensions of climate. All these texts helped to define a socio-economic climatology and to focus the basic purpose of applied climatology as helping society at all scales and levels to achieve a better adjustment to the climatic environment.

Whilst many of the aforementioned texts stressed the resource nature of climate, there was a parallel stream of books that considered climate as a hazard. A wide range of natural, physical and behavioural scientists have investigated the frequencies, magnitude, causes and impacts of climatic hazards. This diversity of approach is exemplified by the recent contributions of Bryant (1991), Alexander (1993) and Smith (1996). The most recent attempt to review the status of applied climatology has been undertaken by Changnon (1995) who envisaged the 'field' as consisting of a core and two concentric and interactive rings as illustrated in Figure 25.1 and was particularly concerned that 'an identity crisis exists because of the diversity and diffuseness of the field'.

It was against this background of questioning and uncertainty that *Applied Climatology: Principles and Practice* was planned and conceived. It grew out of the teaching experiences at Swansea and Reading of the two editors, both of whom had run successful third-year undergraduate option courses for many years. These experiences have taught the editors that whilst the thrust of much recent applied climatology has concerned the cultural environment, the physical and biological environments deserve more attention than they usually receive, particularly as projected environmental changes will have a cascading impact on socio-economic sys-

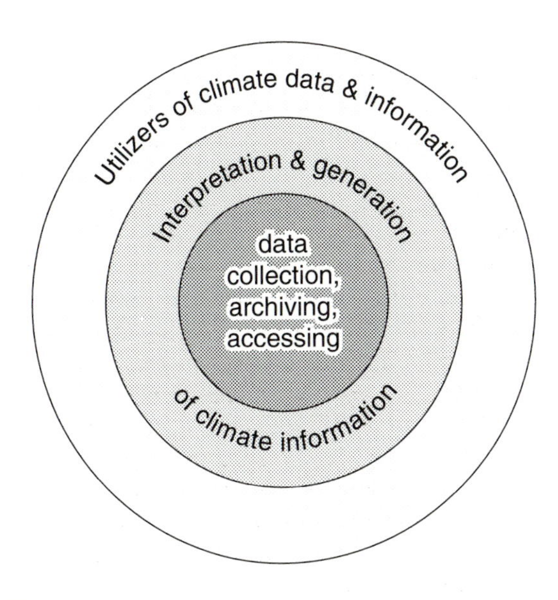

Figure 25.1 The 'field' of applied climatology
Source: Changnon (1995)

tems. Climate is one of the main factors determining the efficiency with which other natural resources can be exploited and this is clearly shown in the chapters on soils and water resources (Chapters 9 and 6). Studying the responses of plants and animals (Chapters 10 and 11) to climate, in a world where food production threatens to be outpaced by population growth, must be a basic thread of applied climatology.

Recently, one of the editors (Perry, 1995) has noted that a 'widespread belief in the reality of impending climate change has clearly been the most influential factor elevating the study of climate and its impacts on natural and socio-economic systems to a new level of interest'. Unfortunately, this interest has not been reflected in major growth in the funding of applied climate research. In the UK, the Natural Environment Research Council acknowledged as long ago as 1976 that 'climatology is becoming increasingly concerned with man's activities and welfare' and recommended that the funding of at least six climatological groups in universities in the UK be given priority. This has not happened and as a result there is a lack of adequate climate training at all levels. At many US colleges and universities, experiments are being conducted in teaching atmospheric science in the context of societal issues and concerns (Moran, 1991), but much more needs to be done to develop a general weather and climate literacy. A welcome boost has come with the launch in 1994 of a new journal, *Meteorological Applications*, one aim of which is to 'open up' the views of the private and public sectors of meteorological activity. This will allow interaction and collaboration between a whole range of specialists in different fields, who use meteorological data.

The future direction of applied climatology is more likely to be determined by the users of weather and climate information and their changing needs and requirements, than it is by academics. If the expected greenhouse warming is accompanied by a destabilized climate, with abrupt and unexpected climate lurches, then the socio-economic impacts are likely to be even greater than they are at present. At the same time, many meteorological services around the world are becoming less government controlled and more open to, and aware of, the needs of the commercial market. Hence, there is a growth in the development of niche, value-added products for private sector companies (WMO, 1994). User requirements will become more demanding and sophisticated and the challenge will be for applied climatologists to develop the methodology and techniques to exploit the commercial opportunities. Underfunded state meteorological services will increasingly be replaced by private commercial forecasting groups which are eager and willing to innovate the development and presentation of weather and climate information more effectively and to maximize the opportunities that will allow the development and planning of the human and natural environments.

REFERENCES

Alexander, D. (1993) *Natural Disasters*, London: UCL Press.

Bryant, E.A. (1991) *Natural Hazards*, London: Cambridge University Press.

Changnon, S.A. (1995) 'Applied climatology: a glorious past – an uncertain future', *American Meteorological Society 9th Conference on Applied Meteorology*, Boston: American Meteorological Society, pp. 1–5.

Griffiths, J.F. (1966) *Applied Climatology*, London: Oxford University Press.

Hobbs, J.E. (1980) *Applied Climatology*, London: Butterworths.

Jenny, H. (1941) *Factors of Soil Formation*, New York: McGraw-Hill.

Landsberg, H.E. (1973) 'The meteorologically utopian city', *Bulletin of the American Meteorological Society* 51: 86–9.

Landsberg, H.E. and Jacobs, W.C. (1951) 'Applied climatology', in T. Malone (ed.) *Compendium of Meteorology*, Boston: American Meteorological Society, pp. 976–92.

Maunder, J. (1970) *The Value of the Weather*, London: Methuen.

Maunder, J. (1986) *The Uncertainty Business*, London: Methuen.

Maunder, J. (1989) *The Human Impact of Climate Uncertainty*, London: Routledge.

Maunder, J. (1994) 'Climate, climate variations and climate change – the Irish response', in J. Freeham (ed.) *Climate Variation and Climate Change in Ireland*, Dublin: Environmental Institute University College, pp. 14–24.

McGillivray, C.W. and Grime, J.P. (1995) 'Testing predictions of the resistance and resilience of vegetation subjected to extreme events', *Functional Ecology* 1.

Moran, J.M. (1991) 'Teaching meteorology to non-science majors', *2nd International Conference on School and Popular Meteorology, American Meteorological Society*.

Perry, A.H. (1995) 'New climatologists for a new climatology', *Progress in Physical Geography* 19: 280–8.

Santer, B.D., Taylor, K.E., Wigley, T.M.L., Johns, T.C., Jones, P.D., Karoly, D.J. Mitchell, J.F.B., Oort, A.H., Penner, J.E., Ramaswamy, V., Schwarzkopf, M.D., Stouffer, R.J. and Tett, S. (1996) 'A search for human influences on the thermal structure of the atmosphere', *Nature* 382: 39–47.

Smith, K. (1975) *Principles of Applied Climatology*, London: McGraw-Hill.

Smith, K. (1996) *Environmental Hazards*, London: Routledge.

UK Climate Change Impacts Review Group (1996) *The Potential Effects of Climate Change in the United Kingdom*, London: HMSO.

WMO (1994) *The Economic Benefits of Meteorological and Hydrological Services*, Geneva: WMO/TD, No 630.

INDEX

Page numbers in italics refer to figures and tables.